CAMBRIDGE STUDIES IN ADVANCED MATHEMATICS 220

Editorial Board

J. BERTOIN, B. BOLLOBÁS, W. FULTON, B. KRA, I. MOERDIJK,
C. PRAEGER, P. SARNAK, B. SIMON, B. TOTARO

GAUSSIAN FREE FIELD AND LIOUVILLE QUANTUM GRAVITY

In this comprehensive volume, the authors introduce some of the most important recent developments at the intersection of probability theory and mathematical physics, including the Gaussian free field, Gaussian multiplicative chaos and Liouville quantum gravity.

This is the first book to present these topics using a unified approach and language, drawing on a large array of multidisciplinary techniques. These range from the combinatorial (discrete Gaussian free field, random planar maps) to the geometric (culminating in the path integral formulation of Liouville conformal field theory on the Riemann sphere) via the complex analytic (based on the couplings between Schramm–Loewner evolution and the Gaussian free field).

The arguments (currently scattered over a vast literature) have been streamlined and the exposition very carefully thought out to present the theory as much as possible in a reader-friendly, pedagogical yet rigorous way, suitable for graduate students as well as researchers.

Nathanaël Berestycki has held the Chair of Stochastics at the University of Vienna since 2018. He is a Fellow of the Institute of Mathematical Statistics and has been an associate editor of numerous journals, including the *Annals of Probability*. He was an invited speaker at the International Congress of Mathematical Physics (ICMP) in 2024.

Ellen Powell is a Professor in the probability group at Durham University, where she has been since 2019. She is currently a UKRI Future Leader's Fellow.

"Beautifully written and illustrated, this is a perfect introduction for anyone with a graduate-level probability background who wants to learn more about Gaussian free fields, random surfaces, conformal field theory, Liouville quantum gravity, SLE and the surrounding circle of ideas. The authors have spent years perfecting this exposition. I highly recommend this book to my own graduate students – and to interested researchers at any level".

Scott Sheffield, Massachusetts Institute of Technology

"For a technically difficult research area with a large body of literature, it is essential to have a clean and pedagogical introductory text. Berestycki and Powell's book fills exactly this need, and it has been a tremendous help for all my students while learning this research area. Additionally, the book has quickly become a standard reference and it's an excellent text for looking up key facts of our field".

Nina Holden, New York University

"Berestycki and Powell succeed brilliantly in making the rapidly evolving field of Liouville Quantum Gravity accessible to the probability community. There is no better place to learn the basics of the theories of Gaussian free field and multiplicative chaos than this book, and no better introduction to the recent research in Liouville CFT and its connections to SLE".

Antti Kupiainen, University of Helsinki

CAMBRIDGE STUDIES IN ADVANCED MATHEMATICS

Editorial Board
J. Bertoin, B. Bollobás, W. Fulton, B. Kra, I. Moerdijk, C. Praeger, P. Sarnak, B. Simon, B. Totaro

All the titles listed below can be obtained from good booksellers or from Cambridge University Press.
For a complete series listing, visit www.cambridge.org/mathematics.

Already Published
179 N. Nikolski *Hardy Spaces*
180 D.-C. Cisinski *Higher Categories and Homotopical Algebra*
181 A. Agrachev, D. Barilari & U. Boscain *A Comprehensive Introduction to Sub-Riemannian Geometry*
182 N. Nikolski *Toeplitz Matrices and Operators*
183 A. Yekutieli *Derived Categories*
184 C. Demeter *Fourier Restriction, Decoupling and Applications*
185 D. Barnes & C. Roitzheim *Foundations of Stable Homotopy Theory*
186 V. Vasyunin & A. Volberg *The Bellman Function Technique in Harmonic Analysis*
187 M. Geck & G. Malle *The Character Theory of Finite Groups of Lie Type*
188 B. Richter *Category Theory for Homotopy Theory*
189 R. Willett & G. Yu *Higher Index Theory*
190 A. Bobrowski *Generators of Markov Chains*
191 D. Cao, S. Peng & S. Yan *Singularly Perturbed Methods for Nonlinear Elliptic Problems*
192 E. Kowalski *An Introduction to Probabilistic Number Theory*
193 V. Gorin *Lectures on Random Lozenge Tilings*
194 E. Riehl & D. Verity *Elements of ∞-Category Theory*
195 H. Krause *Homological Theory of Representations*
196 F. Durand & D. Perrin *Dimension Groups and Dynamical Systems*
197 A. Sheffer *Polynomial Methods and Incidence Theory*
198 T. Dobson, A. Malnič & D. Marušič *Symmetry in Graphs*
199 K. S. Kedlaya *p-adic Differential Equations*
200 R. L. Frank, A. Laptev & T. Weidl *Schrödinger Operators:Eigenvalues and Lieb–Thirring Inequalities*
201 J. van Neerven *Functional Analysis*
202 A. Schmeding *An Introduction to Infinite-Dimensional Differential Geometry*
203 F. Cabello Sánchez & J. M. F. Castillo *Homological Methods in Banach Space Theory*
204 G. P. Paternain, M. Salo & G. Uhlmann *Geometric Inverse Problems*
205 V. Platonov, A. Rapinchuk & I. Rapinchuk *Algebraic Groups and Number Theory, I (2nd Edition)*
206 D. Huybrechts *The Geometry of Cubic Hypersurfaces*
207 F. Maggi *Optimal Mass Transport on Euclidean Spaces*
208 R. P. Stanley *Enumerative Combinatorics, II (2nd Edition)*
209 M. Kawakita *Complex Algebraic Threefolds*
210 D. Anderson & W. Fulton *Equivariant Cohomology in Algebraic Geometry*
211 G. Pineda Villavicencio *Polytopes and Graphs*
212 R. Pemantle, M. C. Wilson & S. Melczer *Analytic Combinatorics in Several Variables (2nd Edition)*
213 A. Yadin *Harmonic Functions and Random Walks on Groups*
214 Y. Kawamata *Algebraic Varieties: Minimal Models and Finite Generation*
215 J. Gillespie *Abelian Model Category Theory*
216 L. Anderson *Oriented Matroids*
217 Y. Motohashi *Essays in Classical Number Theory*
218 H. L. Montgomery & R. C. Vaughan *Multiplicative Number Theory II*
219 E. Rijke *Introduction to Homotopy Type Theory*

Gaussian Free Field and Liouville Quantum Gravity

NATHANAËL BERESTYCKI
University of Vienna

ELLEN POWELL
Durham University

CAMBRIDGE
UNIVERSITY PRESS

Shaftesbury Road, Cambridge CB2 8EA, United Kingdom

One Liberty Plaza, 20th Floor, New York, NY 10006, USA

477 Williamstown Road, Port Melbourne, VIC 3207, Australia

314–321, 3rd Floor, Plot 3, Splendor Forum, Jasola District Centre, New Delhi – 110025, India

103 Penang Road, #05–06/07, Visioncrest Commercial, Singapore 238467

Cambridge University Press is part of Cambridge University Press & Assessment, a department of the University of Cambridge.

We share the University's mission to contribute to society through the pursuit of education, learning and research at the highest international levels of excellence.

www.cambridge.org
Information on this title: www.cambridge.org/9781009405508

DOI: 10.1017/9781009405492

© Nathanaël Berestycki and Ellen Powell 2026

This publication is in copyright. Subject to statutory exception and to the provisions of relevant collective licensing agreements, no reproduction of any part may take place without the written permission of Cambridge University Press & Assessment.

When citing this work, please include a reference to the DOI 10.1017/9781009405492

First published 2026

A catalogue record for this publication is available from the British Library

A Cataloging-in-Publication data record for this book is available from the Library of Congress

ISBN 978-1-009-40550-8 Hardback

Cambridge University Press & Assessment has no responsibility for the persistence or accuracy of URLs for external or third-party internet websites referred to in this publication and does not guarantee that any content on such websites is, or will remain, accurate or appropriate.

For EU product safety concerns, contact us at Calle de José Abascal, 56, 1°, 28003 Madrid, Spain, or email eugpsr@cambridge.org

For our families

Contents

Preface		*page* xiii
Acknowledgements		xvii
1	**Definition and Properties of the GFF**	**1**
1.1	Discrete case	1
1.2	Continuous Green function	9
1.3	GFF as a stochastic process	20
1.4	Random variables and convergence in the space of distributions	26
1.5	Integration by parts and Dirichlet energy	27
1.6	Reminders about function spaces	29
1.7	GFF as a random distribution	32
1.8	Itô's isometry for the GFF	37
1.9	Cameron–Martin space of the Dirichlet GFF	38
1.10	Markov property	40
1.11	Conformal invariance	43
1.12	Circle averages	44
1.13	Thick points	46
1.14	Scaling limit of the discrete GFF	49
1.15	Exercises	54
2	**Liouville Measure**	**57**
2.1	Preliminaries	59
2.2	Convergence and uniform integrability in the L^2 phase	60
2.3	Weak convergence to Liouville measure	62
2.4	The GFF viewed from a Liouville typical point	63
2.5	The full L^1 phase	67
2.6	Change of coordinate formula and conformal covariance	70
2.7	Random surfaces	72
2.8	Exercises	73

Contents

3 Gaussian Multiplicative Chaos — 75
- 3.1 Motivation and background — 75
- 3.2 Set-up for Gaussian multiplicative chaos — 77
- 3.3 Construction of Gaussian multiplicative chaos — 81
- 3.4 Shamov's approach to Gaussian multiplicative chaos — 90
- 3.5 Rooted measures and Girsanov lemma for GMC — 94
- 3.6 Kahane's convexity inequality — 96
- 3.7 Scale-invariant fields — 99
- 3.8 Multifractal spectrum — 105
- 3.9 Positive moments of Gaussian multiplicative chaos (Lebesgue case) — 107
- 3.10 Degeneracy of multiplicative chaos for $\gamma > \sqrt{2d}$ — 112
- 3.11 Positive moments for general reference measures — 115
- 3.12 Negative moments of Gaussian multiplicative chaos — 117
- 3.13 Exercises — 130

4 Statistical Physics on Random Planar Maps — 133
- 4.1 Fortuin–Kasteleyn weighted random planar maps — 133
- 4.2 Conjectured connection with Liouville quantum gravity — 138
- 4.3 Mullin–Bernardi–Sheffield's bijection in the case of spanning trees — 141
- 4.4 The loop-erased random walk exponent — 147
- 4.5 Sheffield's bijection in the general case — 152
- 4.6 Infinite volume limit — 156
- 4.7 Scaling limit of the two canonical trees — 158
- 4.8 Exponents associated with FK-weighted random planar maps — 160
- 4.9 Exercises — 164

5 Introduction to Liouville Conformal Field Theory — 168
- 5.1 Preliminary background — 169
- 5.2 Spherical GFF — 175
- 5.3 Defining the Polyakov measure — 187
- 5.4 Weyl anomaly formula — 190
- 5.5 Convergence of correlation functions within Seiberg bounds — 193
- 5.6 An alternative choice of background metric — 205
- 5.7 Geometric and probabilistic interpretation of Seiberg bounds — 209
- 5.8 Liouville fields — 212
- 5.9 Unit volume Liouville sphere — 214

		Contents	xi
	5.10	Some integrability results	216
	5.11	Exercises	219
6	**Gaussian Free Field with Neumann Boundary Conditions**		220
	6.1	The Neumann GFF as a random distribution	222
	6.2	Covariance formula: the Neumann Green function	230
	6.3	Neumann GFF as a stochastic process	234
	6.4	Other boundary conditions	239
	6.5	Semicircle averages and boundary Liouville measure	247
	6.6	Finiteness of the GMC on a disc with Neumann boundary conditions	251
	6.7	Exercises	254
7	**Quantum Wedges and Scale-Invariant Random Surfaces**		256
	7.1	Convergence of random surfaces	256
	7.2	Quantum cones	272
	7.3	Thin quantum wedges	275
	7.4	Quantum discs	280
	7.5	Quantum spheres	285
	7.6	Special cases	288
	7.7	Equivalence of quantum and Liouville spheres	289
	7.8	Exercises	297
8	**SLE and the Quantum Zipper**		299
	8.1	SLE and GFF coupling: domain Markov property	300
	8.2	Quantum length of SLE	309
	8.3	Proof of Theorem 8.9	311
	8.4	Slicing a wedge with an SLE	327
9	**Liouville Quantum Gravity as a Mating of Trees**		334
	9.1	Orientation	334
	9.2	Whole plane SLE_κ and $SLE_\kappa(\rho)$	335
	9.3	Space-filling SLE in the case $\kappa' \geq 8$	343
	9.4	Space-filling SLE for $\kappa' \in (4, 8)$	351
	9.5	Cutting and welding theorems	358
	9.6	Statement of the mating of trees theorem	362
	9.7	Discussion and uniqueness	364
	9.8	Some elements of the proof of Theorem 9.33	368
Appendix A	**Chordal Loewner Chains and Chordal SLE**		373
	A.1	Chordal Loewner chains	373
	A.2	Chordal SLE_κ	375
	A.3	Chordal $SLE_\kappa(\underline{\rho})$	376

Appendix B	**Reverse Loewner Flow and Reverse SLE**	382
B.1	Definitions	382
Appendix C	**Radial Loewner Chains and Radial SLE**	391
C.1	Radial Loewner chains	391
C.2	Radial SLE_κ and $\text{SLE}_\kappa(\rho)$	392
Appendix D	**Convergence of Random Variables in the Space of Distributions**	396
	References	399
	Notation and Symbols	411
	Subject Index	414

Preface

Over 40 years ago, the physicist Polyakov [Pol81] proposed a bold framework for string theory, in which the problem was reduced to the study of certain "random surfaces". He further made the tantalising suggestion that this theory could be explicitly solved.

Recent breakthroughs from the last 15 years such as (among many other works) [DKRV16, KRV20, DS11, DMS21, MS20, HS23, LG13, Mie13, GM21c, DDDF20, GM21d, GM21b, BGK+24] have not only given a concrete mathematical basis for this theory but also verified some of its most striking predictions – as well as Polyakov's original vision. This theory, now known in the mathematics literature either as **Liouville quantum gravity (LQG)** or **Liouville conformal field theory (CFT)**, is based on a remarkable combination of ideas coming from different fields, above all probability and geometry. At its heart is the planar Gaussian free field (GFF) h, a random distribution on a given reference surface or domain of $\mathbb{R}^2$ whose covariance involves the Green function. A key role is played by the family of measures $\mathcal{M}^\gamma$ (sometimes referred to as Liouville measures, although this should not be confused with the notion of Liouville measure arising for instance in Hamiltonian dynamics) defined formally as $\mathcal{M}^\gamma(dx) = \exp(\gamma h(x)) \, dx$, for a parameter γ known as the coupling constant.

This book is intended to be an introduction to these developments assuming as few prerequisites as possible. Our starting point is a self-contained and thorough introduction to the two-dimensional continuum **Gaussian free field** (GFF). Although surveys and overviews of this object have been written before (notably [She07, WP21]), which give plenty of context, both historical and in relation to other topics, the presentation here gives a comprehensive and systematic treatment of some of the analytic subtleties that arise. Many of the details given here for the construction and basic properties of the GFF have perhaps surprisingly not appeared anywhere else before, to the best of our knowledge.

The second basic ingredient and main building block for subsequent chapters is the theory of **Gaussian multiplicative chaos**. Historically, this theory was first proposed by Høegh-Krohn in [HK71] with motivations from constructive quantum field theory not too dissimilar from the ones in this book. In the mathematical literature however it was Kahane, in his seminal contribution [Kah85], who introduced it, independently of (and going considerably beyond) [HK71]. Kahane was for his part initially motivated by the description of turbulence. In addition to these two distinct motivations, the theory has since found numerous applications in seemingly unrelated areas, such as random matrices, number theory, mathematical finance and planar Brownian motion. A useful and early survey of this theory was written in [RV14] which sketched some of the arguments of the best results available at the time and also outlined some of these applications. However the state of the art has evolved considerably since then; as a result ours is probably the first unified, systematic and self-contained presentation of this theory.

With these tools to hand, the second part of our book is devoted to an exposition of some aspects of Liouville quantum gravity as well as Liouville conformal field theory. These two topics are closely related to one another and they describe, roughly speaking, the same physical theory but with somewhat different perspectives. Essentially, we use the label "Liouville quantum gravity" for a random geometric approach highlighting connections with Schramm–Loewner Evolution (SLE). By contrast, we use the label "Liouville conformal field theory" for an approach based on the path integral formulation. We cover topics such as correlation functions and the so-called Seiberg bounds, Weyl anomaly formula, quantum cones and wedges, quantum zipper and mating of trees, as well as discrete counterparts to this theory in the form of random planar maps decorated by a model of statistical mechanics (namely, the self-dual Fortuin–Kasteleyn percolation model).

These developments require us to work with variants of the (Dirichlet) GFF, namely the GFF on a Riemannian surface and GFF with Neumann boundary conditions, respectively, to which we also provide a systematic introduction. In fact, to the best of our knowledge, this is the first place where the analytic details of their construction are given in full.

More specifically, the topics covered include:

- **Chapter 1:** the definition and main properties of the GFF with Dirichlet (or zero) boundary conditions;

- **Chapter 2:** the construction of the Liouville measure (in the GFF case), its non-degeneracy and change of coordinate formula;
- **Chapter 3:** a comprehensive exposition of the construction and properties of general Gaussian multiplicative chaos measures;
- **Chapter 4:** an introduction to statistical mechanics on random planar maps – the discrete counterparts of Liouville quantum gravity – and Sheffield's bijection with pairs of trees [She16b];
- **Chapter 5:** an introduction to Liouville conformal field theory, as developed in a series of papers starting with [DKRV16] by David, Kupiainen, Rhodes and Vargas;
- **Chapter 6:** the definition, construction and main properties of the GFF with Neumann boundary conditions;
- **Chapter 7:** an account of the notion of quantum surfaces, and the theory of special quantum surfaces enjoying scale invariance, such as quantum spheres, discs, wedges and cones, and a proof of equivalence with aspects of the theory developed in Chapter 5;
- **Chapter 8:** an exposition of Sheffield's quantum zipper theorem (with novel additional details) and its relation with conformal welding [She16a];
- **Chapter 9:** an introduction to the powerful mating-of-trees theory of Duplantier, Miller and Sheffield [DMS21]. This includes an extensive and partly novel treatment of space-filling and whole-plane SLE.

The final three topics above are rather technical, and readers are advised that it will be of most use to people who are actively working in this area. See also Figure 1 for a reading guide.

The theory is in full blossom and attempting to make a complete survey of the field would be hopeless, so quickly is it developing. Nevertheless, as the theory grows in complexity and applicability, it has appeared useful to summarise some of its basic and foundational aspects in one place, especially since complete proofs of some facts can be spread over a multitude of papers.

Clearly, the main drawback of this approach is that many of the important subsequent developments and alternative points of view are not included. For instance: the expansive body of work on random planar maps and their rigorous connections with Liouville quantum gravity, the Brownian map, Liouville Brownian motion, imaginary geometry, imaginary chaos and the Liouville quantum gravity metric, do not feature in this book. For all this, we apologise in advance.

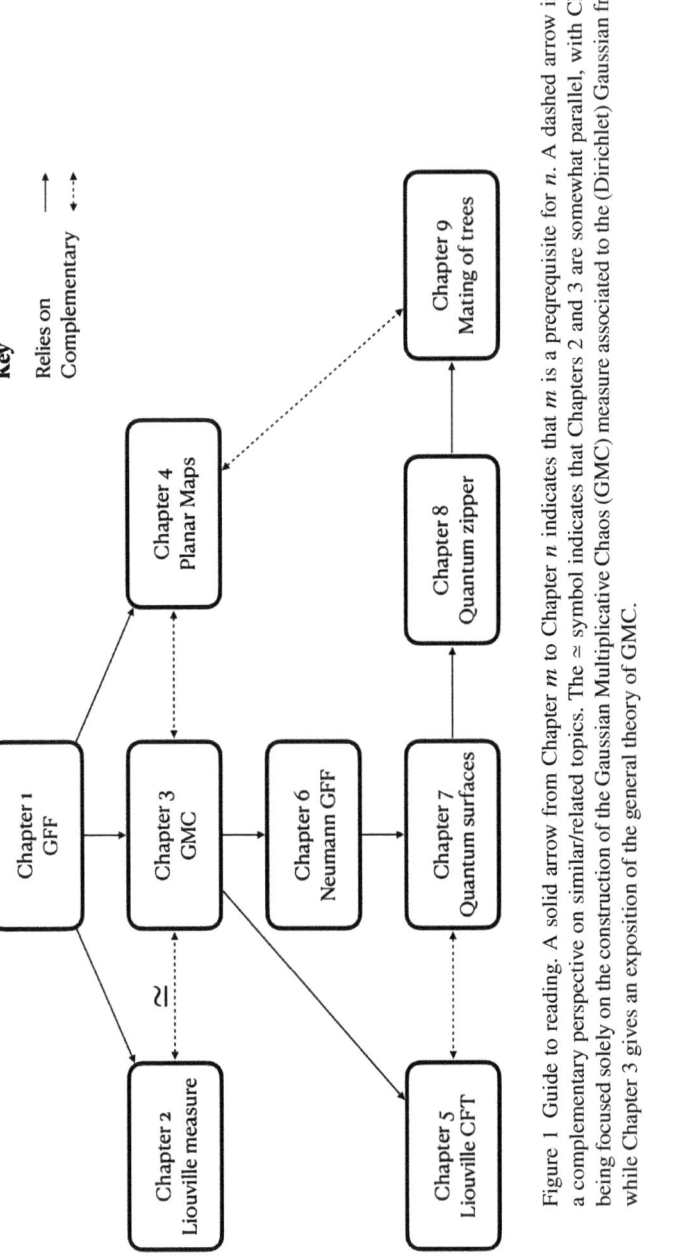

Figure 1 Guide to reading. A solid arrow from Chapter m to Chapter n indicates that m is a prerequisite for n. A dashed arrow indicates a complementary perspective on similar/related topics. The $\simeq$ symbol indicates that Chapters 2 and 3 are somewhat parallel, with Chapter 2 being focused solely on the construction of the Gaussian Multiplicative Chaos (GMC) measure associated to the (Dirichlet) Gaussian free field, while Chapter 3 gives an exposition of the general theory of GMC.

Acknowledgements

An initial draft was written by the first-named author, in preparation for the LMS/Clay Institute research school on *Modern Developments in Probability* taking place in Oxford, July 2015. The draft was subsequently revised on the occasion of several courses given on this material: at the Spring School on *Geometric Models in Probability* in Darmstadt, then in July 2016 for the Probability Summer School at Northwestern (for which the chapter on statistical physics on random planar maps was added), in December 2017 for the Lectures on Probability and Statistics (LPS) at ISI Kolkata and in Berlin at the Stochastic Analysis in Interaction summer school, in 2023. The second-named author lectured on parts of this material in Helsinki in 2022, Santiago de Chile in 2023 and Guanajuato (CIMAT) in 2023.

In all cases, we thank the organisers (Christina Goldschmidt and Dmitry Beliaev; Volker Betz and Matthias Meiners; Antonio Auffinger and Elton Hsu; Arijit Chakrabarty, Manjunath Krishnapur, Parthanil Roy and especially Rajat Subhra Hazra; Peter Friz and Peter Bank; Eero Saksman and Eveliina Peltola; Avelio Sepúlveda and Daniel Remenik; and Daniel Kious, Andreas Kyprianou, Sandra Palau and Juan Carlos Pardo) for their invitations and superb organisation. Thanks also to Benoit Laslier for agreeing to run the exercise sessions accompanying the lectures at the initial school.

The Isaac Newton Institute's semester on *Random Geometry* in 2015 was another important influence and motivation for this book, and we would like to thank the INI for its hospitality. In fact, this semester served as the second author's initiation into the world of the Gaussian free field and Liouville quantum gravity. The resulting years of discussions between us has led to the present expanded and revised version. We would like to thank many of the INI programme participants for enlightening discussions related to aspects of the book; especially, Omer Angel, Juhan Aru, Stéphane Benoît, Bertrand

Duplantier, Ewain Gwynne, Nina Holden, Henry Jackson, Benoit Laslier, Jason Miller, James Norris, Gourab Ray, Scott Sheffield, Xin Sun, Wendelin Werner and Ofer Zeitouni. Special thanks in particular to Juhan Aru, Ewain Gwynne, Nina Holden and Xin Sun for many inspiring discussions over the years over a broad range of topics.

We would also like to thank the participants of two reading groups at the University of Bonn and ETH Zürich/University of Zürich, respectively (particularly the organisers Nina Holden and Eveliina Peltola), which followed these notes and led to many helpful comments. We also received important feedback following graduate courses based on this material which took place at MIT, University of Washington and University of Vienna. Participants and organisers (Scott Sheffield and Zhen-Qing Chen) are warmly acknowledged. We would particularly like to thank Scott Sheffield for his constant encouragement throughout.

A substantial part of the writing took place while the authors were invited participants to the semester on *The Analysis and Geometry of Random Spaces* which took place at MSRI in 2022 (now Simons–Laufer Mathematical Sciences Institute) in Berkeley, California. We are very grateful to the organisers of the programme (Mario Bonk, Steffen Rohde, Joan Lind, Eero Saksman, Fredrik Viklund and Jang-Mei Wu) for this amazing opportunity. We also thank the many other participants of this programme for the pleasant atmosphere and the many comments we received while working on Chapter 5 of this book in connection with the reading group on Liouville CFT.

Finally, comments on versions of this draft have been received at various stages from Morris Ang, Juhan Aru, Jacopo Borga, Zhen-Qing Chen, William Da Silva, Nina Holden, Henry Jackson, Jakob Klein, Aleksandra Korzhenkova, Benoit Laslier, Joona Oikarinen, Léonie Papon, Eveliina Peltola, Gourab Ray, Mark Sellke, Huy Tran, Joonas Turunen, Fredrik Viklund, Mo-Dick Wong, Henrik Weber and Dapeng Zhan. We are grateful for their input which helped to correct minor problems, as well as to emphasise some of the subtle aspects of the arguments; as a result, this text is much better than it would have been otherwise. Of course we hasten to add that we retain the entire responsibility for any remaining typo, error, omission or lack of clarity.

This book has been greatly enhanced by a number of beautiful simulations reproduced here with the permission of their authors. These are, respectively: Jason Miller (Figure 4.3, which can also be seen on the cover, and Figure 4.9); Jérémie Bettinelli and Benoit Laslier (Figure 4.2), Henry Jackson (Figure B.2) and Oskar-Laurin Koiner (Figures 1.1 and 1.2). We thank them wholeheartedly.

Acknowledgements

The work of the first author was supported during various stages of the writing by EPSRC (via grants EP/I03372X/1 and EP/L018896/1) and the FWF (via grants 10.55776/P33083 and 10.55776/F1002), while the second author has been supported by the SNF grant 175505 and the UKRI Future Leader's Fellowship MR/W008513/1. This support is gratefully acknowledged.

1
Definition and Properties of the GFF

1.1 Discrete case

Note: The discrete case is included here only for the purpose of guiding intuition when we come to work in the continuum.

Consider a finite, weighted, undirected graph $\mathcal{G} = (V, E)$ (with weights $(w_e)_{e \in E}$ on the edges). For instance, $\mathcal{G}$ could be a finite portion of the Euclidean lattice $\mathbb{Z}^d$ with weights $w_e \equiv 1$. Let ∂ be a distinguished set of vertices, called the boundary of the graph, and set $\hat{V} = V \setminus \partial$. Let $(X_t)_{t \geq 0}$ be the random walk on $\mathcal{G}$ in continuous time, meaning that it jumps from x to y at rate $w_{x,y}$, and let τ be the first time that X hits ∂ (which we assume to be finite almost surely for every starting point).

Write $Q = (q_{x,y})_{x,y \in V}$ for the Q-matrix of X. That is, its infinitesimal generator, so that for each $x \in V$, $q_{x,y} = w_{x,y}$ for $y \neq x$ and $q_{x,x} = -\sum_{y \sim x} w_{x,y} < \infty$, where $y \sim x$ means that x and y are connected by an edge in E. Note that the uniform measure $\pi(x) \equiv 1$ is reversible for X. We write $\mathbb{P}_x$ for the law of the random walk started and $x \in V$ and $\mathbb{E}_x$ for the corresponding expectation.

Definition 1.1 (Green function) The Green function $G(x, y)$ is defined for any $x, y \in V$ by setting

$$G(x, y) = \mathbb{E}_x \left(\int_0^\infty \mathbf{1}_{\{X_t = y; \tau > t\}} \, dt \right).$$

In other words, $G(x, y)$ is the expected time that X spends at y, when started from x, before hitting the boundary. Note that with this definition, we have $G(x, y) = G(y, x)$ for all $x, y \in \hat{V}$, since $\mathbb{P}_x(X_t = y; \tau > t) = \mathbb{P}_y(X_t = x; \tau > t)$ by reversibility of X with respect to π.

An equivalent expression for the Green function when working with the random walk in discrete time $Y = (Y_n)_{n \geq 0}$ (which jumps from x to y with probability proportional to $w_{x,y}$) is

$$G(x, y) = \frac{1}{q_y} \mathbb{E}_x \left(\sum_{n=0}^{\infty} \mathbf{1}_{\{Y_n = y; \tau(Y) > n\}} \right), \quad (1.1)$$

where $q_y = \sum_{y \sim x} w_{x,y} = -q_{y,y}$ and $\tau(Y)$ is the first time that Y hits ∂D. Indeed, X and Y can be coupled in such a way that for each $y \in \hat{V}$ and each visit of Y to y, X stays at y for an exponentially distributed time with mean $1/q_y$.

The Green function is a basic ingredient in the definition of the Gaussian free field, so the following elementary properties will be important to us.

Proposition 1.2 *Let $\hat{Q}$ denote the restriction of Q to $\hat{V} \times \hat{V}$. Then*

1. $(-\hat{Q})^{-1}(x, y) = G(x, y)$ *for all $x, y \in \hat{V}$.*
2. *G is a symmetric and non-negative definite function. That is, one has*

$$G(x, y) = G(y, x)$$

for all $x, y \in V$, and if $(\lambda_x)_{x \in V}$ is any vector of length $|V|$, then

$$\sum_{x, y \in V} \lambda_x \lambda_y G(x, y) \geq 0.$$

Equivalently, G is symmetric and therefore diagonalisable (when viewed as a matrix), and all of the eigenvalues of G are non-negative. Furthermore, restricted to $\hat{V}$, G is a positive definite function (i.e. its eigenvalues are strictly positive).

3. *$G(x, \cdot)$ is discrete harmonic in $\hat{V} \setminus \{x\}$; more precisely G is the unique function of $x, y \in V$ such that $\hat{Q}G(x, \cdot) = -\delta_x(\cdot)$ for all $x \in \hat{V}$, and satisfies the "boundary condition" $G(x, \cdot) = 0$ on ∂ for all $x \in V$.*

Here, $\delta_x(\cdot)$ denotes the Dirac function at x, namely $\delta_x(\cdot) = \mathbf{1}_{\{\cdot = x\}}$. We also use the natural notation $Qf(x) = \sum_{y \sim x} q_{xy}(f(y) - f(x))$ for the action of the generator Q on functions. Viewed as an operator in this way, Q is often referred to as the discrete Laplacian in continuous time. (Note that by definition, $Qf(x)$ measures the infinitesimal expected change in $f(X_t)$ if the chain starts at x.)

Remark 1.3 The proof below is written in the formalism of continuous time Markov chains, which is a little more natural. However, it can equivalently be written using discrete-time Markov chains and the definition of the Green function in (1.1).

1.1 Discrete case

Proof Note that since $\hat{Q}$ is symmetric, it is diagonalisable, and that all its eigenvalues are negative (this is true of the infinitesimal generator of any Markov chain in continuous time, and here $\hat{Q}$ is nothing else but the infinitesimal generator of the Markov chain absorbed at ∂). Since the chain is absorbed at ∂, 0 is not an eigenvalue and all the eigenvalues of $\hat{Q}$ are therefore strictly negative.

Furthermore, if $\hat{P}^t(x, y) = \mathbb{P}_x(X_t = y, \tau > t)$, then $\hat{P}_t$ satisfies the backward Kolmogorov equation, namely

$$(d/dt)\hat{P}^t(x, y) = \hat{Q}\hat{P}^t(x, y),$$

so that $\hat{P}^t(x, y) = e^{\hat{Q}t}(x, y) = 1 + \sum_{j \geq 1} \frac{1}{j!}(\hat{Q})^j(x, y)$. It then follows, by Fubini, that

$$\begin{aligned} G(x, y) &= \mathbb{E}_x\left(\int_0^\infty \mathbf{1}_{\{X_t = y; \tau > t\}} \, dt\right) \\ &= \int_0^\infty \hat{P}^t(x, y) \, dt \\ &= \int_0^\infty e^{\hat{Q}t}(x, y) \, dt \\ &= (-\hat{Q})^{-1}(x, y). \end{aligned} \quad (1.2)$$

The justification for the last equality comes from thinking about the action of the operator $\int_0^\infty e^{\hat{Q}t} \, dt$ on a single eigenfunction of $\hat{Q}$ (recalling that the corresponding eigenvalue is negative). Since there is a basis of eigenfunctions of $\hat{Q}$ by symmetry, this suffices to prove the last equality.

For the second point, we have already mentioned that $G(x, y) = G(y, x)$. Since $G(x, y) = 0$ whenever $y \in \partial$ it suffices to show that the restriction of G to $\hat{V}$ is positive definite. For this, we can use again that all the eigenvalues of $-\hat{Q}$, and hence of $(-\hat{Q})^{-1}$ are positive. This gives that G is positive definite when restricted to $\hat{V}$, by (1.2).

Let us finally check the third point. This can be seen as a straightforward consequence of the first point, but we prefer to also include a probabilistic proof which is based on the Markov property; effectively, we decompose according to the first jump of the chain.[1] Let $L(x) = \int_0^\infty \mathbf{1}_{\{X_t = x; \tau > t\}} \, dt$. Suppose that $y \neq x$ and $t \geq 0$. If $X_0 = y$ and J is the first time that X jumps away from y (so J is an exponential random variable with rate $q_y = -q_{y,y}$), we can decompose $\mathbb{E}_y(L(x))$ according to whether $\{J > t\}$ or $\{J = s \text{ for some } 0 \leq s \leq t\}$). Also

---

[1] The analogous derivation of the same fact using the discrete-time chain Y instead of the continuous time chain X is in fact slightly simpler – we recommend this as an exercise for the reader!

applying the Markov property at time J, we obtain that

$$G(x,y) = G(y,x) = \mathbb{E}_y(L(x))$$
$$= \mathbb{E}_y(L(x)|J > t)\mathbb{P}_y(J > t)$$
$$+ \int_0^t \sum_{z \neq y} \mathbb{P}_y(J \in ds, X_J = z)\mathbb{E}_y(L(x)|J = s, X_J = z)$$
$$= G(y,x)e^{-q_y t} + \int_0^t q_y e^{-q_y s} ds \sum_{z \neq y} \frac{q_{y,z}}{q_y} \mathbb{E}_z(L(x)).$$

Taking the time derivative on both sides at $t = 0$ and again invoking symmetry, we arrive at the equality

$$0 = -q_y G(y,x) + \sum_{z \neq y} q_{y,z} G(z,x) = -q_y G(x,y) + \sum_{z \neq y} q_{y,z} G(x,z).$$

This means (for fixed x, viewing $G(x,y)$ as a function $g(y)$ of y only) that $Qg(y) = \sum_z q_{y,z} g(z) = 0$. Hence $G(x,\cdot)$ is harmonic in $\hat{V} \setminus \{x\}$.

When $y = x$, a similar argument can be made, but now the event $\{J > t\}$ contributes to $L(x)$, namely,

$$G(x,x) = \mathbb{P}(J > t)(t + G(x,x)) + \int_0^t q_x e^{-q_x s} ds \sum_{z \neq x} \frac{q_{x,z}}{q_x} \mathbb{E}_z(L(x))$$
$$= e^{-q_x t}(t + G(x,x)) + \int_0^t e^{-q_x s} \sum_{z \neq x} q_{x,z} G(x,z) ds.$$

Taking the derivative of both sides at $t = 0$ gives

$$0 = -q_x G(x,x) + 1 + \sum_{z \neq x} q_{x,z} G(x,z),$$

and hence

$$\sum_z q_{xz} G(x,z) = -1.$$

The uniqueness comes from the invertibility of $-\hat{Q}$. □

Remark 1.4 An alternative proof of the first point (i.e. of (1.2)) uses the transition matrix $\hat{R}^n(x,y) = \mathbb{P}_x(Y_n = y, \tau(Y) > n)$ of the jump chain. Indeed, we have already noted that

$$G(x,y) = \frac{1}{q_y} \mathbb{E}_x \left(\sum_{n=0}^{\infty} \mathbf{1}_{\{Y_n = y, \tau(Y) > n\}} \right)$$
$$= \frac{1}{q_y} \sum_{n=0}^{\infty} \hat{R}^n(x,y)$$

1.1 Discrete case

$$= \frac{1}{q_y}(I - \hat{R})^{-1}(x, y)$$
$$= (-\hat{Q})^{-1}(x, y),$$

where in jumping from the second to the third line, we used the fact that $\sum_{n=0}^{\infty} \hat{R}^n = (I - \hat{R})^{-1}$, an identity valid for any matrix of spectral radius (i.e. largest eigenvalue modulus) strictly smaller than 1, which is the case here. An alternative proof that G is non-negative definite can be obtained using the same argument in the proof of Lemma 1.28 (this is stated in the continuous case but can also easily be adapted to this discrete setting).

Definition 1.5 (Discrete GFF) The (zero boundary) discrete Gaussian free field on $\mathcal{G} = (V, E)$ is the centred Gaussian vector $(h(x))_{x \in V}$ with covariance given by the Green function G.

Remark 1.6 This definition is justified. Indeed, suppose that $(C(x, y))_{x,y \in V}$ is a given function. Then there exists a centred Gaussian vector X having covariance matrix C if and only if C is symmetric and non-negative definite (in the sense of property 2 above).

Note that if $x \in \partial$, then $G(x, y) = 0$ for all $y \in V$ and hence $h(x) = 0$ almost surely.

In fact, it is possible to provide a concrete construction of the discrete Gaussian free field, in terms of i.i.d. standard Gaussian random variables. This construction has the advantage that it is very easy to implement on a computer to produce simulations, such as the one in Figure 1.1. We first introduce some notations. Set $N = |\hat{V}|$ and consider the space of functions $f : \hat{V} \to \mathbb{R}$, equipped with the inner product

$$(f, g) = \sum_{x \in \hat{V}} f(x)g(x). \tag{1.3}$$

For this reason (and even though $\hat{V}$ is finite), we denote this space of functions by $\ell^2(\hat{V})$. Any function in $\ell^2(\hat{V})$ can canonically be extended to a function on V by setting it to zero on ∂. Recall that a function $f \in \ell^2(\hat{V})$ is an eigenfunction of $-\hat{Q}$ with eigenvalue λ (necessarily positive) if for all $x \in \hat{V}$, $-\hat{Q}f(x) = \lambda f(x)$, that is,

$$-\sum_{y \in \hat{V}} q_{x,y} f(y) = \lambda f(x).$$

As already mentioned in the proof of Proposition 1.2, since $-\hat{Q}$ is symmetric, it is diagonalisable in an orthonormal basis of $\ell^2(\hat{V})$. Let $e_1, \ldots, e_N$ denote the orthonormal eigenfunctions and let $0 < \lambda_1 \leq \cdots \leq \lambda_N$ denote the corresponding eigenvalues (with multiplicities).

Theorem 1.7 *Let $(e_m)_{m=1}^N$ and $(\lambda_m)_{m=1}^N$ be as above. Then for $x, y \in \hat{V}$ we have the expansion*

$$G(x,y) = \sum_{m=1}^N \frac{1}{\lambda_m} e_m(x) e_m(y). \tag{1.4}$$

This also extends to ∂ if we extend e_m by zero on ∂.

Furthermore, let $(X_m)_{m=1}^N$ be a sequence of i.i.d. standard Gaussians. Set

$$h(x) := \sum_{m=1}^N \frac{X_m}{\sqrt{\lambda_m}} e_m(x); \quad x \in V \tag{1.5}$$

Then h is a discrete GFF on $\mathcal{G}$.

Proof Since $-\hat{Q}$ is invertible, it suffices to check that for $x \in \hat{V}$, if we define $g(y) = g_x(y)$ as on the right-hand side of (1.4), namely $g(y) = \sum_{m=1}^N (1/\lambda_m) e_m(x) e_m(y)$ viewed as a function of $y \in \hat{V}$, one has

$$-\hat{Q}g = \delta_x. \tag{1.6}$$

By linearity and since e_m is an eigenfunction of $-\hat{Q}$, we see that (recall that $x \in \hat{V}$ is fixed)

$$-\hat{Q}g = \sum_{m=1}^N \frac{1}{\lambda_m} e_m(x)(-\hat{Q}e_m) = \sum_{m=1}^N e_m(x) e_m.$$

On the other hand, expanding δ_x in the basis $(e_m)_{m=1}^N$, we also find that

$$\delta_x = \sum_{m=1}^N (\delta_x, e_m) e_m = \sum_{m=1}^N e_m(x) e_m,$$

which is indeed the same as the right-hand side of the previous equation. This proves (1.6) and thus (1.4).

Turning to (1.5), we simply note that $(h(x))_{x \in \hat{V}}$ is clearly a centred Gaussian vector, whose covariance is given by

$$\mathbb{E}[h(x)h(y)] = \mathbb{E}\left(\sum_{m,m'=1}^N \frac{X_m X_{m'}}{\sqrt{\lambda_m \lambda_{m'}}} e_m(x) e_{m'}(y) \right)$$

$$= \sum_{m=1}^N \frac{1}{\lambda_m} e_m(x) e_m(y) = G(x,y)$$

by (1.4), as desired. □

1.1 Discrete case

Figure 1.1 A discrete Gaussian free field. Simulation by Oskar-Laurin Koiner.

Usually for Gaussian fields, looking at the covariance structure is the most useful way of gaining intuition. However, in this case, the joint probability density function of the $|V|$ components of h is perhaps more illuminating.

Theorem 1.8 (Law of the GFF and Dirichlet energy) *The law of the discrete GFF is absolutely continuous with respect to Lebesgue measure on $\mathbb{R}^{\hat{V}}$, with joint density proportional to*

$$\exp\left(-\frac{1}{4}\sum_{x,y\in V} q_{x,y}(h(x)-h(y))^2\right)$$

at any point $(h(x))_{x\in V}$ with $h(x) = 0$ for $x \in \partial$, viewed as a fixed element of $\mathbb{R}^{\hat{V}}$. (Note that the sum includes the vertices $v \in \partial$.)

Remark 1.9 The previous formula might seem a little confusing at first, since we are using $(h(x))_{x\in\hat{V}}$ both to denote the random vector consisting of the values of the discrete Gaussian free field, and for a fixed (deterministic) element in $\mathbb{R}^{\hat{V}}$ at which we evaluate the density of this random vector. To avoid any confusion, the formula above means the following: if we write Y_v for the random variable $Y_v := h(v)$, then

$$\mathbb{P}((Y_v)_{v\in V} \in A) = \int_A \frac{1}{Z}\exp(-\frac{1}{4}\sum_{v,w\in V} q_{v,w}(x_v - x_w)^2)\prod_{v\in\hat{V}}dx_v,$$

where $Z = \int_{\mathbb{R}^N}\exp(-\frac{1}{4}\sum_{v,w\in V} q_{v,w}(x_v - x_w)^2)\prod_{v\in\hat{V}}dx_v$, where $N = |\hat{V}|$. This holds for any Borel set A contained in the hyperplane $\{(x_v)_{v\in V}: x_v = 0 \text{ for all } v \in \partial\}$ of $\mathbb{R}^V$.

For a given function $f: V \to \mathbb{R}$, the quantity

$$\mathcal{E}(f,f) := \frac{1}{2}\sum_{x,y\in V} q_{x,y}(f(x)-f(y))^2 \tag{1.7}$$

8 Definition and Properties of the GFF

is known as the **Dirichlet energy** of f: it is a discrete analogue of $(1/2) \int_D |\nabla f|^2$.

Proof of Theorem 1.8 The result follows from the fact that for a centred Gaussian vector $(Y_1, \ldots, Y_N)$ with invertible covariance matrix Σ, the joint probability density function on $\mathbb{R}^N$ is proportional to

$$f(x_1, \ldots, x_N) = \exp(-\frac{1}{2} x^T \Sigma^{-1} x).$$

For us, the vertices $v \in \hat{V}$ play the roles of the indices $1 \le i \le N$ above with $N = |\hat{V}|$, and the values $h(v)$ for $v \in V$ play the roles of the x_i (to get a non-degenerate covariance matrix we restrict ourselves to vertices in $\hat{V}$, in which case G is invertible by Proposition 1.2). Note that since we are only considering h with $h(v) = 0$ for $v \in \partial$, it suffices to show that

$$-\frac{1}{2} h(\hat{\mathbf{v}})^T G^{-1} h(\hat{\mathbf{v}}) = -\frac{1}{4} \sum_{x,y \in V} q_{x,y} (h(x) - h(y))^2, \quad \text{for } h(\hat{\mathbf{v}}) = (h(v))_{v \in \hat{V}}.$$

Recall that $(-\hat{Q})^{-1}(x, y) = G(x, y)$ for $x, y \in \hat{V}$, so that $G^{-1}(x, y) = -q_{xy}$. Hence

$$h(\hat{\mathbf{v}})^T G^{-1} h(\hat{\mathbf{v}}) = \sum_{x,y \in \hat{V}} G^{-1}(x, y) h(x) h(y) = \sum_{x,y \in \hat{V}} -q_{x,y} h(x) h(y).$$

Moreover, as we only consider h with $h(x) = 0$ for $x \in \partial$, this can be rewritten as

$$-\sum_{x,y \in V} q_{x,y} h(x) h(y) = \frac{1}{2} \sum_{x,y \in V} q_{x,y} (h(x) - h(y))^2 - \frac{1}{2} \sum_{x,y \in V} h(x)^2 q_{x,y}$$
$$- \frac{1}{2} \sum_{x,y \in V} h(y)^2 q_{x,y},$$

where since $\sum_{y \in V} q_{x,y} = 0$ and $q_{x,y} = q_{y,x}$ for all x, y, the terms

$$\sum_{x,y \in V} h(x)^2 q_{x,y} \quad \text{and} \quad \sum_{x,y \in V} h(y)^2 q_{x,y}$$

are both equal to 0. Note that in this final line of reasoning, it is important to sum over all of V and not just $\hat{V}$. Thus

$$-\frac{1}{2} h(\hat{\mathbf{v}})^T G^{-1} h(\hat{\mathbf{v}}) = -\frac{1}{2} \times \frac{1}{2} \sum_{x,y \in V} q_{x,y} (h(x) - h(y))^2,$$

as required. □

Notice that the Dirichlet energy of functions is minimised by harmonic functions. This means that the Gaussian free field can be viewed as a "Gaussian perturbation of a harmonic function": as much as possible, it "tries" to be harmonic. In fact, this is a little ironic, given that in the continuum it is not even a function (see Sections 1.3 and 1.7).

This heuristic is at the heart of the Markov property, which is without a doubt the most useful property of the GFF. We state it here without proof, as we will soon prove its (very similar) continuum counterpart.

Theorem 1.10 *[Markov property of the discrete GFF] Fix $U \subset V$. The discrete GFF $h = (h(x))_{x \in V}$ can be decomposed as*

$$h = h_0 + \varphi,$$

where h_0 is Gaussian free field on U and φ is harmonic in U. Moreover, h_0 and φ are independent.

By a Gaussian free field in U, we mean the GFF on the graph (V, E) but now with $\partial = V \setminus U$, in particular $h_0 = 0$ outside of U.

In other words, this theorem says that conditionally on the values of h outside of U, the field can be written as the sum of two independent terms. One of these is a zero boundary GFF in U, and the other is just the harmonic extension into U of the values of h outside U. To see this, note that the information about the values of h outside of U is completely contained in φ, since h_0 is zero outside of U. Thus conditioning on the values of h outside of U is the same as conditioning on φ. Since h_0 is independent of φ, the conditional law of h given φ is as described.

1.2 Continuous Green function

We will follow a route that is similar to the previous discrete case. First we need to recall the definition of the Green function. We will only cover the basics here, and readers who want to know more are advised to consult, for instance, Lawler's book [Law05] which reviews important facts in a very accessible way. The presentation here will be somewhat different.

Let $d \geq 1$. Let $p_t(x, y)$ denote the transition probability of a Brownian motion B in $\mathbb{R}^d$ with "speed" two (i.e. $B_t = (X^1_{2t}, \ldots, X^d_{2t})$ for $t \geq 0$, where $X^1, \ldots, X^d$ are independent standard Brownian motions[2] in $\mathbb{R}$). Then

$$p_t(x, y) = (4\pi t)^{-d/2} \exp(-|x - y|^2/(4t)), \tag{1.8}$$

which by the Markov property is also the density, with respect to Lebesgue

---

[2] This choice ensures that the infinitesimal generator of B is the Laplace operator Δ instead of $\Delta/2$.

measure on $\mathbb{R}^d$, of the law of B_t (when started from x). For $D \subset \mathbb{R}^d$ an open set, we define $p_t^D(x, y)$ to be the transition probability of Brownian motion with speed two, killed when leaving D, which is defined as the density, with respect to Lebesgue measure on $\mathbb{R}^d$, of the law of B_t, but restricted to the event $\{\tau_D > t\}$, where

$$\tau_D = \inf\{t > 0 \colon B_t \notin D\}.$$

In other words, for any Borel set A in $\mathbb{R}^d$, it satisfies

$$\mathbb{P}_x(B_t \in A, \tau_D > t) = \int_{\mathbb{R}^d} 1_A(y) p_t^D(x, y) \, dy. \tag{1.9}$$

The (almost everywhere, for a fixed $t \geq 0$) existence of a function satisfying (1.9) follows directly from the Radon–Nikodym derivative theorem, since it is clear that if A has zero Lebesgue measure, then $\mathbb{P}_x(B_t \in A, \tau_D > t) \leq \mathbb{P}_x(B_t \in A) = 0$.

By conditioning on the position at time t of B_t, it is not hard to check that $p_t^D(x, y)$ can be expressed rather simply in terms of the whole space transition probabilities in (1.8) and the so-called (speed two) Brownian bridge[3] $(b_s)_{0 \leq s \leq t}$ of duration t from x to y, which describes the law of B, conditionally given $B_0 = x$ and $B_t = y$. Namely, if we denote by $\mathbb{P}_{x \to y;t}$ this law, then we see that

$$\mathbb{P}_x(B_t \in A; \tau_D > t) = \int_{\mathbb{R}^d} \mathbb{P}_{x \to y;t}(b_t \in A; \tau_D > t) p_t(x, y) \, dy$$

$$= \int_{\mathbb{R}^d} 1_A(y) \mathbb{P}_{x \to y;t}(\tau_D > t) p_t(x, y) \, dy.$$

Comparing with (1.9), we deduce that, for every fixed $t \geq 0$ and almost every y,

$$p_t^D(x, y) = \pi_t^D(x, y) p_t(x, y); \text{ where } \pi_t^D(x, y) = \mathbb{P}_{x \to y;t}(\tau_D > t). \tag{1.10}$$

The right-hand side is easily seen to be a jointly continuous function in $t > 0$ and $x, y \in \overline{D}$, as this is clearly satisfied by both $\pi_t^D(x, y)$ and $p_t(x, y)$ separately. This defines the transition probability function $p_t^D(x, y)$ of Brownian motion killed when leaving D uniquely.

Clearly, by the Markov property of Brownian motion, the transition probabilities satisfy the Chapman–Kolmogorov equation:

$$p_{t+s}^D(x, y) = \int_{\mathbb{R}^d} p_t^D(x, z) p_s^D(z, y) \, dz \quad \text{for } s, t \geq 0, x, y \in D. \tag{1.11}$$

Note also immediately for future reference that, by definition of $p_t^D(x, y)$

---

[3] A reader who is unfamiliar with the notion of Brownian bridge may without danger skip to the conclusion immediately following (1.10).

1.2 Continuous Green function

and the monotone class theorem, if ϕ is any non-negative Borel function and $t \geq 0$, then

$$\mathbb{E}_x(\phi(B_t)1_{\{\tau_D>t\}}) = \int_{\mathbb{R}^d} \phi(y) p_t^D(x,y) \, dy.$$

Consequently, by Fubini's theorem,

$$\begin{aligned}\mathbb{E}_x(\int_0^{\tau_D} \phi(B_s) \, ds) &= \mathbb{E}_x(\int_0^{\infty} \phi(B_s)1_{\{\tau_D>s\}} \, ds) \\ &= \int_0^{\infty} \int_{\mathbb{R}^d} \phi(y) p_s^D(x,y) \, dy \, ds \\ &= \int_{\mathbb{R}^d} \phi(y)(\int_0^{\infty} p_s^D(x,y) \, ds) \, dy. \end{aligned} \quad (1.12)$$

The time integral in brackets in (1.12) plays a crucial role in this book and is called the (continuous) Green function. Note the parallel with Definition 1.1; intuitively, as in the discrete case, the Green function measures the expected amount of time spent "at" a point y (i.e. near y) before exiting D.

Definition 1.11 (Continuous Green function)   Let $D \subset \mathbb{R}^d$ be an open set. The Green function $G_0(x,y) = G_0^D(x,y)$ is defined by

$$G_0(x,y) = \int_0^{\infty} p_t^D(x,y) \, dt \quad (1.13)$$

for $x \neq y$ in D.

Note in particular that, combining our definition of the Green function with (1.12), we obtain the following:

$$\mathbb{E}_x(\int_0^{\tau_D} \phi(B_s) \, ds) = \int_{\mathbb{R}^d} G_0^D(x,y)\phi(y) \, dy. \quad (1.14)$$

This agrees with our intuition that the Green function measures the expected amount of time spent by a Brownian motion near a point y before leaving D.

Remark 1.12 (Normalisation)   We call the attention of the reader to the fact that the normalisation of the Green function is a little arbitrary. We have chosen to normalise it so that G, as we will soon see, is the inverse of (minus) the Laplacian, with no multiplicative constant in front. This choice is consistent with say [WP21]. In particular, in two dimensions, our normalisation is chosen so that for $D \subset \mathbb{C}$ simply connected, say, we will have

$$G_0^D(x,y) \sim \frac{1}{2\pi} \log(|x-y|^{-1})$$

as $y \to x$ (see Proposition 1.18). This is however *not* the standard set-up for

12   *Definition and Properties of the GFF*

Gaussian multiplicative chaos (see Chapter 2 and 3) or in papers on Liouville quantum gravity, where the Green function is often normalised so that it blows up like $\log(|x-y|^{-1})$ (i.e. it differs from our choice by a factor of 2π). This means that the Gaussian free field we are about to define will differ by a factor of $\sqrt{2\pi}$ from the field usually considered in the Gaussian multiplicative chaos literature, and which we will also switch to in Chapters 2 and 3. While from the point of view of Gaussian multiplicative chaos, it is more natural to define the Gaussian free field as a log-correlated field rather than a $(2\pi)^{-1}$-log-correlated field, we have chosen the above normalisation of the Green function for this chapter, since it is more natural from an analytic perspective. In particular, it saves us many tedious powers of 2π in our subsequent considerations involving Sobolev spaces.

Another commonly used normalisation of the Green function corresponds to the integral of the transition density for Brownian motion with speed 1 rather than speed 2. This Green function differs from ours by a factor of 2, and the resulting Gaussian free field by a factor of $\sqrt{2}$.

We will sometimes drop the notational dependence of G_0^D on D when it is clear from the context. The subscript 0 refers to the fact that G has **zero boundary conditions**; equivalently, that G is defined from a Brownian motion killed when leaving D.

When $d \geq 2$, it is easy to see that $G_0^D(x,x)$ is typically ill-defined ($= \infty$) for all $x \in D$. This is because $\pi_t^D(x,x) \to 1$ as $t \to 0$ and so $(4\pi t)^{-d/2}\pi_t^D(x,x)$ cannot be integrable. When $x \neq y$, the bound $p_t^D(x,y) \leq p_t(x,y)$ suffices to show there can be no problem of integrability near $t = 0$. Furthermore, at least if $d \geq 3$, $p_t(x,y)$ is also integrable near $t = \infty$, so that $G_D(x,y) < \infty$.

The case $d = 2$ is more delicate. In fact, it can be shown that in that case, then $G_0^D(x,y) < \infty$ if $x \neq y \in D$ and D has at least one **regular** boundary point, that is, a point $b \in \partial D$ such that $\mathbb{P}_b(\tau_D = 0) = 1$ (in other words, starting from a boundary point, a Brownian motion leaves D instantaneously – equivalently, $b \in \partial D$ is regular if for all $\varepsilon > 0$, $\mathbb{P}_z(\tau_D > \varepsilon) \to 0$ as $z \in D$ converges to b). This follows for instance from Lemma 2.32 in [Law05], which gives a uniform bound of the form

$$p_t^D(x,y) \leq ct^{-1}(\log t)^{-2} \quad (1.15)$$

uniformly over $x, y \in D$ and $t > 1$ for a regular domain D. Any proper simply connected open set in two dimensions is easily seen to be regular, that is, every point on the boundary of D is regular. (When the domain is simply connected, an even stronger bound than (1.15) holds, and is in fact easier to prove, see [BN11].) We also note for further reference that (1.15) implies

1.2 Continuous Green function

(through dominated convergence and the continuity of both $\pi_t^D(x, y)$ and that of $p_t(x, y)$) that $(x, y) \in D^2 \mapsto G_0^D(x, y)$ is continuous away from the diagonal $\{(x, y) \in D^2 \colon y = x\}$. (This continuity could also be proved differently from Proposition 1.18 and using elliptic regularity arguments.) Although we will not use this, we note that [BN11] proves that the continuity of $y \in D \setminus \{x\} \mapsto G_0^D(x, y)$ extends to regular boundary points: that is, as $y \to b \in \partial D$ regular, $G_0^D(x, y) \to 0$.

Finally, in dimension $d = 1$, we will see that $G_0^D(x, y)$ is actually finite even when $x = y$. In this case, $p_t^D(x, y)$ is zero as soon as x or y are in D^c (including on the boundary of D), for any $t > 0$.

Example 1.13 Suppose $D = \mathbb{H} \subset \mathbb{C}$ is the upper half plane. Then it is not hard to see that $p_t^{\mathbb{H}}(x, y) = p_t(x, y) - p_t(x, \overline{y})$ by a reflection argument (in fact, by the reflection principle of ordinary one-dimensional Brownian motion). Hence one can deduce that

$$G_0^{\mathbb{H}}(x, y) = \frac{1}{2\pi} \log \left| \frac{x - \overline{y}}{x - y} \right| \qquad (1.16)$$

for $x \neq y$ (see Exercise 1.5 for a hint on the proof).

In the special case $d = 2$, a fundamental property of Brownian motion (also with speed 2) is that it is **conformally invariant**. That is, suppose that $(B_s)_{s \geq 0}$ is a Brownian motion in the plane with speed 2, and T is an analytic map defined on a simply connected open set D with $T' \neq 0$ on D (at this stage, we do not require T to be one to one). Then $T(B_s)$, considered up until the exit time τ_D from D by B, is a Brownian motion in the image domain $D' = T(D)$, up to a time change. More precisely, if we define

$$F(t) = \int_0^{\tau_D \wedge t} |T'(B_s)|^2 \, ds,$$

then we can talk about its right continuous inverse (which is also simply its inverse here)

$$F^{-1}(s) = \inf\{t > 0 \colon F(t) > s\}.$$

The conformal invariance of Brownian motion states that if we set

$$B'_s := T(B_{F^{-1}(s)}); \qquad \text{for } 0 \leq s < \tau' = F(\tau_D) \qquad (1.17)$$

then $(B'_s)_{0 \leq s \leq \tau'}$ is another Brownian motion with speed 2, stopped at the time τ' when it first leaves $D' = T(D)$. This fundamental property, predicted by Lévy in the 1940s, can be proved relatively easily using the Cauchy–Riemann equations satisfied by T and an application of Itô calculus (both Itô's formula and the Dubins–Schwarz theorem).

14 *Definition and Properties of the GFF*

Although the conformal invariance of Brownian motion is only up to a time change, and the Green function G_0 measures the expected time spent by Brownian motion close to a location before leaving the domain, a remarkable property of the Green function is that it is *completely* invariant under conformal isomorphisms, in the following sense.

We say that $D \subset \mathbb{R}^d$ or $D \subset \mathbb{C}$ is a **domain** if it is *open* and *connected*.

Proposition 1.14 (Conformal invariance of the Green function) *Let $D, D' \subset \mathbb{C}$ be regular domains. Suppose that $T \colon D \to D'$ is a conformal isomorphism (i.e. analytic with non-vanishing derivative and one to one). Then*

$$G_0^{T(D)}(T(x), T(y)) = G_0^D(x, y).$$

Note that together with (1.16) and the Riemann mapping theorem, this allows us to determine G_0^D in any simply connected proper domain $D \subset \mathbb{C}$.

Proof The proof is a simple application of the change of variable formula. Let ϕ be a test function and let $x' = T(x)$. Then, by (1.14),

$$\int_{D'} G_0^{D'}(x', y') \phi(y') \, dy' = \mathbb{E}_{x'} \Big(\int_0^{\tau'} \phi(B'_s) \, ds \Big),$$

where B' is a Brownian motion, and τ' is its exit time from D'. On the other hand, the change of variable formula applied to the left-hand side gives us, letting $y' = T(y)$ (a change of variable whose Jacobian derivative evaluates to $dy' = |T'(y)|^2 \, dy$):

$$\int_{D'} G_0^{D'}(x', y') \phi(y') \, dy' = \int_D G_0^{D'}(T(x), T(y)) \phi(T(y)) |T'(y)|^2 \, dy. \quad (1.18)$$

Now let us compute the right-hand side of the initial equation in a different way, using the conformal invariance of Brownian motion discussed above. This allows us to write $B'_s = T(B_{F^{-1}(s)})$; moreover, in this correspondence one has $\tau' = F^{-1}(\tau_D)$. We apply the change of variable formula, but now to the time parameter $t = F^{-1}(s)$, or (since F^{-1} is the inverse of F), $s = F(t)$. The Jacobian derivative is thus

$$ds = F'(t) \, dt = |T'(B_t)|^2 \, dt,$$

by definition of F and the fundamental theorem of calculus. Thus,

$$\mathbb{E}_{x'} \Big(\int_0^{\tau'} \phi(B'_s) \, ds \Big) = \mathbb{E}_x \Big(\int_0^{F^{-1}(\tau)} \phi(T(B_{F^{-1}(s)})) \, ds \Big)$$

$$= \mathbb{E}_x \Big(\int_0^{\tau} \phi(T(B_t)) |T'(B_t)|^2 \, dt \Big)$$

1.2 Continuous Green function

$$= \int_D G_0^D(x,y) \phi(T(y)) |T'(y)|^2 \, dy. \tag{1.19}$$

Identifying the right-hand sides of (1.18) and (1.19), since the test function ϕ is arbitrary, we conclude that

$$G_0^{D'}(T(x), T(y)) |T'(y)|^2 = G_0^D(x,y) |T'(y)|^2$$

first as distributions, and thus by continuity as functions defined for $x \neq y$. The result follows by cancelling the factors of $|T'(y)|^2$ on both sides. □

Remark 1.15 We have already mentioned that the conformal invariance of the Green function is at first a little surprising, since conformal invariance of Brownian motion holds only up to a time change, whereas the Green function, which measures time spent in a neighbourhood of a point, is a priori very sensitive to the time parametrisation. Having done the proof, we can now *a posteriori* explain this remarkable fact. When we apply the change of variables spatially, we pick up a term $|T'(y)|^2$ because we are in dimension $d = 2$. When we apply it temporally, we pick up another term $|T'(y)|^2$ from Itô's formula. The fact that these two factors match exactly is what gives the conformal invariance of the Green function.

From this perspective, conformal invariance of the Green function is a miraculous property, unique to the case $d = 2$. In higher dimensions, it is not simply a problem of defining conformal maps: if we consider scalings $z \mapsto rz$ (note that this leaves Brownian motion invariant up to time change in any dimension), it is only in dimension $d = 2$ that such scalings leave the Green function invariant.

Remark 1.16 We will make use of conformal invariance to analyse the Green function in dimension $d = 2$, as it often suffices to prove some desired property in a concrete domain (where we have explicit formulae, such as the upper half plane), and use conformal invariance to deduce the desired property in an arbitrary simply connected domain. We believe this to be an elegant approach, appropriate for many potential readers of this book. However, it has the drawback that it does not apply in other dimensions, where instead one must usually rely on hands-on estimates. The latter approach also works in dimension $d = 2$ of course and might be more appropriate for readers who do not have a background in complex analysis – after all, the theory which will be developed throughout the first four chapters of this book depends only very tangentially on complex analysis arguments, and can mostly be read without any such background. For more on this hands-on approach to properties of the Green function, we refer potentially interested readers to Chapter 2.4 in [Law05], where none of the arguments appeal to conformal invariance.

Definition and Properties of the GFF

Example 1.17 Having identified the Green function in one simply connected domain (the upper half plane $\mathbb{H}$), the conformal invariance of the Green function can be used in conjunction with the Riemann mapping theorem to evaluate it on an arbitrary simply connected domain D. Here is an example in which the Green function becomes very simple. Let $D = \mathbb{D}$ be the unit disc. We can find a Möbius transformation

$$T(z) = \frac{i-z}{i+z},$$

which maps $\mathbb{H}$ to $\mathbb{D}$. (To check this, note that any Möbius map, that is, any function of the form $z \mapsto (az + b)/(cz + d)$ with $ad - bc \neq 0$, is always a homeomorphism of the extended plane $\mathbb{R}^2 \cup \{\infty\}$ onto itself, and maps circles to circles – where by circles we also allow for infinite lines. Here it is easy to check that if $z \in \mathbb{R}$ then $|T(z)| = 1$, so T maps the real line to the unit circle; since $T(i) = 0$, the image of $\mathbb{H}$ must be the unit disc.) From the explicit form of $G_0^{\mathbb{H}}$ obtained in (1.16), we deduce

$$G_0^{\mathbb{D}}(0, z) = -\frac{1}{2\pi} \log |z|. \tag{1.20}$$

We state below some basic and fundamental properties of the Green function in two dimensions, which will be used throughout.

Proposition 1.18 *For any regular, simply connected domain $D \subset \mathbb{R}^2$, and any $x \in D$:*

1. $G_0^D(x, y) \to 0$ as $y \to y_0 \in \partial D$;
2. $G_0^D(x, y) = -\frac{1}{2\pi} \log(|x - y|) + O(1)$ as $y \to x$.
3. $G_0^D(x, \cdot)$ is harmonic in $D \setminus \{x\}$; and as a distribution

$$\Delta G_0^D(x, \cdot) = -\delta_x(\cdot); \tag{1.21}$$

Proof For the first point, observe that on the unit disc $\mathbb{D}$ with $x = 0$, $|G_0^{\mathbb{D}}(0, y)| \leq C \operatorname{dist}(y, \partial \mathbb{D})$ for all y with $|y| \geq 1/2$ (say), so converges to zero, uniformly as y approaches $\partial \mathbb{D}$. Now suppose D is an arbitrary regular simply connected domain and $x \in D$. Fix a conformal isomorphism f from $\mathbb{D}$ to D with $f(0) = x$. Let $y_n \in D$ be a sequence such that $y_n \to y \in \partial D$. Then we claim that $w_n := f^{-1}(y_n) \in \mathbb{D}$ is a sequence approaching the boundary of $\mathbb{D}$, in the sense that $\operatorname{dist}(w_n, \partial \mathbb{D}) \to 0$ (note however that there is no guarantee, without additional assumptions on D, that w_n will converge to a point on $\partial \mathbb{D}$). Indeed, since $w_n \in \mathbb{D}$ is a bounded sequence, it suffices to check that no subsequence can converge to a point $w \in \mathbb{D}$. But if that were the case, then $f(w_n)$ would converge to $f(w)$ along that subsequence, which contradicts the fact that y_n converges to y. Hence, by conformal invariance of the Green

1.2 Continuous Green function

function and the uniformity of the convergence to zero in $\mathbb{D}$, we deduce that $G_0^D(x, y_n) = G_0^{\mathbb{D}}(0, w_n) \to 0$, as required.

The second point also follows from the explicit form of the Green function on the unit disc and conformal invariance. In particular, integrals of the form $\int_D G_0^D(x, y) f(y) \, dy$ are well defined for any test function $f \in \mathcal{D}_0(D)$, so $G_0^D(x, \cdot)$ may be viewed as a distribution.

For the final point, we can again use the explicit form of $G_0^{\mathbb{D}}$ on $\mathbb{D}$, which shows that $G_0^{\mathbb{D}}(0, \cdot)$ is a harmonic function away from 0 (as the real part of a holomorphic function). Furthermore, harmonicity is preserved under conformal isomorphisms. This shows that $G_0^D(x, \cdot)$ is harmonic away from x. To prove (1.21) requires a little more. For instance, one can reduce (1.21) by conformal invariance to showing that $\Delta \log |z| = 2\pi \delta_0$ in the sense of distributions. This follows from explicit computations of $\Delta f_\varepsilon(z)$, where $f_\varepsilon(z) = \log(|z| \vee \varepsilon)$, and the fact that $f_\varepsilon(z)$ converges to $\log |z|$ in the sense of distributions, hence $\Delta f_\varepsilon(z)$ converges to $\Delta \log |z|$ in the sense of distributions.

However, perhaps the simplest argument for (1.21) is as follows: since we already know by the second point that $G_0^D(x, \cdot)$ is a distribution, it suffices to show that for each test function $f \in \mathcal{D}_0(D)$,

$$\int_D G_0^D(x, y) \Delta f(y) \, dy = -f(x). \tag{1.22}$$

Using (1.14) (and using the consequence of the second point above that the integral in (1.22) is well defined) the left-hand side can be rewritten as

$$\mathbb{E}_x \left(\int_0^{\tau_D} \Delta f(B_s) \, ds \right).$$

On the other hand, by Itô's formula, $M_t^f = f(B_{t \wedge \tau_D}) - \int_0^{t \wedge \tau_D} \Delta f(B_s) \, ds$ is a martingale with initial value $M_0^f = f(x)$, hence

$$\mathbb{E}_x \left(\int_0^{t \wedge \tau_D} \Delta f(B_s) \, ds \right) = \mathbb{E}_x(f(X_{t \wedge \tau_D})) - f(x).$$

The result thus follows by letting $t \to \infty$: in the right-hand side, the first term tends to zero since f has compact support and $\tau_D < \infty$ almost surely. In the left-hand side, we apply the dominated convergence theorem, together with the fact that the expected occupation measure up to time t is dominated by the total expected occupation measure $G_0^D(x, y) \, dy$, which integrates Δf by the second point as Δf is itself a smooth compactly supported function. This proves (1.22) and thus (1.21).

Instead of such computations, one could also argue that p_t^D solves the heat

18   *Definition and Properties of the GFF*

equation

$$\frac{\partial}{\partial t} p_t^D(x,y) = \Delta p_t^D(x,y).$$

Integrating this identity over time gives, at least informally,

$$\Delta G_0^D(x,\cdot) = p_\infty^D(x,\cdot) - p_0^D(x,\cdot) = -\delta_x(\cdot)$$

as desired, where p_∞^D denotes the limit as $t \to \infty$ of $p_t^D(x,y)$, which is zero because Brownian eventually leaves D in finite time. Of course, justifying this requires some careful arguments too, so the exact computations using the form of $G_0^\mathbb{D}$ on $\mathbb{D}$ are more direct. □

Remark 1.19 In fact, the above result also holds in other dimensions with appropriate changes. One can check that in any dimension $d \geq 1$, for any regular domain D such that the Green function $\int_0^\infty p_t^D(x,y) \, dt$ in D is finite for all $x \neq y$ (note that D does not need to be simply connected), and for any fixed $x \in D$:

1. $G_0^D(x,y) \to 0$ as $y \to y_0 \in \partial D$;
2. $G_0^D(x,\cdot)$ is harmonic in $D \setminus \{x\}$ with $\Delta G_0^D(x,\cdot) = -\delta_x(\cdot)$ as distributions;
3.

$$G_0^D(x,y) = \begin{cases} G_0^D(x,x) + o(1) & d = 1 \\ -(2\pi)^{-1} \log(|x-y|) + O(1) & d = 2 \\ \frac{1}{A_d} |x-y|^{2-d} + O(1) & d \geq 3 \end{cases}$$

as $y \to x$, where A_d is the $(d-1)$-dimensional surface area of the unit ball in d dimensions.

Remark 1.20 One can in fact say more than what is contained in Proposition 1.18 or Remark 1.19. Firstly, in Theorem 1.23, the on-diagonal behaviour of the Green function will be estimated more sharply. Secondly, the properties contained in Proposition 1.18 are in fact sufficient to characterise the Green function, and thus may be used to identify it explicitly. See Exercise 1.7 where such a characterisation will be proved, in fact under even weaker assumptions: if $\phi \colon D \setminus \{z_0\} \to \mathbb{R}$ is harmonic, converges to 0 near the boundary, and blows up logarithmically near z_0, in the sense that $\phi(z) = (1+o(1))/(2\pi) \log(|z-z_0|^{-1})$, then ϕ coincides with the Green function. See also [WP21, Lemma 3.7] for more on this, and below for two examples (in dimensions $d = 2$ and $d = 1$).

Example 1.21 Using this characterisation, we obtain another, more conceptual, proof of (1.20). Indeed, it is clear that $z \in \mathbb{D} \setminus \{0\} \mapsto -\log|z|$ is a harmonic function, as the real part of the holomorphic function $\log(z)$ (defined locally,

1.2 Continuous Green function

say, which is sufficient for harmonicity). The logarithmic blow-up near $z = 0$ is of course obvious in this case.

As another example, this time in dimension $d = 1$, one can show:

Example 1.22 If $d = 1$ and $D = (0, 1)$, then

$$G_0^D(s, t) = s(1 - t) \tag{1.23}$$

for $0 < s \leq t < 1$ (a more symmetric expression, independent of the relative position of s and t, is $(s, t) \in (0, 1)^2 \mapsto s \wedge t - st$). Note that this does not blow up on the diagonal.

Actually, one can be slightly more precise about the behaviour of the Green function near the diagonal; that is, one can find a sharper estimate for the error term $O(1)$ in Proposition 1.18:

Theorem 1.23

$$G_0^D(x, y) = -\frac{1}{2\pi} \log(|x - y|) + \frac{1}{2\pi} \log R(x; D) + o(1) \tag{1.24}$$

as $y \to x$, where $R(x; D)$ is the **conformal radius** of x in D. That is, $R(x; D) = |f'(0)|$ for f any conformal isomorphism taking $\mathbb{D}$ to D and satisfying $f(0) = x$. Furthermore, we may write

$$G_0^D(x, \cdot) = -\frac{1}{2\pi} \log |x - \cdot| + \xi_x(\cdot), \tag{1.25}$$

where $\xi_x(\cdot)$ is a harmonic function over all of D, which equals the harmonic extension to D of the function $1/(2\pi) \log(|x - \cdot|)$ on ∂D. (Combining with (1.24), we must have $\xi_x(x) = 1/(2\pi) \log R(x; D)$.)

Proof Recall from (1.20) that if $D = \mathbb{D}$ is the unit disc, we have

$$G_0^{\mathbb{D}}(0, z) = -\frac{1}{2\pi} \log |z|.$$

This makes (1.24) obvious for $D = \mathbb{D}$ and $x = 0$, and so (1.24) follows immediately in the general case by conformal invariance, Taylor approximation, and definition of the conformal radius. To prove (1.25), we note that the difference $G_0^D(x, \cdot) + 1/(2\pi) \log |x - \cdot|$ is bounded and harmonic in the sense of distributions in all of D. Elliptic regularity arguments, or direct argumentation with planar Brownian motion, imply that this is a smooth function which is harmonic in the usual sense. $\square$

Remark 1.24 Note that the conformal radius is unambiguously defined: the value $|f'(0)|$ does not depend on the choice of f (f is unique up to rotation, which does not affect the value of the modulus derivative). Although we will not use this, we note also that by the classical **Köbe quarter theorem**, we have

$$\text{dist}(x, \partial D) \leq R(x; D) \leq 4\,\text{dist}(x, \partial D),$$

so the conformal radius is essentially a measure of the Euclidean distance to the boundary.

The conformal radius appears in Liouville quantum gravity in various formulae which will be discussed later on in the course. The reason it shows up in these formulae is usually because of (1.24).

The last property of G_0^D that we will need, as in the discrete case, is that it is a non-negative definite function. We will see this in Section 1.3.

1.3 GFF as a stochastic process

From now on, we will always assume that $D \subset \mathbb{R}^d$ is a regular domain; that is, an open connected set with regular boundary.

Essentially, as in the discrete case, we would like to define the GFF as a Gaussian "random function" with mean zero and covariance given by the Green function. However (when $d \geq 2$), the divergence of the Green function on the diagonal means that the GFF cannot be defined pointwise, as the variance at any point would have to be infinite. So instead, we define it as a random distribution, or generalised function in the sense of Schwartz.[4] More precisely, we will take the point of view that it assigns values to certain measures with finite Green energy. In doing so, we follow the approach in the two sets of lecture notes [BN11] and [WP21]. The latter in particular contains a great deal more about the relationship between the GFF, SLE, Brownian loop soups and conformally invariant random processes in the plane, which will not be discussed in this book. The foundational paper by Dubédat [Dub09b] is also an excellent source of information regarding basic properties of the Gaussian free field. We should point out that the rest of this text is particularly focused on the case $d = 2$, but we will include results relevant to other dimensions when there is no cost in doing so.

Recall that if I is an index set, a **stochastic process indexed by** I is just a collection of random variables $(X_i)_{i \in I}$, defined on some given probability space. The **law** of the process is a measure on $\mathbb{R}^I$, endowed with the product topology. It is uniquely characterised by its finite-dimensional marginals, that

---
[4] This conflicts with the usage of distribution to mean the law of a random variable but is standard and should not cause confusion.

1.3 GFF as a stochastic process

is, the law of $(X_{i_1}, \ldots, X_{i_n})$ for arbitrary $i_1, \ldots, i_n$ in I, via Kolmogorov's extension theorem.

Given $n \geq 1$, a random vector $X = (X_i)_{1 \leq i \leq n}$ is called **Gaussian** if any linear combination of its entries is real Gaussian; that is, if $\langle \lambda, X \rangle$ is a real Gaussian random variable for any $\lambda = (\lambda_1, \ldots, \lambda_n) \in \mathbb{R}^n, n \in \mathbb{N}$. The law of X is uniquely specified by its mean vector $\mu = \mathbb{E}(X) \in \mathbb{R}^n$, that is, $\mu_i = \mathbb{E}(X_i)$ for each $1 \leq i \leq n$, and its covariance matrix $\Sigma \in \mathcal{M}(\mathbb{R}^n)$ given by $\Sigma_{i,j} = \text{Cov}(X_i, X_j)$, $1 \leq i, j \leq n$. Conversely, given a vector $\mu \in \mathbb{R}^n$ and a symmetric, non-negative[5] matrix $\Sigma \in \mathcal{M}(\mathbb{R}^n)$, there exists a (unique) law on $\mathbb{R}^n$ which is that of a Gaussian vector with mean μ and covariance matrix Σ.

Fix a set I and suppose we are given a function $C \colon I \times I \to \mathbb{R}$, symmetric and non-negative in the sense that

$$\sum_{i,j=1}^n \lambda_i \lambda_j C(t_i, t_j) \geq 0 \quad \text{for all } n \geq 1, t_1, \ldots, t_n \in I \text{ and } \lambda_1, \ldots, \lambda_n \in \mathbb{R}.$$
(1.26)

Then associated to this function C, for each $t_1, \ldots, t_n \in I$, we can define a centred Gaussian vector $(X_{t_1}, \ldots, X_{t_n})$ with covariance matrix $\Sigma_{i,j} = C(t_i, t_j)$, $1 \leq i, j \leq n$. The resulting laws are automatically consistent, in the sense of Kolmogorov as the parameters $t_1, \ldots, t_n \in I$ and $n \geq 1$ are varied. Therefore, by Kolmogorov's extension theorem, the function C defines a unique law on $\mathbb{R}^I$. This is the law of a stochastic process $(X_t)_{t \in I}$ indexed by I such that the restriction of $(X_t)_{t \in I}$ to any n tuple of indices $t_1, \ldots, t_n \in I$ gives us a centred Gaussian vector $(X_{t_1}, \ldots, X_{t_n})$ with the above covariance matrix. The process $(X_t)_{t \in I}$ is called the (centred) **Gaussian stochastic process on I with covariance function** C. Given a real valued function $(\mu(t), t \in I)$, we can also define a Gaussian stochastic process Y on I with mean function μ and covariance function C, simply by shifting X by $\mu(t)$ at each $t \in I$, that is, setting $(Y_t)_{t \in I} := (X_t + \mu(t))_{t \in I}$.

Now, let $D \subset \mathbb{R}^d$ be an open set with **regular** boundary and recall from Section 1.2 that for such D the Green function G_0^D is finite away from the diagonal: $G_0^D(x,y) < \infty$ for $x \neq y$. We will define the Gaussian free field in D (with zero boundary conditions) as a centred Gaussian stochastic process indexed by the set $\mathfrak{M}_0$ (defined below) of signed Borel measures with finite logarithmic energy.

Definition 1.25 (Index set for the GFF) Let $\mathfrak{M}_0^+$ denote the set of (non-negative) Radon measures supported in D, such that $\int \rho(\mathrm{d}x)\rho(\mathrm{d}y) G_0^D(x,y) < \infty$.

---

[5] Here non-negative is in the sense of matrices, that is, $\sum_{i,j} \lambda_i \lambda_j \Sigma_{i,j} \geq 0$ for each $\lambda \in \mathbb{R}^n$, or equivalently, the eigenvalues of Σ are all non-negative.

22 Definition and Properties of the GFF

Denote by $\mathfrak{M}_0$ the set of signed measures of the form $\rho = \rho^+ - \rho^-$ with $\rho^\pm \in \mathfrak{M}_0^+$.

Note that when $d = 2$, due to the logarithmic divergence of the Green function on the diagonal, $\mathfrak{M}_0^+$ includes the case where $\rho(\mathrm{d}x) = f(x)\,\mathrm{d}x$ and f is continuous, but does not include Dirac point masses.

For test functions $\rho_1, \rho_2 \in \mathfrak{M}_0$, we set

$$\Gamma_0(\rho_1, \rho_2) := \int_{D^2} G_0^D(x,y)\rho_1(\mathrm{d}x)\rho_2(\mathrm{d}y) \qquad (1.27)$$

and also define $\Gamma_0(\rho) = \Gamma_0(\rho, \rho)$. We will see below why these quantities are in fact well defined, but note for now that this is not immediately obvious.

Essentially, our definition will be that the Gaussian free field on D with zero boundary conditions is the centred Gaussian stochastic process $(\Gamma_\rho)_{\rho \in \mathfrak{M}_0}$ indexed by $\mathfrak{M}_0$ such that for $\rho_1, \rho_2 \in \mathfrak{M}_0$ we have

$$\mathrm{Cov}(\Gamma_{\rho_1}, \Gamma_{\rho_2}) = \Gamma_0(\rho_1, \rho_2) = \int_{D^2} G_0^D(x,y)\rho_1(\mathrm{d}x)\rho_2(\mathrm{d}y).$$

However in order to do so, a few things need to be checked. Namely:

- $\Gamma_0(\rho_1, \rho_2)$ is well defined whenever $\rho_1, \rho_2 \in \mathfrak{M}_0$. In fact this is not obvious,[6] even if we assume $\rho_1, \rho_2 \in \mathfrak{M}_0^+$.
- The function $\Gamma_0(\cdot, \cdot)$ is symmetric and non-negative on $\mathfrak{M}_0 \times \mathfrak{M}_0$, in the sense of (1.26) with $I = \mathfrak{M}_0$, so is a valid covariance function.

As we will see, these properties will follow rather easily from the following lemma.

Lemma 1.26 *If $\rho_1, \rho_2 \in \mathfrak{M}_0^+$ then $\Gamma_0(\rho_1, \rho_2) < \infty$. Furthermore $\rho_1 + \rho_2 \in \mathfrak{M}_0^+$.*

Proof By the Markov property, we have

$$p_t^D(x,y) = \int_D p_{t/2}^D(x,z) p_{t/2}^D(z,y)\,\mathrm{d}z,$$

and hence by symmetry (i.e. $p_t^D(x,y) = p_t^D(y,x)$, which follows from the same symmetry in the full plane, and the fact that a Brownian bridge from x to y has as much chance to stay in D as one from y to x, as one is the time reversal of the other), we can deduce that

$$G_0^D(x,y) = 2\int_D \mathrm{d}z \int_0^\infty p_u^D(x,z) p_u^D(y,z)\,\mathrm{d}u.$$

---

[6] The necessity of such an argument (in the absence of any form of Cauchy–Schwarz inequality at this stage) seems to not have been noticed before; correspondingly Lemma 1.26, although not difficult, is new.

1.3 GFF as a stochastic process

Consequently, if $\rho_1, \rho_2 \in \mathfrak{M}_0^+$ are arbitrary,

$$\begin{aligned}
\Gamma_0(\rho_1, \rho_2) &= \iint G_0^D(x,y) \rho_1(dx) \rho_2(dy) \\
&= \int_D 2\,dz \int_0^\infty \iint \rho_1(dx) \rho_2(dy) p_u^D(x,z) p_u^D(y,z)\,du \\
&= \int_D 2\,dz \int_0^\infty \left(\int \rho_1(dx) p_u^D(x,z) \right) \times \left(\int \rho_2(dx) p_u^D(x,z) \right) du.
\end{aligned} \tag{1.28}$$

In particular, if $\rho_1 = \rho_2 \in \mathfrak{M}_0^+$ then

$$\Gamma_0(\rho_1, \rho_1) = \int_D 2\,dz \int_0^\infty \left(\int \rho_1(dx) p_u^D(x,z) \right)^2 du < \infty. \tag{1.29}$$

Hence using the inequality $2ab \le a^2 + b^2$, valid for any real numbers a and b, we deduce that $\Gamma_0(\rho_1, \rho_2) < \infty$ whenever $\rho_1, \rho_2 \in \mathfrak{M}_0^+$. This proves the first point.

For the second point, observe that *a priori* $\Gamma_0(\rho_1 + \rho_2) = \Gamma_0(\rho_1) + 2\Gamma_0(\rho_1, \rho_2) + \Gamma_0(\rho_2)$. This is an equality between terms which are non-negative but might be infinite. Nevertheless, from what we have just seen, if $\rho_1, \rho_2 \in \mathfrak{M}_0^+$, all three terms on the right-hand side are finite. Thus the left-hand side is finite too, which concludes the proof of Lemma 1.26. □

Lemma 1.26 allows us to extend the notion of energy $\Gamma_0(\rho_1, \rho_2)$ onto $\mathfrak{M}_0 \times \mathfrak{M}_0$ and not just $\mathfrak{M}_0^+ \times \mathfrak{M}_0^+$, justifying the definition in (1.27). Indeed writing $\rho_i = \rho_i^+ - \rho_i^-$ for $i = 1, 2$, we have

$$\Gamma_0(\rho_1, \rho_2) = \Gamma_0(\rho_1^+, \rho_2^+) + \Gamma_0(\rho_1^-, \rho_2^-) - \Gamma_0(\rho_1^+, \rho_2^-) - \Gamma_0(\rho_1^-, \rho_2^+),$$

where the finiteness of all four terms on the right-hand side is guaranteed by Lemma 1.26. Note also that $\mathfrak{M}_0$ is a vector space (again by Lemma 1.26), with Γ_0 a **bilinear form** on $\mathfrak{M}_0$.

Remark 1.27 In fact, we will soon see as a consequence of Lemma 1.43 that $\mathfrak{M}_0$ is the intersection of the Sobolev space $H_0^{-1}(D)$ with the set of signed measures on D.

Lemma 1.28 *The bilinear form Γ_0 is symmetric and non-negative (in the sense of covariance functions) on $\mathfrak{M}_0 \times \mathfrak{M}_0$. That is, for every $n \ge 1$ and every $\rho_1, \ldots, \rho_n \in \mathfrak{M}_0$, for every $\lambda_1, \ldots, \lambda_n \in \mathbb{R}$,*

$$\sum_{i,j=1}^n \lambda_i \lambda_j \Gamma_0(\rho_i, \rho_j) \ge 0.$$

24 Definition and Properties of the GFF

In particular, Γ_0 is a valid covariance function for a Gaussian stochastic process on $\mathfrak{M}_0$.

Proof Since Γ_0 is a bilinear form, we have:

$$\sum_{i,j=1}^{n} \lambda_i \lambda_j \Gamma_0(\rho_i, \rho_j) = \Gamma_0(\rho),$$

where

$$\rho = \sum_{i=1}^{n} \lambda_i \rho_i \in \mathfrak{M}_0.$$

The desired non-negativity therefore follows directly from (1.29). □

As a consequence of Lemma 1.28, we can now finally give the definition of a Gaussian free field (with zero boundary conditions) as a stochastic process.

Theorem 1.29 (Zero boundary or Dirichlet GFF) *There exists a unique stochastic process $(\mathbf{h}_\rho)_{\rho \in \mathfrak{M}_0}$, indexed by $\mathfrak{M}_0$, such that for every choice of $\rho_1, \ldots, \rho_n$, $(\mathbf{h}_{\rho_1}, \ldots, \mathbf{h}_{\rho_n})$ is a centred Gaussian vector with covariance structure $\mathrm{Cov}(\mathbf{h}_{\rho_i}, \mathbf{h}_{\rho_j}) = \Gamma_0(\rho_i, \rho_j)$.*

Let us emphasise that for a stochastic process, the index set I does not a priori *need* to be a vector space, although this is the case when $I = \mathfrak{M}_0$. Similarly, the covariance function of a Gaussian stochastic process indexed by I does not *need* to be a bilinear non-negative form on I, although again this is true for Γ_0 on $\mathfrak{M}_0$, and this helped us to prove its validity as a covariance function.

Definition 1.30 The process $(\mathbf{h}_\rho)_{\rho \in \mathfrak{M}_0}$ is called the Gaussian free field in D (with Dirichlet or zero boundary conditions). We write GFF as shorthand for Gaussian free field.

Note that in such a setting, it might not be possible to "simultaneously observe" more than a countable number of random variables, because our σ-algebra for the stochastic process $(\mathbf{h}_\rho)_{\rho \in \mathfrak{M}_0}$ is the product σ-algebra, which is generated by the random variables of the form $(\mathbf{h}_{\rho_1}, \ldots, \mathbf{h}_{\rho_n})$, $n \geq 1$, $\rho_1, \ldots, \rho_n \in \mathfrak{M}_0$. A good analogy is with the construction of one-dimensional Brownian motion $(B_t, t \geq 0)$: so long as it is constructed as a Gaussian stochastic process indexed by time, numerical quantities such as $\sup_{s \in [t_1, t_2]} B_s$ are not measurable with respect to the product σ algebra and so are not random variables. In the case of Brownian motion, it is not until a continuous modification is constructed that such quantities can be seen as (measurable) random variable. Likewise, in the case of the GFF, we will have to rely on the existence

1.3 GFF as a stochastic process

of suitable modifications with nice continuity properties. More precisely, this modification will be a random distribution living in a certain Sobolev space of negative index, see Section 1.4, whose law as a stochastic process indexed by $\mathfrak{M}_0$ has the same finite-dimensional marginals as the GFF $(\mathbf{h}_\rho)_{\rho \in \mathfrak{M}_0}$. In other words, this random distribution defines a *version* of the GFF.

Remark 1.31 (Terminology.) We will use the terminology "Dirichlet GFF", "zero boundary GFF" and "GFF with zero/Dirichlet boundary conditions" interchangeably throughout. With a slight abuse of vocabulary, some authors use the term "Dirichlet boundary condition" to indicate that the field has *some* specified (deterministic) boundary conditions, which however may not be identically zero. It will be made clear in the sequel if we wish to talk about anything other than the zero boundary condition case.

Remark 1.32 In Liouville quantum gravity and in Gaussian multiplicative chaos, it is more convenient (as mentioned previously) to work directly with a field which is logarithmically correlated (as opposed to $(2\pi)^{-1}$-logarithmically correlated), that is, with

$$h = \sqrt{2\pi}\mathbf{h}. \quad (1.30)$$

We will use the notations h and $\mathbf{h}$ throughout to make the distinction between these two different conventions.

The following property of "almost sure" linearity is a consequence of the fact that the covariance function Γ_0 is a bilinear form on $\mathfrak{M}_0$; its proof is left as an exercise.

Proposition 1.33 (Linearity.) *If $\lambda, \lambda' \in \mathbb{R}$ and $\rho, \rho' \in \mathfrak{M}_0$ then $\mathbf{h}_{\lambda\rho+\lambda'\rho'} = \lambda\mathbf{h}_\rho + \lambda'\mathbf{h}_{\rho'}$ almost surely.*

In the rest of this text, we will abuse notation slightly and write $(\mathbf{h}, \rho)$ for $\mathbf{h}_\rho$ when $\rho \in \mathfrak{M}_0$. We will think of $(\mathbf{h}, \rho)$ as "$\mathbf{h}$ integrated against ρ", as if $\mathbf{h}$ were an actual distribution, and ρ was a test function. In Section 1.4, we will see that a version of $\mathbf{h}$ can be defined as a random variable taking values in the space of distributions.

At this stage, simply note that if $\mathbf{h}$ is a GFF, it cannot be evaluated pointwise (because $\rho = \delta_x$ does not lie in $\mathfrak{M}_0$). However, it may be tested against smooth, compactly supported test functions $\rho \in \mathcal{D}_0(D)$. In fact, h may be tested against relatively more singular measures: for instance, the "integral" (in the above sense) of $\mathbf{h}$ along a one dimensional segment or a circular arc is always well defined, since the Lebesgue measure on such a one dimensional smooth curve is an element of $\mathfrak{M}_0$. Indeed, one can deduce this from the fact that the divergence of the Green function is only logarithmic, and that in one dimension, $\int_0^1 \log(r^{-1})dr < \infty$.

By Proposition 1.33, an alternative definition of the GFF is simply as the unique stochastic process $(\mathbf{h}, \rho)$ which:

- is almost surely linear in ρ (in the sense that for every $\lambda_1, \lambda_2 \in \mathbb{R}$ and $\rho_1, \rho_2 \in \mathfrak{M}_0$, $(\mathbf{h}, \lambda_1\rho_1 + \lambda_2\rho_2) = \lambda_1(\mathbf{h}, \rho_1) + \lambda_2(\mathbf{h}, \rho_2)$ almost surely); and
- is such that $(\mathbf{h}, \rho)$ is a centred Gaussian random variable with variance $\Gamma_0(\rho)$ for every $\rho \in \mathfrak{M}_0$.

Example Suppose that $d = 1$ and $D = (0, 1)$. Then by (1.23), we know that $G_0^D(x, y) = x(1 - y)$ for $0 < x \le y < 1$, and this turns out to be the covariance of a (speed one) **Brownian bridge** ($b_s, 0 \le s \le 1$) (see Chapter 1.3 of [RY99]). So, a zero boundary Gaussian free field in one dimension is simply a (speed one) Brownian bridge, at least in the sense of stochastic processes indexed by, say, test functions.

Other boundary conditions than zero will also be relevant in practice. For this, we make the following definition (in the case $d = 2$ for simplicity). Suppose that f is a (possibly random) continuous function on the *conformal boundary* of a simply connected domain $D \subset \mathbb{C}$ (equivalent to the Martin boundary of the domain for Brownian motion). Then the GFF with boundary data given by f is the random variable $\mathbf{h} = \mathbf{h}_0 + \varphi$, where $\mathbf{h}_0$ is an independent Dirichlet GFF, and φ is the harmonic extension of f to D.

The reason for this definition will become clear in light of the Markov property discussed in Section 1.10. Alternatively, it can be justified by the fact that if one defines a discrete GFF with prescribed boundary condition f by modifying Theorem 1.8 in the natural way (i.e. taking the same definition but setting $h(y) = f(y)$ for y on the boundary), then for an appropriate sequence of approximating graphs, the discrete GFF with boundary condition f converges to $\mathbf{h}_0 + \varphi$ as defined above. See Section 1.14 for the proof of such a statement in the case $f \equiv 0$.

If we do not specify the boundary conditions, we always mean a Gaussian free field with zero (or Dirichlet) boundary conditions.

1.4 Random variables and convergence in the space of distributions

As we will soon see, the Gaussian free field can be understood as a random distribution. However, since the space of distributions is not metrisable, we first need to address a few foundational issues related to measurability and convergence.

Let $\mathcal{D}_0(D)$ denote the set of compactly supported, C^∞ functions in D, also known as **test functions**. The set $\mathcal{D}_0(D)$ is equipped with a topology in which convergence is characterised as follows. A sequence $(f_n)_{n\geq 0}$ converges to 0 in $\mathcal{D}_0(D)$ if and only if there is a compact set $K \subset D$ such that $\operatorname{supp} f_n \subset K$ for all n and f_n and all its derivatives converge to 0 uniformly on K. A continuous linear map $u\colon \mathcal{D}_0(D) \to \mathbb{R}$ is called a **distribution** on D. Thus, the set of distributions on D is the dual space of $\mathcal{D}_0(D)$. It is denoted by $\mathcal{D}'_0(D)$ and is equipped with the weak-$*$ topology. In particular, $u_n \to u$ in $\mathcal{D}'_0(D)$ if and only if $u_n(\rho) \to u(\rho)$ for all $\rho \in \mathcal{D}_0(D)$.

Let $(\Omega, \mathcal{F}, \mathbb{P})$ be a probability space. A random variable X in the space of distributions is, as always, a function $X\colon \Omega \to \mathcal{D}'_0(D)$ which is measurable with respect to the Borel σ-field on $\mathcal{D}'_0(D)$ induced by the weak-$*$ topology.

Let $(X_n)_{n\geq 1}$ be a sequence of random variables in $\mathcal{D}'_0(D)$. We will often ask ourselves whether this sequence converges in $\mathcal{D}'_0(D)$. However, since the topology of convergence on $\mathcal{D}'_0(D)$ is not metrisable, it is not clear *a priori* if the event (or rather the subset of Ω)

$$E = \{\omega \in \Omega \colon X_n(\omega) \text{ is weak-} * \text{ convergent}\}$$

is measurable. We show here that it is.

Lemma 1.34 *Let D be a domain of $\mathbb{R}^d$. Let* Conv *denote the set of sequences in $\mathcal{D}'_0(D)$ which are weak-$*$ convergent. Then* Conv *is a Borel set in $\mathcal{D}'_0(D)^\mathbb{N}$ equipped with the product Borel σ-algebra.*

See Appendix D for the proof.

1.5 Integration by parts and Dirichlet energy

In order to do view the Gaussian free field as a random variable in the space of distributions, our first step is to relate the covariance of the GFF to the Dirichlet energy of a function (as in the discrete case). The following **Gauss–Green formula**, which is really just an integration by parts formula, will allow us to do so.

Lemma 1.35 (Gauss–Green formula) *Suppose that D is a C^1 smooth domain. If f, g are smooth functions on $\overline{D}$, then*

$$\int_D \nabla f \cdot \nabla g = -\int_D f \Delta g + \int_{\partial D} f \frac{\partial g}{\partial n}, \tag{1.31}$$

28   *Definition and Properties of the GFF*

where $\frac{\partial g}{\partial n}$ denotes the (exterior) normal derivative.

Remark 1.36 For general D, the formula holds whenever $g \in \mathcal{D}_0(D)$ and f is continuously differentiable on D (with the boundary term on the right equal to zero). When $f \in \mathcal{D}'_0(D)$ is a distribution, the distributional derivative ∇f is defined to be the distribution such that (1.31) holds (again with zero boundary term) for all $g \in \mathcal{D}_0(D)$.

With Lemma 1.35 in hand, we can now rewrite the variance $\Gamma_0(\rho, \rho)$ of $(\mathbf{h}, \rho)$ in terms of the Dirichlet energy of an appropriate function f. This Dirichlet energy is of course the continuous analogue of the discrete Dirichlet energy which we encountered in Theorem 1.8 for instance.

Lemma 1.37 *Suppose that D is a regular domain, $f \in \mathcal{D}_0(D)$ and that ρ is a smooth function such that $-\Delta f = \rho$. Then $\rho \in \mathfrak{M}_0$ and*

$$\Gamma_0(\rho, \rho) = \int_D |\nabla f|^2. \tag{1.32}$$

Proof By the Gauss–Green formula (Lemma 1.35), noting that there are no boundary terms arising in each application, we have that

$$\Gamma_0(\rho) = -\int_x \rho(x) \int_y G_0^D(x, y) \Delta_y f(y) \, dy \, dx$$
$$= -\int_x \rho(x) \int_y \Delta_y G_0^D(x, y) f(y) \, dy \, dx.$$

Then using that $\Delta G_0^D(x, \cdot) = -\delta_x(\cdot)$ (in the distributional sense, see Proposition 1.18), we conclude that this is equal to

$$\int_x \rho(x) f(x) \, dx = -\int_D (\Delta f(x)) f(x) \, dx = \int_D |\nabla f(x)|^2 \, dx$$

as required. □

Note that this gives another proof that $\Gamma_0(\rho, \rho) \geq 0$, and therefore that the GFF is well defined as a Gaussian stochastic process (at least when indexed by smooth functions ρ). Indeed, when ρ is smooth one can always find a smooth function f such that $-\Delta f = \rho$: simply define

$$f(x) = \int G_0^D(x, y) \rho(y) \, dy. \tag{1.33}$$

The following lemma will also be useful.

Lemma 1.38 *Suppose that $\rho \in \mathfrak{M}_0$ and $g \in \mathcal{D}_0(D)$. Then*

$$\left| \int_D g(x)\rho(dx) \right|^2 \le \Gamma_0(\rho) \int_D |\nabla g(x)|^2 \, dx.$$

Proof It is a simple exercise to check, using dominated convergence, that if $\rho_\varepsilon \in \mathcal{D}_0(D)$ is defined by $\rho_\varepsilon(x) = \int_D \varepsilon^{-d} \varphi(\varepsilon^{-1}(x-z)) \mathbf{1}_{\{d(z,\partial D) > 2\varepsilon\}} \rho(dz)$ for some smooth positive function φ supported in the unit ball of $\mathbb{R}^d$ with $\int \varphi(y) \, dy = 1$, then $\Gamma_0(\rho_\varepsilon) \to \Gamma_0(\rho)$ and also $\int_D g(x) \rho_\varepsilon(dx) \to \int_D g(x) \rho(dx)$ as $\varepsilon \to 0$. Hence, it suffices to prove the inequality for $\rho(dx) = \rho(x) \, dx$ with $\rho \in \mathcal{D}_0(D)$. In this case, we have that

$$\int_D g(x)\rho(x) \, dx = \int_D \int_D G_0^D(x,y)(-\Delta g(y)) \, dy \rho(x) \, dx$$
$$= \int_D (-\Delta g(y)) f(y) \, dy,$$

where f is defined by (1.33) and satisfies $\Delta f = -\rho$. Applying Gauss–Green, we see that this is equal to $\int_D \nabla g(y) \nabla f(y) \, dy$, whose modulus is bounded above by the square root of $\int_D |\nabla g(y)|^2 \, dy \int_D |\nabla f(y)|^2 \, dy$ using Cauchy–Schwarz. Since $\int_D |\nabla f(y)|^2 \, dy = \Gamma_0(\rho)$ by Lemma 1.37, this concludes the proof. □

1.6 Reminders about function spaces

As we have already mentioned, one drawback of defining the GFF as a stochastic process is that we cannot realise (h, ρ) for all $\rho \in \mathfrak{M}_0$ simultaneously. For example, it will not always be possible to define (h, ρ) when $\rho \in \mathfrak{M}_0$ is random.

With this in mind, it is often useful to work with versions of the GFF that almost surely live in some "function" space. For example, it turns out to be possible to define a version of the GFF that is a random variable taking values in the space of distributions, or generalized functions. In fact, versions of the GFF taking values in much nicer Sobolev spaces (with negative index) can also be defined.

For completeness, we include some brief reminders on function spaces here. We continue to assume that D is a regular domain, unless stated otherwise.

Definition 1.39 (Dirichlet inner product) We define the Dirichlet inner product

$$(f, g)_\nabla := \int_D \nabla f(x) \cdot \nabla g(x) \, dx \quad (1.34)$$

for $f, g \in \mathcal{D}_0(D)$. It is straightforward to see that $(\cdot, \cdot)_\nabla$ is a valid inner product.

Definition 1.40 (The space H_0^1) We define the space $H_0^1(D)$ to be the completion of $\mathcal{D}_0(D)$ with respect to the Dirichlet inner product.

By definition $H_0^1(D)$ is a separable Hilbert space with inner product $(\cdot, \cdot)_\nabla$.

Remark 1.41 Observe that since any element of $H_0^1(D)$ corresponds, by definition, to (the limit of) a Cauchy sequence of functions $f_n \in \mathcal{D}_0(D)$ with respect to the Dirichlet inner product, it can be identified with a distribution $f \in \mathcal{D}_0'(D)$ via $f(\rho) := \lim_{n \to \infty} f_n(\rho) := \int_D f_n(x) \rho(dx)$ for each $\rho \in \mathcal{D}_0(D)$. In fact, due to Lemma 1.38, this limit exists whenever $\rho \in \mathfrak{M}_0$. In this case, we also have $|f(\rho)| \le (f, f)_\nabla \Gamma_0(\rho)$.

Remark 1.42 For general D, the standard definition of the Sobolev space $H_0^1(D)$ (see for example [AF03]) is the completion of $\mathcal{D}_0(D)$ with respect to the inner product $(f, g) := (f, g)_{L^2(D)} + (f, g)_\nabla$. When D is bounded, this coincides with Definition 1.40; indeed, by the Poincaré inequality, the norms $\|u\| := (u, u)$ and $\|u\|_\nabla = (u, u)_\nabla$ are equivalent in this case.

Eigenbasis of $H_0^1(D)$. When D is bounded, it is easy to find a suitable orthonormal eigenbasis for $H_0^1(D)$. Indeed in this case, $H_0^1(D)$ is compactly embedded in $L^2(D)$ by Rellich's embedding theorem, which implies that the resolvent of minus the Laplacian with Dirichlet boundary conditions is a compact operator. Note that this does not require any assumption of smoothness on the boundary of D. Consequently, there exists an orthonormal basis $(f_n)_{n \ge 1}$ of eigenfunctions of $-\Delta$ on D, with zero (Dirichlet) boundary conditions, having eigenvalues $(\lambda_n)_{n \ge 1}$. That is, f_n, λ_n satisfy

$$\begin{cases} -\Delta f_n = \lambda_n f_n & \text{in } D \\ f_n = 0 & \text{on } \partial D \end{cases}$$

for each $n \ge 1$. The $(\lambda_n)_{n \ge 1}$ are positive, ordered in non-decreasing order and $\lambda_n \to \infty$ as $n \to \infty$. Moreover, the Gauss–Green formula (1.31) implies that for $\lambda_n \ne \lambda_m$,

$$(f_n, f_m)_\nabla = \lambda_m \int_D f_n f_m = \lambda_n \int_D f_n f_m.$$

Hence $(f_n, f_m)_\nabla = 0$, and the eigenfunctions corresponding to different eigenvalues are orthogonal with respect to $(\cdot, \cdot)_\nabla$.

Often, the eigenfunctions of $-\Delta$ are normalised to have unit L^2 norm, since they also form an orthogonal basis of $L^2(D)$ for the standard L^2 inner product

1.6 Reminders about function spaces

(again by the Gauss–Green formula). If $(e_j)_{j \geq 1}$ are normalised in this way, then the above considerations imply that setting

$$f_j = \frac{e_j}{\sqrt{\lambda_j}} \qquad (1.35)$$

for each j, we get an orthonormal basis $(f_j)_j$ of $H_0^1(D)$.

In particular, $f \in L^2(D)$ is an element of $H_0^1(D)$ if and only if

$$(f, f)_\nabla = \sum_{j \geq 1} (f, f_j)_\nabla^2 = \sum_{j \geq 1} \lambda_j (f, e_j)_{L^2(D)}^2 < \infty. \qquad (1.36)$$

Sobolev spaces of general index $H_0^s(D), s \in \mathbb{R}$. The above leads us to define H_0^s for general $s \in \mathbb{R}$, and bounded D, to be the Hilbert space completion of $\mathcal{D}_0(D)$ with respect to the inner product

$$(f, g)_s = \sum_{j \geq 1} \lambda_j^s (f, e_j)_{L^2(D)} (g, e_j)_{L^2(D)}. \qquad (1.37)$$

Note that the above series does converge for $f, g \in \mathcal{D}_0(D)$: this can be seen by applying Cauchy–Schwarz, using that $\mathcal{D}_0(D) \subset L^2(D)$, and that all derivatives of functions in $\mathcal{D}_0(D)$ are again elements $\mathcal{D}_0(D)$, with $(\Delta f, e_n) = -\lambda_n (f, e_n)$ for $f \in \mathcal{D}_0(D)$ and $n \geq 1$. We have also seen in (1.36) that it agrees with the previous definition of $H_0^1(D)$ when $s = 1$.

Let us make a few more straightforward observations.

- When $s = 0$ the above space is equivalent, by definition, to $L^2(D)$.
- In general, when $s \geq 0$, it is simple to check that $L^2(D) \supset H_0^s(D)$, and that $f \in L^2(D)$ is an element of $H_0^s(D)$ if and only if $\sum_{j \geq 1} \lambda_j^s (f, e_j)_{L^2(D)}^2 < \infty$.
- If $s \leq 0$, then an element of $H_0^s(D)$ is by definition the limit of a sequence $\{f_n\}_n \in \mathcal{D}_0(D)$ for which $\sum_{j \geq 1} \lambda_j^s (f_n, e_j)_{L^2(D)}^2$ has a limit as $n \to \infty$. In particular, for any $\phi \in H_0^{-s}(D)$, $\lim_{n \to \infty} (f_n, \phi)_{L^2} =: f(\phi)$ exists by Cauchy–Schwarz, and we can identify our element of $H_0^s(D)$ with the distribution $f \in \mathcal{D}_0(D)'$, $\phi \mapsto f(\phi)$. Moreover, this distribution f extends to a continuous linear functional on $H^{-s}(D)$.

   In summary: for $s \leq 0$, $H_0^s(D)$ can be identified with a subspace of $\mathcal{D}_0'(D)$, and is the dual space[7] of $H_0^{-s}(D)$.
- It is also clear from the above that convergence in any negative index Sobolev space implies convergence in the space of distributions $\mathcal{D}_0'(D)$.

---

[7] The space $H_0^s(D)$ for $s < 0$ is usually referred to in the literature as simply $H^s(D)$, but we use the notation H_0^s to emphasise that it is the dual of $H_0^{-s}(D)$ rather than $H^{-s}(D)$. When $-s \in \mathbb{Z}_{\geq 0}$, the latter is the space of L^2 functions with $|s|$ derivatives in $L^2(D)$ and is a strict superspace of $H_0^{-s}(D)$.

It will be useful in what follows to rephrase the expression for $\text{Var}(\mathbf{h}, \rho)$ (when $\mathbf{h}$ is a GFF) in terms of Sobolev norms. Recall that by (1.32), if $-\Delta f = \rho$ for $f, \rho \in \mathcal{D}_0(D)$, then

$$\Gamma_0(\rho) = (f, f)_\nabla = \sum_{j \geq 1} \lambda_j (f, e_j)^2_{L^2(D)}. \tag{1.38}$$

On the other hand, by Gauss–Green we have $(\rho, e_j)_{L^2(D)} = -\lambda_j (f, e_j)_{L^2(D)}$ for every j, so that $(\rho, \rho)_{-1} = \sum_{j \geq 1} \lambda_j^{-1} (\lambda_j (f, e_j)_{L^2(D)})^2 = (f, f)_\nabla$. In other words:

Lemma 1.43 *Suppose that $D \subset \mathbb{R}^n$ is bounded and $\rho \in \mathcal{D}_0(D)$. Then*

$$\text{Var}(\mathbf{h}, \rho) = \Gamma_0(\rho) = (\rho, \rho)_{-1}. \tag{1.39}$$

1.7 GFF as a random distribution

At this stage, we do not yet know that the GFF may be viewed as a random distribution (i.e. as a random variable in $\mathcal{D}'_0(D)$). The goal of this section will be to prove that such a representation exists. Guided by (1.32) (and by Theorem 1.7), we will find an expression for the GFF as a random series, which we will show converges in the distribution space $\mathcal{D}'_0(D)$. In fact, we will show that it converges in a Sobolev space of appropriate index.

The property (1.39) suggests that $\mathbf{h}$ is formally the canonical Gaussian random variable "in" the dual space to $H_0^{-1}(D)$, i.e. in $H_0^1(D)$ (the quotation marks are added since in fact $\mathbf{h}$ does not live in $H_0^1(D)$). It should thus have the expansion

$$\mathbf{h} = \sum_{n=1}^{\infty} X_n g_n = \lim_{N \to \infty} \sum_{n=1}^{N} X_n g_n, \tag{1.40}$$

where X_n are i.i.d. standard Gaussian random variables and $(g_n)_{n \geq 1}$ is an arbitrary orthonormal basis of $H_0^1(D)$. (See for example [Jan97] for more about the general theory of Gaussian Hilbert spaces, and associated series such as the one above.)

It is not clear at this point in what sense (if any) this series converges. We will see in Theorem 1.45 that when D is bounded, it converges in an appropriate Sobolev space and hence in the space of distributions. Note however that the series does **not** converge almost surely in $H_0^1(D)$, since the H_0^1 norms of the partial sums tend to infinity almost surely as $N \to \infty$ (by the law of large numbers).

We start with the following observation, where now D can be any open set

1.7 GFF as a random distribution

with regular boundary. Set $\mathbf{h}_N := \sum_{n=1}^{N} X_n g_n$, and let $f \in \mathcal{D}_0(D)$ or more generally let $f \in H_0^1(D)$. Then

$$(\mathbf{h}_N, f)_\nabla = \sum_{n=1}^{N} X_n (g_n, f)_\nabla \tag{1.41}$$

does converge almost surely and in $L^2(\mathbb{P})$, by the martingale convergence theorem. Its limit is a Gaussian random variable with variance $\sum_{n \geq 1} (g_n, f)_\nabla^2 = \|f\|_\nabla^2$ by Parseval's identity. This defines a random variable which we call $(\mathbf{h}, f)_\nabla$, which has the law of a mean zero Gaussian random variable with variance $\|f\|_\nabla^2$. Hence while the series (1.40) does *not* converge in H_0^1, when we take the inner product with a given $f \in H_0^1$, then this does converge almost surely.

By a density argument, we can extend this to the following theorem. We use the notation (f, φ) for the action of a distribution f on a smooth function φ.

Theorem 1.44 (GFF as a random Fourier series) *Let D be a regular domain and let $\mathbf{h}_N = \sum_{n=1}^{N} X_n g_n$ be the truncated series in (1.40). Then for any $\rho \in \mathfrak{M}_0$,*

$$\lim_{N \to \infty} (\mathbf{h}_N, \rho) =: (\mathbf{h}, \rho)$$

exists in $L^2(\mathbb{P})$ (and hence in probability as well). The limit $(\mathbf{h}, \rho)$ is a Gaussian random variable with variance $\Gamma_0(\rho, \rho)$.

Observe that since $\mathbf{h}_N$ is an element of $H_0^1(D)$, $(\mathbf{h}_N, \rho)$ is well defined for every N by Remark 1.41.

Proof We will first show that for any $\nu \in \mathfrak{M}_0$, we have the upper bound

$$\text{Var}(\mathbf{h}_N, \nu) \leq \Gamma_0(\nu) \tag{1.42}$$

for all $N \geq 1$. To see this, recall from (the argument of) Lemma 1.38 that for any $\nu \in \mathfrak{M}_0$, there exists a sequence $\nu_k \in \mathcal{D}_0(D)$ with $\Gamma(\nu_k) \to \Gamma(\nu)$ as $k \to \infty$. Furthermore, for this sequence, it holds by Remark 1.41 that for each fixed N, $\text{Var}(\mathbf{h}_N, \nu_k) = \sum_{n=1}^{N} (g_n, \nu_k)^2 \to \sum_{n=1}^{N} (g_n, \nu)^2 = \text{Var}(\mathbf{h}_N, \nu)$ as $k \to \infty$. Finally, if we define f_k such that $-\Delta f_k = \nu_k$ for each k, then the discussion just above implies that $\text{Var}(\mathbf{h}_N, \nu_k) = \text{Var}(\mathbf{h}_N, f_k)_\nabla \leq \text{Var}(\mathbf{h}, f_k)_\nabla = (f, f)_\nabla = \Gamma_0(\nu_k)$ for each k. Combining these observations gives the upper bound for any $\nu \in \mathfrak{M}_0$ and $N \geq 1$

$$\text{Var}(\mathbf{h}_N, \nu) = \lim_{k \to \infty} \text{Var}(\mathbf{h}_N, \nu_k) \leq \lim_{k \to \infty} \Gamma_0(\nu_k) = \Gamma_0(\nu),$$

as desired.

We will now use this to prove the result. Take $\rho \in \mathfrak{M}_0$ and choose a sequence

$\rho_\varepsilon \in \mathcal{D}_0(D)$ approximating ρ in the sense that $\Gamma_0(\rho_\varepsilon) \to \Gamma_0(\rho)$ (again using Lemma 1.38). Set $v_\varepsilon = \rho - \rho_\varepsilon$ for each ε. Then $\Gamma_0(v_\varepsilon) \to 0$, so that applying (1.42) to $v = v_\varepsilon$ we deduce that $(\mathbf{h}_N, v_\varepsilon)$ converges to 0 in $L^2(\mathbb{P})$ and in probability as $\varepsilon \to 0$, uniformly in N. The result then follows since for smooth ρ_ε, we already know (as a consequence of the martingale convergence argument before the statement of the theorem) that $(\mathbf{h}_N, \rho_\varepsilon)$ converges to a limit in L^2, and that this limit has the same law as $(\mathbf{h}, \rho_\varepsilon)$. □

We finally address convergence of the series (1.40):

Theorem 1.45 (GFF as a random variable in a Sobolev space) *Suppose D is a regular, bounded domain. If $(X_n)_{n \geq 1}$ are i.i.d. standard Gaussian random variables and $(g_n)_{n \geq 1}$ is **any** orthonormal basis of $H_0^1(D)$, then the series $\sum_{n \geq 1} X_n g_n$ converges almost surely in $H_0^s(D)$, where*

$$s = 1 - \frac{d}{2} - \varepsilon,$$

for any $\varepsilon > 0$. In particular, for $d = 2$, the series converges in $H_0^{-\varepsilon}(D)$ for any $\varepsilon > 0$.

Observe that by Theorem 1.44, the law of the limit is uniquely defined, and coincides with the Gaussian free field $\mathbf{h}$ when its index set is restricted to $H_0^{-s}(D)$.

Proof Let us take $(e_m)_{m \geq 1}$ to be an orthonormal basis of $L^2(D)$ which are eigenfunctions for $-\Delta$, as in Chapter 1.6. This is possible since D is bounded. As usual we write λ_m for the eigenvalue corresponding to e_m; so that $(\lambda_m^{-s/2} e_m)_{m \geq 1}$ is an orthonormal basis of $H_0^s(D)$ and $(\lambda_m^{-1/2} e_m)_{m \geq 1}$ is an orthonormal basis of $H_0^1(D)$. In some cases, λ_m can be computed explicitly: for example, when D is a rectangle or the unit disc. In general, we will make use of the following fundamental estimate due to Weyl (see for example [Cha84, VI.4, page 155] for a proof):

Lemma 1.46 *We have*

$$\lambda_m \sim cm^{2/d}$$

as $m \to \infty$, in the sense that the ratio of the two sides tends to 1 as $m \to \infty$, where $c = (2\pi)^2/(a_d \operatorname{Leb}(D))^{2/d}$, where a_d is the volume of the unit ball in $\mathbb{R}^d$.

Now let $(g_n)_{n \geq 1}$ be *any* orthonormal basis of $H_0^1(D)$. We start by observing that

$$\mathbb{E}\Big(\Big\|\sum_{n=1}^N X_n g_n\Big\|_{H_0^s}^2\Big) = \mathbb{E}\Big(\sum_{n=1}^N \sum_{m=1}^N X_n X_m (g_n, g_m)_{H_0^s}\Big) = \sum_{n=1}^N \|g_n\|_{H_0^s}^2. \quad (1.43)$$

1.7 GFF as a random distribution

In fact, by the same argument, if $n_0 \geq 1$ is fixed and $N \geq n_0$ then, setting $S_N = \|\sum_{n=n_0}^N X_n g_n\|_{H_0^s}^2$, we see that $(S_N)_{N \geq n_0}$ is a submartingale (with respect to the natural filtration generated by $(X_n)_{n \geq 1}$.) Furthermore, by applying Parseval's identity, we have that

$$\sum_{n \geq 1} \|g_n\|_{H_0^s}^2 = \sum_{n \geq 1} \sum_{m \geq 1} (g_n, \lambda_m^{-s/2} e_m)_{H_0^s}^2$$

$$= \sum_{m \geq 1} \lambda_m^{-1+s} \sum_{n \geq 1} (g_n, \lambda_m^{-1/2} e_m)_{H_0^1}^2$$

$$= \sum_{m \geq 1} \lambda_m^{-1+s} < \infty,$$

where we have used positivity and Fubini to interchange the order of summation. The finiteness of the last sum follows by Lemma 1.46, since $(2/d)(-1+s) = -1 - \varepsilon(2/d) < -1$. Thus $(S_N)_{N \geq n_0}$ is bounded in $L^1(\mathbb{P})$. In fact,

$$\mathbb{E}(S_N^2) = \sum_{n=n_0}^N \mathbb{E}(X_n^4) \|g_n\|_{H_0^s}^4 + 3 \sum_{n_0 \leq m \neq n \leq N} \mathbb{E}(X_n^2 X_m^2) \|g_n\|_{H_0^s}^2 \|g_m\|_{H_0^s}^2$$

$$= 3 \sum_{n=n_0}^N \|g_n\|_{H_0^s}^4 + 3 \sum_{n_0 \leq m \neq n \leq N} \|g_n\|_{H_0^s}^2 \|g_m\|_{H_0^s}^2$$

$$\leq 3 \left(\sum_{n_0 \leq n \leq N} \|g_n\|_{H_0^s}^2 \right)^2,$$

so $(S_N)_{N \geq n_0}$ is bounded in $L^2(\mathbb{P})$. Furthermore, by Doob's maximal inequality, and monotone convergence, for every $t > 0$,

$$\mathbb{P}(\sup_{N \geq n_0} S_N \geq t) \leq \frac{4}{t^2} \sup_{N \geq n_0} \mathbb{E}(S_N^2) \leq \frac{12}{t^2} \left(\sum_{n=n_0}^\infty \|g_n\|_{H_0^s}^2 \right)^2. \quad (1.44)$$

We now show how this implies almost sure convergence of the series $\sum_{n \geq 1} X_n g_n$ in $H_0^s(D)$. Set $U(n_0) = \sup_{N \geq n_0} \|\sum_{n=m}^N X_n g_n\|_{H^s}^2$, which is the random variable on the left-hand side of (1.44).

Fix $k \geq 1$. Since $\sum_{n \geq 1} \|g_n\|_{H^s}^2 < \infty$, there exists $n_k \geq 1$ such that $\sum_{n \geq n_k} \|g_n\|_{H^s}^2 \leq 4^{-k}$. By (1.44),

$$\mathbb{P}[U(n_k) \geq 2^{-k}] \leq 12 \cdot 4^{-k}.$$

As the right-hand side above is summable, we deduce that $U(n_k) \leq 2^{-k}$ for k sufficiently large and hence $U(n_k) \to 0$ almost surely. On the other hand if $n \geq n_k$, since $\|a+b\|^2 \leq 2(\|a\|^2 + \|b\|^2)$ on any Hilbert space, we deduce that if $n \geq n_k$ then $U(n) \leq 4 U(n_k)$. Thus $U(n) \to 0$ almost surely. Hence

36   *Definition and Properties of the GFF*

$\sum_{n=1}^{N} X_n g_n$ is Cauchy in H_0^s and therefore converges. This completes the proof of the theorem. □

Remark 1.47   The above theorem implies that the series $\sum_{n=1}^{N} X_n g_n$ converges almost surely in the space of distributions $\mathcal{D}'_0(D)$ whenever D is bounded. Recall that the measurability of the event

$$E = \left\{ \omega \in \Omega \colon \sum_{n=1}^{N} X_n(\omega) g_n \text{ converges in the space } \mathcal{D}'_0(D) \right\}$$

is provided by Lemma 1.34. However, even if we do not appeal to this lemma, the statement "the series $\sum_{n=1}^{N} X_n g_n$ converges almost surely in the space of distributions $\mathcal{D}'_0(D)$ whenever D is bounded" would still be meaningful. Indeed, in Theorem 1.45 we have checked that this series converges almost surely in the space $H_0^s(D)$ for some $s < 0$ (an event which is clearly measurable since $H_0^s(D)$ is a metric and indeed Hilbert space). On that (measurable) event of probability one, say E_s, it is clear that convergence in the space of distribution holds. Thus $E \supseteq E_s$ where E_s has probability one. Another way to state this is that, given Theorem 1.45 (and independently of Lemma 1.34), the event E is measurable on the completed σ-field $\mathcal{F}^*$ of the probability space (the completed σ-field $\mathcal{F}^*$ is the σ-field generated by $\mathcal{F}$ and the null sets).

Furthermore, by Theorem 1.44, this means that the GFF as a stochastic process, when its index set is restricted to smooth test functions, has a version that is almost surely a random element of $\mathcal{D}'_0(D)$. Moreover in two dimensions, the boundedness assumption can be removed using conformal invariance, see Theorem 1.57.

Let us reiterate one of the important conclusions from Theorems 1.45 and 1.44.

Corollary 1.48   *Suppose that $D \subset \mathbb{R}^d$ is a bounded domain, and $s = 1 - \frac{d}{2} - \varepsilon$ for some $\varepsilon > 0$. Then there exists a version of the Dirichlet GFF as a stochastic process $(\mathbf{h}, \rho)_{\rho \in H_0^{-s}}$ (with restricted index set) that is almost surely an element of $H_0^s(D)$.*

1.8 Itô's isometry for the GFF

Note: This section will not be used in the rest of the text and the reader may wish to skip it on a first reading.

In this section, we describe an observation which emerged from joint discussions with James Norris. It is closely linked to Lemma 1.43, which implies that for a zero boundary GFF **h** in a bounded domain D, and for any $f \in H_0^{-1}(D)$, the quantity $(\mathbf{h}, f)$ makes sense almost surely. That is, as the almost sure (and $L^2(\mathbb{P})$) limit of $(\mathbf{h}, f_n)$ for any sequence f_n converging to f in $H_0^{-1}(D)$.

In other words, even though h is only almost surely defined as a continuous linear functional on $H^{d/2-1+\varepsilon}(D)$ for $\varepsilon > 0$ (Theorem 1.45), we can actually test it against fixed functions that are much less regular. Namely, we can test it against any fixed function $H_0^{-1}(D)$. Note that this agrees with (in fact slightly extends) our previous definition of h as a stochastic process, since we have seen that $\mathfrak{M}_0$ is precisely the set of signed measures that are elements of H_0^{-1}, a consequence of Lemma 1.43.

In this section, we will essentially formulate the above discussion in terms of an isometry. To motivate this, it is useful to recall the following well known analogy within Itô's theory of stochastic integration. Let B be a standard Brownian motion. Even though dB does not have the regularity of a function in L^2 (in fact, it is essentially an element of $H^{-1/2-\varepsilon}$ for any $\varepsilon > 0$), it makes perfect sense to integrate it against a test function in L^2. This is thanks to the fact that the map

$$f \mapsto \int f_s \, dB_s$$

defines an isometry of suitable Hilbert spaces. Thus much flexibility has been gained: *a priori* we don't even have the right to integrate against functions in $H^{1/2}$, and yet, taking advantage of some almost sure properties of Brownian motion – namely, quadratic variation – it is possible to integrate against functions in L^2 (and actually much more).

A similar gain can be seen in the context of the GFF: *a priori*, as an element of $H_0^{1-d/2-\varepsilon}$ ($\varepsilon > 0$), it would seem that integrating against an arbitrary test function $f \in L^2$ is not even allowed when $d \geq 2$. Yet, as discussed above, we can almost surely integrate against much rougher objects, namely distributions in H_0^{-1}:

Theorem 1.49 (Itô isometry) *The map X sending $f \in \mathcal{D}_0(D)$ to the random variable $X_f = (\mathbf{h}, f)$ can be viewed as a linear map between $\mathcal{D}_0(D)$ and the* set of random variables $L^2(\Omega, \mathcal{F}, \mathbb{P})$ viewed as a function space. If we endow $\mathcal{D}_0(D)$ with the $H_0^{-1}(D)$ norm and $L^2(\Omega, \mathcal{F}, \mathbb{P})$ with its L^2 norm then X is an isometry:

$$\|f\|_{H_0^{-1}(D)} = \|X_f\|_2 = \mathbb{E}((\mathbf{h}, f)^2)^{1/2}.$$

In particular, since $\mathcal{D}_0(D)$ is dense in $H_0^{-1}(D)$, X_f extends uniquely as an isometry from $H_0^{-1}(D)$ into $L^2(\Omega, \mathcal{F}, \mathbb{P})$. Hence if $f \in H_0^{-1}(D)$, then we can set $(\mathbf{h}, f)$ to be the unique limit in $L^2(\mathbb{P})$ of $(\mathbf{h}, f_n)$, where f_n is any sequence of test functions that converge in $H_0^{-1}(D)$ to f.

Proof This is a direct consequence of Lemma 1.43. □

Remark 1.50 Note that although $(\mathbf{h}, f)$ makes sense as an almost sure limit for any fixed $f \in H_0^{-1}(D)$, or indeed for any countable collection of such f, this does not mean that $\mathbf{h}$ is an element of H_0^1 or that we can test $\mathbf{h}$ against every element of H_0^{-1} simultaneously. For example, writing

$$\mathbf{h} = \lim_{N \to \infty} \mathbf{h}_N := \lim_{N \to \infty} \sum_{n=1}^{N} \frac{X_n}{\sqrt{\lambda_n}} e_n$$

with $(X_n)_n \sim \mathcal{N}(0, 1)$ i.i.d. and $(e_n)_n$ an orthonormal basis of Laplacian eigenfunctions for $L^2(D)$, we have $\mathbf{h}_N \to \mathbf{h}$ almost surely in $H_0^{-1}(D)$ but $\text{Var}(\mathbf{h}, \mathbf{h}_N) = \sum_{n=1}^{N} \lambda_n^{-1} \to \infty$ (at least when $d \geq 2$). So there do exist random elements of $H_0^{-1}(D)$ that cannot be tested against $\mathbf{h}$.

1.9 Cameron–Martin space of the Dirichlet GFF

Note: This section will not be used until Chapter 7 and the reader may wish to skip it on a first reading.

In this section, we will address the following question:

- for $\mathbf{h}$ a Dirichlet (zero) boundary condition GFF in D and F a (deterministic) function on D, when are $\mathbf{h} + F$ and $\mathbf{h}$ mutually absolutely continuous?[8]

The answer is that this holds whenever $F \in H_0^1(D)$. This question can be phrased for general Gaussian processes, and the space of "F" for which absolute continuity holds is known as the **Cameron–Martin space** of the process. Thus,

[8] As stochastic processes indexed by $\mathfrak{M}_0$.

1.9 Cameron–Martin space of the Dirichlet GFF

the lemma below says that $H_0^1(D)$ is the Cameron–Martin space of the (Dirichlet boundary condition) GFF.

Proposition 1.51 *Let $\mathbf{h}$ be a GFF in a bounded domain D with Dirichlet (zero) boundary conditions. Then $\mathbf{h}$ and $\mathbf{h} + F$ are mutually absolutely continuous, as stochastic processes indexed by $\mathfrak{M}_0$, if and only if $F \in H_0^1(D)$. When this holds, the Radon–Nikodym derivative of $(\mathbf{h} + F)$ with respect to $\mathbf{h}$ is given by*

$$\frac{\exp((\mathbf{h}, F)_\nabla)}{\exp((F, F)_\nabla/2)}.$$

Proof Let $(e_i)_i$ be an orthonormal basis of $L^2(D)$ consisting of eigenfunctions of the Laplacian, with associated eigenvalues $(\lambda_i)_i$. We write $g_i := (\sqrt{\lambda_i})e_i$, so that the $(g_i)_i$ form an orthonormal basis of $H_0^{-1} \supset \mathfrak{M}_0$. Recall that

$$F \in H_0^1(D) \iff \sum (F, g_i)^2 < \infty.$$

For $n \in \mathbb{N}$, we consider the finite vector $((\mathbf{h}, g_i))_{1 \le i \le n}$, which by definition of the GFF is just a vector of independent $\mathcal{N}(0, 1)$ random variables.

This is convenient to work with because of the following elementary fact: if $(X_i)_{1 \le i \le n}$ are i.i.d. standard normals and $(a_i)_{1 \le i \le n}$ are real numbers, then $(X_1, X_2, \ldots, X_n)$ and $(X_1 + a_1, X_2 + a_2, \ldots, X_n + a_n)$ are mutually absolutely continuous. Moreover, the Radon–Nikodym derivative of the latter with respect to the former is given by $e^{\sum a_i X_i}/e^{\sum a_i^2/2}$.

In our context, this means that the law of $((\mathbf{h}, g_i))_{1 \le i \le n}$ is mutually absolutely continuous with that of $((\mathbf{h} + F, g_i))_{1 \le i \le n}$, if and only if $|(F, g_i)| < \infty$ for $1 \le i \le n$. Furthermore when this does hold, the Radon–Nikodym derivative of $((\mathbf{h} + F, g_i))_{1 \le i \le n}$ with respect to $((\mathbf{h}, g_i))_{1 \le i \le n}$ is equal to

$$\frac{\exp(\sum_{i=1}^n (\mathbf{h}, g_i)(F, g_i))}{\mathbb{E}(\exp(\sum_{i=1}^n (\mathbf{h}, g_i)(F, g_i)))} = \frac{\exp(\sum_{i=1}^n (\mathbf{h}, g_i)(F, g_i))}{\exp(\sum_{i=1}^n (F, g_i)^2/2)}. \tag{1.45}$$

Now, for $\mathbf{h}$ and $\mathbf{h} + F$ to be mutually absolutely continuous, the family of random variables on the right-hand side of (1.45) must be uniformly integrable (in n). Indeed, they should be the conditional expectations, with respect to a family of sub σ-algebras, of the Radon–Nikodym derivative of $(\mathbf{h} + F)$ with respect to $\mathbf{h}$. This family is *not* uniformly integrable if $F \notin H_0^1(D)$, that is, $\sum_{i \ge 1}(F, g_i)^2 = \infty$. Hence we obtain the necessity of the condition $F \in H_0^1(D)$ in the proposition.

For the sufficiency, we observe that when $F \in H_0^1(D)$, the random variables on the right-hand side of (1.45) converge in $L^1(\mathbb{P})$ to

$$\frac{\exp(\sum_{i \ge 1}(\mathbf{h}, g_i)(F, g_i))}{\exp(\sum_{i \ge 1}(F, g_i)^2/2)} = \frac{\exp((\mathbf{h}, F)_\nabla)}{\exp((F, F)_\nabla/2)}$$

as $n \to \infty$. We also know by Theorem 1.44 that whenever $\rho \in \mathfrak{M}_0$,

$$\sum_{i=1}^{n} \lambda_i^{-1}(\rho, g_i)(\mathbf{h}, g_i) \quad \text{converges to } (\mathbf{h}, \rho) \text{ almost surely}.$$

This implies that for any $\rho_1, \ldots, \rho_m \in \mathfrak{M}_0$ and any $\psi \colon \mathbb{R}^m \to \mathbb{R}$ continuous and bounded:

$$\mathbb{E}\left(\psi((\mathbf{h} + F, \rho_1), \ldots, (\mathbf{h} + F, \rho_m))\right)$$
$$= \mathbb{E}\left(\frac{\exp((\mathbf{h}, F)_\nabla)}{\exp((F, F)_\nabla/2)} \psi((\mathbf{h}, \rho_1), \ldots, (\mathbf{h}, \rho_m))\right).$$

But this is exactly the statement that $\mathbf{h} + F$ is absolutely continuous with respect to $\mathbf{h}$, as a stochastic process indexed by $\mathfrak{M}_0$, with the desired Radon–Nikodym derivative. Since the inverse of the Radon–Nikodym derivative is also in L^1, we obtain the mutual absolute continuity. □

1.10 Markov property

We are now ready to state one of the main properties of the GFF, which is the (domain) Markov property. As in the discrete case, informally speaking, it states that conditionally on the values of $\mathbf{h}$ outside of a given subset U, the free field inside U is obtained by harmonically extending $\mathbf{h}|_{D \setminus U}$ into U and then adding an independent GFF with Dirichlet boundary conditions in U. Note that in this case, however, it is not at all clear that such a harmonic extension is well defined.

Theorem 1.52 (Markov property) *Fix $U \subset D$ a regular subdomain. Let $\mathbf{h}$ be a GFF (with zero boundary conditions on D). Then we may write*

$$\mathbf{h} = \mathbf{h}_0 + \varphi,$$

where

1. $\mathbf{h}_0$ *is a zero boundary condition GFF in U, and is zero outside of U;*
2. φ *is harmonic in U; and*
3. $\mathbf{h}_0$ *and φ are independent.*

This makes sense whether we view $\mathbf{h}$ as a random distribution or a stochastic process indexed by $\mathfrak{M}_0$. Note that since $\mathbf{h}_0 = 0$ on U^c, φ coincides with $\mathbf{h}$ on U^c. See Figure 1.2 for an illustration.

Corollary 1.53 *By Remark 1.47, when $D \subset \mathbb{R}^2$ is an arbitrary domain (i.e. potentially unbounded) this Markov property implies that the random distribution $\mathbf{h}$ almost surely defines a random element of the local Sobolev*

1.10 Markov property

(a) (b)

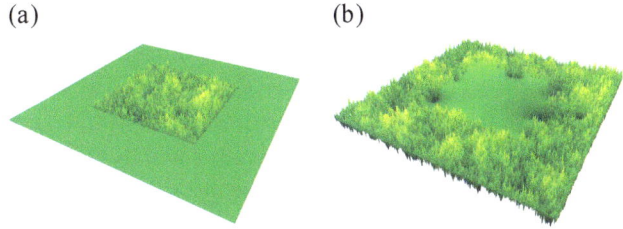

Figure 1.2 The Markovian decomposition of the GFF: here D is a square, and $U \subset D$ a slightly smaller square. (a) shows $\mathbf{h}_0$, and (b) shows φ. Their sum $\mathbf{h}$ is a GFF in D, shown in Figure 1.1. Simulations by Oskar-Laurin Koiner.

space $H^{-1}_{\mathrm{loc}}(D)$.[9] *In general dimension $d \geq 3$, it follows from the above Markov property that there is a version of the stochastic process $\mathbf{h}$ that is almost surely a random distribution (in fact, a random element of $H^{1-d/2-\varepsilon}_{\mathrm{loc}}(D)$ for any $\varepsilon > 0$).*

Proof The key point is the following Hilbertian decomposition:

Lemma 1.54 *Let U be as in Theorem 1.52. We have*
$$H^1_0(D) = H^1_0(U) \oplus \mathrm{Harm}(U),$$
where $\mathrm{Harm}(U)$ consists of harmonic functions in U – that is, elements of $H^1_0(D)$ whose restriction to U coincide with a harmonic function in U.

Proof We first prove orthogonality. Let $f \in H^1_0(U) \subset H^1_0(D)$. Then there exists $f_n \in \mathcal{D}_0(U) \subset \mathcal{D}_0(D)$ such that $f_n \to f$ in the $H^1_0(D)$ sense. Now let $g \in H^1_0(D)$ such that g coincides in U with a harmonic function. Note that
$$(f_n, g)_\nabla = \int_D (\nabla f_n) \cdot (\nabla g) = \int_{U_n} (\nabla f_n) \cdot (\nabla g),$$
where $U_n \subset U$ is chosen to be compactly contained in U, have smooth boundary, and contain the (closure of the) support of f_n (in particular $\nabla f_n = 0$ outside of U_n). Since U_n is smooth, we can apply the Gauss–Green formula (Lemma 1.35) in U_n with boundary term, because g is non-zero on ∂U_n, this boundary term does need to be considered. However, the boundary term vanishes because $\partial g/\partial n$ is a smooth function on ∂U_n and $f_n = 0$ on ∂U_n.

Therefore $(f_n, g)_\nabla = -\int_{U_n} f_n \Delta g$, which is clearly 0 because in U_n, $\Delta g = 0$ and f_n is a smooth function. This shows that f_n and g are orthogonal in $H^1_0(D)$. Then by taking a limit (since f_n approximates f in the $H^1_0(D)$ sense), f and g must also be orthogonal.

[9] $H^{-1}_{\mathrm{loc}}(D)$ is the space of distributions whose restriction to any $U \Subset D$ (i.e. such that $\overline{U}$ is a compact subset of D) is an element of $H^{-1}_0(U)$.

Now let us show that the sum of the two spaces spans $H_0^1(D)$. Let us suppose to begin with that U is C^1 smooth. For $f \in H_0^1(D)$, let f_0 denote the orthogonal projection of f onto $H_0^1(U)$. Set $\varphi = f - f_0$: our aim is to show that φ is harmonic in U. Note that φ is (by definition) orthogonal to $H_0^1(U)$. Hence for any test function $\psi \in \mathcal{D}_0(U)$, we have that $(\varphi, \psi)_\nabla = 0$. By the Gauss–Green formula (and since U is C^1 smooth), we deduce that

$$\int_D (\Delta \varphi) \psi = \int_U (\Delta \varphi) \psi = 0$$

and hence $\Delta \varphi = 0$ as a distribution in U. Elliptic regularity arguments (going beyond the scope of these notes) show that a distribution which is harmonic in the sense of distributions must in fact be a smooth function, harmonic in the usual sense. Therefore $\varphi \in \mathrm{Harm}(U)$ and we are done.

If U does not have C^1 boundary, let $(U_n)_{n\in\mathbb{N}}$ be a sequence of increasing open subsets of U with C^1 boundaries, such that $\cup U_n = U$. For $f \in H_0^1(D)$, by the previous paragraph, we can write $f = f_0^n + \varphi^n$ for each $n \in \mathbb{N}$, where f_0^n is the projection of f onto $H_0^1(U_n)$ and $\varphi^n \in \mathrm{Harm}(U_n)$. Then we just need to show that: (a) $f_0^n \to f_0$ as $n \to \infty$ for some $f_0 \in H_0^1(U)$; and (b) that $f - f_0$ is harmonic in U. For (a), we observe that $H_0^1(U) = \overline{\cup_n H_0^1(U_n)}$ (by definition of $H_0^1(U)$ as the closure of $\mathcal{D}_0(U)$ with respect to the Dirichlet inner product) and so the projections f_0^n of f onto $H_0^1(U_n)$ converge, with respect to $\|\cdot\|_\nabla$, to $f_0 \in H_0^1(U)$. For (b), notice that by definition of f_0, $f - f_0$ is the limit of φ_n as $n \to \infty$, with respect to $\|\cdot\|_\nabla$. In particular, it is clear that when restricted to any U_n, $f - f_0 = \lim_n \varphi_n$ is harmonic in the distributional sense, and thus harmonic by elliptic regularity. Since this holds for any n, it follows that $f - f_0$ is harmonic in U. □

Having this decomposition in hand, we may deduce the Markov property in a rather straightforward way. Indeed, let $(f_n^0)_n$ be an orthonormal basis of $H_0^1(U)$, and let $(\phi_n)_n$ be an orthonormal basis of $\mathrm{Harm}(U)$. For $((X_n, Y_n))_n$ an i.i.d. sequence of independent standard Gaussian random variables, set $\mathbf{h}_0 = \sum_n X_n f_n^0$ and $\varphi = \sum_n Y_n \phi_n$. Then the first series converges in $\mathcal{D}_0'(D)$ since it is a series of a GFF in U. The sum of the two series gives $\mathbf{h}$ by construction, and so the second series also converges in the space of distributions. In the space of distributions, the limit of harmonic distributions must be harmonic as a distribution, and hence (by the same elliptic regularity arguments as above) a true harmonic function. This proves the theorem. □

Remark 1.55 It is worth pointing out an important message from the proof above: any orthogonal decomposition of $H_0^1(D)$ gives rise to a decomposition of the GFF into independent summands.

Example When $d = 1$, this is the statement that if $(b_s)_{s \in [0,1]}$ is a Brownian bridge from 0 to 0 and $[a, b] \subset [0, 1]$, then conditionally on $(b_s)_{s \in [0,a] \cup [b,1]}$, the law of $(b_s)_{s \in [a,b]}$ is given by a the linear interpolation of b_a and b_b, plus an independent Brownian bridge from 0 to 0 on $[a, b]$.

Remark 1.56 In the case when D is an unbounded domain of $\mathbb{R}^d$ with $d \neq 2$, applying the Markov property in bounded subdomains shows that the GFF, viewed as a stochastic process with restricted index set $\mathcal{D}_0(D)$, has a version that almost surely defines a distribution on D.

1.11 Conformal invariance

In the remainder of this chapter, we restrict ourselves to dimension $d = 2$.

In this case the GFF possesses the important additional property of **conformal invariance**, which follows almost immediately from the construction in Section 1.10. Indeed, a straightforward change of variable formula shows that the Dirichlet inner product is conformally invariant: if $\varphi \colon D \to D'$ is a conformal isomorphism, then

$$\int_{D'} \nabla(f \circ \varphi^{-1}) \cdot \nabla(g \circ \varphi^{-1}) = \int_D \nabla f \cdot \nabla g.$$

Consequently, if $(f_n)_n$ is an orthonormal basis of $H_0^1(D)$, then $(f_n \circ \varphi^{-1})_n$ defines an orthonormal basis of $H_0^1(D')$. (Watch out however, that eigenfunctions of $-\Delta$ are not conformally invariant in any sense). So by Theorem 1.45 we have:

Theorem 1.57 (Conformal invariance of the GFF) *If $\mathbf{h}$ is a random distribution on $\mathcal{D}_0'(D)$ with the law of the Gaussian free field on D, then the distribution $\mathbf{h} \circ \varphi^{-1}$, defined by setting $(\mathbf{h} \circ \varphi^{-1}, f) = (\mathbf{h}, |\varphi'|^2 (f \circ \varphi))$ for $f \in \mathcal{D}_0'(D')$, has the law of a GFF on D'.*

Recently, a kind of converse was shown in [BPR21, BPR20]: if a field $\mathbf{h}$ with zero boundary conditions satisfies conformal invariance and the domain Markov property, as well as a moment condition ($\mathbb{E}((\mathbf{h}, \phi)^{1+\varepsilon}) < \infty$ for some $\varepsilon > 0$ and all $\phi \in \mathcal{D}_0(D)$), then $\mathbf{h}$ must be a multiple of the Gaussian free field. In fact, one can reduce the conformal invariance assumption to scale invariance, and obtain the result in all dimensions, [AP22]. See [BPR21, BPR20, AP22] for details.

1.12 Circle averages

An important tool for studying the GFF is the process which describes its average values on small circles centred around a point $z \in D$. This is known as the **circle average process** around z.

More precisely, fix $z \in D$ and let $0 < \varepsilon < \text{dist}(z, \partial D)$. Let $\rho_{z,\varepsilon}$ denote the uniform distribution on the circle of radius ε around z. Note that $\rho_{z,\varepsilon} \in \mathfrak{M}_0$. This follows from the fact that $G_0^D(x, y) \leq -(2\pi)^{-1} \log |x - y| + O(1)$, and the fact, when we fix one of the variables x on the circle, the integral over the circle of $-\log |x - y|$ with respect to y is finite (just like the integral of $-\log r$ with respect to r is finite in one dimension). More generally, this argument shows that the Lebesgue measure on any smooth curve is an element of $\mathfrak{M}_0$.

We set $\mathbf{h}_\varepsilon(z) = (\mathbf{h}, \rho_{z,\varepsilon})$. The following theorem is a consequence of the Kolmogorov–Čentsov continuity theorem (a multidimensional generalisation of the more classical Kolmogorov continuity criterion), and will not be proved here. The interested reader is directed to Proposition 3.1 of [DS11] for a proof.

Proposition 1.58 (Circle average is jointly Hölder) *There exists a modification of $\mathbf{h}$ such that $(\mathbf{h}_\varepsilon(z), z \in D, 0 < \varepsilon < \text{dist}(z, \partial D))$ is almost surely jointly Hölder continuous of order $\eta < 1/2$ on all compact subsets of $\{z \in D \text{ s.t. } 0 < \varepsilon < \text{dist}(z, \partial D)\}$.*

In fact, it can be shown that this version of the GFF is the same as the version which turns h into a random distribution in Theorem 1.44. The reason circle averages are so useful is because of the following result.

Theorem 1.59 (Circle average is a Brownian motion) *Let $\mathbf{h}$ be a GFF on D. Fix $z \in D$ and let $0 < \varepsilon_0 < \text{dist}(z, \partial D)$. For $t \geq t_0 = \log(1/\varepsilon_0)$, set*

$$B_t = \sqrt{2\pi} \mathbf{h}_{e^{-t}}(z).$$

Then $(B_t, t \geq t_0)$ has the law of a Brownian motion started from B_{t_0}: in other words, $(B_{t+t_0} - B_{t_0}, t \geq 0)$ is a standard Brownian motion.

Proof In order to avoid factors of $\sqrt{2\pi}$ everywhere, we use $h = \sqrt{2\pi}\mathbf{h}$ as defined in (1.30), and call $h_\varepsilon(z) = (h, \rho_{z,\varepsilon})$. The theorem statement is then equivalent to saying that $(B_t = h_{e^{-t}}, t \geq t_0)$ is a Brownian motion starting from B_{t_0}. Various proofs can be given. For instance, the covariance function can be computed explicitly (this is a good exercise)! Alternatively, we can use the Markov property of the GFF to see that B_t must have stationary and independent increments. Indeed, suppose $\varepsilon_1 > \varepsilon_2$, and we condition on h outside of $B(z, \varepsilon_1)$. That is, we write $h = h^0 + \varphi$, where φ is harmonic in $U = B(z, \varepsilon_1)$ and h^0 is a GFF in U that is independent of $(h_\varepsilon(z))_{\varepsilon \geq \varepsilon_1}$ (scaled in the same manner as

1.12 Circle averages

(1.30)). Then $h_{\varepsilon_2}(z)$ is the sum of two terms: $h^0_{\varepsilon_2}(z)$; and the circle average of φ on $\partial B(z, \varepsilon_2)$. By harmonicity of φ, the latter is nothing else than $h_{\varepsilon_1}(z)$. This gives that the increment can be expressed as

$$h_{\varepsilon_2}(z) - h_{\varepsilon_1}(z) = h^0_{\varepsilon_2}(z),$$

and hence, since h^0 is independent of $(h_\varepsilon(z))_{\varepsilon \geq \varepsilon_1}$, the increments are independent. Moreover, by applying the change of scale $w \mapsto (w-z)/\varepsilon_1$, so that the outer circle is mapped to the unit circle, we see that the distribution of $h_{\varepsilon_2}(z) - h_{\varepsilon_1}(z)$ depends only on $r = \varepsilon_2/\varepsilon_1$. This means that they are also stationary.

To show from here that $h_{e^{-t}}(z)$ is a Brownian motion, it suffices to compute its variance. That is (by the Markov property), to check that if h is a GFF in the unit disc $\mathbb{D}$ and $r < 1$, then $h_r(0)$ has variance $-\log r$.

For this, let ρ denote the uniform distribution on the circle $\partial(r\mathbb{D})$ at distance r from the origin, so that

$$\text{Var}(h_r(0)) = 2\pi \int_{\mathbb{D}^2} G^{\mathbb{D}}_0(x, y)\rho(dx)\rho(dy). \tag{1.46}$$

The point is that by harmonicity of $G^{\mathbb{D}}_0(x, \cdot)$ in $\mathbb{D} \setminus \{x\}$ and the mean value property, the above integral is simply

$$\text{Var}(h_r(0)) = 2\pi \int_{\mathbb{D}} G^{\mathbb{D}}_0(x, 0)\rho(dx), \tag{1.47}$$

which completes the proof since $G^{\mathbb{D}}_0(x, 0) = -(2\pi)^{-1} \log |x| = -(2\pi)^{-1} \log r$ on $\partial(r\mathbb{D})$.

To check (1.47) rigorously, first consider for a fixed $\eta > 0$, the double integral

$$I_\eta = 2\pi \int_{\mathbb{D}^2} G^{\mathbb{D}}_0\big((1+\eta)x, y\big)\rho(dx)\rho(dy).$$

Then I_η converges clearly to the right-hand side of (1.46) as $\eta \to 0$, and it is now rigorous to exploit the mean value property for the harmonic function $G^{\mathbb{D}}_0((1+\eta)x, \cdot)$ in the entire ball $B(0, r)$ to deduce that

$$I_\eta = 2\pi \int_{\mathbb{D}} G^{\mathbb{D}}_0\big((1+\eta)x, 0\big)\rho(dx).$$

Letting $\eta \to 0$ proves (1.47). □

So, as we "zoom in" towards a point, the average values of the field oscillate like those of a Brownian motion. This gives us a very precise sense in which the field cannot be defined pointwise.

1.13 Thick points

An important notion in the study of Liouville quantum gravity is that of thick points of the Gaussian free field. Indeed, although these points are atypical from the point of view of Euclidean geometry, we will see that they are typical from the point of view of the associated quantum geometry. In order to be consistent with its applications in Gaussian multiplicative chaos and Liouville quantum gravity, we will once again here mostly work with the normalisation $h = \sqrt{2\pi}\mathbf{h}$ from (1.30).

Definition 1.60 Let $\mathbf{h}$ be a GFF in $D \subset \mathbb{C}$ open and simply connected, let $h = \sqrt{2\pi}\mathbf{h}$, and let $\alpha > 0$. We say a point $z \in D$ is α-thick if

$$\liminf_{\varepsilon \to 0} \frac{h_\varepsilon(z)}{\log(1/\varepsilon)} = \alpha.$$

In fact, the lim inf in the definition could be replaced with a lim sup or lim. It is also clear by symmetry that the set of $(-\alpha)$-thick points with $\alpha > 0$ has the same law as the set of α-thick points; hence we restrict to the case $\alpha > 0$ for simplicity.

Note that a given point $z \in D$ is almost surely not thick: the typical value of $h_\varepsilon(z)$ is of order $\sqrt{\log 1/\varepsilon}$ since $h_\varepsilon(z)$ is a Brownian motion at scale $\log 1/\varepsilon$. At this stage, the most relevant result is the following fact due to Hu, Miller and Peres [HMP10] (though it was independently and earlier proved by Kahane in the context of his work on Gaussian multiplicative chaos).

Theorem 1.61 *Let $\mathcal{T}_\alpha$ denote the set of α-thick points. Then almost surely, the Hausdorff dimension $d_H(\mathcal{T}_\alpha)$ of $\mathcal{T}_\alpha$ satisfies*

$$d_H(\mathcal{T}_\alpha) = (2 - \frac{\alpha^2}{2})_+$$

and $\mathcal{T}_\alpha$ is almost surely empty if $\alpha > 2$.

Heuristics The value of the dimension of $\mathcal{T}_\alpha$ is easy to understand and to guess. Indeed, for a given $\varepsilon > 0$,

$$\mathbb{P}(h_\varepsilon(z) \geq \alpha \log(1/\varepsilon)) = \mathbb{P}(\mathcal{N}(0, \log(1/\varepsilon) + O(1)) \geq \alpha \log(1/\varepsilon))$$
$$= \mathbb{P}(\mathcal{N}(0,1) \geq \alpha \sqrt{\log(1/\varepsilon) + O(1)}) \leq \varepsilon^{\alpha^2/2}$$

using scaling and the standard bound $\mathbb{P}(X > t) \leq \mathrm{const} \times t^{-1} e^{-t^2/2}$ for $X \sim \mathcal{N}(0,1)$. Suppose without loss of generality that $D = (0,1)^2$ is the unit square. Then the expected number of squares of size ε such that the centre z satisfies $h_\varepsilon(z) \geq \alpha \log 1/\varepsilon$ is bounded by $\varepsilon^{-2+\alpha^2/2}$. This suggests that the Minkowski

1.13 Thick points

dimension is less or equal to $2 - \alpha^2/2$ when $\alpha < 2$ and that $\mathcal{T}_\alpha$ is empty if $\alpha > 2$. □

Rigorous proof of upper bound. We now turn the above heuristics into a rigorous proof that $d_H(\mathcal{T}_\alpha) \le (2 - \alpha^2/2) \vee 0$, which follows closely the argument given in [HMP10]. The lower bound given in [HMP10] is more complicated, but we will obtain an elementary proof in Chapter 2, via the Liouville measure: see Exercise 2.4.

To start the proof of the upper bound, we begin by stating an improvement of Theorem 1.58, which is Proposition 2.1 in [HMP10]. This is the circle average analogue of Lévy's modulus of continuity for Brownian motion.

Lemma 1.62 *Suppose D is bounded with smooth boundary. Then there exists a version of the circle average process $(h_r(z))_{r<1, z \in D}$, such that for every $\eta < 1/2, \zeta > 0$ and $\varepsilon > 0$, there exists $M = M(\eta, \zeta, \varepsilon)$ which is finite almost surely and such that*

$$|h_r(z) - h_s(w)| \le M \left(\log \frac{1}{r}\right)^\zeta \frac{(|z - w| + |r - s|)^\eta}{r^{\eta + \varepsilon}}$$

holds for every $z, w \in D$ and for all $r, s \in (0, 1)$ such that $r/s \in [1/2, 2]$ and $B(z, r), B(w, s) \subset D$.

See Proposition 2.1 in [HMP10] for a proof.

Without loss of generality, we will now work in the case where D is bounded with smooth boundary. This yields the proof in the general case by the domain Markov property, and since $d_H(\mathcal{T}_\alpha) = \lim_{n \to \infty} d_H(\mathcal{T}_\alpha \cap D_n)$ for a sequence of smooth, bounded domains D_n with $\bigcup D_n = D$.

In this setting, the above lemma allows us to "discretise" the set of ε and z on which it suffices to check thickness. More precisely, set $\varepsilon > 0, K > 0$ and consider the sequence of scales $r_n = n^{-K}$. Fix $\zeta < 1$, and $\eta < 1/2$ arbitrarily (say $\zeta = 1/2, \eta = 1/4$), and let $M = M(\eta, \zeta, \varepsilon)$ be as in the lemma. Then for any $z \in D$, we have that if $r_{n+1} \le r \le r_n$,

$$|h_r(z) - h_{r_n}(z)| \le MK^\zeta (\log n)^\zeta \frac{(r_{n+1} - r_n)^\eta}{r_n^{\eta(1+\varepsilon)}}$$

$$\lesssim (\log n)^\zeta n^{K\eta(1+\varepsilon) - (K+1)\eta} \lesssim (\log n)^\zeta$$

if we choose $\varepsilon = K^{-1}$. Thus any point $z \in D$ is in $\mathcal{T}_\alpha$ if and only if

$$\lim_{n \to \infty} \frac{h_{r_n}(z)}{\log 1/r_n} = \alpha.$$

Now for any $n \ge 1$, let $\{z_{n,j}\}_j = D \cap r_n^{1+\varepsilon} \mathbb{Z}^2$ be a set of discrete points spaced

by $r_n^{1+\varepsilon}$ within D. Then if $z \in B(z_{n,j}, r_n^{1+\varepsilon})$ we have, for the same reasons,
$$|h_{r_n}(z) - h_{r_n}(z_{n,j})| \lesssim (\log n)^{\zeta}.$$
Thus for fixed $\delta > 0$, we let
$$I_n = \{j \colon h_{r_n}(z_{n,j}) \geq (\alpha - \delta)\log(1/r_n)\}.$$
Then for each $N \geq 1$, and each $\delta > 0$,
$$\mathcal{T}_\alpha' = \bigcup_{n>N} \bigcup_{j \in I_n} B(z_{n,j}, r_n^{1+\varepsilon})$$
is a cover of $\mathcal{T}_\alpha$. Consequently, if $\mathcal{H}_q$ denotes q-dimensional Hausdorff measure for $q > 0$,

$$\mathbb{E}(\mathcal{H}_q(\mathcal{T}_\alpha)) \leq \mathbb{E}\left(\sum_{n>N} \sum_{j \in I_n} \operatorname{diam} B(z_{n,j}, r_n^{1+\varepsilon})^q\right)$$
$$\lesssim \sum_{n>N} r_n^{-2-2\varepsilon} r_n^{q(1+\varepsilon)} \max_j \mathbb{P}(j \in I_n).$$

For a fixed n and a fixed j, as argued in the heuristics,
$$\mathbb{P}(j \in I_n) \lesssim \exp(-\frac{(\alpha-\delta)^2}{2}\log(1/r_n)) = r_n^{(\alpha-\delta)^2/2},$$
where the implied constants are uniform over D. We deduce
$$\mathbb{E}(\mathcal{H}_q(\mathcal{T}_\alpha)) \lesssim \sum_{n>N} r_n^{-2-2\varepsilon+(\alpha-\delta)^2/2+q(1+\varepsilon)}.$$

As $r_n = n^{-K}$ and K can be chosen arbitrarily large, the right-hand side tends to zero as $N \to \infty$ as soon as the exponent of r_n in the above sum is positive, or, in other words, if q is such that
$$-2 - 2\varepsilon + (\alpha - \delta)^2/2 + q(1+\varepsilon) > 0.$$
Thus we deduce that $\mathcal{H}_q(\mathcal{T}_\alpha) = 0$ almost surely (and hence $d_H(\mathcal{T}_\alpha) \leq q$ whenever $q(1+\varepsilon) > 2 + 2\varepsilon - (\alpha-\delta)^2/2$. So
$$d_H(\mathcal{T}_\alpha) \leq \frac{2 + 2\varepsilon - (\alpha-\delta)^2/2}{1+\varepsilon},$$
almost surely. Since $\varepsilon > 0, \delta > 0$ are arbitrary, we deduce
$$d_H(\mathcal{T}_\alpha) \leq 2 - \alpha^2/2,$$
almost surely, as desired. □

The value $\alpha = 2$ corresponds informally to the maximum of the free field, and the study of the set $\mathcal{T}_2$ is, informally at least, related to the study of extremes in a branching Brownian motion (see [ABBS13, ABK13]).

1.14 Scaling limit of the discrete GFF

In this short section, we briefly explain why the discrete GFF on appropriate sequences of planar graphs converges in the scaling limit to the continuum GFF. Before we give general arguments, let us point out a situation in which this is relatively straightforward to see.

Let $D = (0, 1)^2$ be the unit square, and $V_N = D \cap (\mathbb{Z}^2/N)$ be the portion of the square lattice (scaled to have mesh size $1/N$) that intersects D, and let E_N be the edges of the whole square lattice scaled by $1/N$. Let ∂_N denote the set of vertices $v \in V_N$ with at least one neighbour outside of V_N, which is the natural boundary of this graph. Let h_N be the discrete Gaussian free field associated with V_N, ∂_N (and with $q_{x,y} = 1$ for every pair of neighbouring vertices x, y in V_N). In order to discuss convergence to the continuum GFF, it is useful to extend the definition of h_N to all of $\mathbb{R}^2$: namely, we extend h_N to be constant on each face of the dual graph of (V_N, E_N); that is, for $x \in V_N$, and $y \in (-1/(2N), +1/(2N)]^2$, we set $h_N(y) = h_N(x)$.

We then claim that for a fixed $k \geq 1$ and fixed test functions $\phi_1, \ldots, \phi_k \in \mathcal{D}_0(D)$, the law of the vector $(h_N, \phi_i)_{i=1}^k$ converges (without scaling) as $N \to \infty$ to the law of $(h, \phi_i)_{i=1}^k$, where h is a continuum GFF. In fact, we will check the following stronger convergence.

Proposition 1.63 *Consider the above discrete GFF h_N in the unit square. We then have the convergence in distribution:*

$$h_N \to h \quad (1.48)$$

as random variables on $H_0^s(D)$ for any $s < 0$, where h is a continuum GFF with zero boundary conditions on $D = (0, 1)^2$.

Note that the choice of normalisation is consistent from discrete to continuum, in that the discrete random walk associated to the graph $G_N = (V_N, E_N)$ converges, after speeding up time by a factor N^2, to a *speed two* Brownian motion.

Proof For $k, m \geq 1$ let

$$f_{k,m}(x, y) = 2 \sin(\pi k x) \sin(\pi m y).$$

It is elementary that $f_{k,m}$ is an eigenfunction of $-\Delta$ in $D = (0, 1)^2$ (with Dirichlet boundary conditions), corresponding to the eigenvalue $\lambda_{k,m} = \pi^2(k^2 + m^2)$, and has unit $L^2(D)$ norm; an elementary fact from Fourier analysis is that $(f_{k,m})_{k,m \geq 1}$ form an orthonormal basis of $L^2(D)$.

Furthermore, on the unit square, a minor miracle happens: namely, if $1 \leq$

50   *Definition and Properties of the GFF*

$k \leq N$ and $1 \leq m \leq N$, then $f_{k,m}$ is *also* a discrete eigenfunction of the negative discrete Laplacian $-Q_N$, with associated eigenvalue

$$\lambda_{k,m}^N = 2 - 2\cos\left(\frac{\pi k}{N}\right) + 2 - 2\cos\left(\frac{\pi m}{N}\right).$$

In particular, letting $N \to \infty$ but keeping $k, m \geq 1$ fixed, we see that

$$\lambda_{k,m}^N \sim \frac{1}{N^2} \lambda_{k,m}.$$

We denote by $\bar{f}_{k,m}^N$ the eigenfunction $f_{k,m}$, normalised to have unit (discrete) L^2 norm, that is,

$$\bar{f}_{k,m}^N(\cdot) = \frac{1}{c_{m,k}^N} f_{k,m}(\cdot); \quad \text{with } c_{k,m}^N = \left(\sum_{z \in V_N} f_{k,m}(z)^2\right)^{1/2}.$$

Clearly, as $N \to \infty$ with $k, m \geq 1$ fixed,

$$(c_{k,m}^N)^2 \sim 4N^2 \iint_D \sin^2(\pi k x) \sin^2(\pi m y) \, dx \, dy = N^2.$$

In fact, using simple trigonometric identities, we can check that for all $N \geq 1$ and all $1 \leq k, m \leq N$ we have $c_{k,m}^N = N$ exactly.

The functions $f_{k,m}$ are linearly independent and thus $(\lambda_{k,m}^N)_{1 \leq k \leq N, 1 \leq m \leq N}$ give us all possible eigenvalues of $-Q_N$ (counted with multiplicity in case of repetition).

We deduce, using Theorem 1.7, that the discrete GFF can be written as

$$h_N(\cdot) = \sum_{k,m=1}^N \frac{1}{\sqrt{\lambda_{k,m}^N}} X_{k,m} \bar{f}_{k,m}^N(\cdot) = \sum_{k,m=1}^N \frac{1}{c_{k,m}^N \sqrt{\lambda_{k,m}^N}} X_{k,m} f_{k,m}(\cdot),$$

where $(X_{k,m})_{1 \leq k,m \leq N}$ are independent standard Gaussian random variables. By Theorem 1.45, to deduce (1.48), it remains to check that

- when $k, m \geq 1$ are fixed and $N \to \infty$, $c_{k,m}^N \sqrt{\lambda_{k,m}^N} \to \pi\sqrt{\lambda_{k,m}}$; and
- the expected H_0^s square norm of the remainder series is controlled uniformly in N.

The first point is elementary given the above asymptotics. For the second point, we need to show that for all $\varepsilon > 0$ we can find $A \geq 1$ large but fixed (i.e. independent of N) such that,

$$\mathbb{E}\left(\left\|\sum_{k=1}^N \sum_{m=A}^N \frac{1}{c_{k,m}^N \sqrt{\lambda_{k,m}^N}} X_{k,m} f_{k,m}(\cdot)\right\|_{H_0^s}^2\right) \leq \varepsilon \qquad (1.49)$$

1.14 Scaling limit of the discrete GFF

for all $N \geq 1$. Note that the left-hand side above is equal to

$$\sum_{k=1}^{N}\sum_{m=A}^{N} \frac{1}{(c_{k,m}^{N})^2 \lambda_{k,m}^{N}} \lambda_{k,m}^{s}$$

by definition of the H_0^s norm. To conclude, we simply observe that $1 - \cos(x) \geq ax^2/2$ for all $x \in [0,1]$ and some $a > 0$. Hence

$$\lambda_{k,m}^{N} \geq \frac{a}{N^2} \lambda_{k,m},$$

and since $c_{k,m}^{N} = N$, we see that

$$\frac{1}{c_{k,m}^{N} \lambda_{k,m}^{N}} \lambda_{k,m}^{s} \leq C \lambda_{k,m}^{-1+s}$$

for some constant $C > 0$. We conclude that (1.49) holds as in the proof of Theorem 1.45, that is, using Weyl's law. This proves (1.48). □

If we take a general bounded domain $D \subset \mathbb{R}^2$, the argument above can no longer be applied because there is no exact relation between the discrete and continuous eigenfunctions. A different argument is therefore needed.

We fix a bounded domain $D \subset \mathbb{R}^2$. Let $(G_\delta)_{\delta > 0}$ denote a sequence of undirected graphs (with weights on the edges) embedded in D. Denote their vertex sets by $v(G_\delta)$ and prespecified boundaries by $\partial_\delta \subset v(G_\delta)$. Let $\mathbb{P}_x^\delta$ denote the law of continuous time random walk on G_δ starting from some vertex x of G_δ, killed when it reaches the boundary ∂_δ, and let $\mathbb{E}_x^\delta$ denote the associated expectation. Our main assumption is that the random walk under $\mathbb{P}_x^\delta$ converges to (speed two) Brownian motion as $\delta \to 0$, uniformly on compact time intervals and uniformly in space, in the sense that for any smooth test function $\phi \in \mathcal{D}_0(D)$, we have

$$\left| \mathbb{E}_x^\delta [\phi(X_{s\delta^{-2}})] - \mathbb{E}_x[\phi(B_s) 1_{\{\tau > s\}}] \right| \to 0, \tag{1.50}$$

as $\delta \to 0$, uniformly over $s \in [0,T]$ for every $T > 0$ and $x \in V(G_\delta)$. We also suppose that if τ_δ is the time that the random walk under $\mathbb{P}_x^\delta$ first hits the boundary ∂_δ, then $\delta^2 \tau_\delta$ is uniformly integrable: that is, for every $\varepsilon > 0$ we can choose $K < \infty$ such that

$$\mathbb{E}_x^\delta (\delta^2 \tau_\delta 1_{\{\tau_\delta \geq K\delta^{-2}\}}) \leq \varepsilon, \tag{1.51}$$

uniformly in the vertices x of G_δ. Finally, we assume that the vertices of G_δ have density asymptotically uniform, in the following sense: for any open set A such that $A \subset D$,

$$\frac{\#v(G_\delta) \cap A}{\delta^{-2}} \to \text{Leb}(A) \tag{1.52}$$

as $\delta \to 0$.

Theorem 1.64 *Let h_δ denote the discrete GFF associated to the graph G_δ and ∂_δ, as above, and suppose that (1.50), (1.51) and (1.52) hold. For a test function $\phi \in \mathcal{D}_0(D)$, let $h_\delta(\phi) = \delta^2 \sum_{x \in v(G_\delta)} h_\delta(x)\phi(x)$. Then for every $k \geq 1$, and for every set of test functions $\phi_1, \ldots, \phi_k \in \mathcal{D}_0(D)$, we have*

$$(h_\delta(\phi_i))_{i=1}^k \to (h, \phi_i)_{i=1}^k$$

in distribution as $\delta \to 0$, where h is a continuum GFF with zero boundary conditions in D.

Proof Since the variables $h_\delta(\phi_i)$ are Gaussian and linear in ϕ_i, it once again suffices to prove the statement for $k = 1$, in which case we write ϕ instead of ϕ_1. Having fixed ϕ, we observe that $h_\delta(\phi)$ is a centred Gaussian random variable with variance

$$\sigma_\delta^2 = \delta^4 \sum_{x,y \in v(G_\delta)} G_\delta(x,y)\phi(x)\phi(y). \tag{1.53}$$

In order to show that $h_\delta(\phi)$ converges to (h, ϕ), it suffices to check that

$$\sigma_\delta^2 \to \sigma^2 = \iint_{D^2} G_0^D(x,y)\phi(x)\phi(y)\,dx\,dy$$

as $\delta \to 0$. Our goal is therefore to show that the Green function for the random walk (in continuous time) on G_δ converges to the continuous Green function, in the integrated sense above. Under the sole assumption that random walk converges to Brownian motion, showing pointwise convergence of the Green functions is not completely straightforward; in fact if one wants any kind of uniformity in the arguments x and y this will typically be false close to the diagonal. Showing the integrated convergence of the Green function, which is what we require here, is fortunately much simpler.

Indeed, fix $x \in v(G_\delta)$. Observe that by definition of the Green function as an occupation measure,

$$\delta^2 \sum_{y \in v(G_\delta)} G_\delta(x,y)\phi(y) = \delta^2 \mathbb{E}_x^\delta \left(\int_0^{\tau_\delta} \phi(X_s)\,ds \right),$$

where X is the continuous time random walk associated to G_δ, as explained before the statement of the theorem. We first change variables $s = u\delta^{-2}$ to get

$$\delta^2 \sum_{y \in v(G_\delta)} G_\delta(x,y)\phi(y) = \int_0^\infty \mathbb{E}_x^\delta(\phi(X_{u\delta^{-2}})1_{\{\tau_\delta > u\delta^{-2}\}})\,du.$$

Fix $\varepsilon > 0$. Choose $K > 0$ sufficiently large that $\int_K^\infty \mathbb{P}_x^\delta(\tau_\delta > u\delta^{-2})\,du \leq \varepsilon$ for

1.14 Scaling limit of the discrete GFF

all $x \in D$, which is possible since we assumed in (1.51) that $\delta^2 \tau_\delta$ is uniformly integrable (uniformly in space). We may also assume without loss of generality that $\int_K^\infty \mathbb{P}(\tau > u)\,du \leq \varepsilon$ (where τ is the first hitting time of ∂D by B), because $\sup_{x \in D} \mathbb{E}_x(\tau) < \infty$ as D is bounded. We then observe, letting $M = \|\phi\|_\infty$, that

$$\sup_{x \in v(G_\delta)} \left| \int_0^\infty \mathbb{E}_x^\delta(\phi(X_{u\delta^{-2}})1_{\{\tau_\delta > u\delta^{-2}\}})\,du - \int_0^\infty \mathbb{E}_x(\phi(B_u)1_{\{\tau > u\}})\,du \right|$$

$$\leq \int_0^K \sup_{x \in v(G_\delta)} \left| \mathbb{E}_x^\delta(\phi(X_{u\delta^{-2}})1_{\{\tau_\delta > u\delta^{-2}\}}) - \mathbb{E}_x(\phi(B_u)1_{\{\tau > u\}}) \right| du + 2M\varepsilon.$$

Since the position of the random walk on G_δ killed at ∂_δ converges uniformly on compact time intervals and uniformly in space to Brownian motion killed when leaving D by (1.50), we deduce that

$$\limsup_{\delta \to 0} \sup_{x \in v(G_\delta)} \left| \int_0^\infty \mathbb{E}_x^\delta(\phi(X_{u\delta^{-2}})1_{\{\tau_\delta > u\delta^{-2}\}})\,du - \int_0^\infty \mathbb{E}_x(\phi(B_u)1_{\{\tau > u\}})\,du \right|$$

$$\leq 2M\varepsilon.$$

Since $\varepsilon > 0$ is arbitrary, we deduce (using (1.14)),

$$\int_0^\infty \mathbb{E}_x^\delta(\phi(X_{u\delta^{-2}})1_{\{\tau_\delta > u\delta^{-2}\}})\,du$$

$$\to \int_0^\infty \mathbb{E}_x(\phi(B_u)1_{\{\tau > u\}})\,du = \int_D G_0^D(x,y)\phi(y)\,dy, \quad (1.54)$$

as $\delta \to 0$, uniformly in $x \in v(G_\delta)$.

It remains to sum over $x \in v(G_\delta)$. Since the above convergence is uniform, and the right-hand side of (1.54) is a continuous function of $x \in \overline{D}$, we deduce, using (1.52), that

$$\delta^2 \sum_{x \in v(G_\delta)} \phi(x) \int_0^\infty \mathbb{E}_x^\delta(\phi(X_{u\delta^{-2}})1_{\{\tau_\delta > u\delta^{-2}\}})\,du$$

$$\to \iint_D G_0^D(x,y)\phi(x)\phi(y)\,dy\,dx,$$

as desired in (1.53). This completes the proof of the theorem. □

Remark 1.65 Let us conclude with some remarks on this theorem.

1. When the area of each face f surrounding a given $x \in v(G_\delta)$ is constant as a function of x (and of order δ^2) then the quantity $h_\delta(\phi) = \delta^2 \sum_{x \in v(G_\delta)} h(x)\phi(x)$ may be viewed up to a multiple factor (coming from the area of each cell) as the integral of h_δ against the test function ϕ, provided that we extend h_δ to all of $\mathbb{R}^2$ by setting it equal to $h(x)$ in the face

f. In that case, Theorem 1.64 says that h_δ, thus extended and viewed as a stochastic process indexed by $\mathcal{D}_0(D)$, converges in the sense of finite-dimensional distributions to a (multiple of) the continuum Gaussian free field. This applies in particular to any periodic lattice such as the square, triangular or hexagonal lattices.

2. In situations where a stronger convergence is desired (such as convergence in the Sobolev space $H_0^s(D)$ for some given $s < 0$, as in Proposition 1.63), the Rellich–Kondrachov embedding theorem is a useful criterion which can be used to establish relative compactness (and hence tightness) in such a Sobolev space. In particular, assuming that the boundary of D is at least C^1, if h_δ is a family of random variables in $H_0^s(D)$ such that $\mathbb{E}(\|h_\delta\|_{H_0^{s'}}^2) \le C$ for some $s' > s$ and $C < \infty$ independent of δ, then $(h_\delta)_{\delta>0}$ is tight in $H_0^s(D)$.

This criterion is particularly simple to use in combination with Lemma 1.43 in order to show convergence in $H^{-1-\varepsilon}$ for any $\varepsilon > 0$. Indeed, once we extend h_δ from the vertices $v(G_\delta)$ to a function defined on all D (for instance we extend h_δ to be constant on each face, and suppose as above that each face has equal area) then by Lemma 1.43,

$$\|h_\delta\|_{H^{-1}}^2 = \iint_D G_D(x,y) h_\delta(x) h_\delta(y) \, dx \, dy.$$

Taking the expectation, and using similar estimates on the discrete Green function as the ones obtained in the proof of Theorem 1.64, it is not hard to see that

$$\mathbb{E}(\|h_\delta\|_{H^{-1}}^2) \to \text{const.} \iint G_D(x,y)^2 \, dx \, dy < \infty,$$

with the above constant related to the area per face. Thus by the Rellich–Kondrachov criterion, $(h_\delta)_{\delta>0}$ is tight in the Sobolev space $H^{-1-\varepsilon}$, for any $\varepsilon > 0$. By Theorem 1.64, the unique limit point is an appropriate multiple of the Gaussian free field. Hence convergence takes place in distribution in the space $H^{-1-\varepsilon}$, for any $\varepsilon > 0$.

1.15 Exercises

Discrete GFF

1.1 Describe the GFF on a binary tree of depth n, where ∂ is the root of the tree.

1.15 Exercises

1.2 Using an orthonormal basis of eigenfunctions for $-\hat{Q}$, show that the partition function Z in Proposition 1.8 is given by

$$Z = \det(-\hat{Q})^{-1/2},$$

1.3 In this exercise, we will show that the minimiser of the discrete Dirichlet energy is discrete harmonic. Fix $U \subset V$ and fix a function $g: V \setminus U \to \mathbb{R}$. Consider

$$\inf\{\mathcal{E}(f, f), \text{ over } f: V \to \mathbb{R}; f|_{V \setminus U} = g\},$$

where $\mathcal{E}(f, f)$ is defined in (1.7).

  (a) Show that the inf is attained at a function f_0.

  (b) Show that f_0 is harmonic in U: that is, $Qf_0(x) = 0$ for all $x \in U$. To see this, it may be helpful to note that, for every function φ supported in U, and for every $\varepsilon > 0$, $\mathcal{E}(f_0 + \varepsilon\varphi, f_0 + \varepsilon\varphi) \geq \mathcal{E}(f_0, f_0)$, and to use the following integration by parts formula: if $u, v: V \to \mathbb{R}$ with v supported on U,

$$\mathcal{E}(u, v) = -(Qu, v).$$

where the inner product on the right-hand side is defined in (1.3).

1.4 Prove the spatial Markov property of the discrete GFF (Theorem 1.10). One way to do this is to consider the harmonic extension φ to U of the boundary data (i.e. $h|_{U^c}$) and check that $h - \varphi$ and φ are jointly Gaussian vectors indexed by U, so the desired property follows by computing suitable covariances.

Continuum GFF

1.5 Show that on the upper half plane,

$$G_0^{\mathbb{H}}(x, y) = \frac{1}{2\pi} \log \left| \frac{x - \bar{y}}{x - y} \right|.$$

Hint: use that $p_t^{\mathbb{H}}(x, y) = p_t(x, y) - p_t(x, \bar{y})$ by symmetry, and use the formula $e^{-a/t} - e^{-b/t} = t^{-1} \int_a^b e^{-x/t} \, dx$.

Deduce the value of $G_0^{\mathbb{D}}(0, \cdot)$ on the unit disc.

1.6 Let $p_t(x, y)$ be the transition function of Brownian motion on the whole plane (with diffusivity 2). Show that $\int_0^1 p_t(x, y) \, dt = -(2\pi)^{-1} \log |x - y| + O(1)$ as $x \to y$. Then use this to argue that if D is connected and bounded (for simplicity), then $G_0^D(x, y) = -(2\pi)^{-1} \log |x - y| + O(1)$ as $x \to y$, recovering the third property of Proposition 1.18.

1.7 Let D be a bounded domain and $z_0 \in D$. Suppose that $\phi(z)$ is harmonic in $D \setminus \{z_0\}$ and that

$$\phi(z) = -(2\pi)^{-1}(1 + o(1)) \log |z - z_0| \text{ as } z \to z_0;$$
$$\phi(z) \to 0 \text{ as } z \to w \in \partial D.$$

Show that $\phi(z) = G_0^D(z_0, z)$ for all $z \in D \setminus \{z_0\}$. (*Hint*: use the optional stopping theorem.)

1.8 Let $\mathbf{h}$ be a GFF in a domain D. Consider $\tilde{\mathbf{h}}_\varepsilon(z)$, the average value of $\mathbf{h}$ on a square of side length ε centred at z. Let $\bar{h}_\varepsilon(z) = \sqrt{2\pi}\tilde{\mathbf{h}}_\varepsilon(z)$. Is this a Brownian motion as a function of $t = \log 1/\varepsilon$? If not, how can you modify it so that it becomes a Brownian motion? More generally, what about the average of the field on a scaled contour $\varepsilon\lambda$, where λ is a piecewise smooth loop (the so-called potato average...)?

1.9 **Radial decomposition.** Suppose $D = \mathbb{D}$ is the unit disc and $\mathbf{h}$ is a GFF in D. Show that $\mathbf{h}$ can be written as the sum

$$\mathbf{h} = \mathbf{h}_{\text{rad}} + \mathbf{h}_{\text{circ}},$$

where $\mathbf{h}_{\text{rad}}$ is a radially symmetric function, $\mathbf{h}_{\text{circ}}$ is a distribution with zero average on each disc, and the two parts are independent. Specify the law of each of these two parts.

1.10 Let D be a proper simply connected domain and let $z \in D$.
(a) Show that

$$\log R(z; D) = \mathbb{E}_z(\log |B_T - z|),$$

where $T = \inf\{t > 0 : B_t \notin D\}$. (*Hint*: let g be a map sending D to $\mathbb{D}$ and z_0 to 0. Let $\phi(z) = \frac{g(z)}{z - z_0}$ for $z \neq z_0$ and $\phi(z_0) = g'(z_0)$; and consider $\log |\phi|$.)

(b) Deduce the following useful formula: let $D \subset \mathbb{C}$ be as above, let $U \subset D$ be a subdomain, and for $z \in U$, let ρ_z be the harmonic measure on ∂U as seen from z. Then show that $\rho \in \mathfrak{M}_0$ and that

$$\text{Var}(\mathbf{h}, \rho) = \frac{1}{2\pi} \log \frac{R(z; D)}{R(z, U)}.$$

1.11 Show that the constraints in Remark 1.19 uniquely identify G^D when $d \geq 3$. For $x \in D$, defining $H_x(y) = (1/A_d)|x - y|^{2-d}$, let h_x be the unique harmonic extension of $H_x|_{\partial D}$ into D. Show that the function $H(x, y) = H_x(y) - h_x(y)$, defined for $x \neq y \in D$, satisfies the constraints of Remark 1.19. Deduce that $G^D = H$. Show this directly by proving that the transition probability $p_t^D(x, y)$ solves the heat equation in D.

2
Liouville Measure

In this chapter, we fix $\gamma > 0$ (the **coupling constant**) and introduce the Liouville measure. Informally speaking, this measure $\mathcal{M}$ (depending on γ) takes the form

$$\mathcal{M}(\mathrm{d}z) = e^{\gamma h(z)} \, \mathrm{d}z, \qquad (2.1)$$

where $h = \sqrt{2\pi}\mathbf{h}$ is a GFF in two dimensions (normalised according to (1.30)). The scaling factor $\sqrt{2\pi}$ is introduced so that (formally) $\mathbb{E}[h(x)h(y)] = -\log|x-y| + O(1)$; that is, h is logarithmically correlated. The construction will be generalised in Chapter 3 which is devoted to **Gaussian multiplicative chaos**, which are measures of the form (2.1) but for generic log-correlated Gaussian fields h. While the Gaussian free field in two dimensions is of course an example of such a field, so that Liouville measure really is just a particular case of the theory of Gaussian multiplicative chaos, some arguments specific to the GFF can be used to simplify the presentation and introduce relevant ideas in a clean way, without the need to introduce too much machinery. This is the reason why we have chosen to do the construction of Liouville measure (i.e. in the case of the GFF) in this separate chapter.

Heuristics. The informal definition (2.1) should be interpreted as follows. Some abstract Riemann surface has been parametrised, after Riemann uniformisation, by a domain of our choice – perhaps the disc, assuming that it has a boundary, or perhaps the unit sphere in three dimensions if it doesn't. In this parametrisation, the conformal structure is preserved: that is, curves crossing at an angle θ at some point in the domain would also correspond to curves crossing at an angle θ in the original surface. However, in this parametrisation, the metric and the volume do not correspond to the ambient volume and metric of Euclidean space. Namely, a small element of Euclidean volume dz in D parametrises a small element of volume $e^{\gamma h(z)} dz$ in terms of the intrinsic geometry of the original surface. Hence points where h is *very big* (for example, thick points) correspond in reality to relatively big portions of the surface; while points where h is very low are points which correspond to small portions of the surface. The first points will tend to be typical from the point of view of sampling from the volume measure, while the second points will be where geodesics tend to travel.

Rigorous approach. Let $D \subset \mathbb{R}^2$ be an open set and let h be a Dirichlet (or zero boundary) GFF on D. When we try to give a precise meaning (2.1), we immediately run into a serious problem: the exponential of a distribution (such as h) is not *a priori* defined. This corresponds to the fact that while h is regular enough to be a *distribution*, so small rough oscillations cancel each other when we average h over macroscopic regions of space, these oscillations become highly magnified when we take the exponential, and they can no longer cancel each other out. In fact, giving a meaning to (2.1) will require non-trivial work, and will be done via an approximation procedure, using

$$\mathcal{M}_\varepsilon(\mathrm{d}z) := e^{\gamma h_\varepsilon(z)} \varepsilon^{\gamma^2/2} \, \mathrm{d}z, \qquad (2.2)$$

for $\varepsilon > 0$, where $h_\varepsilon(z)$ is a jointly continuous version of the circle average. (More general regularisations will be considered in Chapter 3.) It is straightforward to see that $\mathcal{M}_\varepsilon$ is a (random) Radon measure on D for every ε. Our goal will be to prove the following theorem.

Theorem 2.1 *Suppose $0 \le \gamma < 2$. If D is bounded, then the random measure $\mathcal{M}_\varepsilon$ converges weakly almost surely to a random measure $\mathcal{M}$, the* (bulk) *Liouville measure, along the subsequence $\varepsilon = 2^{-k}$. $\mathcal{M}$ almost surely has no atoms, and for any $A \subset D$ open, we have $\mathcal{M}(A) > 0$ almost surely. In fact, $\mathbb{E}(\mathcal{M}(A)) = \int_A R(z, D)^{\gamma^2/2} \, \mathrm{d}z \in (0, \infty)$.*

We remind the reader that the notation $R(z, D)$ above stands for the conformal radius of D seen from z. That is, $R(z, D) = |f'(0)|$ where f is (any) conformal isomorphism taking $\mathbb{D}$ to D and 0 to z. If D is unbounded then weak convergence can be replaced by vague convergence with exactly the same proof.

In this form, the result is due to Duplantier and Sheffield [DS11]. It could also have been deduced from earlier work of Kahane [Kah85] who used a different approximation procedure, together with results of Robert and Vargas [RV10b] showing universality of the limit with respect to the approximating procedure. (In fact, these two results would have given convergence in the distribution of $\mathcal{M}_\varepsilon$ rather than in probability; and hence would not show that the limiting measure $\mathcal{M}$ depends solely on the free field h. However, a strengthening of the arguments of Robert and Vargas due to Shamov [Sha16] has recently yielded convergence in probability.) Earlier, Høegh–Krohn [HK71] had introduced a similar model in the context of quantum field theory and analysed it in the relatively easy L^2 phase when $0 \le \gamma < \sqrt{2}$. Here we will follow the elementary approach developed in [Ber17], which works in the more general context of Gaussian multiplicative chaos (see Chapter 3), but with the simplifications that are allowed by taking the underlying field to be the GFF.

2.1 Preliminaries

Before we start the proof of Theorem 2.1, we first observe that this is the right normalisation.

Lemma 2.2 *We have that* $\mathrm{Var}(h_\varepsilon(x)) = \log(1/\varepsilon) + \log R(x, D)$. *As a consequence,*

$$\mathbb{E}(M_\varepsilon(A)) = \int_A R(z, D)^{\gamma^2/2}\, dz \in (0, \infty).$$

Proof The proof is very similar to the argument in Theorem 1.59 and is a good exercise. Fix $x \in D$. By definition,

$$\mathrm{Var}(h_\varepsilon(x)) = 2\pi \Gamma(\rho_{x,\varepsilon}) = 2\pi \int \rho_{x,\varepsilon}(dz)\rho_{x,\varepsilon}(dw) G_0^D(z,w).$$

For a fixed z, $G_0^D(z, \cdot)$ is harmonic on $D \setminus \{z\}$ and so $\int \rho_{x,\varepsilon}(dw) G_0^D(w,z) = G_0^D(x,z)$ by the mean value property and an approximation argument similar to (1.47). Therefore,

$$\mathrm{Var}(h_\varepsilon(x)) = 2\pi \Gamma(\rho_{x,\varepsilon}) = 2\pi \int \rho_{x,\varepsilon}(dz) G_0^D(z,x).$$

Now, observe that $2\pi G_0^D(x, \cdot) = -\log|x - \cdot| + \xi(\cdot)$ where $\xi(\cdot)$ is harmonic and $\xi(x) = \log R(x; D)$. Indeed let $\xi(\cdot)$ be the harmonic extension of $-\log|x - \cdot|$ on ∂D. Then $2\pi G_0^D(x, \cdot) + \log|x - \cdot| - \xi(\cdot)$ has zero boundary values on ∂D, and is bounded and harmonic in $D \setminus \{x\}$. Hence it must be zero in all of D by the uniqueness of solutions to the Dirichlet problem among bounded functions (for example, by the optional stopping theorem). Note that $\xi(x) = \log R(x; D)$ by (1.24). Therefore, by harmonicity of ξ and the mean value property,

$$\mathrm{Var}(h_\varepsilon(x)) = 2\pi \int G_0^D(x,z)\rho_{x,\varepsilon}(dz) = \log(1/\varepsilon) + \xi(x)$$

as desired. □

We now make a couple of remarks:

1. Not only is the expectation constant, but we have that for each fixed z, $e^{\gamma h_\varepsilon(z)} \varepsilon^{\gamma^2/2}$ forms a martingale as a function of ε. This is nothing but the exponential martingale of a Brownian motion.
2. However, the integral $\mathcal{M}_\varepsilon(A)$ is **not** a martingale. This is because the underlying filtration in which $e^{\gamma h_\varepsilon(z)} \varepsilon^{\gamma^2/2}$ is a martingale depends on z. If we try to condition on $(h_\varepsilon(z), z \in D)$, then this gives too much information, and we lose the martingale property.

2.2 Convergence and uniform integrability in the L^2 phase

The bulk of the proof consists in showing that for any fixed bounded Borel subset S of D (including possibly D itself), we have that $\mathcal{M}_\varepsilon(S)$ converges almost surely along the subsequence $\varepsilon = 2^{-k}$ to a non-degenerate limit. We will then explain in Section 2.3, using fairly general arguments, why this implies the almost sure weak convergence of the sequence of measures $\mathcal{M}_\varepsilon$ along the same subsequence.

Let us now fix S and set $I_\varepsilon = \mathcal{M}_\varepsilon(S)$. We first suppose that $\gamma \in [0, \sqrt{2})$. In this case, the so-called L^2 phase, it is relatively easy to check the convergence (which actually holds in L^2), but difficulties arise when $\gamma \in [\sqrt{2}, 2)$. (As luck would have it, this coincides precisely with the phase which is interesting from the point of view of statistical physics on random planar maps.)

Proposition 2.3 *If $\gamma \in [0, \sqrt{2})$ and $\varepsilon > 0$, $\delta = \varepsilon/2$, then we have the estimate $\mathbb{E}((I_\varepsilon - I_\delta)^2) \leq C\varepsilon^{2-\gamma^2}$. In particular, I_ε is a Cauchy sequence in $L^2(\mathbb{P})$ and so converges to a limit in probability as $\varepsilon \to 0$. Along the sequence $\varepsilon = 2^{-k}$, this convergence occurs almost surely, and the limit is almost surely strictly positive.*

Proof For ease of notations, let $\bar{h}_\varepsilon(z) = \gamma h_\varepsilon(z) - (\gamma^2/2) \operatorname{Var}(h_\varepsilon(z))$, and let

$$\sigma(\mathrm{d}z) = R(z, D)^{\gamma^2/2} \mathrm{d}z.$$

The idea is to say that if we consider the Brownian motions $h_\varepsilon(x)$ and $h_\varepsilon(y)$ (viewed as functions of $\varepsilon = e^{-t}$), then they are (approximately) identical until $\varepsilon \leq |x - y|$, after which time they evolve (exactly) independently.

Observe that by Fubini's theorem,

$$\mathbb{E}((I_\varepsilon - I_\delta)^2)$$
$$= \int_{S^2} \mathbb{E}\left((e^{\bar{h}_\varepsilon(x)} - e^{\bar{h}_\delta(x)})(e^{\bar{h}_\varepsilon(y)} - e^{\bar{h}_\delta(y)})\right) \sigma(\mathrm{d}x)\sigma(\mathrm{d}y)$$

2.2 Convergence and uniform integrability in the L^2 phase

$$= \int_{S^2} \mathbb{E}\left(e^{\overline{h}_\varepsilon(x)+\overline{h}_\varepsilon(y)}(1 - e^{\overline{h}_\delta(x)-\overline{h}_\varepsilon(x)})(1 - e^{\overline{h}_\delta(y)-\overline{h}_\varepsilon(y)})\right) \sigma(dx)\sigma(dy).$$

By the Markov property, $\overline{h}_\varepsilon(x) + \overline{h}_\varepsilon(y)$, $\overline{h}_\varepsilon(x) - \overline{h}_\delta(x)$ and $h_\varepsilon(y) - h_\delta(y)$ are independent as soon as $|x - y| \geq 2\varepsilon$. Indeed, we can apply the Markov property in $U = B(x, \varepsilon)$, which allows us to write $h = \tilde{h} + \varphi$, where φ is harmonic in U and $\tilde{h}$ is an independent GFF in U. Since $\tilde{h}$ is zero outside of U, the increment $\overline{h}_\delta(y) - \overline{h}_\varepsilon(y)$ and the term $\overline{h}_\varepsilon(y) + \overline{h}_\varepsilon(x)$ depend only on φ, and are therefore independent of the increment $\overline{h}_\delta(x) - \overline{h}_\varepsilon(x)$ (which, as noted in Theorem 1.59, depends only on $\tilde{h}$.) Applying the same argument with $U = B(y, \varepsilon)$ gives that $\overline{h}_\delta(y) - \overline{h}_\varepsilon(y)$ is independent of the pair $\{\overline{h}_\delta(x) - \overline{h}_\varepsilon(x), \overline{h}_\varepsilon(y) + \overline{h}_\varepsilon(x)\}$.

Hence if $|x - y| \geq 2\varepsilon$, we can factorise the expectation in the above integral as

$$= \mathbb{E}\left(e^{\overline{h}_\varepsilon(x)+\overline{h}_\varepsilon(y)}\right)\mathbb{E}\left(1 - e^{\overline{h}_\delta(x)-\overline{h}_\varepsilon(x)}\right)\mathbb{E}\left(1 - e^{\overline{h}_\delta(y)-\overline{h}_\varepsilon(y)}\right),$$

where both second and third terms are equal to zero, because of the pointwise martingale property. Therefore, the expectation is just 0 as soon as $|x - y| > 2\varepsilon$.

Also note that by the martingale property for a fixed x,

$$\mathbb{E}((e^{\overline{h}_\varepsilon(x)} - e^{\overline{h}_\delta(x)})^2) = \mathbb{E}(e^{2\overline{h}_\delta(x)} - e^{2\overline{h}_\varepsilon(x)})$$

$$\leq \mathbb{E}(e^{2\overline{h}_\delta(x)}) \leq C\mathbb{E}(e^{2\overline{h}_\varepsilon(x)})$$

for some $C > 0$. Hence using Cauchy–Schwarz in the case where $|x - y| \leq 2\varepsilon$,

$$\mathbb{E}((I_\varepsilon - I_\delta)^2)$$

$$\leq \int_{|x-y|\leq 2\varepsilon} \sqrt{\mathbb{E}((e^{\overline{h}_\varepsilon(x)} - e^{\overline{h}_\delta(x)})^2)\mathbb{E}((e^{\overline{h}_\varepsilon(y)} - e^{\overline{h}_\delta(y)})^2)}\sigma(dx)\sigma(dy)$$

$$\leq C\int_{|x-y|\leq 2\varepsilon} \sqrt{\mathbb{E}(e^{2\overline{h}_\varepsilon(x)})\mathbb{E}(e^{2\overline{h}_\varepsilon(y)})}\sigma(dx)\sigma(dy) \tag{2.3}$$

$$\leq C\int_{|x-y|\leq 2\varepsilon} \varepsilon^{\gamma^2} e^{\frac{1}{2}(2\gamma)^2 \log(1/\varepsilon)}\sigma(dx)\sigma(dy)$$

$$\leq C\varepsilon^{2+\gamma^2-2\gamma^2} = C\varepsilon^{2-\gamma^2}. \tag{2.4}$$

Thus I_ε is a Cauchy sequence in $L^2(\mathbb{P})$. To check almost sure convergence along the subsequence $\varepsilon = 2^{-k}$, we just note that since γ is assumed to be smaller than $\sqrt{2}$, the exponent $2 - \gamma^2$ is positive, and hence the sum $\sum_{k\geq 1} 2^{-k(2-\gamma^2)} < \infty$. The almost sure convergence thus follows from the Borel–Cantelli lemma.

It remains to check that $\mathbb{P}(\lim_{\varepsilon\to 0} I_\varepsilon > 0) = 1$. We will appeal to Kolmogorov's $0 - 1$ law. We already know that $\mathbb{P}(\lim_{\varepsilon\to 0} I_\varepsilon > 0) > 0$, since $\mathbb{E}(\lim_{\varepsilon\to 0} I_\varepsilon) = \lim_{\varepsilon\to 0} \mathbb{E}(I_\varepsilon) > 0$. Moreover, notice that if $(f_i)_{i\geq 1}$ is an orthonormal basis of $H_0^1(D)$, then $\{h_\varepsilon(x) : x \in D, \varepsilon > 0\}$ and therefore

62   *Liouville Measure*

$\lim_{\varepsilon \to 0} I_\varepsilon$, is a function of the sequence of coefficients $(h, f_i)_\nabla$. Now, we have seen that these coefficients are independent standard Gaussians, and it is clear that the event $\{\lim_{\varepsilon \to 0} I_\varepsilon > 0\}$ is in the tail σ-algebra generated by the sequence (since this event is invariant under resampling any finite number of terms). Thus it has probability zero or one, and since the probability is positive, it must be one. This concludes the proof of the proposition. □

The moral of this proof is the following: while I_ε is not a martingale in ε (because there is no filtration common to all points x such that $e^{\bar{h}_\varepsilon(x)}$ forms a martingale), we can use the pointwise martingales to estimate the second moment of the increment $I_\varepsilon - I_\delta$. Only for points x, y which are very close (of order ε) do we get a non-trivial contribution.

We defer the proof of the general case $\gamma \in [0, 2)$ until slightly later (see Section 2.5); and for now show how convergence of masses $\mathcal{M}_\varepsilon(S)$ towards some non-negative limit implies the almost sure weak convergence of the sequence of measures $\mathcal{M}_\varepsilon$.

2.3 Weak convergence to Liouville measure

Proof of Theorem 2.1 continued. We now finish the proof of the weak convergence in Theorem 2.1 (assuming convergence of masses of fixed bounded Borel subsets $S \subseteq D$ toward some non-negative limit with probability 1) by showing that the sequence of measures $\mathcal{M}_\varepsilon$ converges in probability for the weak topology towards a measure $\mathcal{M}$. This measure will be defined by the limits of quantities of the form $\mathcal{M}_\varepsilon(S)$, where S is a cube such that $\bar{S} \subset D$. These arguments are borrowed from [Ber17].

Note that since $\mathcal{M}_\varepsilon(D)$ converges almost surely, we have that the measures $\mathcal{M}_\varepsilon$ are almost surely tight in the space of Borel measures on $\bar{D}$ with the topology of weak convergence (along the subsequence $\varepsilon = 2^{-k}$, which we will not repeat). Let $\tilde{\mathcal{M}}$ be any weak limit.

Let $\mathcal{A}$ denote the π-system of subsets of $\mathbb{R}^2$ of the form $A = [x_1, y_1] \times [x_2, y_2]$ where $x_i, y_i \in \mathbb{Q}, i = 1, 2$ and such that $\bar{A} \subset D$, and note that the σ-algebra generated by $\mathcal{A}$ is the Borel σ-field on D. Observe that $\mathcal{M}_\varepsilon(A)$ converges almost surely to a limit (which we call $\mathcal{M}(A)$) for any $A \in \mathcal{A}$, by the part of the theorem which is already proved (or assumed in the case $\gamma \geq \sqrt{2}$). Observe that this convergence holds almost surely simultaneously for all $A \in \mathcal{A}$, since $\mathcal{A}$ is countable.

Let $A = [x_1, y_1) \times [x_2, y_2) \in \mathcal{A}$. We first claim that

$$\mathcal{M}(A) = \sup_{x'_i, y'_i} \{\mathcal{M}([x'_1, y'_1] \times [x'_2, y'_2])\}, \qquad (2.5)$$

where the sup is over all $x'_i, y'_i \in \mathbb{Q}$ with $x'_i > x_i$ and $y'_i < y_i$, $1 \le i \le 2$. Clearly the left-hand side is almost surely greater or equal to the right-hand side, but both sides have the same expectation by monotone convergence (for $\mathbb{E}$). Likewise, it is easy to check that

$$\mathcal{M}(A) = \inf_{x'_i, y'_i} \{\mathcal{M}((x'_1, y'_1) \times (x'_2, y'_2))\}, \qquad (2.6)$$

where now the infimum is over all $x'_i, y'_i \in \mathbb{Q}$ with $x'_i < x_i$ and $y'_i > y_i$, $1 \le i \le 2$.

We aim to check that $\tilde{\mathcal{M}}(A) = \mathcal{M}(A)$, which uniquely identifies the weak limit $\tilde{\mathcal{M}}$ and hence proves the desired weak convergence.

Note that by the portmanteau lemma, for any $A = [x_1, y_1) \times [x_2, y_2)$, and for any $x'_i, y'_i \in \mathbb{Q}$ with $x'_i < x_i$ and $y'_i > y_i$, $1 \le i \le 2$, we have:

$$\tilde{\mathcal{M}}(A) \le \tilde{\mathcal{M}}((x'_1, y'_1) \times (x'_2, y'_2))$$
$$\le \liminf_{\varepsilon \to 0} \mathcal{M}_\varepsilon((x'_1, y'_1) \times (x'_2, y'_2))$$
$$= \mathcal{M}((x'_1, y'_1) \times (x'_2, y'_2)).$$

(The portmanteau lemma is classically stated for probability measures, but there is no problem in using it here since we already know convergence of the total mass, and thus can equivalently work with the normalised measures $\mathcal{M}_\varepsilon / \mathcal{M}_\varepsilon(D)$.)

Since the x'_i, y'_i are arbitrary, taking the infimum over the admissible values and using (2.6), we get

$$\tilde{\mathcal{M}}(A) \le \mathcal{M}(A).$$

The converse inequality follows in the same manner, using (2.5). We deduce that $\tilde{\mathcal{M}}(A) = \mathcal{M}(A)$, almost surely, as desired. As already explained, this uniquely identifies the limit $\tilde{\mathcal{M}}$. Hence $\mathcal{M}_\varepsilon$ converges almost surely, weakly, to $\mathcal{M}$ on D. □

2.4 The GFF viewed from a Liouville typical point

Let h be a Gaussian free field on a domain D, with associated Liouville measure $\mathcal{M}$ for some $\gamma < 2$. An interesting question is the following: if z is a random point sampled according to the Liouville measure, normalised to be a probability distribution (this is possible when D is bounded), then what does

h look like near the point z? This gives rise to the concept of *rooted measure* in the terminology of [DS11] or to the Peyrière measure in the terminology of Gaussian multiplicative chaos.

We expect some atypical behaviour: after all, for any given fixed $z \in D$, $e^{\gamma h_\varepsilon(z)} \varepsilon^{\gamma^2/2}$ converges almost surely to 0, so the only reason $\mathcal{M}$ could be non-trivial is if there are enough points on which h is atypically big. Of course this leads us to suspect that $\mathcal{M}$ is in some sense carried by certain thick points of the GFF. It remains to identify the level of thickness. As mentioned before, a simple back of the envelope calculation (made slightly more rigorous in the next result) suggests that these points should be γ-thick. As we will see, this is in fact a simple consequence of Girsanov's lemma: essentially, when we bias h by $e^{\gamma h(z)}$, we shift the mean value of the field by $\gamma G_0^D(\cdot, z) = \gamma \log 1/|\cdot - z| + O(1)$, thereby resulting in a γ-thick point.

Theorem 2.4 *Suppose D is bounded. Let z be a point sampled according to the Liouville measure $\mathcal{M}$, normalised to be a probability measure. Then, almost surely,*

$$\lim_{\varepsilon \to 0} \frac{h_\varepsilon(z)}{\log(1/\varepsilon)} = \gamma.$$

In other words, z is almost surely a γ-thick point ($z \in \mathcal{T}_\gamma$).

When D is not bounded, we can simply say that $\mathcal{M}(\mathcal{T}_\gamma^c) = 0$, almost surely. In particular, $\mathcal{M}$ is singular with respect to Lebesgue measure, almost surely.

Proof The proof is elegant and at its core relatively simple, but requires a conceptual shift. This comes with the following important but elementary lemma, which can be seen as a (completely elementary) version of Girsanov's theorem. Because it is Gaussian in nature rather relying on stochastic calculus, it is appropriate to also credit Cameron and Martin.

Lemma 2.5 (Tilting lemma/Girsanov/Cameron–Martin) *Let $X = (X_1, \ldots, X_n)$ be a Gaussian vector under the law $\mathbb{P}$, with mean μ and covariance matrix V. Let $\alpha \in \mathbb{R}^n$ and define a new probability measure by*

$$\frac{d\mathbb{Q}}{d\mathbb{P}} = \frac{e^{\langle \alpha, X \rangle}}{Z},$$

where $Z = \mathbb{E}(e^{\langle \alpha, X \rangle})$ is a normalising constant. Then under $\mathbb{Q}$, X is still a Gaussian vector, with covariance matrix V and mean $\mu + V\alpha$.

It is worth rephrasing this lemma in plain words. Suppose we weigh the law of a Gaussian vector by some linear functional. Then the process remains

2.4 The GFF viewed from a Liouville typical point

Gaussian, with unchanged covariances, however the mean is shifted, and the new mean of the variable X_i say, is

$$\mu'_i = \mu_i + \mathrm{Cov}(X_i, \langle \alpha, X \rangle).$$

In other words, the mean is shifted by an amount which is simply the covariance of the quantity we are considering and what we are weighting by.

Proof Assume for simplicity (and in fact without loss of generality) that $\mu = 0$. It is simple to check it with Laplace transforms: indeed if $\lambda \in \mathbb{R}^n$, then

$$\mathbb{Q}(e^{\langle \lambda, X \rangle}) = \frac{1}{Z}\mathbb{E}(e^{\langle \lambda + \alpha, X \rangle}) = \frac{1}{e^{\frac{1}{2}\langle \alpha, V\alpha \rangle}} e^{\frac{1}{2}\langle \alpha + \lambda, V(\alpha + \lambda) \rangle} = e^{\frac{1}{2}\langle \lambda, V\lambda \rangle + \langle \lambda, V\alpha \rangle}$$

The first term in the exponent $\langle \lambda, V\lambda \rangle$ is the Gaussian term with variance V, while the second term $\langle \lambda, V\alpha \rangle$ shows that the mean is now $V\alpha$, as desired. □

Let $\mathbb{P} = \mathbb{P}(\mathrm{d}h)$ be the law of the GFF, and let Q_ε denote the joint law on (z, h) defined by:

$$Q_\varepsilon(\mathrm{d}z, \mathrm{d}h) = \frac{1}{Z} e^{\gamma h_\varepsilon(z)} \varepsilon^{\gamma^2/2} \, \mathrm{d}z \mathbb{P}(\mathrm{d}h). \tag{2.7}$$

Here Z is a normalising (non-random) constant depending solely on ε. Note that the marginal law of h is weighted by $\mathcal{M}_\varepsilon(D)$ under Q_ε, and given h, the point z is sampled proportionally to $\mathcal{M}_\varepsilon$.

Also define $Q(\mathrm{d}z, \mathrm{d}h) = (\mathbb{E}(\mathcal{M}_h(D))^{-1} \mathcal{M}_h(\mathrm{d}z) \mathbb{P}(\mathrm{d}h)$ where by $\mathcal{M}_h$ we mean the Liouville measure which is almost surely defined by h. Note that Q_ε converges to Q weakly with respect to the product topology induced by the Euclidean metric for z and the Sobolev H^{-1} norm for h, say, or, if we prefer the point of view that h is a stochastic process indexed by $\mathfrak{M}_0$, then the meaning of this convergence is with respect to the infinite product $D \times \mathbb{R}_0^{\mathfrak{M}}$: that is, for any fixed $m \geq 1$ and $\rho_1, \ldots, \rho_m \in \mathfrak{M}_0$, and any continuous bounded function f on D,

$$\mathbb{E}\left((h, \rho_1) \ldots (h, \rho_m) \int f(z) \varepsilon^{\gamma^2/2} e^{\gamma h_\varepsilon(z)} \, \mathrm{d}z\right)$$
$$\to \mathbb{E}\left((h, \rho_1) \ldots (h, \rho_m) \int f(z) \mathcal{M}_h(\mathrm{d}z)\right).$$

This can be verified exactly with the same argument which shows the weak convergence of the approximate Liouville measures. For simplicity, we will keep the point of view of a stochastic process for the rest of the proof.

Recall that under the law Q_ε, the marginal law of h is simply that of a GFF biased by its total mass, so that in particular, $Z = \mathbb{E}(\mathcal{M}_\varepsilon(D))$ is (up to some small effects from the boundary, which we freely ignore from now on) equal to

66   *Liouville Measure*

$\int_D R(z, D)^{\gamma^2/2} \, dz$, and does not depend on ε. Furthermore, the marginal law of z is

$$Q_\varepsilon(dz) = \frac{1}{Z} dz \mathbb{E}(e^{\gamma h_\varepsilon(z)} \varepsilon^{\gamma^2/2}) = \frac{dz}{Z} R(z, D)^{\gamma^2/2}.$$

Here again, the law does not depend on ε and is nice, that is, absolutely continuous, with respect to Lebesgue measure. Finally, it is clear that under Q_ε, given h, the conditional law of z is just given by a sample from $\mathcal{M}_\varepsilon$.

We will simply reverse the procedure and focus instead on the *conditional distribution of h* given z. We start by explaining the argument without worrying about its formal justification, and add the justifications where needed afterwards.

Note that by definition,

$$Q_\varepsilon(dh|z) = \frac{1}{Z(z)} e^{\gamma h_\varepsilon(z)} \varepsilon^{\gamma^2/2} \mathbb{P}(dh),$$

where $Z(z) := R(z, D)^{\gamma^2/2}$. In other words, the law of the Gaussian field h has been reweighted by an exponential linear functional. By Girsanov's lemma, we deduce that under $Q_\varepsilon(dh|z)$, h is a field with the same covariances as under $\mathbb{P}$, and *non-zero* mean at point w given by

$$\mathrm{Cov}(h(w), \gamma h_\varepsilon(z)) = \gamma \log(1/|w - z|) + O(1).$$

(More rigorously, we apply Girsanov's lemma to the Gaussian stochastic process $(h, \rho)_{\rho \in \mathfrak{M}_0}$ and find that under Q_ε, its covariance structure remains unchanged, while its mean has been shifted by $\mathrm{Cov}((h, \rho); \gamma h_\varepsilon(z))$.)

In the limit as $\varepsilon \to 0$, this amounts to adding the function $\gamma G_0^D(\cdot, z)$ to the field $h(\cdot)$. We now argue that this must coincide with the law of $Q(dh|z)$. To see this, we use the previous paragraph to write for any $\varepsilon > 0$, and for any $m \geq 1, \rho_1, \ldots, \rho_m \in \mathfrak{M}_0, \psi \in C_b(D)$:

$$\mathbb{E}_{Q^\varepsilon}((h, \rho_1) \ldots (h, \rho_m) \psi(z)) =$$

$$\int_D dz\, \psi(z) R(z, D)^{\frac{\gamma^2}{2}} \mathbb{E}_h \left(\prod_{i=1}^m ((h, \rho_i) + \mathrm{Cov}((h, \rho_i), \gamma h_\varepsilon(z))) \right).$$

Invoking the weak convergence of Q_ε to Q, we see that the left-hand side of the above equality converges to $\mathbb{E}_Q((h, \rho_1) \cdots (h, \rho_m) \psi(z))$ as $\varepsilon \to 0$. At the same time, an application of the dominated convergence theorem shows that the right-hand side converges as $\varepsilon \to 0$ to

$$\int_D dz\, \psi(z) R(z, D)^{\frac{\gamma^2}{2}} \mathbb{E}_h \big((h + \gamma G_0^D(z, \cdot), \rho_1) \cdots (h + \gamma G_0^D(z, \cdot), \rho_m) \big). \quad (2.8)$$

Hence the law of $Q(dh \mid z)$ is as claimed.

To summarise, under Q and given z, a logarithmic singularity of strength γ has been introduced at the point z. Hence we find that under $Q(\mathrm{d}h \mid z)$, almost surely,

$$\lim_{\delta \to 0} \frac{h_\delta(z)}{\log(1/\delta)} = \gamma,$$

so $z \in \mathcal{T}_\gamma$, almost surely as desired. In other words, $Q(\mathcal{M}_h(\mathcal{T}_\gamma^c) = 0) = 1$.

We conclude the proof of the theorem by observing that the marginal laws $Q(\mathrm{d}h)$ and $\mathbb{P}(\mathrm{d}h)$ are mutually absolutely continuous with respect to one another, so any property which holds almost surely under Q holds also almost surely under $\mathbb{P}$. (This absolute continuity follows simply from the fact that $\mathcal{M}(S) \in (0, \infty)$, $\mathbb{P}$–almost surely). □

2.5 The full L^1 phase

To address the difficulties that arise when $\gamma \geq \sqrt{2}$, we proceed as follows. Roughly, we claim that the second moment of I_ε blows up because of rare points which are *too thick* and which do not contribute to the integral in an almost sure sense, but nevertheless inflate the value of the second moment. So we will remove these points by hand. To see which points to remove, we appeal the considerations of the previous section: this suggests that we should be safe to get rid of points that are strictly more than γ-thick.

Let $\alpha > 0$ be fixed (it will be chosen $> \gamma$ and very close to γ soon). We define a good event $G_\varepsilon^\alpha(x) = \{\overline{h}_\varepsilon(x) \leq \alpha \log(1/\varepsilon)\}$, for which the point x is not too thick at scale ε.

Lemma 2.6 (Liouville points are no more than γ-thick) *For $\alpha > \gamma$, we have*

$$\mathbb{E}(e^{\overline{h}_\varepsilon(x)} 1_{G_\varepsilon^\alpha(x)}) \geq 1 - p(\varepsilon),$$

where the function p may depend on α and for a fixed $\alpha > \gamma$, $p(\varepsilon) \to 0$ as $\varepsilon \to 0$, polynomially fast. The same estimate holds if $\overline{h}_\varepsilon(x)$ is replaced with $\overline{h}_{\varepsilon/2}(x)$.

Proof Note that

$$\mathbb{E}(e^{\overline{h}_\varepsilon(x)} 1_{\{G_\varepsilon^\alpha(x)\}}) = \tilde{\mathbb{P}}(G_\varepsilon^\alpha(x)), \text{ where } \frac{\mathrm{d}\tilde{\mathbb{P}}}{\mathrm{d}\mathbb{P}} = e^{\overline{h}_\varepsilon(x)}.$$

By Girsanov's lemma, under $\tilde{\mathbb{P}}$, the process $X_s = h_{e^{-s}}(x)$ has the same covariance structure as under $\mathbb{P}$ and its mean is now $\gamma \operatorname{Cov}(X_s, X_t) = \gamma s + O(1)$ for $s \leq t$. Hence it is a Brownian motion with drift γ, and the lemma follows from the fact that such a process does not exceed αt at time t with high probability

68   Liouville Measure

when t is large (and the error probability is exponential in t, or polynomial in ε, as desired).

Changing ε into $\varepsilon/2$ means that the drift of X_s is $\gamma s + O(1)$ over a slightly larger interval of time, namely until time $t + \log 2$. In particular, the same argument as above shows that the same estimate holds for $\overline{h}_{\varepsilon/2}(x)$ as well. □

We therefore see that points which are more than γ-thick do not contribute significantly to I_ε in expectation and can be safely removed. To this end, we fix $\alpha > \gamma$ and introduce:

$$J_\varepsilon = \int_S e^{\overline{h}_\varepsilon(x)} \mathbf{1}_{G_\varepsilon(x)} \sigma(dx); \quad \hat{J}_{\varepsilon/2}(x) = \int_S e^{\overline{h}_{\varepsilon/2}(x)} \mathbf{1}_{G_\varepsilon(x)} \sigma(dx) \quad (2.9)$$

with $G_\varepsilon(x) = G_\varepsilon^\alpha(x)$, and where we recall that $\sigma(dx) = R(x, D)^{\gamma^2/2} dx$. Note that a consequence of Lemma 2.6 is that

$$\mathbb{E}(|I_\varepsilon - J_\varepsilon|) \le p(\varepsilon)|\sigma(S)| \to 0 \text{ and } \mathbb{E}(|I_{\varepsilon/2} - \hat{J}_{\varepsilon/2}|) \le p(\varepsilon)|\sigma(S)| \to 0 \quad (2.10)$$

as $\varepsilon \to 0$.

Lemma 2.7 *We have the estimate $\mathbb{E}((J_\varepsilon - \hat{J}_{\varepsilon/2})^2) \le \varepsilon^r$ for some $r > 0$. In particular, J_ε is a Cauchy sequence in L^2. Along $\varepsilon = 2^{-k}$, this convergence occurs almost surely.*

Proof The proof of this lemma is virtually identical to that in the L^2 phase (see Proposition 2.3). The key observation there was that if $|x - y| \ge 2\varepsilon$, then the increments $h_\varepsilon(x) - h_{\varepsilon/2}(x)$ and $h_\varepsilon(y) - h_{\varepsilon/2}(y)$ are independent of each other, and in fact also of $\mathcal{F}$: the σ-algebra generated by h restricted to the complement of $B(x, \varepsilon) \cup B(y, \varepsilon)$. Since the events $G_\varepsilon(x)$ and $G_\varepsilon(y)$ are both measurable with respect to $\mathcal{F}$, we may therefore deduce from *that* proof (see (2.3)) that

$$\mathbb{E}((J_\varepsilon - \hat{J}_{\varepsilon/2})^2) \le$$

$$C \int_{|x-y| \le 2\varepsilon} \sqrt{\mathbb{E}(e^{2\overline{h}_\varepsilon(x)} \mathbf{1}_{G_\varepsilon(x)}) \mathbb{E}(e^{2\overline{h}_\varepsilon(y)} \mathbf{1}_{G_\varepsilon(y)})} \sigma(dx) \sigma(dy).$$

Now,

$$\mathbb{E}(e^{2\overline{h}_\varepsilon(x)} \mathbf{1}_{G_\varepsilon(x)}) \le \mathbb{E}(e^{2\overline{h}_\varepsilon(x)} \mathbf{1}_{\{h_\varepsilon(x) \le \alpha \log(1/\varepsilon)\}})$$

$$\le O(1)\varepsilon^{-\gamma^2} \mathbb{Q}(h_\varepsilon(x) \le \alpha \log 1/\varepsilon),$$

where by Girsanov's lemma, under the law $\mathbb{Q}$, $h_\varepsilon(x)$ is a normal random variable with mean $2\gamma \log(1/\varepsilon) + O(1)$ and variance $\log 1/\varepsilon + O(1)$. This means that

$$\mathbb{Q}(h_\varepsilon(x) \le \alpha \log 1/\varepsilon) \le O(1) \exp(-\frac{1}{2}(2\gamma - \alpha)^2 \log 1/\varepsilon),$$

2.5 The full L^1 phase

and hence

$$\mathbb{E}((J_\varepsilon - \hat{J}_{\varepsilon/2})^2) \leq O(1)\varepsilon^{2-\gamma^2}\varepsilon^{\frac{1}{2}(2\gamma-\alpha)^2}.$$

Again, choosing $\alpha > \gamma$ sufficiently close to γ ensures that the bound on the right-hand side is at most $O(1)\varepsilon^r$ for some $r > 0$, as desired. This finishes the proof of the lemma. It also concludes the proof of Theorem 2.1 in the general case $\gamma < 2$, by (2.10), and recalling that $p(\varepsilon)$ decays polynomially in ε for fixed α, so we can apply Borel–Cantelli to get almost sure convergence along the sequence $\varepsilon = 2^{-k}$. □

Proof of Theorem 2.1 As a consequence of Lemma 2.7 and (2.10), I_ε is a Cauchy sequence in L^1 and so converges to a limit in probability. The almost sure convergence along the sequence $\varepsilon = 2^{-k}$ follows from the fact that $p(\varepsilon)$ converges to zero polynomially fast by Lemma 2.6 and the Borel–Cantelli lemma. The limit of I_ε is almost surely strictly positive by the same 0–1 argument as in the case $\gamma \leq \sqrt{2}$, and the weak convergence in the sense of measures then follows by the argument detailed in Section 2.3. The formula for the expectation of the limit in Theorem 2.1 is a consequence of Lemma 2.2. Finally, the fact that $\mathcal{M}$ almost surely has no atoms follows from Exercise 2.4.
□

We note that the almost sure convergence over the entire range of ε (not just the dyadic values $\varepsilon = 2^{-k}$) was proved by Sheffield and Wang [SW16].

Understanding the phase transition for the Liouville measure. The description of the Liouville measure viewed from a typical point, explained in Theorem 2.4, was established in the entire L^1 regime and so for all $0 < \gamma < 2$. The fact that the Liouville measure $\mathcal{M} = \mathcal{M}_\gamma$ is supported on the γ-thick points, $\mathcal{T}_\gamma$, is very helpful to get a clearer picture what changes when $\gamma = 2$. Indeed recall from Theorem 1.61 that $\dim(\mathcal{T}_\gamma) = (2 - \gamma^2/2)_+$, and $\mathcal{T}_\gamma$ is empty if $\gamma > 2$. The point is that $\mathcal{M} = \mathcal{M}_\gamma$ does not degenerate when $\gamma < 2$ *because* there are thick points to support it. Once $\gamma > 2$, there are no longer any thick points, and this makes it in some sense "clear" that any approximations to $\mathcal{M}_\gamma$ must degenerate to the zero measure.

When $\gamma = 2$ however, $\mathcal{T}_\gamma$ is not empty, and there is therefore a hope to construct a meaningful critical Liouville measure $\mathcal{M}_2$. Such a construction has indeed been carried out in two separate papers by Duplantier, Rhodes, Sheffield and Vargas [DRSV14b, DRSV14a]. However, the normalisation must be done more carefully – see these two papers for details, as well as the more recent works [JS17, HRV18, Pow18].

2.6 Change of coordinate formula and conformal covariance

Of course, it is natural to wonder in what way the conformal invariance of the GFF manifests itself at the level of the Liouville measure. As it turns out, these measures are not simply conformally invariant. This is easy to believe intuitively, since the total mass of the Liouville measure has to do with total surface area (measured in quantum terms) enclosed in a domain, and so this must grow as the domain grows.

However, the measures are **conformally covariant**: that is, to relate their laws under conformal mappings one must include a correction term accounting for the inflation of the domain under the conformal map. This term is naturally proportional to the derivative of the conformal map.

To formulate the result, it is convenient to use the following notation. Suppose that h is a given distribution – perhaps a realisation of a GFF, but also perhaps one of its close relatives (for example, the GFF plus some smooth deterministic function) – and suppose that its circle average process is well defined. Then we define $\mathcal{M}_h$ to be the measure, if it exists, given by $\mathcal{M}_h(dz) = \lim_{\varepsilon \to 0} e^{\gamma h_\varepsilon(z)} \varepsilon^{\gamma^2/2} dz$. Of course, if h is just a GFF, then $\mathcal{M}_h$ is nothing else but the measure we have constructed in the previous part. If h can be written as $h = h_0 + \varphi$ where φ is deterministic, h_0 is a GFF and $e^{\gamma \varphi} \in L^1(\mathcal{M}_{h_0})$, then $\mathcal{M}_h(dz) = e^{\gamma \varphi(z)} \cdot \mathcal{M}_{h_0}(dz)$ is absolutely continuous with respect to the Liouville measure $\mathcal{M}_{h_0}$.

Theorem 2.8 (Conformal covariance of Liouville measure) *Let $f : D \to D'$ be a conformal isomorphism, and let h be a GFF in D. Then $h' = h \circ f^{-1}$ (where we define this image in the sense of distributions) is a GFF in D', and*

$$\mathcal{M}_h \circ f^{-1} = \mathcal{M}_{h \circ f^{-1} + Q \log |(f^{-1})'|} = e^{\gamma Q \log |(f^{-1})'|} \mathcal{M}_{h'},$$

where

$$Q = \frac{\gamma}{2} + \frac{2}{\gamma}.$$

In other words, pushing forward the Liouville measure $\mathcal{M}_h$ by the map f, we get a measure which is absolutely continuous (with density $|(f^{-1})'(z)|^{\gamma Q}$ at $z \in D'$) with respect to the Liouville measure on D'. The quantity Q plays a very important role in the theory developed in the subsequent chapters. In physics, it is known under the name of **background charge**.

Informal Proof The idea behind this formula may be understood quite easily. Indeed, note that $\gamma Q = \gamma^2/2 + 2$. When we use the map f, a small circle of

2.6 Change of coordinate formula and conformal covariance

radius ε is mapped *approximately* into a small circle of radius $\varepsilon' = |f'(z)|\varepsilon$ around $f(z)$. So $e^{\gamma h_\varepsilon(z)} \varepsilon^{\gamma^2/2}\, dz$ approximately corresponds to

$$e^{\gamma h'_{|f'(z)|\varepsilon}(z')} \varepsilon^{\gamma^2/2} \frac{dz'}{|f'(z)|^2}$$

by the usual change of variable formula. This can be rewritten as

$$e^{\gamma h'_{\varepsilon'}(z')} (\varepsilon')^{\gamma^2/2} \frac{dz'}{|f'(z)|^{2+\gamma^2/2}}.$$

Letting $\varepsilon \to 0$ we get, at least heuristically speaking, the desired result. $\square$

Proof of Theorem 2.8. Of course, the above heuristic is far from a proof, and the main reason is that $h_\varepsilon(z)$ is not a very well-behaved approximation of h under conformal maps. It is better to instead work with a different approximation of the GFF, using an orthonormal basis of $H_0^1(D)$ as in Section 1.7, which has the advantage of being conformally invariant.

In view of this, we make the following definition: suppose $h = \sum_n X_n f_n$, where X_n are i.i.d. standard normal random variables, and f_n is an orthonormal basis of $H_0^1(D)$. Set $h^N(z) = \sum_{i=1}^N X_i f_i$ to be the truncated series, and define

$$\mathcal{M}^N(S) = \int_S \exp\left(\gamma h^N(z) - \frac{\gamma^2}{2}\operatorname{Var}(h^N(z))\right) \sigma(\mathrm{d}z),$$

where we recall that $\sigma(\mathrm{d}z) = R(z,D)^{\gamma^2/2}\, \mathrm{d}z$. Note that $\mathcal{M}^N(S)$ has the same expected value as $\mathcal{M}(S)$. Furthermore, $\mathcal{M}^N(S)$ is a non-negative martingale with respect to the filtration $(\mathcal{F}_N)_N$ generated by $(X_N)_N$, so has an almost sure limit which we will call $\mathcal{M}^*(S)$.

Lemma 2.9 *Almost surely, $\mathcal{M}^*(S) = \mathcal{M}(S)$.*

Proof When we take the circle averages of the series, we obtain

$$h_\varepsilon = h_\varepsilon^N + h'_\varepsilon,$$

where h'_ε is independent from h^N, and h_ε^N denotes the circle average of the function h^N. Hence

$$\varepsilon^{\gamma^2/2} e^{\gamma h_\varepsilon(z)} = e^{\gamma h_\varepsilon^N(z)} \varepsilon^{\gamma^2/2} e^{\gamma h'_\varepsilon(z)}.$$

Consequently, integrating over S and taking the conditional expectation given $\mathcal{F}_N$, we obtain that

$$\mathcal{M}_\varepsilon^N(S) := \mathbb{E}(\mathcal{M}_\varepsilon(S) \mid \mathcal{F}_N) = \int_S \exp\left(\gamma h_\varepsilon^N(z) - \frac{\gamma^2}{2}\operatorname{Var}(h_\varepsilon^N(z))\right) \sigma(\mathrm{d}z).$$

When $\varepsilon \to 0$, the right-hand side converges to $\mathcal{M}^N(S)$, since h^N is a nice smooth function. Consequently,

$$\mathcal{M}^N(S) = \lim_{\varepsilon \to 0} \mathbb{E}(\mathcal{M}_\varepsilon(S) \mid \mathcal{F}_N).$$

Since $\mathcal{M}_\varepsilon(S) \to \mathcal{M}(S)$ in L^1, we have $\mathcal{M}^N(S) = \lim_{\varepsilon \to 0} \mathbb{E}(\mathcal{M}_\varepsilon(S) \mid \mathcal{F}_N) = \mathbb{E}(\mathcal{M}(S) \mid \mathcal{F}_N)$ and so by martingale convergence, $\mathcal{M}^N(S) \to \mathcal{M}(S)$ as $N \to \infty$. Hence $\mathcal{M}(S) = \mathcal{M}^*(S)$, as desired. $\square$

To finish the proof of conformal covariance (Theorem 2.8), we now simply recall that if f_n is an orthonormal basis of $H_0^1(D)$ then $f_n \circ f^{-1}$ gives an orthonormal basis of $H_0^1(D')$. Hence if $h' = h \circ f^{-1}$, then its truncated series h'_N can also simply be written as $h'_N = h^N \circ f^{-1}$. Thus, consider the measure $\mathcal{M}^N$ and apply the map f. We obtain a measure $\tilde{\mathcal{M}}'_N$ in D' such that

$$\tilde{\mathcal{M}}'_N(D') = \int_{D'} \exp\{\gamma h^N(f^{-1}(z')) - \frac{\gamma^2}{2} \text{Var}(h^N(f^{-1}(z')))\} \times$$

$$R(f^{-1}(z'), D)^{\gamma^2/2} \frac{dz'}{|f'(f^{-1}(z'))|^2}$$

$$= \int_{D'} d\mathcal{M}'_N(z') e^{(2+\gamma^2/2) \log |(f^{-1})'(z')|},$$

where $d\mathcal{M}'_N$ is the approximating measure to $\mathcal{M}_{h'}$ in D'. (The second identity is justified by properties of the conformal radius.) Letting $N \to \infty$, and recalling that $d\mathcal{M}'_N$ converges to $d\mathcal{M}_{h'}$ by the previous lemma, we obtain the desired statement of conformal covariance. This finishes the proof of Theorem 2.8. $\square$

2.7 Random surfaces

The notion of **random surface** is a way of identifying Gaussian free field-type distributions that give rise to different "parametrisations" of the same Liouville measure. Essentially, we want to consider the surfaces encoded by $\mathcal{M}_h$ and by $\mathcal{M}_h \circ f^{-1}$ to be "the same" for any given conformal isomorphism $f: D \to D'$. By the conformal covariance formula (Theorem 2.8) if h is a GFF, we have

$$\mathcal{M}_h \circ f^{-1} = \mathcal{M}_{h'} \text{ almost surely, where } h' = h \circ f^{-1} + Q \log |(f^{-1})'|. \quad (2.11)$$

Thus we should think of h and h' as encoding the same quantum surface.

In fact, (when h is a GFF) this equality holds almost surely for *all D'* and **all** conformal isomorphisms $f: D \to D'$ simultaneously. This result was proved by Sheffield and Wang in [SW16].

This motivates the following definition, due to Duplantier and Sheffield

[DS11]. Define an equivalence relation on pairs (D, h), consisting of a simply connected domain D and an element h of $\mathcal{D}'(D)$, by declaring that

$$D_1 \sim D_2$$

if there exists $f: D_1 \to D_2$ a conformal isomorphism such that

$$h_2 = h_1 \circ f^{-1} + Q \log |(f^{-1})'|.$$

It is easy to see that this is an equivalence relation.

Definition 2.10 A (random) surface is a pair (D, h) consisting of a domain and a (random) distribution $h \in \mathcal{D}'(D)$, where the pair is considered modulo the above equivalence relation.

Observe that this definition of (random) surface depends on the parameter $\gamma \ge 0$ of the Liouville measure (since Q depends on γ).

Interesting random surfaces arise, among other things, when we sample a point according to the Liouville measure (either in the bulk, or on the boundary for a free field with a non-trivial boundary behaviour, see later), and we 'zoom in' near this point. Roughly speaking, these are the *quantum cones* and *quantum wedges* introduced by Sheffield in [She16a]. A particular kind of wedge will be studied in a fair amount of detail later on in this book (see Theorem 7.11).

2.8 Exercises

2.1 Explain why Lemmas 2.7 and 2.6 imply uniform integrability of $\mathcal{M}_\varepsilon(S)$.

2.2 Let $\mathcal{M}$ be the Liouville measure with parameter $0 \le \gamma < 2$. Use uniform integrability and the Markov property of the GFF to show that $\mathcal{M}(S) > 0$ almost surely.

2.3 (a) How would you normalise $e^{\gamma h_\varepsilon(z)}$ if you were just aiming to define the Liouville measure on some line segment contained in D, or more generally a smooth simple curve in D? Show that with this normalisation you get a non-degenerate limit.

(b) What is the conformal covariance property in this case?

2.4 Recall the events $G_\varepsilon(z) = \{h_\varepsilon(z) \le \alpha \log 1/\varepsilon\}$ from the proof of uniform integrability of the Liouville measure in the general case. Show that for any $0 \le d < 2 - \gamma^2/2 < 2$,

$$\mathbb{E}\left(\int_{S^2} \frac{1}{|x-y|^d} e^{\bar{h}_\varepsilon(x)} \mathbf{1}_{G_\varepsilon(x)} \sigma(\mathrm{d}x) \, e^{\bar{h}_\varepsilon(y)} \mathbf{1}_{G_\varepsilon(y)} \sigma(\mathrm{d}y)\right) \le C < \infty,$$

where C does not depend on ε. Deduce that
$$\dim(\mathcal{T}_\gamma) \geq 2 - \gamma^2/2$$
almost surely and that $\mathcal{M}$ almost surely has no atoms. Conclude with a proof of Theorem 1.61.

3
Gaussian Multiplicative Chaos

3.1 Motivation and background

In Chapter 2, we constructed the Liouville measure, which is an example of Gaussian multiplicative chaos. Gaussian multiplicative chaos is the theory, developed initially by Kahane in the 1980s, whose goal is the definition and study of random measures of the form

$$\mathcal{M}(\mathrm{d}z) = \exp(\gamma h(z) - \frac{\gamma^2}{2}\mathbb{E}(h(z)^2))\sigma(\mathrm{d}z),$$

where γ is a parameter (the **coupling constant**), h is a centred, logarithmically correlated Gaussian field, and σ is a reference measure. In this chapter, we will give a modern presentation of the general theory. There are two main reasons why we devote an entire chapter to this theory. The first one, continuing on the theme of Chapters 1 and 2, is because the tools we will develop in the process are very useful for the study of Liouville measure: for instance, they will enable us to describe the multifractal spectrum of Liouville measure, leading us to the KPZ relation,[1] which was one of the initial motivations of the seminal work of Duplantier and Sheffield [DS11]. The second, and possibly more important reason, is that Gaussian multiplicative chaos has arisen in many natural models coming from motivations beyond Liouville quantum gravity. For instance, Kahane's original motivation, following the pioneering ideas of Mandelbrot and in particular Kolmogorov, was the description of **turbulence** and especially the phenomenon of intermittency (see [Fri95] for a classical book on the legacy of Kolmogorov in turbulence, and we refer, for instance, to [Che15] and [CGRV19] for a recent survey and article outlining the connection to Gaussian multiplicative chaos). In this case, it is of course more natural

---

[1] Here KPZ stands for Khnizhnik–Polyakov–Zamolodchikov, and should not be confused with the Kardar–Parisi–Zhang equation.

to assume that the field lives in three (rather than two) dimensions. We have however already observed that the Gaussian free field is not logarithmically correlated except in dimension two: indeed the correlations are given by the Green function, which is a multiple of $|y - x|^{2-d}$ in dimensions $d \geq 3$. This is one of the reasons why developing a general theory (going beyond the two-dimensional case of the Gaussian free field) is of great interest. Let us mention briefly a few further topics, where a connection to Gaussian multiplicative chaos has been observed.

- In **random matrix theory**, Gaussian multiplicative chaos describes (or is conjectured to describe) the limiting behaviour of (powers of) the characteristic polynomial of a large random matrix drawn from many of the classical random matrix ensembles. See, for example, [Web15] followed by [NSW20] for the case of CUE, and [BWW18] for a general class of random Hermitian matrices including GUE. Other relevant works include (but are not limited to) [FK21, Kiv22, LOS18].
- In number theory, Gaussian multiplicative chaos describes the (real part) of the **Riemann zeta function** on the critical line, when it is recentred at a random point. See [SW20] and references therein for a long line of works leading to this result.
- In **mathematical finance**, Gaussian multiplicative chaos is used as a model for the stochastic volatility of a financial asset, following some highly influential works of Bacry, Muzy and Delour ([MDB00]), Mandelbrot et al. [MFC97], and also Cont [Con01]).
- In the study of **planar Brownian motion**, a closely related theory has been developed by Jego [Jeg20], [Jeg23], and Aïdékon, Hu and Shi [AHS20]; both of these works generalise the older paper of Bass, Burdzy and Koshnevisan [BBK94] to the full L^1 regime. In the case of the random walk, see [Jeg23] as well as [AB22] (although this requires wiring the boundary); and also [ABJL23] in the context of the loop soup with explicit connections to Liouville measure.

The reader is also invited to consult the survey [RV14], which contains many useful discussions, facts and references.

3.2 Set-up for Gaussian multiplicative chaos

Note: Sections 3.2–3.7 could be skipped by a reader interested only in the GMC measures associated with the Gaussian free field (i.e. the Liouville measures). In this case, the reader may wish to skip to Section 3.8, although tools such as Kahane's inequality (Theorem 3.19) will be needed.

We first describe the set-up in which we will be working. We consider a more general set-up than before and in particular for the rest of this chapter, we do not assume we are working exclusively in two dimensions. Let $D \subset \mathbb{R}^d$ be a bounded domain. Consider a kernel $K(x, y)$ of the form

$$K(x, y) = \log(|x - y|^{-1}) + g(x, y), \qquad (3.1)$$

where g is continuous over $\overline{D} \times \overline{D}$.

We will now define a Gaussian field whose pointwise covariance function is, in some sense given by (3.1). We will proceed in a way that is somewhat inspired by what we did in the case of the Gaussian free field (Chapter 1), defining first a space of measures $\mathfrak{M}$ against which we will be able to test the field, and viewing the field as a stochastic process indexed by this space of measures. However, we will face an immediate difficulty. Namely, if we define $\mathfrak{M}$ to be the space of bounded signed measures[2] such that $\iint |K(x, y)| |\rho|(\mathrm{d}x)|\rho|(\mathrm{d}y) < \infty$, it is hard to show directly that this forms a vector space. (In the case of the GFF, this was a consequence of Lemma 1.26, which has no obvious analogue here.) In two dimensions, this would follow from the theory developed in that chapter, since integrability against K is essentially equivalent to integrability agains G_D. However, we wish to develop the theory in a setting which is more general. Our approach will thus differ a little, and the index set of our stochastic process will first be defined as an abstract Hilbert space induced by K. While this works well, the downside is that it is not easy to check if a concrete bounded signed measure ρ lies in this space. We will essentially check it for bounded measures which can not only integrate K (or, equivalently, can integrate the logarithm), but can do so *continuously*, see the energy condition (3.4).

Let us now explain how to associate a stochastic process to the covariance function K in (3.1). For smooth, compactly supported functions $f, g \in \mathcal{D}(D)$ we write

$$K(f, g) = \int K(x, y) f(x) g(y) \, \mathrm{d}x \, \mathrm{d}y. \qquad (3.2)$$

---

[2] A signed measure $\rho = \rho^+ - \rho^-$ is called bounded if ρ^+ and ρ^- are finite.

Then K is bilinear over the vector space $\mathcal{D}(D)$. We assume that K is positive definite in the sense that $K(f, f) \geq 0$ for all $f \in \mathcal{D}(D)$ and $K(f, f) = 0$ implies $f = 0$. Thus $K(f, g)$ defines an inner product over $\mathcal{D}(D)$ and $f \mapsto \|f\|_K := \sqrt{K(f, f)}$ defines a norm over $\mathcal{D}(D)$. We let $\mathfrak{M}$ be the Hilbert space completion of $\mathcal{D}(D)$ with respect to this norm. By definition, if $\rho \in \mathfrak{M}$, then there exists a sequence of smooth, compactly supported functions $f_n \in \mathcal{D}(D)$ such that $f_n \to \rho$ with respect to the norm induced by K, that is, $\|f_n - \rho, f_n - \rho\|_K^2 \to 0$. Note that if K is the Green function with Dirichlet boundary conditions in a bounded domain $D \subset \mathbb{R}^2$ then $\mathfrak{M} = H_0^{-1}(D)$ with inner product $(\cdot, \cdot)_{-1}$, and its restriction to signed measures coincides with our earlier notation $\mathfrak{M}_0$.

Let h be the centred Gaussian generalised function with covariance K. That is, we view h as a stochastic process indexed by $\mathfrak{M}$, characterised by the two properties that: (h, ρ) is linear in $\rho \in \mathfrak{M}$ in the sense that $(h, \alpha \rho_1 + \beta \rho_2) = \alpha(h, \rho_1) + \beta(h, \rho_2)$ almost surely for $\alpha, \beta \in \mathbb{R}$, $\rho_1, \rho_2 \in \mathfrak{M}$; and for any $\rho \in \mathfrak{M}$,

(h, ρ) is a centred Gaussian random variable with variance $\|\rho\|_K^2$.

Equivalently, $((h, \rho))_{\rho \in \mathfrak{M}}$ is a centred Gaussian stochastic process with covariance

$$(\rho, \rho') \in \mathfrak{M}^2 \mapsto \mathbb{E}[(h, \rho)(h, \rho')] = (\rho, \rho')_K \qquad (3.3)$$

(where the latter denotes the inner product associated to the Hilbert space $\mathfrak{M}$, associated to the norm $\|\cdot\|_K$). We will write $\int h(x)\rho(dx)$ or (h, ρ) with an abuse of notation for the random variable h_ρ. Note that this set-up covers the case of a Gaussian free field in two dimensions with Dirichlet boundary conditions. In fact, it also covers the case of the Gaussian free field with free or Neumann boundary conditions, see Chapter 7. We extend the definition of h outside of D by setting $h|_{D^c} = 0$, so if ρ is, say, a measure such that $\rho|_D \in \mathfrak{M}$, by definition $(h, \rho) = (h, \rho|_D)$.

As mentioned above, it is not straightforward to check if a concrete non-negative measure lies in $\mathfrak{M}$. Let ρ be a fixed non-negative finite Borel measure on $\mathbb{R}^d$ compactly supported in D, and suppose that ρ satisfies the *energy* condition

$$x \mapsto \mathcal{E}_\rho(x) := \int \log\left(\frac{1}{|x - y|}\right)\rho(dy) \text{ is continuous on } \overline{B}(0, R), \qquad (3.4)$$

where $R > 0$ is large enough that $\mathrm{Supp}(\rho) \subset B(0, R)$. In particular, $\mathcal{E}_\rho$ is uniformly bounded on $\overline{B}(0, R)$. It is easy to check that the condition (3.4) is satisfied whenever ρ has an L^p Lebesgue density supported in D) for some $p > 1$, but also in many other cases, for example, when ρ is the uniform distribution on some circle contained in D.

3.2 Set-up for Gaussian multiplicative chaos

For ρ satisfying the energy condition (3.4), it is not hard to see that ρ can be identified with an element of $\mathfrak{M}$ as follows. Let us fix a smooth, symmetric, non-negative function ψ supported in $B(0,1)$ such that $\int \psi(z)\,dz = 1$, and let $\rho_n(y) = \rho * \psi_n(y) = \int \rho(dz)\psi_n(y-z)$, where $\psi_n(z) = n^d\psi(zn)$. Then ρ_n is a smooth, compactly supported function in D for n large enough. Furthermore, one can check that $\|\rho_m - \rho_n\|_K \to 0$ as $n, m \to \infty$, that is, $(\rho_n)_{n \geq 0}$ is a Cauchy sequence with respect to the norm induced by the covariance kernel K. Indeed consider, for instance, $K(\rho_n, \rho_m)$. By definition of ρ_n and ρ_m, using the symmetry of ψ_n and Fubini's theorem (in the non-negative case, as K is bounded below from (3.1) and ρ_n, ρ_m have finite mass)

$$K(\rho_n, \rho_m) = \iint \rho_n(x)K(x,y)\rho_m(y)\,dx\,dy$$
$$= \iint \rho(dz)\rho(dw) \iint \psi_n(z-x)K(x,y)\psi_m(w-y)\,dx\,dy$$
$$= \mathbb{E}\Big[\iint K(z + \frac{X}{n}, w + \frac{Y}{m})\rho(dz)\rho(dw)\Big],$$

where X, Y are independent and distributed according to $\psi(z)\,dz$. Let us check that the above expression converges to $K(\rho, \rho)$. First, as ρ is finite and the function g in the definition of K is bounded and continuous, it suffices to replace $K(z + X/n, w + Y/m)$ by $-\log|z - w - X/n - Y/m|$. Let us fix z and consider the integral in w. This integral can be rewritten as

$$\int -\log(|z + \frac{X}{n} - \frac{Y}{m} - w|)\rho(dw) = \mathcal{E}_\rho(z + \frac{X}{n} - \frac{Y}{m})$$

in the notation of (3.4). Hence as $n, m \to \infty$ this converges to $\mathcal{E}_\rho(z)$ by continuity of $\mathcal{E}_\rho$ in z. Furthermore as $\mathcal{E}_\rho$ is bounded and ρ is a finite measure, we have in turn $\int \mathcal{E}_\rho(z + X/n - Y/m)\rho(dz) \to \int \mathcal{E}_\rho(z)\rho(dz) = K(\rho, \rho)$ by dominated convergence. Using boundedness again, we can take expectations by dominated convergence, and we see that $K(\rho_n, \rho_m) \to K(\rho, \rho)$ as $n, m \to \infty$, as desired. For the same reason, all four terms in the expansion of $K(\rho_n - \rho_m, \rho_n - \rho_m)$ converge to the same limit (i.e. $K(\rho, \rho)$) and thus $\rho_n - \rho_m$ is a Cauchy sequence with respect to the norm $\|\cdot\|_K$. The sequence $(\rho_m)_{m \geq 1}$ therefore has a limit in $\mathfrak{M}$, and this limit does not depend on the choice of smoothing (i.e. on the choice of ψ subject to the above properties). Let us denote this limit by $\rho_\mathfrak{M}$. Note then that for two measures ρ, ρ' satisfying (3.4), we have the following expression for the covariance function (3.3) of the field:

$$(\rho_\mathfrak{M}, \rho'_\mathfrak{M})_K = \lim_{n \to \infty} K(\rho_n, \rho'_n) = K(\rho, \rho') := \iint \rho(dx)K(x,y)\rho'(dy). \quad (3.5)$$

Hence a measure satisfying (3.4) induces an element $\rho_\mathfrak{M}$ of the abstract Hilbert space $\mathfrak{M}$, and the corresponding covariance has the concrete expression (3.5)

80 *Gaussian Multiplicative Chaos*

in this case. As the map $\rho \mapsto \rho_{\mathfrak{M}}$ is linear, we will still write, with a similar abuse of notation as earlier, (h, ρ) for the random variable $h_{\rho_{\mathfrak{M}}} = (h, \rho_{\mathfrak{M}})$. To summarise, for measures ρ and ρ' satisfying (3.4):

$$\mathrm{Cov}((h,\rho);(h,\rho')) = \iint \rho(\mathrm{d}x) K(x,y) \rho'(\mathrm{d}y).$$

This is the expression we will use in practice when constructing Gaussian multiplicative chaos below.

Let σ be a Radon measure[3] on $\overline{D}$ of dimension at least $\mathfrak{d}$ (where $0 \le \mathfrak{d} \le d$), in the sense that,

$$\iint_{\overline{D} \times \overline{D}} \frac{1}{|x-y|^{\mathfrak{d}-\varepsilon}} \sigma(\mathrm{d}x) \sigma(\mathrm{d}y) < \infty \tag{3.6}$$

for all $\varepsilon > 0$ (so for example, if σ is Lebesgue measure, then $d = \mathfrak{d}$). Note that $\mathfrak{d} \ge 0$ and may be equal to 0, but the statement of the theorem below will be empty in that case. In particular, we will only care about the case $\mathfrak{d} > 0$, which prevents σ from having any atoms. Throughout this chapter, when $\mathfrak{d} > 0$ we will fix a number $0 < \mathbf{d} < \mathfrak{d}$, such that

$$\iint_{\overline{D} \times \overline{D}} \frac{1}{|x-y|^{\mathbf{d}}} \sigma(\mathrm{d}x) \sigma(\mathrm{d}y) < \infty. \tag{3.7}$$

Remark 3.1 Many results of this chapter (for example, Theorems 3.2 or 3.30) are stated under an assumption of a strict inequality involving $\mathbf{d}$; since $\mathbf{d}$ can be chosen arbitrarily close to $\mathfrak{d}$, the same results could be stated by replacing $\mathbf{d}$ by $\mathfrak{d}$.

Now, fix θ a non-negative, Borel measure on $\mathbb{R}^d$ which supported on the closed unit ball $\overline{B}(0,1)$ and such that $\theta(\mathbb{R}^d) = 1$. Suppose in addition that $x \mapsto \mathcal{E}_\theta(x)$ is continuous (and hence uniformly bounded) on $\overline{B}(0,5)$, as in (3.4). Thus we may test h against θ, as explained above. More generally, for $\varepsilon > 0$, set $\theta_\varepsilon(\cdot)$ to be the image of the measure θ under the mapping $x \mapsto \varepsilon x$; that is $\theta_\varepsilon(A) = \theta(A/\varepsilon)$ for all Borel sets A. We view this as an approximation of the identity based on θ (and will sometimes write $\theta_\varepsilon(x) \, \mathrm{d}x$ for the measure $\theta_\varepsilon(\mathrm{d}x)$ with an abuse of notation). We also write $\theta_{x,\varepsilon}(\cdot)$ for the measure θ_ε translated by x. For $x \in D$, it is now straightforward to see that $\theta_{x,\varepsilon}$ also satisfies (3.4) and thus can be viewed as an element of $\mathfrak{M}$. So we can define an ε-regularisation of the field h by setting for ε small,

$$h_\varepsilon(x) = h * \theta_\varepsilon(x) = \int h(y) \theta_\varepsilon(x-y) \, \mathrm{d}y = \int h(y) \theta_{x,\varepsilon}(\mathrm{d}y) \, , \ x \in D. \tag{3.8}$$

---

[3] In fact, on $\mathbb{R}^d$, every locally finite Borel measure is Radon, so it would suffice to assume that σ is a locally finite Borel measure.

One can check that $\mathrm{Var}(h_\varepsilon(x) - h_\varepsilon(x')) \to 0$ as $|x - x'| \to 0$ for a fixed ε, so there exists a version of the stochastic process h such that $h_\varepsilon(x)$ is almost surely a Borel measurable function of $x \in S$ (see for example Proposition 2.1.12 in [GN16]). Hence for any Borel set $S \subset D$ and $\gamma \geq 0$, we may define

$$\mathcal{M}_\varepsilon(S) = I_\varepsilon = \int_S e^{\gamma h_\varepsilon(z) - \frac{\gamma^2}{2}\mathbb{E}(h_\varepsilon(z)^2)} \sigma(\mathrm{d}z). \tag{3.9}$$

In Chapter 2, where h was the 2D Dirichlet Gaussian free field (multiplied by $\sqrt{2\pi}$), our choice for σ was $\sigma(\mathrm{d}z) = R(z, D)^{\gamma^2/2} \mathrm{d}z$, and our choice for the measure θ was the uniform distribution on the unit circle (so $h_\varepsilon(z)$ was the usual circle average process of h).

3.3 Construction of Gaussian multiplicative chaos

With these definitions, we can state the result that guarantees the existence of Gaussian multiplicative chaos. For simplicity we assume D bounded, and let $S \subset D$ be a Borel subset (which may be equal to D itself).

Theorem 3.2 *Let* $0 \leq \gamma < \sqrt{2\mathbf{d}}$ *(equivalently,* $0 \leq \gamma < \sqrt{2\mathfrak{d}}$*). Then* $\mathcal{M}_\varepsilon(S)$ *converges in probability and in* $L^1(\mathbb{P})$ *to a limit* $\mathcal{M}(S)$. *The random variable* $\mathcal{M}(S)$ *does not depend on the choice of the regularising kernel* θ *subject to the above assumptions. Furthermore, the collection* $(\mathcal{M}(S))_{S \subset D}$ *defines a Borel measure* $\mathcal{M}$ *on* D, *and* $\mathcal{M}_\varepsilon$ *converges in probability towards* $\mathcal{M}$ *for the topology of weak convergence of measures on* D. *Finally, the measure* $\mathcal{M}$ *almost surely has no atoms.*

Note: In later chapters, we will sometimes also use the notation $\mathcal{M}_h$ or $\mathcal{M}_h^\gamma$ to indicate the dependence of $\mathcal{M}$ on the underlying field h or the field h and the parameter γ.

Let us assume without loss of generality that $\mathbf{d} > 0$, so that σ has no atoms.

As before, the main idea will be to pick $\alpha > \gamma$ and consider the normalised measure $e^{\gamma h_\varepsilon(x)} \mathrm{d}x$, but *restricted to good points*; that is, points that are not too thick. We will check that the L^1 contribution of bad points is negligible (essentially by the above Cameron–Martin–Girsanov observation), while the remaining part is shown to remain bounded and in fact convergent in $L^2(\mathbb{P})$. The key will be to take a good and slightly more subtle definition of the notion of *good points*, which makes the relevant L^2 computation very simple.

In [Ber17], uniqueness of the limit was obtained by comparing to a different approximation of the field, arising from the Karhuhen–Loeve expansion of h. This gives another approximation of the measure which turns out to be a martingale, and hence also has a limit. In the same paper, it was then shown that the two measures must agree, thereby deducing uniqueness. Here we present a slightly simpler argument partly based on a remark made by Hubert Lacoin (private communication).

3.3.1 Uniform integrability

The goal of this section will be to prove:

Proposition 3.3 *I_ε is uniformly integrable.*

Proof Let $\alpha > 0$ be fixed (it will be chosen $> \gamma$ and very close to γ soon). It is helpful in this proof to introduce another regularisation kernel θ^* satisfying the same assumptions as θ, namely (3.4). We then denote by $h_r^*(z) = h \star \theta_{z,r}^*$. For $s \in S$, and $n \in \mathbb{Z}$, set

$$E_n(x) = \{h_{e^{-n}}^*(x) \le \alpha n\}.$$

We define a **good event**

$$G_\varepsilon^\alpha(x) = \bigcap_{n=n_0}^{n(\varepsilon)} E_n(x), \qquad (3.10)$$

where $e^{-n_0} = \varepsilon_0 \le 1$ for instance, and $n(\varepsilon) = \lceil \log(1/\varepsilon) \rceil$. This is the good event that the point x is never too thick (with respect to the regularisation kernel θ^*) up to scale ε. Mostly, we think of the case where $\theta = \theta^*$, but in the proof of uniqueness (independence of the limit with respect to the choice of the regularisation kernel), it will be useful to define the good event in such a way that it does not depend on θ, hence our choice to let it depend on some other fixed arbitrary regularisation kernel θ^*. Further, let $\overline{h}_\varepsilon(x) = \gamma h_\varepsilon(x) - (\gamma^2/2)\mathbb{E}(h_\varepsilon(x)^2)$ to ease notations.

Lemma 3.4 (Ordinary points are not thick) *For any $\alpha > 0$, we have that, uniformly over $x \in S$, $\mathbb{P}(G_\varepsilon^\alpha(x)) \ge 1 - p(\varepsilon_0)$ where the function p may depend on α and for a fixed $\alpha > 0$, $p(\varepsilon_0) \to 0$ as $\varepsilon_0 \to 0$.*

Proof Set $X_t^* = h_\varepsilon^*(x)$ for $\varepsilon = e^{-t}$. Then a direct computation using (3.1) and the covariance formula (3.5) (see Lemma 3.6, and more precisely (3.13)) implies that

$$|\mathrm{Cov}(X_s^*, X_t^*) - s \wedge t| \le O(1), \qquad (3.11)$$

where the implicit constant is uniform. In particular, $\mathrm{Var}(X_t^*) = t + O(1)$.

3.3 Construction of Gaussian multiplicative chaos

Note that for each $k \ge 1$, $\mathbb{P}(X_k^* \ge \alpha k/2) \le e^{-\alpha^2 k^2/(8 \operatorname{Var}(X_k^*))}$ which decays exponentially in k by the above, and so is smaller than $Ce^{-\lambda k}$ for some $\lambda > 0$ (we may take $\lambda = \alpha^2/16$ for instance). Hence

$$\mathbb{P}(\exists k \ge k_0: |X_k^*| \ge \alpha k) \le \sum_{k \ge k_0} Ce^{-\lambda k}.$$

We call $p(\varepsilon_0)$ to be the right-hand side of the above for $k_0 = \lceil -\log(\varepsilon_0) \rceil$ which can be made arbitrarily small by picking ε_0 small enough. This proves the lemma. □

Lemma 3.5 (Liouville points are no more than γ-thick) *For $\alpha > \gamma$, we have*

$$\mathbb{E}(e^{\overline{h}_\varepsilon(x)} 1_{G_\varepsilon^\alpha(x)}) \ge 1 - p(\varepsilon_0).$$

Proof Note that

$$\mathbb{E}(e^{\overline{h}_\varepsilon(x)} 1_{\{G_\varepsilon^\alpha(x)\}}) = \tilde{\mathbb{P}}(G_\varepsilon^\alpha(x)), \text{ where } \frac{d\tilde{\mathbb{P}}}{d\mathbb{P}} = e^{\overline{h}_\varepsilon(x)}.$$

Analogously to X_t^*, set $X_t := h_\varepsilon(x))$ for $\varepsilon = e^{-t}$. By the Cameron–Martin–Girsanov lemma, under $\tilde{\mathbb{P}}$, the process $(X_s^*)_{-\log \varepsilon_0 \le s \le t}$ has the same covariance structure as under $\mathbb{P}$ and its mean is now $\gamma \operatorname{Cov}(X_s^*, X_t) = \gamma s + O(1)$ for $s \le t$ (see again Lemma 3.6, and more precisely (3.13)). Hence

$$\tilde{\mathbb{P}}(G_\varepsilon^\alpha(x)) \ge \mathbb{P}(G_\varepsilon^{\alpha-\gamma}(x)) \ge 1 - p(\varepsilon_0)$$

by Lemma 3.4 since $\alpha > \gamma$. □

We thus see that points which are more than γ-thick do not contribute significantly to I_ε in expectation and so can be safely removed. We therefore fix $\alpha > \gamma$ and introduce:

$$J_\varepsilon = \int_S e^{\overline{h}_\varepsilon(z)} 1_{\{G_\varepsilon(z)\}} \sigma(dz) \qquad (3.12)$$

with $G_\varepsilon(x) = G_\varepsilon^\alpha(x)$. We will show that J_ε is uniformly integrable from which the result follows.

Before we embark on the main argument of the proof, we record here for ease of reference an elementary estimate on the covariance structure of $h_\varepsilon(x)$. Roughly speaking, the role of the first estimate (3.13) is to bound from above (up to an unimportant constant of the form $e^{O(1)}$) the contribution to $\mathbb{E}(J_\varepsilon^2)$ coming from points x, y that are close to each other. That will suffice to prove uniform integrability. The role of the finer estimate (3.14) is to get a more precise estimate to the contribution to $\mathbb{E}(J_\varepsilon^2)$ coming from points x, y which are macroscopically far away, which we will be able to assume thanks to (3.13). This time the error in the covariance up to an additive term $o(1)$ will translate into an error up to a factor $e^{o(1)} = 1 + o(1)$ in the estimation of this contribution. In turn, this will imply convergence.

Lemma 3.6 *We have the following estimate:*

$$\text{Cov}(h_\varepsilon(x), h_r^*(y)) = \log 1/(|x-y| \vee r \vee \varepsilon) + O(1). \tag{3.13}$$

Moreover, if $\eta > 0$ and $|x - y| \geq \eta$, then

$$\text{Cov}(h_\varepsilon(x), h_\delta^*(y)) = \log(1/|x-y|) + g(x, y) + o(1), \tag{3.14}$$

where $o(1)$ tends to 0 as $\delta, \varepsilon \to 0$, uniformly in $|x - y| \geq \eta$.

Note that, as θ and θ^* are completely arbitrary subject to (3.4), the same estimates hold with $h_\varepsilon(x)$ replaced by $h_\varepsilon^*(x)$ and/or $h_r^*(y)$ replaced by $h_r(y)$.

Proof We start with the proof of (3.13). Assume without loss of generality that $\varepsilon \leq r$. Note that, by (3.5),

$$\text{Cov}(h_\varepsilon(x), h_r^*(y)) = \iint K(z, w) \theta_{x,\varepsilon}(\mathrm{d}w) \theta_{y,r}^*(\mathrm{d}z)$$

$$= \iint -\log(|w - z|) \theta_{x,\varepsilon}(\mathrm{d}w) \theta_{y,r}^*(\mathrm{d}z) + O(1) \tag{3.15}$$

We consider the following cases: (a) $r \leq |x - y|/3$, and (b) $r \geq |x - y|/3$.

In case (a), $|x - y| \leq \varepsilon + |w - z| + r \leq 2r + |w - z| \leq (2/3)|x - y| + |w - z|$ by the triangle inequality, so $|w - z| \geq (1/3)|x - y|$ and we get

$$\text{Cov}(h_\varepsilon(x), h_r^*(y)) \leq -\log|x - y| + O(1)$$

as desired in this case.

The second case (b) is when $r \geq |x - y|/3$. Then by translation and scaling so that $B(y, r)$ becomes $B(0, 1)$, the right-hand side of (3.15) is equal to

$$\log(1/r) + \iint -\log|w - z| \theta_{\frac{x-y}{r}, \frac{\varepsilon}{r}}(\mathrm{d}w) \theta^*(\mathrm{d}z).$$

Conditioning on w (which is necessarily in $\overline{B}(0, 4)$ under the assumptions of case (b)), we see that by the assumption (3.4) on θ^*, the second term is bounded by $O(1)$, uniformly, so that

$$\text{Cov}(h_\varepsilon(x), h_r^*(y)) \leq -\log r + O(1)$$

as desired in this case. This proves (3.13).

The proof of (3.14) is similar but simpler. Indeed, we get (as in (3.15)),

$$\text{Cov}(h_\varepsilon(x), h_\delta^*(y)) = \iint -\log|w - z| \theta_{x,\varepsilon}(\mathrm{d}w) \theta_{y,\delta}^*(\mathrm{d}z) + g(x, y) + o(1), \tag{3.16}$$

3.3 Construction of Gaussian multiplicative chaos 85

where the $o(1)$ term tends to 0 as $\varepsilon, \delta \to 0$, coming from the continuity of g, and hence is uniform in x, y (not even assuming $|x - y| \geq \eta$). Now note that

$$\left|\log|w - z| - \log|x - y|\right| \leq \frac{4\max(\varepsilon, \delta)}{|x - y|}$$

as soon as $\max(\varepsilon, \delta) \leq \eta/4 \leq |x - y|/4$. Therefore the right-hand side of (3.16) is $-\log|x - y| + g(x, y) + O(\max(\varepsilon, \delta)) + o(1)$ when $|x - y| \geq \eta$, which proves the claim (3.14). □

Lemma 3.7 *For $\alpha > \gamma$ sufficiently close to γ, J_ε is bounded in $L^2(\mathbb{P})$ and hence uniformly integrable.*

Proof By Fubini's theorem,

$$\mathbb{E}(J_\varepsilon^2) = \int_{S \times S} \mathbb{E}(e^{\overline{h}_\varepsilon(x) + \overline{h}_\varepsilon(y)} \mathbf{1}_{\{G_\varepsilon(x) \cap G_\varepsilon(y)\}}) \sigma(dx) \sigma(dy)$$

$$= \int_{S \times S} e^{\gamma^2 \operatorname{Cov}(h_\varepsilon(x), h_\varepsilon(y))} \tilde{\mathbb{P}}(G_\varepsilon(x) \cap G_\varepsilon(y)) \sigma(dx) \sigma(dy), \quad (3.17)$$

where $\tilde{\mathbb{P}}$ is a new probability measure obtained by the Radon–Nikodym derivative

$$\frac{d\tilde{\mathbb{P}}}{d\mathbb{P}} = \frac{e^{\overline{h}_\varepsilon(x) + \overline{h}_\varepsilon(y)}}{\mathbb{E}(e^{\overline{h}_\varepsilon(x) + \overline{h}_\varepsilon(y)})}.$$

Note that since σ has no atoms, we may assume that $x \neq y$. Also, if $\varepsilon \leq e^{-1}\varepsilon_0$ and $|x - y| \leq e^{-1}\varepsilon_0$ (else we bound the probability below by 1), we have

$$\tilde{\mathbb{P}}(G_\varepsilon(x) \cap G_\varepsilon(y)) \leq \tilde{\mathbb{P}}(h_r^*(x) \leq \alpha \log 1/r),$$

where

$$r = e^{-n}, \text{ where } n = \left\lceil \log\left(\frac{1}{\varepsilon \vee |x - y|}\right) \right\rceil. \quad (3.18)$$

Furthermore, by Cameron–Martin–Girsanov, under $\tilde{\mathbb{P}}$ we have that $h_r^*(x)$ has the same variance as before (therefore $\log 1/r + O(1)$) and a mean given by

$$\operatorname{Cov}_{\mathbb{P}}(h_r^*(x), \gamma h_\varepsilon(x) + \gamma h_\varepsilon(y)) = 2\gamma \log 1/r + O(1), \quad (3.19)$$

again by Lemma 3.6 (more precisely, by (3.13)). Consequently,

$$\tilde{\mathbb{P}}(h_r^*(x) \leq \alpha \log 1/r) = \mathbb{P}(\mathcal{N}(2\gamma \log(1/r), \log 1/r) \leq \alpha \log(1/r) + O(1))$$

$$\leq \exp\left(-\frac{1}{2}(2\gamma - \alpha)^2(\log(1/r) + O(1))\right)$$

$$= O(1)r^{(2\gamma - \alpha)^2/2}. \quad (3.20)$$

We deduce that

$$\mathbb{E}(J_\varepsilon^2) \le O(1) \int_{S\times S} |(x-y) \vee \varepsilon|^{(2\gamma-\alpha)^2/2 - \gamma^2} \sigma(dx)\sigma(dy). \quad (3.21)$$

(We will get a better approximation in Section 3.3.2). Clearly by (3.6), this is bounded if

$$(2\gamma - \alpha)^2/2 - \gamma^2 > -\mathbf{d}$$

and since α can be chosen arbitrarily close to γ, this is possible if

$$\mathbf{d} - \gamma^2/2 > 0 \text{ or } \gamma < \sqrt{2\mathbf{d}}. \quad (3.22)$$

This proves the lemma. □

To finish the proof of Proposition 3.3, observe that $I_\varepsilon = J_\varepsilon + J'_\varepsilon$. We have $\mathbb{E}(J'_\varepsilon) \le p(\varepsilon_0)$, by Lemma 3.5, and, for a fixed ε_0, J_ε is bounded in L^2 (uniformly in ε). Hence I_ε is uniformly integrable. □

3.3.2 Convergence

As before, since $\mathbb{E}(J'_\varepsilon)$ can be made arbitrarily small by choosing ε_0 sufficiently small, it suffices to show that J_ε converges in probability and in L^1. In fact, we will show that it converges in L^2, from which convergence will follow. To do this, we will show that $(J_\varepsilon)_\varepsilon$ forms a Cauchy sequence in L^2, and we start by writing

$$\mathbb{E}((J_\varepsilon - J_\delta)^2) = \mathbb{E}(J_\varepsilon^2) + \mathbb{E}(J_\delta^2) - 2\mathbb{E}(J_\varepsilon J_\delta). \quad (3.23)$$

Our basic approach is thus to estimate, better than before, $\mathbb{E}(J_\varepsilon^2)$ from above and $\mathbb{E}(J_\varepsilon J_\delta)$ from below. Essentially, the idea is that for x, y which are at a small but macroscopic distance, we can identify the limiting distribution of $(h^*_r(x), h^*_r(y))_{r \le \varepsilon_0}$ under the distribution $\mathbb{P}$ biased by $e^{\overline{h}_\varepsilon(x) + \overline{h}_\delta(y)}$. On the other hand, when x, y are closer than that we know from Section 3.3.1 that the contribution is essentially negligible.

Let us write, similar to (3.17),

$$\mathbb{E}(J_\varepsilon J_\delta) = \int_{S^2} e^{\gamma^2 \operatorname{Cov}(h_\varepsilon(x), h_\delta(y))} \tilde{\mathbb{P}}(G_\varepsilon(x) \cap G_\delta(y)) \sigma(dx)\sigma(dy),$$

where

$$d\tilde{\mathbb{P}}(\cdot) = \frac{e^{h_\varepsilon(x) + h_\delta(y)}}{\mathbb{E}(e^{h_\varepsilon(x) + h_\delta(y)})} d\mathbb{P}(\cdot).$$

We fix $\eta > 0$ arbitrarily for now, and suppose that $|x - y| \ge \eta$.

3.3 Construction of Gaussian multiplicative chaos

Observe that for any fixed $\varepsilon_1 \leq \varepsilon_0$, as $\varepsilon \to 0$, and uniformly over $x, y \in S$ with $|x - y| \geq \eta$ and $r \geq \varepsilon_1$, if $z \in \{x, y\}$,

$$\mathrm{Cov}(h_r^*(z), h_\varepsilon(x)) \to \int_D K(w, x)\theta_r^*(w - z)\, dw. \tag{3.24}$$

Likewise, uniformly over $x, y \in S$ such that $|x - y| \geq \eta$, and over $r \geq \varepsilon_1$, if $z \in \{x, y\}$, as $\delta \to 0$:

$$\mathrm{Cov}(h_r^*(z), h_\delta(y)) \to \int_D K(w, y)\theta_r^*(w - z)\, dw. \tag{3.25}$$

(Note that both right-hand sides of (3.24) and (3.25) are finite by (3.13).) Consequently, by Cameron–Martin–Girsanov, as $\varepsilon \to 0, \delta \to 0$, the joint law of the processes $(h_r^*(x), h_r^*(y))_{r \leq \varepsilon_0}$ under $\tilde{\mathbb{P}}$ converges to a joint distribution $(\tilde{h}_r^*(x), \tilde{h}_r^*(y))_{r \leq \varepsilon_0}$ whose covariance is unchanged and whose mean is given by

$$\mathbb{E}(\tilde{h}_r^*(z)) = \gamma \left[\int_D K(w, x)\theta_r^*(w - z)\, dw + \int_D K(w, y)\theta_r^*(w - z)\, dw \right],$$
$$z \in \{x, y\}, r \leq \varepsilon_0. \tag{3.26}$$

This convergence is for the weak convergence on compacts of $r \in (0, \varepsilon_0]$, and is uniform in $|x - y| \geq \eta$.

The core of the argument will be to prove that $\tilde{\mathbb{P}}(G_\varepsilon(x) \cap G_\delta(y))$ converges as $\varepsilon, \delta \to 0$, uniformly over $|x - y| \geq \eta$, to the analogous probabilities for $(\tilde{h}_r^*(x), \tilde{h}_r^*(y))$. To state this more precisely, let us introduce some notation. Define the event $\tilde{E}_n(z)$ (for $z \in \{x, y\}$) in a way analogous to the event $E_n(z)$:

$$\tilde{E}_n(z) = \{\tilde{h}_{e^{-n}}^*(z) \leq \alpha n\}; \quad n \in \mathbb{Z},$$

and consider the corresponding good event for the field $\tilde{h}$,

$$\tilde{G}(z) = \bigcap_{n=n_0}^{\infty} \tilde{E}_n(z).$$

Lemma 3.8 *As $\varepsilon \to 0, \delta \to 0$, uniformly over $|x - y| \geq \eta$,*

$$\tilde{\mathbb{P}}(G_\varepsilon(x) \cap G_\delta(y)) \to g_\alpha(x, y) := \mathbb{P}(\tilde{G}(x) \cap \tilde{G}(y)). \tag{3.27}$$

Proof This will follow from the fact that the joint law of the processes $(h_r^*(x), h_r^*(y))_{r \leq \varepsilon_0}$ under $\tilde{\mathbb{P}}$ converges on compact sets of $(0, \varepsilon_0]$ to the joint distribution of $(\tilde{h}_r^*(x), \tilde{h}_r^*(y))_{r \leq \varepsilon_0}$, but requires an argument since the good events $\tilde{G}(x), \tilde{G}(y)$ do not depend only on the behaviour of $(h_r^*(x), h_r^*(y))$ in some fixed compact of $(0, \varepsilon_0]$.

Let us start with the upper bound for (3.27). For any $n_1 > n_0$, and any ε, δ such that $n(\varepsilon), n(\delta) \geq n_1$,

$$\tilde{\mathbb{P}}(G_\varepsilon(x) \cap G_\delta(y)) \leq \tilde{\mathbb{P}}\left(\bigcap_{n=n_0}^{n_1} E_n(x) \cap E_n(y) \right)$$

so, using the convergence on compact sets,

$$\limsup_{\delta, \varepsilon \to 0} \tilde{\mathbb{P}}(G_\varepsilon(x) \cap G_\delta(y)) \leq \mathbb{P}\left(\bigcap_{n=n_0}^{n_1} \tilde{E}_n(x) \cap \tilde{E}_n(y) \right)$$

and since $n_1 > n_0$ is arbitrary, we get

$$\limsup_{\varepsilon, \delta \to 0} \tilde{\mathbb{P}}(G_\varepsilon(x) \cap G_\delta(y)) \leq \mathbb{P}(\tilde{G}(x) \cap \tilde{G}(y)).$$

Let us now turn to the lower bound. The key is to observe that the constraint $h^*_{e-n}(z) \leq \alpha n$ defining $E_n(z)$ is essentially guaranteed for large n under $\tilde{\mathbb{P}}$, whether z is x or y. This is because $|x - y| \geq \eta$ so x and y are well separated (note that this would be false however if x and y would not be separated: for instance if $x = y$, we would typically expect $h^*_{e-n}(x)$ to be of the order of $2\gamma n$, which is $> \alpha n$.) More precisely, applying the same argument as in Lemma 3.5, no matter how small $a > 0$ is, we can find $n_1 = n_1(\eta, \alpha, \gamma, a)$, but independent of ε or δ, such that for all $\varepsilon, \delta > 0$, under $\tilde{\mathbb{P}}$,

$$\tilde{\mathbb{P}}\left(\bigcup_{n=n_1}^{n(\delta) \vee n(\varepsilon)} \{h^*_{e-n}(z) \geq \alpha n\} \right) \leq a. \quad (3.28)$$

Indeed, consider $z = x$ first. Lemma 3.5 analyses the effect of biasing by $e^{\overline{h}_\varepsilon(x)}$. To analyse the effect of further biasing by $e^{\overline{h}_\delta(y)}$, observe that since x and y are separated by a distance at least η, the resulting additional shift in the mean of $h^*_{e-n}(x)$, which is explicitly given by $\gamma \operatorname{Cov}(h^*_{e-n}(x), h_\delta(y))$, is at most $O(1)$ by (3.13). The same observation applies to $z = y$.

Therefore, using (3.28),

$$\tilde{\mathbb{P}}(G_\varepsilon(x) \cap G_\delta(y)) \geq \tilde{\mathbb{P}}\left(\bigcap_{n=n_0}^{n_1} E_n(x) \cap E_n(y) \right) - 2a.$$

We can take a limit as $\delta, \varepsilon \to 0$ using convergence on compacts to deduce

$$\liminf_{\varepsilon \to 0} \tilde{\mathbb{P}}(G_\varepsilon(x) \cap G_\delta(y)) \geq \mathbb{P}\left(\bigcap_{n=n_0}^{n_1} \tilde{E}_n(x) \cap \tilde{E}_n(y) \right) - 2a$$

$$\geq \mathbb{P}\left(\bigcap_{n=n_0}^{\infty} \tilde{E}_n(x) \cap \tilde{E}_n(y) \right) - 2a.$$

3.3 Construction of Gaussian multiplicative chaos

This completes the proof of (3.27) because $a > 0$ was arbitrary. $\square$

Using this lemma, we can easily conclude:

Lemma 3.9 *We have*

$$\limsup_{\varepsilon \to 0} \mathbb{E}(J_\varepsilon^2) \leq \int_{S \times S} e^{\gamma^2 g(x,y)} \frac{1}{|x-y|^{\gamma^2}} g_\alpha(x,y) \sigma(dx) \sigma(dy),$$

where $g_\alpha(x, y)$ is a non-negative function depending on α, ε_0 and γ such that the above integral is finite.

Proof We fix $\eta > 0$ arbitrarily small (in particular, η may and will be smaller than $e^{-1}\varepsilon_0$). If $|x-y| \leq \eta$, we use the same bound as in (3.21). The contribution coming from the part $|x - y| \leq \eta$ can thus be bounded, uniformly in ε, by $f(\eta)$ (where $f(\eta) \to 0$ as $\eta \to 0$ and the precise order of magnitude of $f(\eta)$ is determined by (3.6), and is at most polynomial in η).

On the other hand, for $|x - y| \geq \eta$, taking $\delta = \varepsilon$ in (3.8), after applying Lemma 3.6 (more specifically (3.14)) for the pointwise limit and (3.13) for the use of dominated convergence):

$$\int_{S^2; |x-y| \geq \eta} e^{\gamma^2 \operatorname{Cov}(h_\varepsilon(x), h_\varepsilon(y))} \tilde{\mathbb{P}}(G_\varepsilon(x), G_\varepsilon(y)) \sigma(dx) \sigma(dy)$$

$$\to \int_{S^2; |x-y| \geq \eta} \frac{e^{\gamma^2 g(x,y)}}{|x-y|^{\gamma^2}} g_\alpha(x,y) \sigma(dx) \sigma(dy). \tag{3.29}$$

Since we already know that the piece of the integral coming from $|x - y| \leq \eta$ contributes at most $f(\eta) \to 0$ when $\eta \to 0$, it remains to check that the integral on the right-hand side of (3.29) remains finite as $\eta \to 0$. But we have already seen in (3.20) that for $|x - y| \leq \varepsilon_0/3$, $\tilde{\mathbb{P}}(G_\varepsilon(x) \cap G_\varepsilon(y)) \leq O(1)|x-y|^{(2\gamma-\alpha)^2/2-\gamma^2}$; hence this inequality must also hold for $g_\alpha(x, y)$. Hence the result follows as in (3.22). $\square$

Lemma 3.10 *We have*

$$\liminf_{\varepsilon, \delta \to 0} \mathbb{E}(J_\varepsilon J_\delta) \geq \int_{S \times S} e^{\gamma^2 g(x,y)} \frac{1}{|x-y|^{\gamma^2}} g_\alpha(x,y) \sigma(dx) \sigma(dy).$$

Proof In fact, the proof is almost exactly the same as in Lemma 3.9 and is even easier, since we get a lower bound by restricting ourselves to $|x - y| \geq \eta$. We deduce immediately from Lemma 3.8 that

$$\liminf_{\varepsilon, \delta \to 0} \mathbb{E}(J_\varepsilon J_\delta) \geq \int_{S^2; |x-y| \geq \eta} e^{\gamma^2 g(x,y)} \frac{1}{|x-y|^{\gamma^2}} g_\alpha(x,y) \sigma(dx) \sigma(dy).$$

Since η is arbitrary, the result follows. $\square$

Proof of convergence in Theorem 3.2 Using (3.23) together with Lemmas 3.9 and 3.10, we see that J_ε is a Cauchy sequence in L^2 for any $\varepsilon_0 > 0$. Combining with Lemma 3.5, it therefore follows that I_ε is a Cauchy sequence in L^1 and hence converges in L^1 (and also in probability) to a limit $I = \mathcal{M}(S)$. The proof of weak convergence follows by the argument in Section 2.3. We leave the proof that $\mathcal{M}$ has no atoms as an exercise: see Exercise 2.4 or Exercise 3.4 for another proof. □

Remark 3.11 Note that $\lim_{\varepsilon \to 0} \mathbb{E}(J_\varepsilon^2)$ depends on the regularisation θ, even though, as we will see next, $\lim_{\varepsilon \to 0} I_\varepsilon$ does not.

Proof of uniqueness in Theorem 3.2 To prove uniqueness, we take $\tilde\theta$ another non-negative Borel measure on $\mathbb{R}^d$ supported in $\overline{B}(0, 1)$ and satisfying (3.4). Let $\tilde h_\delta(x) = h * \tilde\theta_\delta(x)$, and let $\tilde J_\delta$ be defined as J_δ but with $\tilde\theta$ instead of θ: that is,

$$\tilde J_\delta = \int_S e^{\gamma \tilde h_\delta(z) - (\gamma^2/2)\mathbb{E}(\tilde h_\delta(z)^2)} \mathbf{1}_{\{G_\delta(z)\}} \sigma(\mathrm{d}z),$$

where we use the same good event $G_\delta(z)$ as in (3.10), based on the regularisation kernel θ^*. Then the argument of Lemma 3.10 can be used to show that the same conclusion holds for J_δ replace by $\tilde J_\delta$: that is,

$$\liminf_{\varepsilon, \delta \to 0} \mathbb{E}(J_\varepsilon \tilde J_\delta) \geq \int_{S \times S} e^{\gamma^2 g(x,y)} \frac{1}{|x-y|^{\gamma^2}} g_\alpha(x, y) \sigma(\mathrm{d}x) \sigma(\mathrm{d}y).$$

Hence we deduce $\lim_{\varepsilon \to 0, \delta \to 0} \mathbb{E}((J_\varepsilon - \tilde J_\delta)^2) = 0$, and this implies that the limits associated with θ and $\tilde\theta$ are almost surely the same. □

Remark 3.12 The proof of Theorem 3.2 extends without difficulty to a variety of settings going somewhat beyond the stated assumptions. In such cases, the input that is crucially required for the argument to extend without major modifications is Lemma 3.6 which controls the correlations of the regularised Gaussian field. An example would be the setting of a Gaussian, logarithmically correlated field on a Riemannian manifold.

3.4 Shamov's approach to Gaussian multiplicative chaos

An alternative and powerful viewpoint on Gaussian multiplicative chaos was also developed in Shamov [Sha16]. It is closely related to the generalisation of "rooted measures" for the GFF: see Section 2.4. In what follows, h will be a centred Gaussian field with logarithmically diverging covariance kernel K as in (3.1) (although the original paper [Sha16] works in a more general setting).

3.4 Shamov's approach to Gaussian multiplicative chaos

Before stating the result, let us make an observation about changes of measure for the field h. If $\rho \in \mathfrak{M}$ we write $K\rho$ for the linear operator $\rho' \mapsto K(\rho, \rho')$ on $\mathfrak{M}$. Note that if $\rho \in \mathcal{D}_0(D)$, we have

$$K\rho(x) = \int_D K(x,y)\,\rho(\mathrm{d}y). \tag{3.30}$$

Applying Girsanov's Lemma (Lemma 2.5), we see that if $\rho \in \mathfrak{M}$ then the field $h + K\rho$ (viewed as a stochastic process indexed by $\rho' \in \mathfrak{M}$) is absolutely continuous with respect to h, with associated Radon–Nikodym derivative

$$\frac{\exp((h,\rho))}{\exp(\frac{1}{2}(\rho,\rho)_K)}. \tag{3.31}$$

Note the connection with Section 1.9 in the case of the zero boundary GFF: when $\rho \in \mathfrak{M}_0$ ($\mathfrak{M}_0$ corresponding to the zero boundary condition Green function) then $(h,\rho) = (h,F)_\nabla$, where F is defined by $-\Delta F = 2\pi\rho$ and is an element of $H_0^1(D)$. By (1.33), this is exactly the statement that $F = K\rho$, and the above expression is equal to $\exp((h,F)_\nabla)/\exp(\frac{1}{2}(F,F)_\nabla)$ as in Lemma 1.51. See [Aru20] for more on Shamov's approach when the field is the planar Dirichlet GFF.

Definition 3.13 (Shamov's definition of GMC)   Let h be as above and σ as in (3.6). Let $\gamma \in (0, 2)$. A measure $\mathcal{M}^\gamma$ is a γ-multiplicative chaos measure for h, with background measure σ if:

- $\mathcal{M}^\gamma$ is measurable with respect to h as a stochastic process indexed by $\mathfrak{M}$ (note that this allows us to write $\mathcal{M}^\gamma(\mathrm{d}x) = \mathcal{M}^\gamma(h, \mathrm{d}x)$);
- $\mathbb{E}(\mathcal{M}^\gamma(S)) = \sigma(S)$ for all Borel sets $S \subset D$;
- For every fixed (deterministic) function $\rho \in \mathcal{D}_0(D)$ (i.e. ρ is smooth and compactly supported), if $\xi = K\rho$ is given by (3.30), then

$$\mathcal{M}^\gamma(h + \xi, \mathrm{d}x) = \exp(\gamma\xi(x))\mathcal{M}^\gamma(h, \mathrm{d}x) \text{ almost surely.} \tag{3.32}$$

We use the notation $\mathcal{M}^\gamma$ above to distinguish it from $\mathcal{M}$ in Sections 3.1–3.3. However, we will see just below that $\mathcal{M}^\gamma$ exists, and in fact must be equal to $\mathcal{M}$.

Theorem 3.14 (Shamov, [Sha16])   *Assume the set-up of Definition 3.13. Then a multiplicative chaos measure for h with parameter γ and background measure σ exists. Moreover, it is unique.*

We note that the uniqueness part of Theorem 3.14 may be particularly useful if one wants to identify some limit as being a GMC measure, since the conditions are in many contexts relatively easy to check.

Remark 3.15 As we will see in the proof below, the condition (3.32) ensures that the effect of weighting the law of the field by $\mathcal{M}^\gamma(D)$ is to add the singularity $\gamma K(x, \cdot)$ to the field at a point x chosen from $\mathcal{M}^\gamma$, a property which we will shall see amounts to Girsanov's transform for the field reweighted by the mass of $\mathcal{M}^\gamma(D)$. So essentially, Shamov's approach characterises the GMC measure as a certain Radon–Nikodym derivative for the field.

Proof (i). *Proof of existence.* For the existence part we will show that the GMC measure constructed in Theorem 3.2 does satisfy the stated conditions. The first two properties are an obvious consequence of the construction: in particular, since the GMC measure $\mathcal{M}$ is obtained as a limit in probability as $\varepsilon \to 0$, with respect to the weak convergence of measures, of a sequence of measures $\mathcal{M}_\varepsilon$ that are obviously measurable with respect to h, so is their limit $\mathcal{M}$. The third property requires some additional arguments. First, since $\rho \in \mathcal{D}_0(D)$, by (3.31), we have that $h + \xi = h + K\rho$ is absolutely continuous with respect to h. In particular, it makes sense to apply Theorem 3.2 to $\tilde{h} = h + \xi$ and we have $\mathcal{M}^\gamma(h+\xi, dx) = \lim_{\varepsilon \to 0} \varepsilon^{\gamma^2/2} e^{\gamma(h_\varepsilon + \xi_\varepsilon)(x)} \, dx = \lim_{\varepsilon \to 0} e^{\gamma \xi_\varepsilon(x)} \mathcal{M}_\varepsilon(dx)$ and $\mathcal{M}^\gamma(h, dx) = \lim_{\varepsilon \to 0} e^{\gamma h_\varepsilon(x)} \, dx = \lim_{\varepsilon \to 0} \mathcal{M}_\varepsilon(dx)$, where the limits are in probability and $f \mapsto f_\varepsilon = f * \theta_\varepsilon$ represents smoothing with some fixed mollifier at scale ε. Hence, it suffices to show that $e^{\gamma \xi_\varepsilon} \mathcal{M}_\varepsilon \to e^{\gamma \xi} \mathcal{M}$ in probability, with respect to the topology of weak convergence as $\varepsilon \to 0$. This follows by the triangle inequality since $e^{\gamma \xi_\varepsilon} \to e^{\gamma \xi}$ uniformly on compacts of D as $\varepsilon \to 0$, as $\xi = K\rho$ is continuous on $\overline{D}$, by a simple application of dominated convergence (recall that $\rho \in \mathcal{D}_0(D)$).

(ii). *Proof of uniqueness.* Suppose that a measure $\mathcal{M}^\gamma$ satisfying the constraints of Definition 3.13 exists. We will consider the probability measure (often called the **rooted measure**, and already encountered in Chapter 2 in the context of the Liouville measure associated to the Dirichlet GFF in (2.7)):

$$Q(dx, dh) = \frac{\mathcal{M}^\gamma(h, dx)}{\mathbb{E}(\mathcal{M}^\gamma(h, D))} \mathbb{P}(dh). \qquad (3.33)$$

Note that

$$\int_D \int Q(dx, dh) = \mathbb{E}_h(\int_D \mathcal{M}^\gamma(h, dx))/\mathbb{E}_h(\mathcal{M}^\gamma(h, D)) = 1$$

so that Q is a probability law on pairs (x, h). We will show that:

1. Under Q, the marginal law of x has density proportional to $\sigma(dx)$,
2. Given x, the conditional law of the field (viewed as a stochastic process indexed by $\mathfrak{M}$) is that of h plus the deterministic function $\gamma K(x, \cdot)$.

Observe that these two properties completely characterise the law Q and thus,

3.4 Shamov's approach to Gaussian multiplicative chaos

by disintegration, the conditional law $Q(dx \mid h)$ of x given h under Q. On the other hand, the definition of Q means that this conditional law is exactly $\mathcal{M}^\gamma(h, dx)$ and so we have identified $\mathcal{M}^\gamma$ uniquely (note that this doesn't identify only the law of $\mathcal{M}^\gamma$ but really the joint law of h and $\mathcal{M}^\gamma$).

To show the claim concerning Q, it is enough to prove that the Q marginal law of x is equal to $\sigma(dx)/\sigma(D)$, and that for any $\rho_1, \ldots, \rho_m \in \mathfrak{M}$ and $a_1, \ldots, a_m \in \mathbb{R}$ the Q conditional law of $(a_1(h, \rho_1) + \cdots + a_m(h, \rho_m))$ given x is a normal random variable with the correct mean and covariance. In other words (using linearity of h on the space $\mathfrak{M}$), it suffices to show that for any $g \in L^1(\sigma)$ on D, and $\rho \in \mathfrak{M}$

$$\mathbb{E}_Q(e^{(h,\rho)}g(x)) = \mathbb{E}\left(\int_D e^{(h+\gamma K(x,\cdot),\rho)} g(x) \frac{\sigma(dx)}{\sigma(D)}\right). \tag{3.34}$$

In fact, it follows from the definition of $\mathfrak{M}$ that for $\rho \in \mathfrak{M}$, (h, ρ) is the limit in probability (under $\mathbb{P}$ but therefore also under Q by absolute continuity) of (h, ρ_n) as $n \to \infty$, for some appropriate sequence ρ_n with $\rho_n \in \mathcal{D}_0(D)$ for all n. To characterise the law of h under Q, it therefore suffices to show (3.34) for $\rho \in \mathcal{D}_0(D)$.

Let us conclude the proof by showing this. Note that by Fubini's theorem, when $\rho \in \mathcal{D}_0(D)$ the right-hand side of (3.34) is equal to

$$\sigma(D)^{-1} \int_D e^{\frac{1}{2} \operatorname{Var}((h,\rho)) + \gamma \int K(x,y)\rho(dy)} g(x) \sigma(dx)$$

$$= \sigma(D)^{-1} \int_D e^{\frac{1}{2}(\rho,\rho)_K + \gamma K\rho(x)} g(x) \sigma(dx)$$

(recalling the notation $K\rho$ in (3.30)). Furthermore, the left-hand side of (3.34) (using the assumption that $\mathbb{E}(\mathcal{M}^\gamma(h, D)) = \sigma(D)$ and the definition of Q) is equal to

$$\sigma(D)^{-1} \mathbb{E}\left(\int_D e^{(h,\rho)} g(x) \mathcal{M}^\gamma(h, dx)\right).$$

However, using the observation (3.31) and the property (3.32), we have

$$\mathbb{E}\left(\int_D e^{(h,\rho)} g(x) \mathcal{M}^\gamma(h, dx)\right) = \mathbb{E}\left(\int_D e^{\frac{1}{2}(\rho,\rho)_K} g(x) \mathcal{M}^\gamma(h + K\rho, dx)\right)$$

$$= \mathbb{E}\left(\int_D e^{\frac{1}{2}(\rho,\rho)_K} e^{\gamma K\rho(x)} g(x) \mathcal{M}^\gamma(h, dx)\right)$$

$$= \int_D e^{\frac{1}{2}(\rho,\rho)_K + \gamma K\rho(x)} g(x) \sigma(dx), \tag{3.35}$$

where, in the last line, we again used the assumption that $\mathbb{E}(\mathcal{M}^\gamma(h, dx)) = \sigma(dx)$. Dividing by $\sigma(D)$ this is the same as the right-hand side of (3.34), so we get the desired result. $\square$

3.5 Rooted measures and Girsanov lemma for GMC

We now return to the notation $\mathcal{M}$ where we suppress the dependence on γ.

Let h be as in Section 3.2 and let σ be as in (3.6). Closely related to the previous theorem (and in particular the rooted measure appearing in its proof) is a description of the law of h after reweighting by $\mathcal{M}(D)/\sigma(D)$. In the case of the two-dimensional Gaussian free field with Dirichlet boundary conditions, this has already been described in (2.8), which is a consequence of Lemma 2.5. The result in this case is that the law of the field, when biased by the total mass $\mathcal{M}(D)$, can simply be described by first sampling a point z with an appropriate (deterministic) law (corresponding to the appropriate multiple of $\sigma(\mathrm{d}z) = R(z,D)^{\gamma^2/2}\,\mathrm{d}z$), and then adding to h a function of the form $\gamma G_D(z,\cdot)$. Since G_D is nothing but the covariance of the field, it is easy to guess that such a description generalises to the broader Gaussian multiplicative chaos framework. This is indeed what the next theorem shows, which is essentially a reformulation of the work done in Theorem 3.14.

Theorem 3.16 (Girsanov's lemma for GMC) *Let h be as in Section 3.2 and σ as in (3.6), $\mathcal{M}$ the γ-multiplicative chaos measure for h with reference measure σ. Then for any $\rho \in \mathfrak{M}$, and any non-negative Borel function g on D,*

$$\mathbb{E}\left[e^{(h,\rho)}\int_D g(x)\mathcal{M}(\mathrm{d}x)\right] = \int_D \sigma(\mathrm{d}x)g(x)\mathbb{E}\left[e^{(h+\gamma K(x,\cdot),\rho)}\right].$$

Proof We note that this could be proved using the same argument explained in (2.8). However, given Theorem 3.14, it is simpler to proceed as follows. We recall that $\mathcal{M}$ is a Gaussian multiplicative chaos for h in the sense of Shamov (Definition 3.13), as explained in Theorem 3.14. Therefore, it satisfies (3.34), as shown in Theorem 3.14. This implies the result. □

Since $\rho \in \mathfrak{M}$ is arbitrary and the law of (h,ρ) for arbitrary ρ characterises the law of a Gaussian additive process $((h,\rho))_{\rho \in \mathfrak{M}}$ uniquely, or (alternatively) applying the argument used in the proof of Theorem 3.14 since we already know that $\mathcal{M}$ is also a GMC in the sense of Definition 3.13, we deduce:

Corollary 3.17 *Define a law $\mathbb{Q}$ on pairs (x,h) through the formula*

$$\mathbb{Q}(\mathrm{d}x, \mathrm{d}h) = \frac{1}{\sigma(D)}\mathcal{M}_h(\mathrm{d}x)\mathbb{P}(\mathrm{d}h),$$

*where $\mathcal{M}_h$ is the GMC measure almost surely associated with h. Then $\mathbb{Q}$ is indeed a probability measure (the **rooted** or Peyrière measure associated to h) and satisfies:*

3.5 Rooted measures and Girsanov lemma for GMC

1. The marginal law of h is given by $\mathbb{Q}(dh) = (\mathcal{M}_h(D)/\sigma(D))\mathbb{P}(dh)$. (In particular, the law of h under $\mathbb{Q}$ is absolutely continuous with respect to $\mathbb{P}$.)
2. Given h, the point $x \in D$ is sampled according to the normalised GMC measure associated to h, $\mathcal{M}_h(\cdot)/\mathcal{M}_h(D)$.
3. The marginal law of x is $\mathbb{Q}(dx) = \sigma(dx)/\sigma(D) =: \hat{\sigma}(dx)$.
4. Finally, the law of (x, h) under $\mathbb{Q}$ is the same as the law of $(x, h + \gamma K(x, \cdot))$, under $\hat{\sigma}(dx) \otimes \mathbb{P}(dh)$.

Corollary 3.18 *Under $\hat{\sigma} \otimes \mathbb{P}$, $\int_D e^{\gamma^2 K(x,y)} \mathcal{M}_h(dy) < \infty$ almost surely. Furthermore, the law of the measure $e^{\gamma^2 K(x,y)} \mathcal{M}_h(dy)$ under $\hat{\sigma} \otimes \mathbb{P}$ is equal to the $\mathbb{Q}$-law of $\mathcal{M}_h(dy)$.*

Proof of Corollary 3.18 If we show the second point, then the first one follows automatically since $\mathbb{Q}$ is absolutely continuous with respect to $\mathbb{P}$ and thus $\mathcal{M}(D) < \infty$ almost surely under $\mathbb{Q}$. We therefore concentrate on the second point. To this end, we note that since $\mathbb{Q}$ is absolutely continuous with respect to $\hat{\sigma} \otimes \mathbb{P}$, the measure $\mathcal{M}_h$ almost surely does not have any atoms under $\mathbb{Q}$. In particular, it almost surely does not have an atom at the marked point x and therefore,

$$\mathcal{M}_h(D \setminus B_\eta(x)) \uparrow \mathcal{M}_h(D) < \infty \tag{3.36}$$

almost surely (under $\mathbb{Q}$) as $\eta \downarrow 0$. On the other hand, by monotone convergence, as $\eta \to 0$,

$$\int_{D \setminus B_\eta(x)} e^{\gamma^2 K(x,y)} \mathcal{M}_h(dy) \uparrow \int_D e^{\gamma^2 K(x,y)} \mathcal{M}_h(dy).$$

Therefore it suffices to show that for each $\eta > 0$, the law of $\mathcal{M}_h$ restricted to $D \setminus B_\eta(x)$ under $\mathbb{Q}$ coincides with the law of $e^{\gamma^2 K(x,y)} \mathcal{M}_h(dy)$ restricted to $D \setminus B_\eta(x)^c$ under $\hat{\sigma} \otimes \mathbb{P}$.

For this, we first note that $K(x, \cdot)$ is bounded and continuous on $\overline{D} \setminus B_\eta(x)$. Let $\mathcal{M}_\varepsilon = \mathcal{M}_{h,\varepsilon}$ denote the approximation to the Gaussian multiplicative chaos measure in Theorem 3.2 associated to the field h. Since $\mathcal{M}_{\varepsilon,h} \Rightarrow \mathcal{M}_h$ in probability as $\varepsilon \to 0$ (where $\Rightarrow$ denotes weak convergence), we deduce immediately that

$$e^{\gamma^2 K(x,y)} \mathcal{M}_{\varepsilon,h}(dy) \Rightarrow e^{\gamma^2 K(x,y)} \mathcal{M}_h(dy) \text{ on } D \setminus B_\eta(x) \tag{3.37}$$

in probability (under $\hat{\sigma} \otimes \mathbb{P}$). On the other hand, when $\varepsilon > 0$ is fixed, it is obvious that

$$\mathcal{M}_{h+\gamma K(x,\cdot),\varepsilon}(dy) = e^{\gamma^2 K_\varepsilon(x,y)} \mathcal{M}_{\varepsilon,h}(dy),$$

where $K_\varepsilon(x, \cdot) = K(x, \cdot) * \theta_\varepsilon$ is the ε-regularisation of $K(x, \cdot)$. By the triangle

96   *Gaussian Multiplicative Chaos*

inequality, and the uniform convergence of $K_\varepsilon(x,\cdot)$ to $K(x,\cdot)$ on $\overline{D} \setminus B_\eta(x)$, and the fact that $\mathbb{E}(\mathcal{M}_h(\mathrm{d}y)) = \sigma(\mathrm{d}y)$ under $\mathbb{P}$, we deduce from (3.37) that

$$\mathcal{M}_{h+\gamma K(x,\cdot),\varepsilon} \Rightarrow e^{\gamma^2 K(x,y)} \mathcal{M}_h(\mathrm{d}y) \text{ on } D \setminus B_\eta(x)$$

weakly in probability under $\hat{\sigma} \otimes \mathbb{P}$. Using Corollary 3.17 (item 4.), the left-hand side however has the same law as $\mathcal{M}_{h,\varepsilon}$ under $\mathbb{Q}$. Taking limits in distribution as $\varepsilon \to 0$, we deduce that the restriction of $\mathcal{M}_h$ to $D \setminus B_\eta(x)$ under $\mathbb{Q}$ has the same law as the restriction of $e^{\gamma^2 K(x,y)} \mathcal{M}_h(\mathrm{d}y)$ to $D \setminus B_\eta(x)$ under $\hat{\sigma} \otimes \mathbb{P}$, as desired. □

As will be illustrated below, Girsanov's theorem (either Theorem 3.16, Corollary 3.17 or Corollary 3.18) is the basis of many calculations for GMC.

3.6 Kahane's convexity inequality

We now present a fundamental tool in the study of Gaussian multiplicative chaos, which is Kahane's convexity inequality. Essentially, this is an inequality that will allow us to "compare" the GMC measures associated with two slightly different fields. Such comparison arguments are very useful in order to do scaling arguments and so compute moments and multifractal spectra, which is our next goal. This inequality was actually crucial to Kahane's construction of Gaussian multiplicative chaos [Kah85], although modern approaches such as the one presented just above (coming from [Ber17]) do not rely on this.

More precisely, the content of Kahane's inequality is to say that given a **convex** function f, and two centred Gaussian fields $X = (X_s)_{s \in T}$ and $Y = (Y_s)_{s \in T}$ with covariances Σ_X and Σ_Y such that $\Sigma_X(s,t) \leq \Sigma_Y(s,t)$ pointwise, we have

$$\mathbb{E}(f(\mathcal{M}_X(D))) \leq \mathbb{E}(f(\mathcal{M}_Y(D)))$$

for $\mathcal{M}_X, \mathcal{M}_Y$ the GMC measures associated with X and Y. The precise statement of the inequality comes in different flavours depending on what one is willing to assume about f and the fields. A statement first appeared in [Kah86], which had an elegant proof but relied on the extra assumption that f is increasing. As we will see, this assumption is crucially violated for us (for example, in the proof of Theorem 3.27, we will use $f(x) = -x^q$ with $q < 1$, so f is convex but decreasing). The assumption of increasing f is removed in [Kah85], whose proof we will follow roughly here.

Theorem 3.19 (Kahane's convexity inequality) *Suppose that $D \subset \mathbb{R}^d$ is*

3.6 Kahane's convexity inequality

bounded and that $(X(x))_{x \in D}$, $(Y(x))_{x \in D}$ *are almost surely continuous centred Gaussian fields with*

$$K_X(x, y) := \mathbb{E}(X(x)X(y)) \le \mathbb{E}(Y(x)Y(y)) =: K_Y(x, y) \text{ for all } x, y \in D.$$

Assume that $f : (0, \infty) \to \mathbb{R}$ *is convex with at most polynomial growth at 0 and* ∞, *and* σ *is a Radon measure as in* (3.6). *Then*

$$\mathbb{E}\left(f\left(\int_D e^{X(x) - \frac{1}{2}\mathbb{E}(X(x)^2)} \sigma(\mathrm{d}x)\right)\right) \le \mathbb{E}\left(f\left(\int_D e^{Y(x) - \frac{1}{2}\mathbb{E}(Y(x)^2)} \sigma(\mathrm{d}x)\right)\right).$$

Proof The proof is closely related to a Gaussian Integration by Parts formula (see for example [Zei15]). Define, for $t \in [0, 1]$:

$$Z_t = \sqrt{1-t}X + \sqrt{t}Y.$$

Thus $Z_0 = X$ and $Z_1 = Y$. Since the fields X and Y are assumed to be continuous, the maxima and minima of X and Y on D have sub-Gaussian tails by Borell's inequality (see for example [Zei15, Theorem 2]). This means that if f is as in the statement of theorem, we have that

$$h(t) := \mathbb{E}\left(f\left(\int_D Q_t(x) \, \sigma(\mathrm{d}x)\right)\right) := \mathbb{E}\left(f\left(\int_D e^{Z_t(x) - \frac{1}{2}\mathbb{E}((Z_t(x))^2)} \sigma(\mathrm{d}x)\right)\right)$$

is well defined for all $t \in [0, 1]$.

Suppose first that f is smooth. This means we can actually differentiate the above expression and obtain that

$$\frac{\mathrm{d}h}{\mathrm{d}t} = \frac{1}{2}\mathbb{E}\left(f'\left(\int_D Q_t(x) \, \sigma(\mathrm{d}x)\right) \times \right.$$
$$\left. \int_D \sigma(\mathrm{d}y) \left(\frac{-X(y)}{\sqrt{1-t}} + \frac{Y(y)}{\sqrt{t}} + K_X(y,y) - K_Y(y,y)\right) Q_t(y)\right).$$

Here we have "differentiated under the integral sign" twice (once for the derivative of the integral $\int_D Q_t(x)\sigma(\mathrm{d}x)$ and once for the expectation) which is permitted since $\int_D Q_t(x)\sigma(\mathrm{d}x)$ has sub-Gaussian tails and f has at most polynomial growth at 0 and ∞ (since f' is increasing this means that f' also has at most polynomial growth at 0 and ∞).

Consequently, by Fubini's theorem, it suffices to show that for any fixed y :

$$\mathbb{E}\left(\left(\frac{-X(y)}{\sqrt{1-t}} + \frac{Y(y)}{\sqrt{t}} + K_X(y,y) - K_Y(y,y)\right) \times \right.$$
$$\left. Q_t(y) f'\left(\int_D Q_t(x) \, \sigma(\mathrm{d}x)\right)\right) \ge 0. \tag{3.38}$$

Indeed, this then implies that h is increasing and so

$$h(0) = \mathbb{E}\left(f\left(\int_D e^{X(x)-(1/2)\mathbb{E}(X(x)^2)}\sigma(\mathrm{d}x)\right)\right)$$

$$\leq h(1) = \mathbb{E}\left(f\left(\int_D e^{Y(x)-(1/2)\mathbb{E}(Y(x)^2)}\sigma(\mathrm{d}x)\right)\right),$$

as desired.

To show (3.38), we fix y and write

$$U_t(y) := \frac{-X(y)}{\sqrt{1-t}} + \frac{Y(y)}{\sqrt{t}},$$

so that $U_t(y)$ is the time derivative of the interpolation $Z_t(y)$. Note that $\mathbb{E}(U_t(y)Z_t(x)) = K_Y(x,y) - K_X(x,y) \geq 0$ for all x. This means that we can decompose

$$Z_t(x) = A_t(x)U_t(y) + V_t(x)$$

for each $x \in D$, where $A_t(x) = (K_Y(x,y) - K_X(x,y))/\mathbb{E}(U_t(y)^2) \geq 0$ and $V_t(x)$ is centred, Gaussian and independent of $U_t(y)$. This corresponds to writing the conditional law of $Z_t(x)$ given $U_t(y)$. Let us rewrite the expectation in (3.38) in terms of $U_t(y)$ and $V_t(y)$. To start with, we decompose

$$Q_t(x) = e^{A_t(x)U_t(y) - \frac{1}{2}A_t(x)^2 \mathbb{E}(U_t(y)^2)} e^{V_t(x) - \frac{1}{2}\mathbb{E}(V_t(x)^2)} \qquad (3.39)$$

for each $x \in D$. Thus applying (3.39) with $x = y$, the expectation in (3.38) can be rewritten as

$$\mathbb{E}\left(\left(U_t(y) - A_t(y)\mathbb{E}(U_t(y)^2)\right) e^{A_t(y)U_t(y) - \frac{1}{2}A_t(y)^2 \mathbb{E}(U_t(y)^2)} \right.$$

$$\left. e^{V_t(y) - \frac{1}{2}\mathbb{E}(V_t(y)^2)} f'\left(\int_D Q_t(x)\sigma(\mathrm{d}x)\right)\right).$$

Now, in order to write this, an expectation involving the single Gaussian random variable $U_t(y)$, we consider the conditional expectation (now expanding $Q_t(x)$ as in (3.39) for clarity):

$$\mathbb{E}\left(e^{V_t(y) - \frac{1}{2}\mathbb{E}(V_t(y)^2)} \times \right.$$

$$\left. f'\left(\int_D e^{A_t(x)U_t(y) - \frac{1}{2}A_t(x)^2 \mathbb{E}(U_t(y)^2)} e^{V_t(x) - \frac{1}{2}\mathbb{E}(V_t(x)^2)} \sigma(\mathrm{d}x)\right) \,\bigg|\, U_t(y)\right).$$

Since $U_t(y)$ is independent of $V_t(x)$ for each $x \in D$ (and thus, by Gaussianity, of $(V_t(x), x \in D)$), and since $A_t(x) \geq 0$ and f' is increasing, we see that the above conditional expectation is an almost surely increasing function of $U_t(y)$. Hence (3.38) can be written as

$$\mathbb{E}\left(g(U_t(y))\left(U_t(y) - A_t(y)\mathbb{E}(U_t(y)^2)\right) e^{A_t(y)U_t(y) - \frac{1}{2}A_t(y)^2 \mathbb{E}(U_t(y)^2)}\right),$$

where g is an increasing function. Approximating g by a positive linear combination of step functions and writing $a = A_t(y)$, $\sigma^2 = \mathbb{E}(U_t(y)^2)$ it therefore suffices to prove that

$$\int_x^\infty e^{-z^2/2\sigma^2}(z-a\sigma^2)e^{az-\frac{a^2\sigma^2}{2}}\,dz \geq 0$$

for any $x \in \mathbb{R}$.

If $x \geq a\sigma^2$ then the above clearly holds by positivity of the integrand. On the other hand, if $x \leq a\sigma^2$ then the integral is greater than

$$\int_{-\infty}^\infty e^{-z^2/2\sigma^2}(z-a\sigma^2)e^{az-\frac{a^2\sigma^2}{2}}\,dz = \frac{d}{da}\int_{-\infty}^\infty e^{-z^2/2\sigma^2}e^{az-\frac{a^2\sigma^2}{2}} = \frac{d}{da}(1) = 0.$$

This concludes the proof when f is smooth.

In the general case of a convex function f, we approximate f by smooth convex functions $f_n \to f$ pointwise with (uniform) polynomial growth at zero and infinity, and then apply dominated convergence, using again the fact that $\sup_x X(x)$ has Gaussian tails; such approximations are easily obtained by approximating the weak derivative of f (a measure) by smooth functions via convolution. □

3.7 Scale-invariant fields

When we apply Kahane's convexity inequality, we will want to compare our Gaussian field with an auxiliary Gaussian field enjoying an exact scaling relation. In this section, we explain a modification, due to Rhodes and Vargas ([RV10a]) of a construction due to Bacry and Muzy ([BM03]) that will give us the desired scale-invariant field. (In the case of the two-dimensional GFF, the Markov property gives a close analogue but would lead to extra technicalities.) The main result of this section is Theorem 3.23, which does not seem to appear in the literature in this form.

3.7.1 One-dimensional cone construction

We first explain the construction we will use in one dimension where things are easier. Fix $0 < \varepsilon < R$ and for $x \in \mathbb{R}$, consider the **truncated cone** $C_{\varepsilon,R}(x)$ in $\mathbb{R}^2$ given by

$$C(x) = C_R(x) = \{z = (y,t) \in \mathbb{R} \times [0,\infty) : |y-x| \leq (t \wedge R)/2\}, \quad (3.40)$$

100 Gaussian Multiplicative Chaos

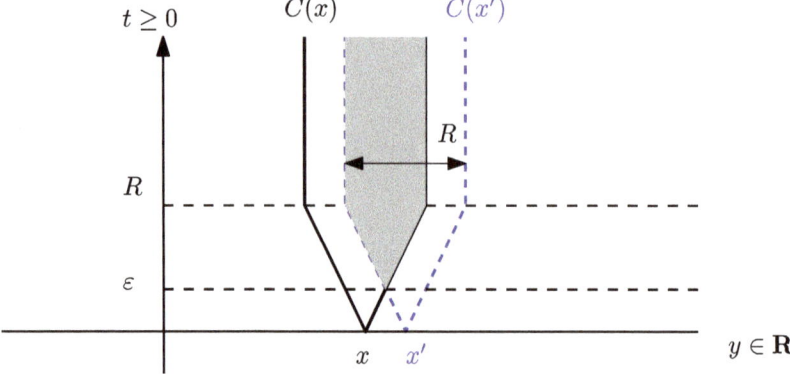

Figure 3.1 The truncated cones in the construction of the scale-invariant auxiliary field. The covariance of the field at (x, x') is obtained by integrating $\mathrm{d}y\,\mathrm{d}t/t^2$ in the shaded area.

where $|y - x|$ denotes Euclidean norm in $\mathbb{R}$ (see Figure 3.1). Define a kernel

$$c_{\varepsilon,R}(x,x') = \int_{y \in \mathbb{R}} \int_{t \in [\varepsilon,\infty)} \mathbf{1}_{\{(y,t) \in C(x) \cap C(x')\}} \frac{\mathrm{d}y\,\mathrm{d}t}{t^2}.$$

Note that since the domain of integration has been truncated at $t = \varepsilon$, the integral is finite. We claim that $c_{\varepsilon,R}$ is non-negative definite and so can be used to define a Gaussian field $X_{\varepsilon,R}$ on $\mathbb{R}$ whose covariance is given by $c_{\varepsilon,R}$. Indeed, for any $n \geq 1$, for any $x_1, \ldots, x_n \in \mathbb{R}^d$ and $\lambda_1, \ldots, \lambda_n \in \mathbb{R}$,

$$\sum_{i,j=1}^{n} \lambda_i \lambda_j c_{\varepsilon,R}(x_i, x_j) = \sum_{i,j=1}^{n} \lambda_i \lambda_j \int_{\mathbb{R}} \int_{\varepsilon}^{\infty} \mathbf{1}_{\{(y,t) \in C(x_i)\}} \mathbf{1}_{\{(y,t) \in C(x_j)\}} \frac{\mathrm{d}y\,\mathrm{d}t}{t^2}$$

$$= \int_{\mathbb{R}^d} \int_{\varepsilon}^{\infty} \left(\sum_{i=1}^{n} \lambda_i \mathbf{1}_{\{(y,t) \in C(x_i)\}} \right)^2 \frac{\mathrm{d}y\,\mathrm{d}t}{t^2} \geq 0. \quad (3.41)$$

As the covariance kernel $c_{\varepsilon,R}$ is a nice continuous function of x, x', we can check (again using for example Proposition 2.1.12 in [GN16]) that there exists a centred Gaussian field $X_{\varepsilon,R}$ whose covariance is given by $c_{\varepsilon,R}$ and which is almost surely Borel measurable as a function on $\mathbb{R}$.

Remark 3.20 This computation showing that $c_{\varepsilon,R}$ is non-negative definite works because the covariance is defined to be of the form $c(x, x') = \int_S f_x(z) f_{x'}(z) \nu(\mathrm{d}z)$, for some fixed function f_x of z associated to each $x \in \mathbb{R}$, where the integral can be on some arbitrary space S with measure ν. Here the space S is $\mathbb{R} \times (0, \infty)$, $\nu(dz) = \mathbf{1}_{\{t \geq \varepsilon\}} dy dt/t^2$, and $f_x(z) = \mathbf{1}_{z \in C_{\varepsilon,R}(x)}$. We will use other choices when considering the higher-dimensional case.

The key property of $X_{\varepsilon,R}$ (and the reason for introducing it) is that its covariance can be computed exactly. This not only shows that the field is logarithmically correlated, but enjoys an exact scaling relation, as follows.

Lemma 3.21 *Define the function*

$$g_{\varepsilon,R}(x) = \begin{cases} \log_+(R/|x|) & \text{if } |x| \geq \varepsilon \\ \log(R/\varepsilon) + 1 - (|x|/\varepsilon) & \text{if } |x| \leq \varepsilon, \end{cases} \quad (3.42)$$

where $\log_+(x) = \log(x) \vee 0$. Then for all $x, y \in \mathbb{R}$,

$$c_{\varepsilon,R}(x, x') = g_{\varepsilon,R}(x - x'). \quad (3.43)$$

In particular, $c_{\varepsilon,R}(x, x') = \log(R/(|x - x'| \vee \varepsilon)) + O(1)$, where the $O(1)$ term does not depend on x, x', ε or R (and is in fact bounded between 0 and 1).

Proof By translation invariance and symmetry we can assume that $x' = 0$ and $x > 0$. If $x \geq R$, there is nothing to prove, so assume first that $\varepsilon \leq x \leq R$. Then the two cones first intersect at height $x \geq \varepsilon$. Moreover, the width of the intersection of these cones at height $t \geq x$ is $(t - x) \wedge (R - x)$, so

$$\begin{aligned} c_{\varepsilon,R}(0, x) &= \int_x^R (t - x) \frac{dt}{t^2} + \int_R^\infty (R - x) \frac{dt}{t^2} \\ &= \log(R/x) - x\left(\frac{1}{x} + \frac{1}{R}\right) + \frac{(R-x)}{R} \\ &= \log(R/x) \end{aligned}$$

as desired. When $x \leq \varepsilon$, the computation is almost the same, but the lower bound of integration for the first integral is ε instead of x, which gives the desired result. $\square$

We now explain why this implies a scaling property. We fix the value R of truncation and write X_ε for $X_{\varepsilon,R}$ (often we will choose $R = 1$ and write Y_ε for $X_{\varepsilon,1}$).

Corollary 3.22 *For $\lambda < 1$,*

$$(X_{\lambda\varepsilon}(\lambda x))_{x \in B(0,R/2)} =_d (\Omega_\lambda + X_\varepsilon(x))_{x \in B(0,R/2)},$$

where Ω_λ is an independent centred Gaussian random variable with variance $\log(1/\lambda)$.

102   Gaussian Multiplicative Chaos

Proof One directly checks that for all $x, x' \in \mathbb{R}$ such that $|x - x'| \leq R$ (and so also $|x - x'| \leq R/\lambda$ automatically),

$$c_{\lambda\varepsilon, R}(\lambda x, \lambda y) = c_{\varepsilon, R}(x, y) + \log(1/\lambda)$$

and hence the result follows. □

3.7.2 Higher-dimensional construction

The one-dimensional Bacry–Muzy construction presented above is beautiful and simple but does not trivially generalise to more than one dimension. This is because if one considers truncated cones in $\mathbb{R}^{d+1}$ (instead of $\mathbb{R}^2$) and integrates with respect to the scale-invariant measure $dy\, dt/t^{d+1}$, the volume of the intersection of two truncated cones based at x and x' does not lead to nice formulae which yield scale invariance in the sense of Corollary 3.22 (see [Cha06] for an article where this model is nevertheless studied).

To overcome this problem, we follow (in a slightly simplified setting) a very nice construction proposed by Rhodes and Vargas [RV10a] in which the exact one-dimensional computation of Bacry and Muzy can be exploited to give a field in any number of dimensions satisfying both logarithmic correlations and exact scaling relations. The basic idea is to define the cones on $\mathbb{R}^d$ based at x and $x' \in \mathbb{R}^d$ by first applying a random rotation in order to preserve isotropy, and then applying the one-dimensional construction to the first coordinates of x and x'.

To be more precise, let $d \geq 1$ and consider $\mathcal{R}$ the orthogonal group of $\mathbb{R}^d$; that is, d-dimensional matrices M such that $MM^t = I$. Let Σ denote Haar measure on $\mathcal{R}$ normalised to be a probability distribution. If $\rho \in \mathcal{R}, x \in \mathbb{R}^d$, let ρx denote the vector of $\mathbb{R}^d$ obtained by applying the isometry ρ to x, and let $(\rho x)_1$ denote its first coordinate. Define the **cone-like** set $\mathbf{C}_R(x)$ as follows:

$$\mathbf{C}_R(x) := \{(\rho, t, y) \in \mathcal{R} \times \mathbb{R} \times (0, \infty) : (t, y) \in C_R((\rho x)_1)\},$$

where if $z \in \mathbb{R}$, $C_R(z)$ is the truncated cone of (3.40). Thus for any given ρ, we first apply ρ to x and consider the truncated cone (in two dimensions) based on the first coordinate of ρx. As in Section 3.7.1, we define a field through its covariance kernel

$$\mathbf{c}_{\varepsilon, R}(x, x') = \int_{\mathcal{R} \times \mathbb{R} \times (0, \infty)} \mathbf{1}_{\{(\rho, t, y) \in \mathbf{C}_R(x) \cap \mathbf{C}_R(x')\}} \mathbf{1}_{\{t \geq \varepsilon\}} \, d\Sigma(\rho) \otimes \frac{dy\, dt}{t^2}.$$

We note that this is non-negative definite for the same reasons as (3.41) (see especially Remark 3.20). Hence, as before we can consider an almost surely Borel measurable function $x \in \mathbb{R}^d \mapsto X_{\varepsilon, R}(x) \in \mathbb{R}$ which is a translation-invariant, centred Gaussian field with $\mathbf{c}_{\varepsilon, R}$ as its covariance kernel.

3.7 Scale-invariant fields

Theorem 3.23 *Fix any $R > 0$. The restriction of the field $(X_{\varepsilon,R})_{x \in \mathbb{R}^d}$ to $x \in B(0, R/2)$, viewed as a function of $\varepsilon > 0$ and $x \in B(0, R/2)$, is scale-invariant in the following sense: for any $\lambda < 1$,*

$$(X_{\lambda\varepsilon,R}(\lambda x))_{x \in B(0,R/2)} =_d (\Omega_\lambda + X_{\varepsilon,R}(x))_{x \in B(0,R/2)}, \qquad (3.44)$$

where Ω_λ is an independent centred Gaussian random variable with variance $\log(1/\lambda)$. Furthermore, its covariance function $\mathbf{c}_{\varepsilon,R}$ satisfies

$$\mathbf{c}_{\varepsilon,R}(x, x') = \log\left(\frac{1}{\|x - x'\| \vee \varepsilon}\right) + O(1), \qquad (3.45)$$

uniformly over $x, x' \in B(0, R/2)$, where the implicit constant $O(1)$ above depends only on the dimension $d \geq 1$.

Proof We start by noticing that we have the following exact expression for the covariance. Recall the function $g_{\varepsilon,R}(t)$ for $t \in \mathbb{R}$ from Lemma 3.21:

$$g_{\varepsilon,R}(t) = \begin{cases} \log_+(R/|t|) & \text{if } |t| \geq \varepsilon \\ \log(R/\varepsilon) + 1 - (|t|/\varepsilon) & \text{if } |t| \leq \varepsilon. \end{cases}$$

Using Fubini's theorem, and since $g_{\varepsilon,R}$ gives the covariance in the one-dimensional case (Lemma 3.21), we have:

$$\mathbf{c}_{\varepsilon,R}(x, x') = \int_{\rho \in \mathcal{R}} g_{\varepsilon,R}((\rho x)_1 - (\rho x')_1) \, d\Sigma(\rho). \qquad (3.46)$$

The scale invariance in (3.44) then follows easily. Indeed, if $x, x' \in \mathbb{R}^d$ are such that $|x - x'| \leq R$ (and so also $|x - x'| \leq R/\lambda$ automatically), then note that

$$g_{\lambda\varepsilon,R}(\lambda x - \lambda y) = g_{\varepsilon,R}(x - y) + \log(1/\lambda)$$

which, as already noticed in Corollary 3.22, immediately implies (3.44).

Let us now turn to the proof of (3.45). Since $g_{\varepsilon,R}(t) = \log_+(R/(|t| \vee \varepsilon)) + O(1)$, we have

$$\mathbf{c}_{\varepsilon,R}(x, x') = \int_{\mathcal{R}} \log\left(\frac{R}{|\rho(x - x')_1| \vee \varepsilon}\right) d\Sigma(\rho) + O(1). \qquad (3.47)$$

Since $|\rho(x - x')_1| \leq |\rho(x - x')| = \|x - x'\|$ for any $\rho \in \mathcal{R}$, we immediately get the lower bound

$$\mathbf{c}_{\varepsilon,R}(x, x') \geq \log\left(\frac{R}{\|x - x'\| \vee \varepsilon}\right) + O(1). \qquad (3.48)$$

To get a bound in the other direction, we observe that for a fixed vector $u \in B(0, R/2)$, the distribution of ρu under the Haar measure $d\Sigma(\rho)$ is uniform on the sphere of radius $\|u\|$. Its first coordinate $(\rho u)_1$ therefore has an absolutely continuous distribution with respect to $(1/\|u\|)$ times (one dimensional) Lebesgue measure. As a consequence, if

$$\mathcal{R}_k(u) = \{\rho \in \mathcal{R} \colon |(\rho u)_1| \in [2^{-(k+1)}\|u\|, 2^{-k}\|u\|]\}$$

then

$$\Sigma(\mathcal{R}_k(u)) \leq O(2^{-k}), \tag{3.49}$$

where the implicit constant depends only on the dimension $d \geq 1$. We note that the right-hand side does not depend on u since the quantity on the left-hand side is clearly scale (and rotation) invariant. Therefore, from (3.47) with $x - x' = u$, and since $\varepsilon \geq 2^{-k-1}\varepsilon$,

$$\mathbf{c}_{\varepsilon,R}(x, x') = O(1) + \sum_{k \geq 0} \int_{\mathcal{R}_k(u)} \log\left(\frac{R}{|(\rho u)_1| \vee \varepsilon}\right) d\Sigma(\rho)$$

$$\leq O(1) + \sum_{k \geq 0} \int_{\mathcal{R}_k(u)} \log\left(\frac{R}{\|2^{-k-1}u\| \vee \varepsilon}\right) d\Sigma(\rho)$$

$$\leq O(1) + \sum_{k \geq 0} \int_{\mathcal{R}_k(u)} \log\left(\frac{R}{\|u\| \vee \varepsilon}\right) d\Sigma(\rho) + O(k)\Sigma(\mathcal{R}_k(u))$$

$$\leq O(1) + \log\left(\frac{R}{\|u\| \vee \varepsilon}\right) + \sum_{k \geq 0} O(k 2^{-k}),$$

where we have used (3.49) in the final line. This proves (3.45). □

Remark 3.24 The covariance kernel takes a particularly nice form in a fixed neighbourhood of a given point when $\varepsilon \to 0$. Indeed, note that if $x \in B(0, R)$ and $|(\rho x)_1| \geq \varepsilon$, then writing $x = \|x\|e_x$ where e_x is the unit vector in the direction of x, we have (letting e_1 denote the unit vector in the first coordinate),

$$g_{\varepsilon,R}((\rho x)_1) = \log(R/\langle \rho x, e_1 \rangle) = \log(R/\|x\|) + \log(R/\langle \rho e_x, e_1 \rangle).$$

When we integrate against $d\Sigma$, we can take advantage of rotational symmetry to note that

$$C = \int_{\rho \in \mathcal{R}} \log(R/\langle \rho e_x, e_1 \rangle) \, d\Sigma(\rho)$$

does not in fact depend on x.

Therefore for *any* $x \in B(0, R)$,

$$\lim_{\varepsilon \to 0} \mathbf{c}_{\varepsilon,R}(x, 0) = \log(R/\|x\|) + C.$$

From this it follows that in $B(0, R)$ that the function $x \mapsto \log(R/\|x\|) + C$ is positive definite in $B(0, R)$. We can get rid of the constant C by changing the value of R, and so we deduce that

$$x \mapsto K(x) := \log(R/\|x\|) \text{ is positive definite in a small neighbourhood of } 0,$$

a fact which appears to have been first proved for all dimensions in [RV10a]. Note that the size of this neighbourhood does depend on the dimension d.

Remark 3.25 It was shown in [JSW19] that if the continuous term g from the decomposition (3.1) of K is an element of $H_{\text{loc}}^{d+\varepsilon}(D \times D)$ for some $\varepsilon > 0$, then K is locally positive definite on D. This provides an alternative justification that $K(x) = \log(R/\|x\|)$ is locally positive definite on $\mathbb{R}^d$.

Remark 3.26 By contrast, note that $x \mapsto \tilde{K}(x) = \log_+(R/\|x\|)$ is positive definite in the *whole space* if and only if $d \leq 3$: see Section 5.2 of [RV10b] for a nice proof based on Fourier transform.

When $d = 1$ or $d = 2$ one can also show that $\tilde{K}(x)$ is not only positive definite but of σ-positive type in the sense of Kahane: that is, it is a sum $\tilde{K}(x) = \sum_{n=1}^{\infty} K_n(x)$ where the summands K_n are not only positive definite functions, but also pointwise non-negative ($K_n(x) \geq 0$). When $d = 3$, it is an open question to determine whether $\tilde{K}(x)$ is σ-positive.

3.8 Multifractal spectrum

We now explain how Kahane's convexity inequality can be used to obtain various estimates on the moments of the mass of small balls, and in turn to the multifractal spectrum of Gaussian multiplicative chaos. We take h, θ as in Section 3.2, and we assume that $d = \mathbf{d}$ and the reference measure σ is Lebesgue measure for simplicity.

Theorem 3.27 (Scaling relation for Gaussian multiplicative chaos) *Let $\gamma \in (0, \sqrt{2d})$. Let $B(r)$ be a ball of radius r in the domain D. Then uniformly over all such balls, and for any $q \in \mathbb{R}$ (including $q < 0$) such that $\mathcal{M}_\varepsilon(B(0,1))^q$ is uniformly integrable in ε,*

$$\mathbb{E}(\mathcal{M}(B(r))^q) \asymp r^{(d+\gamma^2/2)q - \gamma^2 q^2/2}, \qquad (3.50)$$

where $a_r \sim b_r$ if $C^{-1} a_r \leq b_r \leq C a_r$ for some constant C depending only on $\sup_{\overline{D}} |g|$, q, and γ. The function

$$\xi(q) = q(d + \gamma^2/2) - \gamma^2 q^2/2 \qquad (3.51)$$

*is called the **multifractal spectrum** of Gaussian multiplicative chaos.*

Remark 3.28 In Section 3.9, we will see that the assumption on q is equivalent to

$$q < \frac{2d}{\gamma^2}.$$

At this stage, we already know it at least for $0 \le q < 1$.

Remark 3.29 **What is a multifractal spectrum?** The above theorem characterises the multifractal spectrum of Gaussian multiplicative chaos. To explain the terminology, it is useful to consider the opposite case of a *monofractal* object. For instance, Brownian motion is a monofractal because its behaviour is (to first order at least) described by a single fractal exponent, $\alpha = 1/2$. One way to say this is to observe that for all q

$$\mathbb{E}(|B_t|^q) \asymp t^{q/2}.$$

(A variety of exponents can however be obtained by considering logarithmic corrections, see for example [MP10]). The monofractality of Brownian motion is thus expressed through the fact that its moments have a power law behaviour, where the exponent is *linear* in the order of the moment q. By contrast, note that the function ξ in Theorem 3.27 is **non-linear**, which is indicative of multifractal behaviour. That is, several fractal exponents (in fact, a whole spectrum of exponents) are needed to characterise the first-order behaviour of Gaussian multiplicative chaos. Roughly speaking, the multifractal formalism developed among others in [Fal14] is what allows the data of a non-linear function such as the right-hand side of (3.50) to be translated into a knowledge about the various fractal exponents and their relative importance.

Proof of Theorem 3.27 Set $R = 1$ and let $Y_\varepsilon = X_{\varepsilon,1}$ denote the scale-invariant field constructed in Theorem 3.23. As hinted previously, the idea will be to compare h_ε to the scale-invariant field Y_ε. Note that by the estimate (3.45) in Theorem 3.23 on the one hand, and (3.13) on the other hand, there exist constants $a, b > 0$ such that

$$\mathbf{c}_\varepsilon(x, y) - a \le \mathbb{E}(h_\varepsilon(x) h_\varepsilon(y)) \le \mathbf{c}_\varepsilon(x, y) + b, \qquad (3.52)$$

where $\mathbf{c}_\varepsilon = \mathbf{c}_{\varepsilon,1}$ is the covariance function for Y. As a result, it will be possible to estimate the moments of $\mathcal{M}(B(r))$ up to constants by computing those of $\tilde{\mathcal{M}}(B(0,r))$, where $\tilde{\mathcal{M}}$ is the chaos measure associated to Y. More precisely, from (3.52) and Kahane's convexity inequality (applied to the fields h_ε and $Y_\varepsilon + \mathcal{N}(0, a)$ in one direction and to the fields Y_ε and $h_\varepsilon + \mathcal{N}(0, b)$ in the other direction, with the function f taken to be the concave or convex function

$x \mapsto x^q$), we get:
$$\mathbb{E}((\mathcal{M}_\varepsilon(S))^q) \asymp \mathbb{E}((\tilde{\mathcal{M}}_\varepsilon(S))^q) \tag{3.53}$$
for $S \subset D$, where the implicit constants depend only on a, b and $q \in \mathbb{R}$, and not on S or ε, and where
$$\tilde{\mathcal{M}}_\varepsilon(z) = \exp(\gamma Y_\varepsilon(z) - (\gamma^2/2)\mathbb{E}(Y_\varepsilon(z)^2))\, dz.$$
It therefore suffices (also making use of the translation invariance of Y) to study the moments of $\tilde{\mathcal{M}}_\varepsilon(B(0,r))$.

We now turn to the proof of (3.50). Note that $\mathbb{E}(Y_\varepsilon(x)^2) = \log(1/\varepsilon) + O(1)$. Fix $\varepsilon > 0$, and $\lambda = r < 1$. Then
$$\tilde{\mathcal{M}}_{r\varepsilon}(B(0,r)) \asymp \int_{B(0,r)} e^{\gamma Y_{r\varepsilon}(z)} (r\varepsilon)^{\gamma^2/2}\, dz$$
$$= r^{d+\gamma^2/2} \int_{B(0,1)} e^{\gamma Y_{r\varepsilon}(rw)} \varepsilon^{\gamma^2/2}\, dw$$
by the change of variables $z = rw$. Hence by Theorem 3.23,
$$\tilde{\mathcal{M}}_{r\varepsilon}(B(0,r)) \asymp r^{d+\gamma^2/2} e^{\gamma \Omega_r} \tilde{\mathcal{M}}'_\varepsilon(B(0,1)), \tag{3.54}$$
where $\tilde{\mathcal{M}}'$ is a copy of $\tilde{\mathcal{M}}$ and Ω_r is an independent $\mathcal{N}(0, \log(1/r))$ random variable. Raising to the power q, taking expectations and using (3.53), we get:
$$\mathbb{E}(\mathcal{M}_{r\varepsilon}(B(r)^q)) \asymp \mathbb{E}(\tilde{\mathcal{M}}_{r\varepsilon}(B(0,r)^q))$$
$$= r^{q(d+\gamma^2/2)} \mathbb{E}(e^{\gamma q \Omega_r}) \mathbb{E}(\tilde{\mathcal{M}}_\varepsilon(B(0,1))^q)$$
$$\asymp r^{\xi(q)} \mathbb{E}(\mathcal{M}_\varepsilon(B(0,1)^q)), \tag{3.55}$$
where
$$\xi(q) = q(d + \gamma^2/2) - \gamma^2 q^2/2$$
is the multifractal spectrum from the theorem statement. Suppose now that q is such that $\mathcal{M}_\varepsilon(B(0,1))^q$ is uniformly integrable. Then
$$\mathbb{E}(\mathcal{M}(B(r))^q) \asymp r^{\xi(q)},$$
as desired. □

3.9 Positive moments of Gaussian multiplicative chaos (Lebesgue case)

We continue our study of GMC initiated above in $\mathbb{R}^d$ with the reference measure σ taken to be the Lebesgue measure, for a logarithmically correlated field h

108 *Gaussian Multiplicative Chaos*

satisfying the general assumptions of Section 3.2. Let $\mathcal{M}$ be the associated GMC measure. The goal of this section will be to prove the following theorem on its moments. (See Section 3.11 for similar results where σ is allowed to be more general than Lebesgue measure).

Theorem 3.30 *Let $S \subset D$ be bounded and open, and suppose that $\sigma(dx) = dx$ is the Lebesgue measure on $\mathbb{R}^d$. Let $\gamma \in (0, 2)$ and $q > 0$. Then $\mathbb{E}(\mathcal{M}(S)^q) < \infty$ if*

$$q < \frac{2d}{\gamma^2}. \tag{3.56}$$

In fact, the theorem shows that $(\mathcal{M}_\varepsilon(S))^q$ is uniformly integrable in ε, so that Theorem 3.27 applies to this range of values of q.

Before starting the proof of this theorem, we note that from Theorem 3.30,

$$\mathbb{P}(\mathcal{M}(S) > t) \le t^{-2d/\gamma^2 + o(1)}; \quad t \to \infty. \tag{3.57}$$

In fact, much more precise information is known about the tail at ∞: a lower bound matching this upper bound can be obtained so that it becomes an equality. In fact, the $o(1)$ term in the exponent can also be removed and a constant identified: in the case of the two-dimensional GFF, this was done by Rhodes and Vargas [RV19], and the universality of this behaviour (including the calculation of the constant itself) was shown subsequently in a paper by Mo-Dick Wong [Won20].

Proof Note that we already know uniform integrability of $\mathcal{M}_\varepsilon(S)$ (Theorem 3.2) so we can assume $q > 1$. For simplicity (and without loss of generality), we assume that S is the unit cube in $\mathbb{R}^d$. Let $\mathcal{S}_m$ denote the mth level dyadic covering of the domain $\mathbb{R}^d$ by cubes $S_i, i \in \mathcal{S}_m$ of sidelength 2^{-m}. Given $q < 2d/\gamma^2$, we define $n = n(q) \ge 2$ such that $n - 1 < q \le n$. We will show by **induction** on n that

$$M_\varepsilon := \mathbb{E}(\mathcal{M}_\varepsilon(S)^q)$$

is uniformly bounded.

Let us consider the case $n = 2$ first. We first subdivide the cubes of $\mathcal{S}_m$ into 2^d disjoint groups so that no two cubes within any given group touch (including at the boundary); thus any two cubes within a given group are at distance at least 2^{-m} from one another. The reader should convince themselves that this is actually possible (it is a generalisation of the usual checkerboard pattern for $\mathbb{Z}^2$). Let $\mathcal{S}'_m$ denote one of these 2^d groups of cubes of sidelength 2^{-m}.

We will now take advantage of some convexity properties, using the fact that

3.9 Positive moments of GMC (Lebesgue case)

$q/2 \leq 1$ (recall that $n = 2$ and $n - 1 < q \leq n$ by definition). We write, for given m,

$$\left(\sum_{i \in S'_m} \mathcal{M}_\varepsilon(S \cap S_i)\right)^q = \left(\sum_{i,j \in S'_m} \mathcal{M}_\varepsilon(S_i)\mathcal{M}_\varepsilon(S_j)\right)^{q/2}$$
$$\leq \sum_{i,j \in S'_m} \mathcal{M}_\varepsilon(S_i)^{q/2} \mathcal{M}_\varepsilon(S_j)^{q/2}, \qquad (3.58)$$

where we have used the elementary fact that $(x + y)^\alpha \leq x^\alpha + y^\alpha$ if $x, y > 0$ and $\alpha \in (0, 1)$. (This is easily proven by writing $(x + y)^\alpha - x^\alpha = \int_x^{x+y} \alpha t^{\alpha-1} \, dt \leq \int_0^y \alpha t^{\alpha-1} \, dt = y^\alpha$, since the integrand $\alpha t^{\alpha-1}$ is decreasing in t).

We consider the on-diagonal and off-diagonal terms in (3.58) separately. We start with the on-diagonal terms (the estimate in this case works for general $q > 0$ so is not restricted to the case $n = 2$):

Lemma 3.31 *Assume the set-up of Theorem 3.30. Then there exists a constant c_q such that for all sufficiently large m, and for all $\varepsilon > 0$,*

$$\mathbb{E}\left(\sum_{i \in S'_m} \mathcal{M}_\varepsilon(S_i)^q\right) \leq c_q 2^{dm - \xi(q)m} \mathbb{E}(\mathcal{M}_{\varepsilon 2^m}(S)^q). \qquad (3.59)$$

Proof By (3.55), applied with $r = 2^{-m}$, we have for each $i \in S'_m$,

$$\mathbb{E}(\mathcal{M}_\varepsilon(S_i)^q) \leq c_q 2^{-\xi(q)m} \mathbb{E}((\mathcal{M}_{\varepsilon 2^m}(S))^q).$$

Since there are at most 2^{dm} terms in this sum, we deduce the lemma. □

For the off-diagonal terms, we simply observe that in the case where the two indices are distinct:

Lemma 3.32 *Assume the set-up of Theorem 3.30. Then for any fixed m and $q < 2$, there exists a constant $C_{m,q}$ independent of ε such that*

$$\mathbb{E}\left(\sum_{i \neq j \in S'_m} \mathcal{M}_\varepsilon(S_i)^{q/2} \mathcal{M}_\varepsilon(S_j)^{q/2}\right) \leq C_{m,q}.$$

Proof Note that by Jensen's inequality (since $q/2 \leq 1$),

$$\mathbb{E}\left(\mathcal{M}_\varepsilon(S_i)^{q/2} \mathcal{M}_\varepsilon(S_j)^{q/2}\right) \leq \mathbb{E}\left(\mathcal{M}_\varepsilon(S_i) \mathcal{M}_\varepsilon(S_j)\right)^{q/2}$$

for all $i \neq j \in S'_m$. The expectation can easily be computed, and we have for some constant c_m,

$$\lim_{\varepsilon \to 0} \mathbb{E}\left(\mathcal{M}_\varepsilon(S_i) \mathcal{M}_\varepsilon(S_j)\right) \leq \int_{x \in S_i, y \in S_j} e^{\gamma^2 K(x,y)} \, dx \, dy \leq c_m < \infty$$

since the squares S_i and S_j are at distance at least 2^{-m} from one another. Taking the qth power and summing over all terms $i \neq j \in S'_m$ gives the lemma. □

We put these two lemmas together as follows. First, note that for $q < 2d/\gamma^2$, $2d - \xi(q) < 0$. We can therefore choose m large enough that $c_q 2^{dm-\xi(q)m} < 1/(2^d)^{q+1}$, where c_q is as in Lemma 3.31. From (3.58), we obtain that

$$\mathbb{E}\left(\left(\sum_{i \in S'_m} \mathcal{M}_\varepsilon(S_i)\right)^q\right) \leq \frac{1}{(2^d)^{q+1}} \mathbb{E}(\mathcal{M}_{2^m\varepsilon}(S)^q) + C_{m,q}, \qquad (3.60)$$

where $C_{m,q}$ comes from Lemma 3.32. Adding the contributions from all 2^d groups (and using the fact that $(x_1 + \cdots + x_{2^d})^q \leq (2^d)^{q-1}(x_1^q + \cdots + x_{2^d}^q)$ by convexity), and bounding each of the 2^d terms by (3.60),

$$\mathbb{E}\left(\left(\sum_{i \in S_m} \mathcal{M}_\varepsilon(S_i)\right)^q\right) \leq \frac{1}{2^d} \mathbb{E}(\mathcal{M}_{2^m\varepsilon}(S)^q) + (2^d)^q C_{m,q}.$$

Therefore, recalling that $M_\varepsilon = \mathbb{E}(\mathcal{M}_\varepsilon(S)^q)$, we have

$$M_\varepsilon \leq \frac{1}{2^d} M_{2^m\varepsilon} + 2^{dq} C_{m,q}.$$

Taking the sup over $\varepsilon > \varepsilon_0$, and since $2^m \varepsilon \geq \varepsilon$, we get

$$\sup_{\varepsilon > \varepsilon_0} M_\varepsilon \leq \frac{1}{2^d} \sup_{\varepsilon > \varepsilon_0} M_\varepsilon + 2^{dq} C_{m,q}$$

and hence

$$\sup_{\varepsilon > \varepsilon_0} M_\varepsilon \leq \frac{2^{dq}}{1 - 1/2^d} C_{m,q}.$$

We conclude the proof for $q \in (1, 2]$ – that is, $n = 2$ – by letting $\varepsilon_0 \to 0$ and Fatou's lemma.

We now consider the general case, which is in fact very similar to when $n = 2$. We use the fact that $q/n \leq 1$ and thus, arguing as in (3.58),

$$\left(\sum_{i \in S'_m} \mathcal{M}_\varepsilon(S \cap S_i)\right)^q \leq \sum_{i_1,\ldots,i_n \in S'_m} \mathcal{M}_\varepsilon(S_{i_1})^{q/n} \cdots \mathcal{M}_\varepsilon(S_{i_n})^{q/n}. \qquad (3.61)$$

As before, we consider the on-diagonal (when all indices are equal) and off-diagonal terms separately. The on-diagonal terms were already estimated in Lemma 3.31, and we have the same upper bound (3.59) for all sufficiently large m and all $\varepsilon > 0$. For the off-diagonal terms, we obtain the following estimate.

3.9 Positive moments of GMC (Lebesgue case)

Lemma 3.33 *Assume the set-up of Theorem 3.30. Then for any fixed m and $q < 2d/\gamma^2$, there exists a constant $C_{m,q}$ independent of ε such that*

$$\mathbb{E}\left(\sum_{i_1,\ldots,i_n \in S'_m\,:\,i_1 \neq i_2} M_\varepsilon(S_{i_1})^{q/n} \cdots M_\varepsilon(S_{i_n})^{q/n}\right) \leq C_{m,q}.$$

Proof Note that by Jensen's inequality (since $q/n \leq 1$), if $i_1 \neq i_2 \in S'_m$,

$$\mathbb{E}\big(M_\varepsilon(S_{i_1})^{q/n} \cdots M_\varepsilon(S_{i_n})^{q/n}\big) \leq \mathbb{E}\big(M_\varepsilon(S_{i_1}) \cdots M_\varepsilon(S_{i_n})\big)^{q/n}.$$

As before, this expectation can be computed exactly. To begin with, we rewrite the index set $\{i_1,\ldots,i_n\}$ in a way that takes into account which indices are equal and which are distinct. Thus let $\{i_1,\ldots,i_n\} = \{j_1,\ldots,j_p\}$ where the j_k are pairwise distinct and $2 \leq p \leq n$ (since at least two indices are distinct). Call m_k the multiplicity of j_k in $\{i_1,\ldots,i_n\}$ – that is, the number of times j_k is present in that set – so that $m_1 + \cdots + m_p = n$ (with $m_k \geq 1$ by assumption). Then

$$\mathbb{E}\big(M_\varepsilon(S_{i_1}) \cdots M_\varepsilon(S_{i_n})\big) = \int_{x_1 \in S_{i_1}} \cdots \int_{x_n \in S_{i_n}} e^{(\gamma^2/2) \sum_{1 \leq k \neq \ell \leq n} K_\varepsilon(x_k, x_\ell)} \, dx_1 \cdots dx_n.$$

When $x_k \in S_{i_k}, x_\ell \in S_{i_\ell}$ and $S_{i_k} \neq S_{i_\ell}$, the term $K(x_k, x_\ell) = -\log|x_k - x_\ell| + O(1)$ is bounded above by a constant c_m since the cubes are separated by a minimum distance of 2^{-m}. Hence

$$\mathbb{E}\big(M_\varepsilon(S_{i_1}) \cdots M_\varepsilon(S_{i_n})\big)$$
$$\leq c'_m \prod_{k=1}^p \int_{S_{j_k}} \cdots \int_{S_{j_k}} e^{(\gamma^2/2) \sum_{1 \leq k \neq \ell \leq m_k} K_\varepsilon(x_k, x_\ell)} \, dx_1 \cdots dx_{m_k}$$
$$= c'_m \prod_{k=1}^p \mathbb{E}\big((M_\varepsilon(S_{j_k}))^{m_k}\big).$$

Now, since $m_k \leq n-1$ (as there are at least two distinct indices in the set $\{i_1,\ldots,i_n\}$), and since $S_{j_k} \subset S$, we have that

$$\mathbb{E}\big((M_\varepsilon(S_{j_k}))^{m_k}\big) \leq \mathbb{E}(M_\varepsilon(S))^{n-1}$$

which, by the induction hypothesis, is uniformly bounded in ε, by a constant depending only on m and q. This concludes the proof of the lemma. □

Putting together (3.59) and Lemma 3.33, we conclude the proof that M_ε is uniformly bounded for arbitrary $q < 2d/\gamma^2$, as in the case $q < 2 \wedge (2d/\gamma^2)$. This finishes the proof of Theorem 3.30. □

We complement Theorem 3.30 with two results. The first one shows that the condition $q < 2d/\gamma^2$ is sharp for the finiteness of the moment of order $q > 0$. The second will show a partial result in the general framework of Gaussian multiplicative chaos with respect to a d-dimensional reference measure σ – that is, satisfying (3.6). We start with the first result.

Proposition 3.34 *Assume the set-up of Theorem 3.30 (in particular, that $\sigma(dx) = dx$ is the Lebesgue measure on $\mathbb{R}^d$). Let $q > 2d/\gamma^2$. Then*

$$\mathbb{E}(\mathcal{M}(S)^q) = \infty.$$

Proof The proof argues by contradiction, and has the same flavour as Theorem 3.30 but is much simpler (essentially, we can ignore the off-diagonal term). Suppose that for some $q > 2d/\gamma^2$, $\mathbb{E}(\mathcal{M}(S)^q) < \infty$. By Kahane's inequality, there is no loss of generality in assuming that the Gaussian field h is in fact an exactly scale-invariant field X satisfying Theorem 3.23. Then, for any cube S_i of sidelength 2^{-m}, by (3.55) (or more precisely (3.54)),

$$\mathbb{E}\big((\mathcal{M}(S_i))^q\big) \asymp 2^{-m\xi(q)}\mathbb{E}\big((\mathcal{M}(S))^q\big).$$

On the other hand, keeping the same notations as in the proof of Theorem 3.30, and since $(x+y)^q \geq x^q + y^q$ for $q > 1$ and $x, y > 0$,

$$\mathcal{M}(S)^q \geq \sum_{i \in S_m} \mathcal{M}(S_i)^q.$$

Hence, taking expectations,

$$\mathbb{E}(\mathcal{M}(S)^q) \gtrsim 2^{dm-\xi(q)m}\mathbb{E}\big((\mathcal{M}(S))^q\big).$$

However, when $q > 2d/\gamma^2$, we have that $d - \xi(q) > 0$. Since m is arbitrary and the implicit constant does not depend on m, we get the desired contradiction. □

3.10 Degeneracy of multiplicative chaos for $\gamma > \sqrt{2d}$

As in the preceding section, we continue to consider the Lebesgue case, so $\sigma(dx) = dx$ and $d = \mathbf{d}$, while the logarithmically correlated field h satisfies the general assumptions of Section 3.2. In this short section, we give a (non-optimal) bound showing that if $\gamma > \sqrt{2d}$ then the measures $\mathcal{M}_\varepsilon$ converge to zero and so degenerate, in contrast to the case $\gamma < \sqrt{2d}$ studied in Theorem 3.2. This shows that the value $\gamma = \sqrt{2d}$ is indeed a natural (sharp) boundary for the validity of that result.

At the value $\gamma = \sqrt{2d}$ itself, it is still possible with additional renormalisation to define a *critical* Gaussian multiplicative chaos which is a measurable function

3.10 Degeneracy of multiplicative chaos for $\gamma > \sqrt{2d}$

of the field h. See [DRSV14a, DRSV14b] as well as [JS17, Pow18] for a construction, and see [Pow21] for a survey. In the case $\gamma > \sqrt{2d}$, even if we normalise more carefully than above, any limit would only be in law rather than in probability or almost surely; and thus will not be a measurable function of the field h. This is partly related to the phenomenon known as *freezing* in the statistical physics literature. See in particular [MRV16, BJRV13] and references therein for a more precise discussion related to supercritical Gaussian multiplicative chaos.

Theorem 3.35 *Let $D \subset \mathbb{R}^d$ be a bounded domain, let $\sigma(\mathrm{d}x) = \mathrm{d}x$ be the Lebesgue measure and let h be a logarithmically correlated field h satisfying the assumptions of Section 3.2. Let $\gamma > \sqrt{2d}$ and let*

$$\mathcal{M}_\varepsilon(\mathrm{d}x) = \exp(\gamma h_\varepsilon(x) - \tfrac{\gamma^2}{2}\mathrm{Var}[h_\varepsilon(x)])\,\mathrm{d}x,$$

as in (3.9). Then for any compactly supported $A \subset D$, $\mathcal{M}_\varepsilon(A) \to 0$ in probability as $\varepsilon \to 0$. More precisely, for any $0 < \alpha < 1$, there exists $C > 0$ depending only on A, D, γ, d and the law of h (but not on ε), such that

$$\mathbb{E}[(\mathcal{M}_\varepsilon(A))^\alpha] \leq C\varepsilon^{\xi(\alpha)-d}, \tag{3.62}$$

where $\xi(\alpha) = \alpha(d + \gamma^2/2) - \gamma^2\alpha^2/2$. In particular, for $\alpha = \sqrt{2d}/\gamma \in (0,1)$ we have for C depending on A, D, γ, d and the law of h (but not on ε), that

$$\mathbb{E}[(\mathcal{M}_\varepsilon(A))^\alpha]^{1/\alpha} \leq C\varepsilon^\psi, \quad \text{where } \psi = \left(\tfrac{\gamma}{\sqrt{2}} - \sqrt{d}\right)^2 > 0. \tag{3.63}$$

In particular, $\mathcal{M}_\varepsilon$ converges weakly to the null measure in probability.

Remark 3.36 The bound we obtain in (3.63) shows that $\mathcal{M}_\varepsilon(A)$ decays polynomially fast to zero, and is close to optimal except for polylogarithmic terms: indeed, Madaule, Rhodes and Vargas showed in [MRV16, Corollary 2.3] that

$$(\log(1/\varepsilon))^{\frac{3\gamma}{2\sqrt{2d}}} \varepsilon^{-(\frac{\gamma}{\sqrt{2}} - \sqrt{d})^2} \mathcal{M}_\varepsilon(A)$$

converges – necessarily in law, as mentioned above – to a non-degenerate limit. The analysis in [MRV16] comes from a fine understanding of the behaviour of the maximum of the field; this connection to the behaviour of the maximum of the regularised field is at the heart of the freezing transition alluded to above. This is however not entirely captured by the rather rough scaling argument which is given below.

Proof By Kahane's inequality (Theorem 3.19), we may without loss of generality assume that $h_\varepsilon = X_{\varepsilon,10}$ (say) is the scale-invariant field defined on all of $\mathbb{R}^d$ constructed in Theorem 3.23. By increasing A slightly, we can then assume that A is a cube, which we may (still without loss of generality) take to be the unit cube. Since $\mathbb{E}[h_\varepsilon(x)^2] = -\log(1/\varepsilon) + O(1)$, there is also no loss of generality in proving the result for the measure $\mathcal{M}_\varepsilon(dx) = \varepsilon^{\gamma^2/2} e^{\gamma h_\varepsilon(x)} dx$ instead of (3.9). (This turns out to have some advantages in the proof below.)

Let m be such that $r = 2^{-m}$ satisfies $2^{-m-1} \le \varepsilon \le 2^{-m}$. We split the cube A into a union of $r^{-d} = 2^{md}$ cubes of sidelength $r = 2^{-m}$. We take these cubes to be open and non-overlapping and call them $C_i, i \in I$. Then (since the Lebesgue measure of the boundary of these cubes equals zero),

$$\mathcal{M}_\varepsilon(A) = \sum_{i \in I} \mathcal{M}_\varepsilon(C_i).$$

As noted in the proof of Theorem 3.30, the function $x > 0 \mapsto x^\alpha$ is subadditive when $\alpha < 1$. Hence we see that

$$\mathcal{M}_\varepsilon(A)^\alpha \le \sum_{i \in I} \mathcal{M}_\varepsilon(C_i)^\alpha.$$

Now, arguing as in (3.55) using the exact scale invariance and translation invariance of h, we see that

$$\mathbb{E}[\mathcal{M}_\varepsilon(C_i)^\alpha] = r^{\xi(\alpha)} \mathbb{E}[\mathcal{M}_{\varepsilon/r}(A)^\alpha],$$

where $\xi(\alpha) = \alpha(d+\gamma^2/2) - \gamma^2 \alpha^2/2$. (Note that this is an equality rather than an identity valid only up to constants, as is the case in (3.55), thanks to our choice of normalisation of $\mathcal{M}_\varepsilon$ which is slightly different from (3.9).) We deduce that

$$\mathbb{E}[\mathcal{M}_\varepsilon(A)^\alpha] \le r^{\xi(\alpha)} r^{-d} \mathbb{E}[\mathcal{M}_{\varepsilon/r}(A)^\alpha]. \tag{3.64}$$

Now, by Jensen's inequality, $\mathbb{E}[\mathcal{M}_{\varepsilon/r}(A)^\alpha] \le \mathbb{E}[\mathcal{M}_{\varepsilon/r}(A)]^\alpha \le C$, so that (since $r \asymp \varepsilon$),

$$\mathbb{E}[\mathcal{M}_\varepsilon(A)^\alpha] \le C \varepsilon^{\xi(\alpha)-d}.$$

As desired, this shows that the left-hand side decays at least polynomially fast when α is such that $\xi(\alpha) - d > 0$ (note that this is true for some range of α since at $\alpha = 1$ the value of $\xi(\alpha) - d$ is $\xi(1) - d = 0$, whereas its derivative is $\xi'(1) = d+\gamma^2/2-\gamma^2 = d-\gamma^2/2 < 0$). The optimal decay for $\mathbb{E}[(\mathcal{M}_\varepsilon(A))^\alpha]^{1/\alpha}$ is found by choosing α to maximise the function $\alpha \in [0,1] \mapsto (\xi(\alpha)-d)/\alpha$. This is achieved when $\alpha = \sqrt{2d}/\gamma$, and then

$$\frac{(\xi(\alpha)-d)}{\alpha} = \left(\frac{\gamma}{\sqrt{2}} - \sqrt{d}\right)^2,$$

so (3.63) follows. □

3.11 Positive moments for general reference measures

We now introduce the second result complementing Theorem 3.30, which is an extension of Theorem 3.30 to the general set-up of Gaussian multiplicative chaos relative to a **d**-dimensional reference measure σ (satisfying (3.6)). In order to not make the exposition too cumbersome, we limit the proof to the case where $q < (2\mathbf{d}/\gamma^2) \wedge 2$ (hence, at least in the L^1 regime where $\gamma \in [\sqrt{\mathbf{d}}, \sqrt{2\mathbf{d}})$, there is no loss of generality at all).

Before doing so, it may be useful to explain where the previous proof breaks down if σ is not Lebesgue measure. The main issue lies in the scaling argument of Lemma 3.31; when we consider a cube of S_i of S'_m (sidelength 2^{-m}), blowing this up by a factor 2^m will of course still produce a cube of unit sidelength, but the Gaussian multiplicative chaos is now with respect to a reference measure which is no longer σ, but instead reflects the local behaviour of σ near the cube S_i. For very inhomogeneous fractals, this behaviour could be wildly different, and so the inequality in that lemma has no reason to hold true.

Instead, we will need a different approach that accounts for the possible inhomogeneities of the fractal supporting the reference measure σ. The proof below comes from work (written roughly in parallel with this part of the book) in [BSS23]. It is based on Girsanov's lemma (Theorem 3.16).

Proposition 3.37 *Let $S \subset D$ be bounded and open, and suppose that the reference measure σ satisfies the dimensionality condition (3.6). Then if $0 < q < 2 \wedge (2\mathbf{d}/\gamma^2)$,*

$$\mathbb{E}(\mathcal{M}(S)^q) < \infty \tag{3.65}$$

and moreover, $\mathcal{M}_\varepsilon(S)$ converges to $\mathcal{M}(S)$ in L^q.

Proof Again, we can assume without loss of generality that $q > 1$. Set $\delta = q - 1 \in (0, 1)$. Write

$$\mathbb{E}(\mathcal{M}(S)^q) = \mathbb{E}(\mathcal{M}(S)\mathcal{M}(S)^\delta) = \mathbb{E}(\mathcal{M}(S))\mathbb{Q}(\mathcal{M}(S)^\delta) = \sigma(S)\mathbb{E}^*(\mathcal{M}(S)^\delta),$$

where $\mathbb{Q}$ denotes the law of the field biased by $\mathcal{M}(S)$. Using Girsanov's lemma for Gaussian multiplicative chaos, Theorem 3.16 (and Corollary 3.18), we can rewrite this as

$$\mathbb{Q}(\mathcal{M}(S)^\delta) = \int_S \sigma(\mathrm{d}x) \mathbb{E}\left(\left(\int_S e^{\gamma^2 K(x,y)} \mathcal{M}(\mathrm{d}y)\right)^\delta\right).$$

Next, for each $n \geq 0$, let $A_n(x)$ denote the annulus at distance between 2^{-n} and 2^{-n-1} from x; that is, $A_n(x) = \{y : |y - x| \in [2^{-n-1}, 2^{-n}]\}$. Then, using the fact that $K(x, y) = -\log|x - y| + O(1)$ and the inequality $(a_1 + \ldots + a_n)^\delta \leq$

$a_1^\delta + \cdots + a_n^\delta$ for $\delta < 1$ and $a_i > 0$, we see that

$$\mathbb{Q}(\mathcal{M}(S)^\delta) \leq C \sum_{n=0}^\infty \int_S \sigma(dx) \mathbb{E}\left(\left(\int_{A_n(x)} e^{-\gamma^2 \log|x-y|} \mathcal{M}(dy)\right)^\delta\right)$$

$$\leq C \sum_{n=0}^\infty 2^{n\gamma^2 \delta} \int_S \sigma(dx) \mathbb{E}(\mathcal{M}(A_n(x))^\delta).$$

Now fix x, and consider a field X which is exactly scale invariant around x as in Section 3.7.2. Hence for any $\lambda < 1$,

$$(X(x + \lambda z))_{z \in S} = (\tilde{X}(z) + \Omega_{-\log \lambda})_{z \in S},$$

where $\tilde{X}$ has the same law as X and Ω_r is a Gaussian with variance r independent of $\tilde{X}$. Write X_ε for the field truncated at level ε, as in Section 3.7.2.

Set $\lambda = 2^{-n} \leq 1$. Denote by $\sigma_{\lambda,x}(dz)$ the image measure of σ under the map $y = x + \lambda z \mapsto z$ (so that the total mass $\sigma_{\lambda,x}(A_1(0)) = \sigma(A_n(x))$). By applying Theorem 3.23 and changing variables $y \mapsto z$, we obtain:

$$\mathbb{E}\left(\left(\int_{A_n(x)} e^{\gamma X_{\lambda\varepsilon}(y)} (\lambda\varepsilon)^{\frac{\gamma^2}{2}} \sigma(dy)\right)^\delta\right)$$

$$\leq \lambda^{\frac{\gamma^2 \delta}{2}} \mathbb{E}\left(\left(\int_{A_1(0)} e^{\gamma X_\varepsilon(x+\lambda z)} \varepsilon^{\frac{\gamma^2}{2}} \sigma_{\lambda,x}(dz)\right)^\delta\right)$$

$$= \lambda^{\frac{\gamma^2 \delta}{2}} \mathbb{E}\left(e^{\delta\gamma\Omega_{-\log\lambda}} \left(\int_{A_1(0)} e^{\gamma \tilde{X}_\varepsilon(z)} \varepsilon^{\frac{\gamma^2}{2}} \sigma_{\lambda,x}(dz)\right)^\delta\right)$$

$$= \lambda^{\frac{\gamma^2 \delta}{2}} e^{-\frac{\delta^2 \gamma^2 \log(\lambda)}{2}} \mathbb{E}\left(\left(\int_{A_1(0)} e^{\gamma \tilde{X}_\varepsilon(z)} \varepsilon^{\frac{\gamma^2}{2}} \sigma_{\lambda,x}(dz)\right)^\delta\right)$$

$$\leq \lambda^{\frac{\gamma^2 \delta}{2} - \frac{\delta^2 \gamma^2}{2}} \mathbb{E}\left(\left(\int_{A_1(0)} e^{\gamma \tilde{X}_\varepsilon(z)} \varepsilon^{\frac{\gamma^2}{2}} \sigma_{\lambda,x}(dz)\right)\right)^\delta,$$

where the last inequality is by Jensen's inequality since $\delta < 1$.

By Kahane's inequality (Theorem 3.19) and comparing to the trivial field, there exists an absolute constant $C > 0$ such that

$$\mathbb{E}(\mathcal{M}_{\varepsilon 2^{-n}}(A_n(x))^\delta) \leq C \lambda^{\gamma^2 \delta/2 - \delta^2 \gamma^2/2} \sigma(A_n(x))^\delta.$$

Letting $\varepsilon \to 0$, we get that for any $n \geq 0$,

$$\mathbb{E}(\mathcal{M}(A_n(x))^\delta) \leq C 2^{-n(\gamma^2 \delta/2 - \delta^2 \gamma^2/2)} \sigma(A_n(x))^\delta.$$

We deduce that

$$\mathbb{Q}(\mathcal{M}(S)^\delta) \leq C \sum_n 2^{n(\gamma^2 \delta/2 + \delta^2 \gamma^2/2)} \int \sigma(dx) \sigma(A_n(x))^\delta.$$

To estimate the last integral, let $\overline{y}$ and $\overline{x}$ be two independent points distributed according to $\sigma(\cdot \cap S)/\sigma(S)$. Then note that

$$\sigma(A_n(x)) \le \sigma(S)\mathbb{P}(|\overline{y} - x| \le 2^{-n})$$

so that by Jensen's inequality again (as $\delta < 1$),

$$\int_S \sigma(\mathrm{d}x)\sigma(A_n(x))^\delta \le \int_S \sigma(\mathrm{d}x)\sigma(S)^\delta \mathbb{P}(|\overline{y} - \overline{x}| \le 2^{-n} \mid \overline{x} = x)^\delta$$
$$\le \sigma(S)^{\delta+1}\mathbb{E}(\mathbb{P}(|\overline{x} - \overline{y}| < 2^{-n} \mid \overline{x})^\delta)$$
$$\le \sigma(S)^{\delta+1}\mathbb{P}(|\overline{x} - \overline{y}| \le 2^{-n})^\delta$$
$$\le \sigma(S)^{\delta+1}\mathbb{E}(|\overline{x} - \overline{y}|^{-\mathbf{d}})^\delta 2^{-n\mathbf{d}\delta}.$$

by Markov's inequality. Now $\delta < 2\mathbf{d}/\gamma^2 - 1$ implies that

$$\gamma^2\delta/2 + \delta^2\gamma^2/2 - \mathbf{d}\delta = \delta(\gamma^2/2 - \mathbf{d} + \gamma^2\delta/2) < 0.$$

Putting everything together, we can find $c = c(\mathbf{d}, \gamma, \delta)$ such that

$$\mathbb{E}(\mathcal{M}(S)^q) = \sigma(S)\mathbb{Q}(\mathcal{M}(S)^\delta) \le c(\delta)\mathbb{E}(|\overline{x} - \overline{y}|^{-\mathbf{d}})^\delta < \infty,$$

by (3.7). This concludes the proof. $\square$

3.12 Negative moments of Gaussian multiplicative chaos

We now turn our attention to negative moments of the chaos measures. We will first show in Proposition 3.40 that in the general set-up, $\mathcal{M}(S)$ admits moments of order $q \in [q_0, 0]$ for some $q_0 < 0$. This proof is based on a similar argument appearing in [GHSS18]. We will then explain how to bootstrap this to get existence of all negative moments (see Theorem 3.42). Note that, in particular, this implies strict positivity of the measures with probability 1.

We work in the general setting: σ is a Radon measure with dimension at least $\mathbf{d}$ ($0 < \mathbf{d} \le d$) and $0 \le \gamma < \sqrt{2\mathbf{d}}$. The first ingredient we will need concerns the β-dimensional energy of the measure $\mathcal{M}$.

Lemma 3.38 *Assume that $\sigma(D) > 0$. Suppose that $0 \le \beta < \mathbf{d} \vee \sqrt{2\mathbf{d}}\gamma$, and x is a point chosen from the measure $\sigma(\mathrm{d}x)$ in D (normalised to be a probability measure), independently of the field. Then*

$$\int_D |x - y|^{-\beta} \mathcal{M}(\mathrm{d}y) < \infty$$

almost surely and in fact, has finite rth moment for $r > 0$ small enough.

118   *Gaussian Multiplicative Chaos*

Proof If $\sqrt{2\mathbf{d}}\gamma \leq \mathbf{d}$ then $\beta < \mathbf{d}$, in which case this energy will have finite expectation directly by assumption (3.6). So let us assume that $\sqrt{2\mathbf{d}}\gamma > \mathbf{d}$, and thus $\beta < \sqrt{2\mathbf{d}}\gamma$. This means that for $r > 0$ small enough, we will have $1 > 1 - r > 1/2 \vee \beta^2/(2\mathbf{d}\gamma^2)$. For such an r, we bound

$$\mathbb{E}\left(\left(\int_D |x-y|^{-\beta} \mathcal{M}(dy)\right)^r\right) \leq C\mathbb{E}\left(\int_D \left(\int_D e^{\beta K(x,y)} \mathcal{M}(dy)\right)^r \frac{\sigma(dx)}{\sigma(D)}\right)$$

$$\leq C\mathbb{E}(\mathcal{M}(D)^r \mathcal{M}_{\gamma^{-1}\beta}(D))$$

for some finite C, where the last inequality follows from Girsanov's lemma (Theorem 3.16 and Corollary 3.18) and $\mathcal{M}_{\gamma^{-1}\beta}$ is the chaos measure of the field with parameter $\gamma^{-1}\beta$ rather than γ (note that by our assumptions on the parameters, we have $\gamma^{-1}\beta < \sqrt{2\mathbf{d}}$, so this chaos is indeed well defined and non-trivial).

Now by Hölder's inequality with $p = r^{-1}$ and $q = (1-r)^{-1}$, the above is less than or equal to

$$\mathbb{E}(\mathcal{M}(D))^r \mathbb{E}(\mathcal{M}_{\gamma^{-1}\beta}(D)^{\frac{1}{1-r}})^{1-r}.$$

By Proposition 3.37, this is finite as long as $(1-r)^{-1} \leq 2 \wedge 2\mathbf{d}/(\gamma^{-1}\beta)^2$, which is exactly our assumption on r. □

Corollary 3.39 *Take the same set-up as in Lemma 3.38. Then there exists M large enough, depending only on γ and $\mathbf{d}$, such that*

$$\mathbb{P}(E_s) := \mathbb{P}\left(\int_{B(x,s^{-M})} e^{\gamma^2 K(x,y)} \mathcal{M}(dy) \leq \frac{1}{s}\right) \geq \frac{1}{2}$$

for all s sufficiently large.

Proof By the assumptions in Section 3.2 on K, we know that $e^{\gamma^2 K(x,y)} \leq c|x-y|^{-\gamma^2}$ for some $c < \infty$. Writing $|x-y|^{-\gamma^2} = |x-y|^{-(\gamma^2+2/M)}|x-y|^{2/M}$, we therefore have that

$$e^{\gamma^2 K(x,y)} \leq cs^{-2}|x-y|^{-(\gamma^2+2/M)} \text{ for all } y \in B(x, s^{-M}).$$

Hence

$$\mathbb{P}\left(\int_{B(x,s^{-M})} e^{\gamma^2 K(x,y)} \mathcal{M}(dy) \leq \frac{1}{s}\right)$$

$$\geq \mathbb{P}\left(\int_D |x-y|^{-(\gamma^2+2/M)} \mathcal{M}(dy) \leq \frac{s}{c}\right).$$

If M is such that $\beta := \gamma^2 + 2/M < \mathbf{d} \vee \sqrt{2\mathbf{d}}\gamma$ (it is always possible to choose M in this manner, consider separately the cases $\gamma \leq \sqrt{\mathbf{d}}$ and $\mathbf{d} < \gamma < \sqrt{2\mathbf{d}}$), then by Lemma 3.38, the right-hand side converges monotonically to 1 as $s \to \infty$. □

3.12 Negative moments of Gaussian multiplicative chaos

From here, the key observation is the following. If we write $\mathbb{Q}$ for the field biased by $M(D)/\sigma(D)$ as before, then for $s > 0$,

$$\mathbb{Q}(\exp(-sM(D))) = \sigma(D)^{-1}\mathbb{E}(M(D)\exp(-sM(D))) \leq \frac{e^{-1}}{\sigma(D)s}$$

simply because $xe^{-sx} \leq e^{-1}/s$ for all positive x, s. This says that under $\mathbb{Q}$, $M(D)$ is unlikely to be too small. Of course, we would actually like such a statement under $\mathbb{P}$. The trouble is that the field has an extra log singularity under $\mathbb{Q}$, and so it could be this that saves $M(D)$ from being very small. The work now is essentially to rule this out using Corollary 3.39.

To do this, we first claim that if E_s is the event in Corollary 3.39, then

$$\mathbb{E}(\exp(-cs^{1+M\gamma^2}M(D))\mathbf{1}_{E_s}) \leq \frac{C}{s\sigma(D)} \qquad (3.66)$$

for some $c, C, s_0 < \infty$ and all $s \geq s_0$, where these constants depend on $\mathbf{d}, \gamma$ and also K. Indeed, by Girsanov's lemma again (Theorem 3.17),

$$\mathbb{Q}\big(\exp(-sM(D))\big) = \mathbb{E}\bigg(\exp\bigg(-s\int_D e^{\gamma^2 K(x,y)}M(\mathrm{d}y)\bigg)\bigg),$$

where under $\mathbb{P}$, x is a point chosen according to σ, independently of the field. Moreover, on the event E_s,

$$s\int_D e^{\gamma^2 K(x,y)}M(\mathrm{d}y) \leq 1 + cs^{1+M\gamma^2}M(D)$$

for some $c < \infty$. This implies that

$$\exp\big(-cs^{1+M\gamma^2}M(D)\big)\mathbf{1}_{E_s} \leq e^{-1}\exp\bigg(-s\int_D e^{\gamma^2 K(x,y)}M(\mathrm{d}y)\bigg).$$

Taking expectations, we get

$$\mathbb{E}\big(\exp(-cs^{1+M\gamma^2}M(D))\mathbf{1}_{E_s}\big) \leq \mathbb{E}\bigg[\exp\bigg(-s\int_D e^{\gamma^2 K(x,y)}M(\mathrm{d}y)\bigg)\bigg]$$
$$= \mathbb{Q}\big[(\exp(-sM(D)))\big]$$
$$= \frac{1}{s\sigma(D)}\mathbb{E}\big[sM(D)\exp(-sM(D))\big]$$

and so (3.66) holds.

Note that if it weren't for the indicator function in (3.66), this would imply that $M(D)$ has *some* finite negative moments (using the identity

$$y^{-p} = \Gamma(p)^{-1}\int_0^\infty t^{p-1}\exp(-ty)\,\mathrm{d}t,$$

for $p > 0$, this will be detailed below). On the other hand, we have shown in

Corollary 3.39 that the event in the indicator function is rather likely. Putting these ideas together more precisely, we obtain the following.

Proposition 3.40 *Assume that $\sigma(D) > 0$ and $0 \leq \gamma < \sqrt{2\mathbf{d}}$. For some $q_0 < 0$, depending only on γ and $\mathbf{d}$, it holds that $\mathbb{E}(\mathcal{M}(D)^{q_0}) < \infty$.*

Proof Let us first observe that, without loss of generality, we may assume that $K(x, y) \geq 0$ for all $x, y \in D$. Indeed, we can always find some $D' \subset D$ with $\sigma(D') > 0$ and $K(x, y) \geq 0$ for all $x, y \in D'$ (since K diverges logarithmically near the diagonal), and then it clearly suffices to show that $\mathbb{E}(\mathcal{M}(D')^{q_0}) < \infty$. Note that σ also has dimension at least $\mathbf{d}$ when restricted to D'.

The advantage of assuming this set-up is that we can make use of the following tool (see for example, [Pit82]):

Theorem 3.41 (FKG inequality) *Let $(Z(x))_{x \in U}$ be an almost surely continuous centred Gaussian field on $U \subset \mathbb{R}^d$ with $\mathbb{E}(Z(x)Z(y)) \geq 0$ for all $x, y \in U$. Then, if f, g are two bounded, increasing measurable functions,*

$$\mathbb{E}\big(f((Z(x))_{x \in U})g((Z(x))_{x \in U})\big) \geq \mathbb{E}\big(f((Z(x))_{x \in U})\big)\mathbb{E}\big(g((Z(x))_{x \in U})\big).$$

To apply this, we need to work with continuous fields, so let us consider the regularised field h_ε of (3.8) and regularised measure $\mathcal{M}_\varepsilon$ of (3.9), and denote by E_s^ε the event E_s of Corollary 3.39 with $\mathcal{M}$ replaced by $\mathcal{M}_\varepsilon$ (and we still define E_s^ε in terms of a point that is sampled independently of the field and with probability proportional to $\sigma(dx)$). Since h_ε is almost surely continuous and the functions $\mathbf{1}_{E_s^\varepsilon}$ and $\exp(-cs^{1+M\gamma^2}\mathcal{M}_\varepsilon(D))$ are both bounded, decreasing functions of the field h_ε, we can apply Theorem 3.41 to see that

$$\mathbb{E}(\exp(-cs^{1+M\gamma^2}\mathcal{M}_\varepsilon(D))\mathbf{1}_{E_s^\varepsilon}) \geq \mathbb{E}(\exp(-cs^{1+M\gamma^2}\mathcal{M}_\varepsilon(D)))\mathbb{P}(E_s^\varepsilon)$$

for all ε. (Recall that $\mathbb{E}$ is over the field as well as the independent random point x, so we actually apply the FKG inequality conditionally given x, then note that the first term in the right hand does not depend on x.) By dominated convergence, we therefore obtain that

$$\mathbb{E}(\exp(-cs^{1+M\gamma^2}\mathcal{M}(D))\mathbf{1}_{E_s}) \geq \mathbb{E}(\exp(-cs^{1+M\gamma^2}\mathcal{M}(D)))\mathbb{P}(E_s).$$

Hence by Corollary 3.39 and (3.66), for M large enough (depending only on $\gamma, \mathbf{d}$) and s_0 large enough (depending on $\gamma, \mathbf{d}$ and K):

$$\mathbb{E}(\exp(-cs^{1+M\gamma^2}\mathcal{M}(D))) \leq \frac{2C}{s\sigma(D)} \quad \text{for all } s \geq s_0,$$

or to put it another way, for some t_0 sufficiently large,

$$\mathbb{E}(\exp(-t\mathcal{M}(D))) \leq \frac{2C}{(t/c)^{1/(1+M\gamma^2)}\sigma(D)} \quad \text{for all } t \geq t_0. \quad (3.67)$$

3.12 Negative moments of Gaussian multiplicative chaos

Finally, since $y^{-p} = \Gamma(p)^{-1} \int_0^\infty t^{p-1} \exp(-ty) \, dt$ for $p > 0$, this implies that

$$\mathbb{E}(\mathcal{M}(D)^{-p}) = \int_0^\infty t^{p-1} \mathbb{E}(e^{-t\mathcal{M}(D)}) \, dt$$

$$\lesssim 1 + \int_{t_0}^\infty t^{p-1-1/(1+M\gamma^2)} \, dt.$$

The integral in the right-hand side is finite as soon as $p < 1/(1 + M\gamma^2)$, and so for such values of p we get $\mathbb{E}(\mathcal{M}(D)^{-p}) < \infty$. Note that this only depends on M, γ, so since M is a function of γ and $\mathbf{d}$, the obtained q_0 also depends only on γ and $\mathbf{d}$. This completes the proof of Proposition 3.40. □

Now we explain how to extend this to all negative moments, using an iteration procedure. This idea first appeared in the setting of multiplicative cascade measures (a toy model for multiplicative chaos) in [Mol96].

Theorem 3.42 *Suppose that $\sigma(D) > 0$ and $0 \leq \gamma < \sqrt{2\mathbf{d}}$. Then*

$$\mathbb{E}(\mathcal{M}(D)^q) < \infty$$

for all $q < 0$.

We emphasise that we need only our standing assumptions on the measure σ and the field h from Section 3.2 here (as long as $\sigma(D) > 0$), and that there are no restrictions on d or $\mathbf{d} > 0$.

Proof To begin with, note that since σ does not have any atoms, we can find two distinct points x_1, x_2 in the support of σ. Therefore we can find open sets D_1 and D_2 such that $x_1 \in D_1, x_2 \in D_2, \overline{D}_1 \cap \overline{D}_2 = \emptyset$ and $\sigma(D_1)\sigma(D_2) > 0$. Furthermore, by the assumption on K (more precisely, the continuity of g), we may assume that $K(x, y) \leq C$ whenever $x \in D_1, y \in D_2$.

The key point is that by Proposition 3.40, there exists $q_0 < 0$ such that $\mathbb{E}(\mathcal{M}(D_i)^q) < \infty$ for all $q \in [q_0, 0]$ and $i = 1, 2$. Indeed, we have seen that q_0 depends only on $\mathbf{d}, \gamma$, as long as the base measure has strictly positive mass.

The idea to make use of this is to note the trivial bound $\mathcal{M}(D) \geq \mathcal{M}(D_1) + \mathcal{M}(D_2)$, and then apply the AM–GM inequality to see that

$$\mathcal{M}(D) \geq \sqrt{\mathcal{M}(D_1)\mathcal{M}(D_2)}.$$

This gives that

$$\mathbb{E}(\mathcal{M}(D)^q) \leq \mathbb{E}(\mathcal{M}(D_1)^{q/2} \mathcal{M}(D_2)^{q/2})$$

for $q < 0$. If $\mathcal{M}(D_1)$ and $\mathcal{M}(D_2)$ were independent, we could factorise the right-hand side and choose $q = 2q_0$, therefore showing that negative moments exist with orders in the larger interval $[2q_0, 0]$. We could then iterate to get all negative moments.

The problem of course is that they are not actually independent. To get around this we will use the assumption that $K(x, y) \le C$ for $x \in D_1, y \in D_2$, together with Kahane's inequality (Theorem 3.19).

More precisely, let us denote our field restricted to $D_1 \cup D_2$ by X. Let us also define a Gaussian field Y on $D_1 \cup D_2$ by setting it equal to $Y_1 + Y_2 + Z$ where: Y_1, Y_2 are independent; Y_1 has the law of $X|_{D_1}$ on D_1 and is 0 on D_2; Y_2 has the law of $X|_{D_2}$ on D_2 and is 0 on D_1; and Z is an independent normal random variable with variance C. Then the covariance kernel of Y dominates (pointwise) the covariance kernel of X. Since polynomials of negative order are convex, we can apply Kahane's inequality (Theorem 3.19) (and a limiting argument so that we can compare the respective GMC measures) to obtain that

$$\mathbb{E}\big((\mathcal{M}(D_1) + \mathcal{M}(D_2))^q\big) \le \mathbb{E}\big((\mathcal{M}_Y(D_1) + \mathcal{M}_Y(D_2))^q\big)$$
$$\le \mathbb{E}\big(\mathcal{M}_Y(D_1)^{q/2} \mathcal{M}_Y(D_2)^{q/2}\big)$$
$$= \mathbb{E}\big(e^{\frac{q}{2}(\gamma Z - \frac{\gamma^2}{2}C)}\big) \mathbb{E}\big(\mathcal{M}_Y(D_1)^{q/2}\big) \mathbb{E}\big(\mathcal{M}_Y(D_2)^{q/2}\big),$$

where we have applied AM–GM in the second line. By construction, if $q \in [2q_0, 0]$, the right-hand side is finite. So we obtain that $\mathbb{E}(\mathcal{M}(D)^q) < \infty$ for all $q \in [2q_0, 0]$. Repeating the argument, one obtains the existence of any negative moment. □

Since $\mathcal{M}(D)$ has finite negative moments of all orders (as shown by the previous theorem), we deduce that the tail at zero of $\mathcal{M}(D)$ decays faster than any polynomial. It is natural to wonder whether the decay can be characterised precisely. A lognormal upper bound for this decay (meaning, $\mathbb{P}(\mathcal{M}(D) \le \delta) \le \exp(-c(\log 1/\delta)^2)$ for some $c > 0$) was first established in [DS11], see also [Aru20]. In some one-dimensional cases of Gaussian multiplicative chaos, the exact law of the total mass is in fact known (this was obtained by Remy [Rem20], proving a well-known conjecture of Fyodorov and Bouchaud [FB08]). In Exercise 3.7, we propose a lognormal lower bound valid in great generality.

3.12.1 KPZ theorem

In this section, we consider the Gaussian free field with zero boundary conditions in a domain $D \subset \mathbb{R}^2$. The KPZ formula relates the "quantum" and "Euclidean" sizes of a given set A, which is either deterministic, or random but independent of the field. This often has a particularly natural interpretation in the context of discrete random planar maps and critical exponents; see Section 3.12.4. Concrete examples are given in Chapter 4.

3.12 Negative moments of Gaussian multiplicative chaos

We will first formulate the KPZ theorem using the framework of Rhodes and Vargas [RV11]. This article appeared simultaneously with (and independently from) the paper by Duplantier and Sheffield [DS11]. The results of these two papers are similar in spirit, but the version we present here is a bit easier to state, and in fact stronger. The formulation (and sketch of proof) corresponding to [DS11] will be given in Section 3.12.3. We will also include a version, due to Aru [Aru15]. Although this last statement is weaker, its proof is completely straightforward given our earlier work.

We first introduce the notion of *scaling exponent* of a set A (in the sense of [RV11]), starting with the Euclidean version. Let $A \subset D$ be a fixed Borel set and write $d_H(A)$ for the (Euclidean) Hausdorff dimension of A. Since $0 \le d_H(A) \le 2$, we may write

$$d_H(A) = 2(1 - x), \tag{3.68}$$

for $x \in [0, 1]$. The number x is called the **(Euclidean) scaling exponent** of A.

We now define the quantum analogue of the scaling exponent. Let

$$C_\delta(A) := \inf\{\sum_i \mathcal{M}(B(x_i, r_i))^\delta : \{B(x_i, r_i)\}_i \text{ is a cover of } A\},$$

so that $C_\delta(A)$ can be viewed as a (multiple) of the quantum Hausdorff content of A. We now define

$$d_{H,\gamma}(A) = \inf\{\delta > 0 : C_\delta(A) = 0\} \in [0, 1]$$

and call $d_{H,\gamma}(A)$ its "quantum Hausdorff dimension". Finally, we define the **quantum scaling exponent** Δ by

$$\Delta = 1 - d_{H,\gamma}(A).$$

The terms "quantum Hausdorff dimension" and content should perhaps be qualified, for the following reasons.

1. Although it does not feature in this book, a random metric associated with $e^{\gamma h}$ (h a GFF in D) has recently been constructed in a series of works culminating with [DDDF20, GM21c, GM21a]. The Hausdorff dimension d_γ of D equipped with this random metric is currently unknown, except for the special case $\{\gamma = \sqrt{8/3}, d_\gamma = 4\}$. The general bound $d_\gamma > 2$ is also known, as well as more precise estimates: see [DG20, GP19].
2. Under this random metric, the actual value of the Hausdorff dimension of $A \subset D$ is then given by

$$d_\gamma(1 - \Delta).$$

Again it always holds that $\Delta \in [0, 1]$, and note the analogy with (3.68).

124   *Gaussian Multiplicative Chaos*

3. Recently, a metric version of the KPZ formula has been obtained by Gwynne and Pfeffer [GP22]; more details concerning the relation between scaling exponent and Hausdorff dimension can be found there.

Remark 3.43 There is no consensus (even in the physics literature) about the value of d_γ. Until recently, it seemed that the prediction

$$d_\gamma = 1 + \frac{\gamma^2}{4} + \sqrt{\left(1 + \frac{\gamma^2}{4}\right)^2 + \gamma^2}$$

by Watabiki [Wat93] had a reasonable chance of being correct, but it has now been proved false – at least for small γ [DG19]. Simulations are notoriously difficult because of large fluctuations. As mentioned earlier, the only value that is known rigorously is when $\gamma = \sqrt{8/3}$. In this case, the metric space is described by the Brownian map [Mie13, LG13, MS21] and the Hausdorff dimension is equal to 4.

We are now ready to state the KPZ theorem in this set-up.

Theorem 3.44 (Almost sure Hausdorff KPZ formula) *Suppose that A is deterministic and that $\gamma \in (0, 2)$. Then, almost surely it holds that*

$$x = \frac{\gamma^2}{4}\Delta^2 + \left(1 - \frac{\gamma^2}{4}\right)\Delta.$$

We will not prove this result and refer to [RV11] for details. (We will, however, soon see the proof of a closely related result due to Aru [Aru15]). We make a few observations.

1. $x = 0, 1$ if and only if $\Delta = 0, 1$.
2. This is a quadratic relation with positive discriminant so can be inverted.
3. In the particular but important case of uniform random planar map scaling limits (see Chapter 4), $\gamma = \sqrt{8/3}$ and so the relation is

$$x = \frac{2}{3}\Delta^2 + \frac{1}{3}\Delta. \tag{3.69}$$

As we have already mentioned, various forms of the KPZ relation have now been proved; the above statement comes from the work of Rhodes and Vargas [RV11]. Other versions can be found in the works of Aru [Aru15], Duplantier and Sheffield [DS11] which will both be discussed later in this chapter. See also works of Gwynne and Pfeffer [GP22] for a KPZ relation in the sense of metric (Hausdorff) dimensions; Gwynne, Holden and Miller [GHM20] for an effective KPZ formula which can be used rigorously for determining a number of SLE exponents, and Berestycki, Garban, Rhodes and Vargas [BGRV16] for a KPZ relation formulated using the Liouville heat kernel.

3.12.2 Proof in the case of expected Minkowski dimension

We now state Aru's version of the KPZ formula [Aru15] which, as already mentioned, has a straightforward proof given our earlier work. This statement uses an alternative notion of fractal dimension: Minkowski dimension rather than Hausdorff.

We will only state the case $d = 2$ of this result, even though the arguments generalise easily to arbitrary dimensions. We again use the notation S_n for the nth-level dyadic covering of D by squares S_i, $i \in S_n$ of sidelength 2^{-n}. For $\delta > 0$, the (Euclidean) $(\delta, 2^{-n})$-Minkowski content of A is defined by

$$M_\delta(A; 2^{-n}) = \sum_{i \in S_n} \mathbf{1}_{\{S_i \cap A \neq \emptyset\}} \operatorname{Leb}(S_i)^\delta,$$

and the (Euclidean) Minkowski dimension (fraction) of A is then

$$d_M(A) = \inf \left\{ \delta : \limsup_{n \to \infty} M_\delta(A, 2^{-n}) < \infty \right\}.$$

Note that since we used $\operatorname{Leb}(S_i)$ in the definition of the Minkowski content rather than the more standard sidelength 2^{-n} of S_i, the above quantity d_M is in $[0, 1]$ and is related to the more standard notion of Minkowski dimension D_M through the identity $d_M = D_M/2$. Finally, we define the **Minkowski scaling exponent**

$$x_M = 1 - d_M.$$

On the quantum side, we define

$$M_\delta^\gamma(A, 2^{-n}) = \sum_{i \in S_n} \mathbf{1}_{\{S_i \cap A \neq \emptyset\}} \mathcal{M}(S_i)^\delta,$$

and the quantum expected Minkowski dimension by

$$q_M = \inf \left\{ \delta : \limsup_{n \to \infty} \mathbb{E}(M_\delta^\gamma(A, 2^{-n})) < \infty \right\}.$$

The quantum Minkowski scaling exponent is then set to be

$$\Delta_M = 1 - q_M.$$

The KPZ relation for the Minkowski scaling exponents is then

$$x_M = (\gamma^2/4)\Delta_M^2 + (1 - \gamma^2/4)\Delta_M$$

(formally this is the same as the relation in Theorem 3.46). Equivalently, this can be rephrased as follows.

Proposition 3.45 (Expected Minkowski KPZ, [Aru15]) *Suppose $\overline{A}$ lies at a positive distance from ∂D and that A is bounded. Then*

$$d_M = (1 + \gamma^2/4)q_M - \gamma^2 q_M^2/4. \qquad (3.70)$$

Proof First recall Theorem 3.27 from earlier in this chapter, which implies that if $0 \le q \le 1$, then

$$\mathbb{E}(\mathcal{M}(B(r))^q) \asymp r^{(d+\gamma^2/2)q - \gamma^2 q^2/2}$$

for balls $B(r)$ of Euclidean radius r lying strictly within the domain D.

Fix $d \in (0, 1)$ and let q be such that $d = (1 + \gamma^2/4)q - q^2\gamma^2/4$ and note that $q \in (0, 1)$. Therefore,

$$\mathbb{E}\Big(\sum_{i \in S_n} \mathbf{1}_{\{S_i \cap A \ne \emptyset\}} \mathcal{M}(S_i)^q\Big) \asymp \sum_{i \in S_n} \mathbf{1}_{\{S_i \cap A \ne \emptyset\}} \operatorname{Leb}(S_i)^d$$

and consequently the limsup of the left-hand side is infinite if and only if the limsup of the right-hand side is infinite. In other words, d_M and q_M satisfy (3.70). $\square$

3.12.3 Duplantier–Sheffield's KPZ theorem

We end this chapter with a short description of Duplantier and Sheffield's definitions of scaling exponents, as well as a sketch of proof of the resulting KPZ formula [DS11]. (The statement is a bit weaker than Theorem 3.44, since the notions of scaling exponents are slightly harder to use, and the formula holds only in expectation as opposed to almost surely).

In this section, the (Euclidean) scaling exponent of $A \subset D$ is the limit, if it exists, defined by

$$x' = \lim_{\varepsilon \to 0} \frac{\log \mathbb{P}(B(z, \varepsilon) \cap A \ne \emptyset)}{\log(\varepsilon^2)},$$

where $\mathbb{P}$ is the joint law of A (if it is random) and a point z chosen proportionally to Lebesgue measure in D. We will assume that D is bounded.

We need to make a few comments about this definition.

1. First, this is equivalent to saying that the volume of A_ε, the Euclidean ε-neighbourhood of A, decays like $\varepsilon^{2x'}$. In other words, A can be covered with $\varepsilon^{-(2-2x')}$ balls of radius ε, and hence typically the Hausdorff dimension of A is simply

$$d_H(A) = 2 - 2x' = 2(1 - x'),$$

3.12 Negative moments of Gaussian multiplicative chaos

consistent with our earlier definition of Euclidean scaling exponent. In particular, note that $x' \in [0, 1]$ always; $x' = 0$ means that A is practically the full space, $x' = 1$ means it is practically empty.

2. In the definition, we divide by $\log(\varepsilon^2)$, because ε^2 is the volume (with respect to the Euclidean geometry on $\mathbb{R}^2$) of a ball of radius ε. In the quantum world, we would need to replace this by the Liouville area of a ball of radius ε – see below.

The quantum analogue of this is the following. For $z \in D$, we denote by $B^\delta(z)$ the quantum ball of mass δ: that is, the Euclidean ball centred at z whose radius is chosen so that its Liouville area is precisely δ. (In [DS11], this is called the *isothermal* ball of mass δ at z). The quantum scaling exponent of $A \subset D$ is then the limit, if it exists, defined by

$$\Delta' = \lim_{\delta \to 0} \frac{\log \mathbb{P}(B^\delta(z) \cap A \neq \emptyset)}{\log(\delta)},$$

where z is sampled from the Liouville measure $\mathcal{M}$ normalised to be a probability distribution.

Theorem 3.46 (Expected Hausdorff KPZ formula) *Suppose A is independent of the GFF, $\gamma \in (0, 2)$, and D is bounded. Then if A has Euclidean scaling exponent x', it has quantum scaling exponent Δ', where x' and Δ' are related by the formula*

$$x' = \frac{\gamma^2}{4}(\Delta')^2 + \left(1 - \frac{\gamma^2}{4}\right)\Delta'. \tag{3.71}$$

We will now sketch the argument used by Duplantier and Sheffield to prove Theorem 3.46, since it is interesting in its own right and gives a somewhat different perspective (in particular, it shows that the KPZ formula can be seen as a large deviation probability for Brownian motion).

Informal description of the idea of the proof. We wish to evaluate the probability $\mathbb{P}(B^\delta(z) \cap A \neq \emptyset)$, where z is a point sampled from the Liouville measure, and B^δ is the Euclidean ball of Liouville mass δ around z. Of course, the event that this ball intersects A is rather unlikely, since the ball is small. But it can happen for two reasons. The first one is simply that z lands very close (in the Euclidean sense) to A – this has a cost governed by the Euclidean scaling exponent of A, by definition, since we may think of z as being sampled from the Lebesgue measure and then sampling the Gaussian free field given z, as in the description of the rooted measure in Section 2.4. However, it is more economical for z to land relatively further away from A, and instead require that the ball of quantum mass δ have a bigger than usual radius. As the quantum mass of the ball of radius r around z is essentially governed by the size of the circle average $h_r(z)$, which behaves like a Brownian motion plus some drift, we find ourselves computing a large deviation probability for a Brownian motion. The KPZ formula is hence nothing else but the large deviation function for Brownian motion.

Sketch of proof of Theorem 3.46 Now we turn the informal idea above into more concrete mathematics, except for two approximations that we will not justify. Suppose z is sampled according to the Liouville measure $\mathcal{M}$. Then we know from Theorem 3.16 (see also (2.8)) that the joint law of the point z and the free field is absolutely continuous with respect to a point z sampled from Lebesgue measure, together with the field $h^0(\cdot) + \gamma \log|\cdot -z| + O(1)$, where h^0 is a GFF that is independent of z. (See Section 2.4.) Hence the mass of the ball of radius ε about z is approximately given by

$$\mathcal{M}(B(z,\varepsilon)) \approx \varepsilon^{\gamma^2/2} e^{\gamma h_\varepsilon(z)} \times \varepsilon^2 \asymp e^{\gamma(h_\varepsilon^0(z) + \gamma \log 1/\varepsilon)} \varepsilon^{2+\gamma^2/2}$$
$$= \varepsilon^{2-\gamma^2/2} e^{\gamma h_\varepsilon^0(z)}. \qquad (3.72)$$

It takes some time to justify rigorously the approximation in (3.72), but the idea is that the field h_ε fluctuates on a spatial scale of size roughly ε. Hence we are not making a big error by pretending that h_ε is constant on $B(z, \varepsilon)$, equal to $h_\varepsilon(z)$. In a way, making this precise is the most technical part of the paper [DS11]. We will not go through the arguments which do so, and instead we will see how, assuming it, one is naturally led to the KPZ relation.

Working on an exponential scale (which is more natural for circle averages) and writing $B_t = h^0_{e^{-t}}(z)$, we find that

$$\log \mathcal{M}(B(z, e^{-t})) \approx \gamma B_t - (2 - \gamma^2/2)t.$$

We are interested in the maximum radius ε such that $\mathcal{M}(B(z,\varepsilon))$ will be approximately δ: this will give us the Euclidean radius of the quantum ball of mass δ around z. So let

$$T_\delta = \inf\{t \geq 0 : \gamma B_t - (2 - \gamma^2/2)t \leq \log \delta\}$$
$$= \inf\left\{t \geq 0 : B_t + \left(\frac{2}{\gamma} - \frac{\gamma}{2}\right)t \geq \frac{\log(1/\delta)}{\gamma}\right\},$$

where the second equality is in distribution. Note that since $\gamma < 2$ the drift is positive, and hence $T_\delta < \infty$ almost surely.

Now, recall that if $\varepsilon > 0$ is fixed, the probability that z will fall within (Euclidean) distance ε of A is approximately $\varepsilon^{2x'}$. Hence, applying this with $\varepsilon = e^{-T_\delta}$ the probability that $B^\delta(z)$ intersects A is, approximately, given by

$$\mathbb{P}(B^\delta(z) \cap A \neq \emptyset) \approx \mathbb{E}(\exp(-2x' T_\delta)).$$

3.12 Negative moments of Gaussian multiplicative chaos   129

This is the second approximation that we will not seek to justify fully. Consequently, we deduce that

$$\Delta' = \lim_{\delta \to 0} \frac{\log \mathbb{E}(\exp(-2x'T_\delta))}{\log \delta}.$$

For $\beta > 0$, consider the martingale

$$M_t = \exp(\beta B_t - \beta^2 t/2),$$

and apply the optional stopping theorem at the time T_δ (note that this is justified). Then we get, letting $a = 2/\gamma - \gamma/2$, that

$$1 = \exp\left(\beta \frac{\log(1/\delta)}{\gamma}\right) \mathbb{E}(\exp(-(a\beta + \beta^2/2)T_\delta)).$$

Finally set $2x' = a\beta + \beta^2/2$, so that $\mathbb{E}(\exp(-2x'T_A)) = \delta^{\beta/\gamma}$. In other words, $\Delta' = \beta/\gamma$, where β is such that $2x' = a\beta + \beta^2/2$. Equivalently, $\beta = \gamma\Delta'$, and

$$2x' = \left(\frac{2}{\gamma} - \frac{\gamma}{2}\right)\gamma\Delta' + \frac{\gamma^2}{2}(\Delta')^2.$$

This is exactly the KPZ relation.   □

3.12.4 Applications of KPZ to critical exponents

Note: This section explains in a non-rigorous manner how the KPZ relation can be used to compute critical exponents in some models of statistical mechanics in two dimensions. This section can be skipped on a first reading and is only relevant later in connection with the end of Chapter 4. This section also assumes basic familiarity with the notion of random planar maps and the conjectures related to their conformal embeddings, see Section 4.2.

At the discrete level, the KPZ formula can be interpreted as follows. Consider a random planar map M of size N (where 'size' refers indistinctly to the number of faces, vertices or edges). Suppose that a certain subset A within M has a size $|A| \approx N^{1-\Delta}$, so that A is "fractal-like". We have in mind a set A which is defined conditionally independently given the map, and of course depends on N (but we do not indicate this in the notation). For instance, A could be the set of double points of a random walk on the map run until its cover time, or the set of pivotal edges for percolation on the map with respect to some macroscopic event. We may also consider the Euclidean analogue A' of A within a Euclidean box of area N (and thus of side length $n = \sqrt{N}$). Namely, A' is the set that one

130   Gaussian Multiplicative Chaos

obtains when the map M is exactly this subset of the square lattice. In this case we again expect A' to be fractal-like, and so $|A'| \approx N^{1-x} = n^{2-2x}$ for some $x \in [0, 1]$. If A' has a scaling limit, then this x is nothing but its Euclidean scaling exponent (indeed, the discrete size of A' is essentially the number of balls of a fixed radius required to cover a scaled version of it). Likewise, if A has a scaling limit, then Δ is nothing but its quantum scaling exponent.

Hence the KPZ relation suggests that x, Δ should be related by

$$x = \frac{\gamma^2}{4}\Delta^2 + \left(1 - \frac{\gamma^2}{4}\right)\Delta$$

in the limit as $N \to \infty$. Here γ refers to the universality class of the map; this assumes that A is (when embedded suitably in the plane) independent of the field h which represents the embedding of the map in the limit.

In particular, observe that the approximate (Euclidean) Hausdorff dimension of A' is then $2 - 2x$, consistent with our definitions. See Chapter 4 for concrete examples, where this is used, for instance, to guess the loop-erased random walk exponent.

3.13 Exercises

3.1 By considering the set of thick points or otherwise, argue that the KPZ relation does not need to hold if the set A is allowed to depend on the free field; for instance show that we can have $\Delta' = 0$, while $x' > 0$. This type of example was first considered by [Aru15] who also considered the case of flow lines associated to the GFF.

3.2 Suppose $K(x, y) \geq 0$ for all $x, y \in D$. Let $A \subset D$, and let $q \in [0, 1]$. Show that $\mathbb{E}(\mathcal{M}(A)^q)$ is a non-increasing function of $\gamma \in [0, \sqrt{2d})$. (Hint: use Kahane's inequality).

3.3 Let $A \subset D$. For $\gamma < \sqrt{d}$, show that $\mathcal{M}(A)$ admits a continuous modification in γ. (Hint: use the Kolmogorov continuity criterion.)

3.4

(a) Use the scaling invariance properties developed in the proof of the multifractal spectrum to show that $\mathcal{M}$ almost surely has no atoms.

(b) Give an alternative proof, using the energy estimate in Exercise 2.4 of Chapter 2.

3.5 This exercise gives a flavour of Kahane's original pioneering argument for the construction of GMC in [Kah85]. Suppose that K is a covariance kernel

3.13 Exercises

of the form (3.2) that can be written in the form

$$K(x, y) = \sum_{n=1}^{\infty} K_n(x, y)$$

for all $x \neq y$ in $D \subset \mathbb{R}^d$, where for each n, $K_n \colon D \times D \to \mathbb{R}$ is positive definite and satisfies $K_n(x, y) \geq 0$ for all $x, y \in D$. Such a covariance kernel was called σ-**positive** by Kahane. Show that there exists a sequence of centred Gaussian fields $(h^n)_{n \geq 1}$ such that the fields $(h^n - h^{n-1})_{n \geq 1}$ are independent centred Gaussian fields with covariances K_n for each n. Let σ be a reference Radon measure satisfying (3.6) for some $\mathbf{d} > 0$. For $0 \leq \gamma < \sqrt{2\mathbf{d}}$, we use this decomposition to construct a natural sequence of 'chaos approximations' $\mathcal{M}_n$ by setting

$$\mathcal{M}_n(A) = \int_A \exp\{\gamma h_n(x) - \tfrac{\gamma^2}{2} \mathbb{E}(h_n(x)^2)\} \sigma(\mathrm{d}x),$$

for any Borel set A. Prove that $\mathcal{M}_n(A)$ has an almost sure limit $\mathcal{M}(A)$ as $n \to \infty$ which defines a random measure.

Suppose we are given two σ-positive decompositions for K, say

$$K(x, y) = \sum_{n=1}^{\infty} K_n(x, y) = \sum_{n=1}^{\infty} K'_n(x, y),$$

and let $\mathcal{M}$ and $\mathcal{M}'$ be the associated chaos measures constructed above. Using Kahane's inequality (and without using Theorem 3.2), show that for any Borel set A, $\mathbb{E}(\mathcal{M}(A)^q) \leq \mathbb{E}(\mathcal{M}'(A)^q)$ for $q \in (0, 1)$ (note that this argument does not require knowing that either $\mathcal{M}$ or $\mathcal{M}'$ is non-zero). Deduce that the laws of $\mathcal{M}$ and $\mathcal{M}'$ (as random measures) are identical. This is Kahane's theorem on uniqueness of GMC; Kahane's inequality [Kah86] was discovered for the purpose of this proof.

3.6 We now take the same set-up as above, and assume the result of Theorem 3.2. Show that in the case $\gamma < \sqrt{2\mathbf{d}}$, the limit $\mathcal{M}$ constructed above agrees with the GMC measure of Theorem 3.2.

3.7 If K is as in (3.1), define the linear operator T on $L^2(D)$ by setting

$$Tf(x) = \int_D K(x, y) f(y) \, \mathrm{d}y$$

for each $f \in L^2(D)$. When D is bounded, one can show using standard operator theory that there exists an orthonormal basis $\{f_k\}_{k \geq 0}$ of $L^2(D)$, made up of eigenfunctions for T. The ordering can be chosen so that the associated eigenvalues $\{\lambda_k^{-1}\}_{k \geq 0}$ satisfy $0 \leq \lambda_1 \leq \lambda_2 \leq \lambda_3, \ldots$.

(a) Show that for each x in D,

$$\sum_{k=0}^{n} \lambda_k^{-1} f_k(x) f_k(\cdot) \to K(x, \cdot) \text{ in } L^2(D)$$

as $n \to \infty$. Let h be the centred Gaussian field with covariance K. By considering the joint law of $\{\lambda_k^{1/2}(h, f_k)\}_{k \geq 1}$, show that for any smooth compactly supported test function f on D, if $h^n := \sum_{k=0}^{n}(h, f_k) f_k$, then (h^n, f) converges almost surely and in $L^2(\mathbb{P})$ to (h, f) as $n \to \infty$.

Remark This decomposition of h is known as the **Karhuhen–Loeve expansion**.

(b) Show further that for $\gamma \geq 0$, the sequence of measures defined by

$$\mathcal{M}^n(S) := \int_S \exp(\gamma h^n(z) - \gamma^2/2 \operatorname{Var}(h^n(z))) \, dz \quad S \subset D,\, n \geq 0$$

has an almost sure limit with respect to the topology of weak convergence of measures. When $\gamma < \sqrt{2\mathbf{d}}$, show that $\mathcal{M}^n(S)$ is a uniformly integrable family for any fixed S. Use this to show that $\lim_n \mathcal{M}^n(S)$ agrees almost surely with $\mathcal{M}_\gamma(S)$, where $\mathcal{M}_\gamma$ is the GMC measure for h constructed in Theorem 3.2.

(c) Suppose that f_1 is non-negative and bounded. Use (3.67) to show that for $\delta > 0$, $\mathbb{P}(\mathcal{M}_\gamma(D) \leq \delta) \geq c\mathbb{P}(Z \leq \delta)$ where Z is an appropriately chosen lognormal random variable and $c > 0$ does not depend on δ.

4
Statistical Physics on Random Planar Maps

4.1 Fortuin–Kasteleyn weighted random planar maps

In this chapter, we change our focus from the continuous to the discrete and describe the model of random planar maps weighted by self-dual Fortuin–Kasteleyn percolation. As we will see, these maps can be thought of as canonical discretisations of Liouville quantum gravity (but there are in fact many other models of planar maps which are believed to be related to Liouville quantum gravity).

We proceed as follows. We first recall the notion of planar map and **decorated planar map** before defining a probability measure on such maps (maps decorated by self-dual FK loops). In Section 4.2, we discuss aspects of the conjectured connection between this model of planar maps and **Liouville quantum gravity**. In Section 4.3, we focus on the case where the decoration is a spanning tree. Here we describe in detail a powerful **bijection** due (independently) to Mullin, Bernardi and Sheffield, between tree decorated maps and pairs of independent, positive random walk excursions (equivalently, two-dimensional random walk excursions in the positive quadrant). This bijection is a convenient way to approach the question of scaling limits. We use it in Section 4.4 to compute the (quantum) scaling exponent of the **loop-erased random walk** (LERW). Using the KPZ relation of Section 3.12.1, we find that it agrees with various known properties of LERW on the square lattice, including the Hausdorff dimension of its scaling limit SLE_2. In Section 4.5, we discuss Sheffield's bijection, which is a generalisation of the aforementioned bijection to decorations which are no longer spanning trees but **densely packed loop configurations**. Again, this bijection is from decorated maps to pairs of excursions in a suitable sense. In this case, however, the excursions are far from independent; this has an interpretation in terms of a **discrete mating of trees** which will be described in the continuum in Chapter 9. This description is used in Section 4.6 to show the existence of an infinite volume local limit. A scaling limit result is discussed which, roughly speaking, shows that the limiting trees are correlated infinite CRTs.

Planar map, dual map. Recall that a **planar map** is a proper embedding of a (multi) graph with a finite number of edges in the plane $\mathbb{C} \cup \{\infty\}$ (viewed as the Riemann sphere), which is viewed up to orientation preserving homeomorphisms from the sphere to itself. Let m_n be a map with n edges and t_n be a subgraph spanning *all* of its vertices. We call the pair (m_n, t_n) a (subgraph) **decorated map**. Let $m_n^\dagger$ denote the *dual map* of m_n. Recall that the vertices of the dual map correspond to the faces of m_n, and two vertices in the dual map are adjacent if and only if their corresponding faces are adjacent to a common edge in the primal map. Every edge e in the primal map corresponds to an edge $e^\dagger$ in the dual map which joins the vertices corresponding to the two faces adjacent to e. The dual subgraph $t_n^\dagger$ is the graph formed by the subset of edges $\{e^\dagger : e \notin t_n\}$ and all dual vertices. We fix an edge in the map m_n, to which we also assign an orientation, and define it to be the root edge. With an abuse of notation, we will still write m_n for the rooted map; and we let $\mathcal{M}_n$ be the set of maps with n edges together with one distinguished edge called the root.

Canonical triangulation. Given a subgraph decorated map (m_n, t_n) with $m_n \in \mathcal{M}_n$ and a subgraph t_n of m_n, one can associate to it a set of loops where in some sense, each loop forms the interface between two clusters (connected components) associated to t_n and its planar dual. To be more precise, let us say that two vertices x and y of m_n are in the same component if we can travel from x to y using edges in t_n; by convention a vertex x is always in its own component (hence that component will consist only of x if x is not covered by t_n). We can use the same definition to talk about clusters on the planar map $m_n^\dagger$ dual to m_n and the dual configuration of edges $t_n^\dagger$; the loops then separate primal and dual clusters. To define these loops more precisely, we will need to discuss not only the dual planar map but also a couple of related maps that can be constructed from superposing the primal and dual maps.

We first consider an auxiliary map which we call the **Tutte map**, and which is formed by joining the dual vertices in every face of m_n with the primal vertices incident to that face. We call these edges **refinement edges** (drawn in green in Figure 4.1). Thus the vertex set of the Tutte map consists of all primal and dual vertices, but note that its edge set does not contain any of the original edges of m_n or its dual. It is easy to check that this Tutte map is a quadrangulation, meaning each face has exactly four (refinement) edges surrounding it. Each of the original edges of m_n or $m_n^\dagger$ is a diagonal of one of these quadrangles. In other words, every edge in m_n corresponds to a quadrangle in the Tutte map;

4.1 Fortuin–Kasteleyn weighted random planar maps

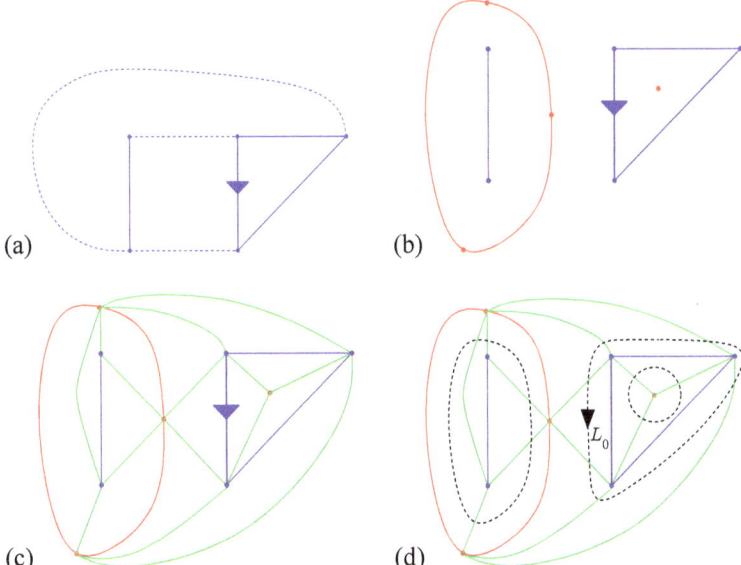

Figure 4.1 A map **m** decorated with loops associated to a set of open edges **t**. (a) The map is in blue, with solid open edges and dashed closed edges. (b) Open clusters and corresponding open dual clusters are shown in blue and red. (c) Every dual vertex is joined to its adjacent primal vertices by a green edge. This results in a refined map $\overline{m}$ which is a triangulation. (d) The primal and dual open clusters are separated by loops, which are drawn in black and are dashed. Each loop is identified with the set of triangles through which it passes: note that it crosses each triangle in the set exactly once. The oriented root edge of the map is indicated with a blue arrow in subfigures (a), (b) and (c). The loop L_0 is marked with an arrow in subfigure (d), and the arrow indicates the orientation of the loop, parallel to the orientation of the root edge.

this quadrangle can be viewed as the union of two triangles, one on either side of the edge.

In fact, this construction defines a bijection between maps with n edges and quadrangulations with n faces, sometimes called the **Tutte bijection**.

Given a subgraph decorated map (m_n, t_n) define the **refinement** map $\overline{m}_n$ to be formed by the union of $t_n, t_n^\dagger$ and the refinement edges, note that its vertex set is the same as the Tutte map, that is, every primal and dual vertex of m_n. The addition of t_n and $t_n^\dagger$ makes the refinement map a triangulation: indeed, every quadrangle from the Tutte map has been split into two (either with a diagonal from t_n or from $t_n^\dagger$). The root edge of m_n induces a **root triangle** on the refinement map, which is taken to be the triangle immediately to the right of the root edge of m_n.

Note that every triangle consists of two refinement edges and one edge from either t_n (primal edge) or $t_n^\dagger$ (dual edge). For future reference, we call such a triangle in $\overline{m}_n$ a **primal triangle** or **dual triangle**, respectively (see Figure 4.4).

Loops. Finally, given (m_n, t_n), we can define the loops induced by t_n as follows. For each connected component C of either t_n or $t_n^\dagger$, we draw a loop surrounding it (meaning a closed curve in the complement of C in the sphere; the complement contains two components, and by convention, we draw it in the "exterior" one that contains the point on the sphere designated to be ∞; note that even in the case where the connected component C is reduced to a single vertex there is still a loop surrounding it which separates the sphere into two components). If this loop is drawn sufficiently close to C it identifies a unique collection of triangles that are adjacent to C (in the sense that they share at least a vertex with it). We view the component C itself as an open cluster for a percolation configuration either on m_n or its dual, and will use the word "cluster" interchangeably from now on.

In what follows, one should visualise the loop of C as being a closed curve drawn sufficiently close to C in its complement, as above. However for precision, we will actually identify the loop with the collection of triangles through which it passes. See Figure 4.1 for an illustration. In this way, each loop is simply a collection of triangles "separating" a primal connected component of t_n from a dual connected component in $t_n^\dagger$, or vice versa. Note that the set of loops is "space filling" in the sense that every triangle of the refined map is contained in a loop. We denote by L_0 the loop that is associated with the root triangle. It comes with a natural orientation, which is parallel to the orientation of the root edge of M_n.

Also, given the Tutte map and the collection of closed curves described above, one can recover the spanning subgraph t_n (hence also $t_n^\dagger$) that generates it. Let $\ell(m_n, t_n)$ denote the number of loops corresponding to a configuration (m_n, t_n). Note that this is equal to the number of clusters in t_n plus the number of clusters in $t_n^\dagger$ minus one; indeed, each new cluster generates a new loop.

Fortuin–Kasteleyn model. The particular distribution on planar maps that we will now consider was introduced in [She16b]. Let $q \geq 0$ and let $n \geq 1$: we will define a random map $M_n \in \mathcal{M}_n$ decorated with a (random) subset T_n of edges. As in the deterministic setting, this induces a dual collection of edges $T_n^\dagger$ on the dual map of M (see Figure 4.1). The law of (M_n, T_n) is defined by declaring that for any fixed planar map m with n edges, and t a given subset of edges of m,

$$\mathbb{P}(M_n = m, T_n = t) \propto \sqrt{q}^\ell, \quad \ell = \ell(m, t). \tag{4.1}$$

4.1 Fortuin–Kasteleyn weighted random planar maps

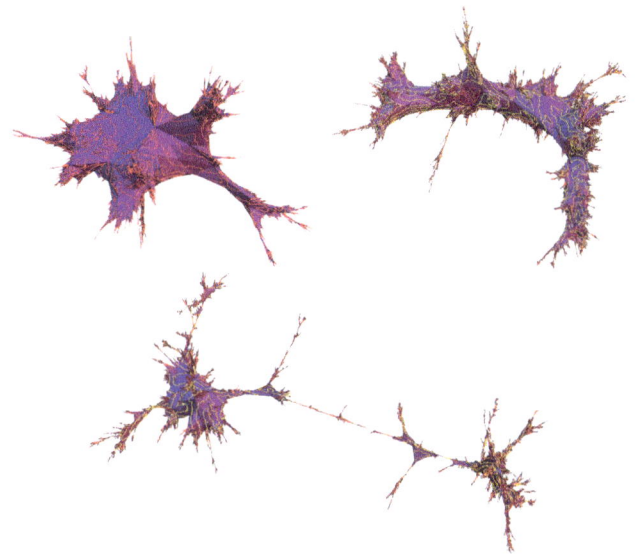

Figure 4.2 A map weighted by the FK model with $q = 0.5$, $q = 2$ (corresponding to the FK-Ising model) and $q = 9$, respectively, together with some of their loops. Simulation by J. Bettinelli and B. Laslier. When $q > 4$, it is believed that the maps become tree-like, and the limiting metric space should be Aldous' continuum random tree.

Recall from above that ℓ is the (total) number of loops separating primal and dual clusters in $(\boldsymbol{m}, \boldsymbol{t})$.

Equivalently, the map M_n is chosen with probability proportional to the "partition function" of the self-dual Fortuin–Kasteleyn model on it, and given the map M_n, the collection of edges T_n is then sampled from this Fortuin–Kasteleyn model. This is in turn closely related to the critical q-state Potts model, see [Bax00]. Note that M_n is actually a rooted map (as all of our maps are) and with this definition, the root edge of the map and its orientation are chosen uniformly at random (given the unrooted version). See Figure 4.2 for simulations of (M_n, T_n) at different values of q.

Uniform random planar maps. Observe that when $q = 1$, the FK model (4.1) has the property that the map M_n is chosen **uniformly** at random among the set $\mathcal{M}_n$ all of (rooted) maps with n edges because the total number of possible configurations for $\boldsymbol{t}_n$ is 2^n *independently* of $\boldsymbol{m}_n$. Furthermore, given $M_n = \boldsymbol{m}_n$, T_n is chosen uniformly at random from the 2^n possibilities: this corresponds to each edge being present (open) with probability $1/2$, independently of one another. In other words, T_n corresponds to bond percolation with parameter $1/2$ given the map M_n. This is in fact the critical parameter for this percolation model, as shown in the work of Angel [Ang03].

The case of a uniformly chosen planar map in $\mathcal{M}_n$ is one in which remarkably detailed information is known about its structure. In particular, a landmark result due to Miermont [Mie13] and Le Gall [LG13] shows that, viewed as a metric space and rescaling edge lengths to be $n^{-1/4}$, the random map converges to a multiple of a certain universal random metric space known as the **Brownian map**. (In fact, the results of Miermont and Le Gall apply, respectively, to uniform quadrangulations with n faces and to p-angulations for $p = 3$ or p even, whereas the convergence result concerning uniform planar maps in $\mathcal{M}_n$ was established a bit later by Bettinelli, Jacob and Miermont [BJM14].) Critical percolation on a related half-plane version of the maps has been analysed in a work of Angel and Curien [AC15], while information on the full plane percolation model was later obtained by Curien and Kortchemski [CK15]. Related works on loop models (sometimes rigorous, sometimes not) appear in [GJSZJ12, BBG12b, EK95, BBM11, BBG12a, CCM20].

One reason for the particular choice of the FK model in (4.1) is the belief that for $q < 4$, after Riemann uniformisation, a large sample of such a map closely approximates a *Liouville quantum gravity* surface. We will try to summarise this conjecture in Section 4.2.

4.2 Conjectured connection with Liouville quantum gravity

The distribution (4.1) gives us a natural family of distributions on planar maps (indexed by the parameter $q \geq 0$). As already mentioned, in this model, the weight of a particular map $m \in \mathcal{M}_n$ is proportional to the *partition function* $Z(m, q)$ of the critical FK model on the map.

Conformal Embedding. Suppose that $q < 4$ in what follows. It is strongly believed that in the limit $n \to \infty$, the geometry of such maps are related to Liouville quantum gravity with parameter γ, where

$$q = 2 + 2\cos\left(\frac{\gamma^2 \pi}{2}\right). \tag{4.2}$$

(Note that this equation has no real solution if $q > 4$.)

To be more precise about this, one must relate the world of planar maps to the world of Liouville quantum gravity by specifying a "natural" embedding of

4.2 Conjectured connection with Liouville quantum gravity

Figure 4.3 Circle packing of a uniform random planar map. Simulation by Jason Miller.

the maps into the plane. There are various ways to do this, and a couple of the simplest are as follows.

- **Via the circle packing theorem.** By a theorem of Koebe–Andreev–Thurston (see the book by K. Stephenson [Ste05] for a comprehensive introduction), any planar map can be represented as a circle packing. A circle packing is a collection of circles in the plane such that any two of the corresponding discs either are tangent to one another, or do not overlap. In the circle packing representation, the vertices of the map are given by the centres of the circles, and the edges correspond to tangent circles. See Figure 4.3. Each circle packing representation of a map gives an embedding in the plane, and when the map is a simple triangulation, this embedding is unique up to Möbius transformations.
- **Via the uniformisation theorem.** In this approach, a given map is viewed as a Riemann surface by declaring that each face of degree p is a regular p-gon of unit area, endowed with the standard metric, and specifying the charts near a vertex in the natural manner. This Riemann surface can then be embedded into the disc (say) by the uniformisation theorem (which is a generalisation of the Riemann mapping theorem from subsets of $\mathbb{C}$ to arbitrary Riemann surfaces).

These embeddings are essentially unique up to Möbius transforms (in the first case, we can circle pack the refinement map $\overline{m}_n$ instead of m_n). The choice of Möbius transform can be fixed by requiring, for instance, that the root edge is mapped to $(0, 1)$.

Once an embedding has been chosen, a natural object to study is the measure μ_n in the plane which puts mass $1/N$ (N being the number of vertices in M_n) at the position of each embedded vertex. The conjecture alluded to above says that in the limit as $n \to \infty$, if M_n is sampled from (4.1), then this measure μ_n should converge to γ-Liouville quantum gravity. More precisely, if γ and q are related by (4.2), it should converge in distribution for the topology of weak convergence, to a *variant* of the Liouville measure μ_γ (this variant will be specified, for example, in Chapter 5).

Remark 4.1 Note that when $q = 1$, which we have already discussed is the case of uniformly chosen random planar maps, we have $\cos(\gamma^2 \pi/2) = -1/2$, that is, $\gamma = \sqrt{8/3}$. Consequently, the limit of a (conformally embedded) uniformly chosen map should be related to Liouville quantum gravity with this parameter. This has been verified for a slightly different type of conformal embedding called the **Cardy embedding** in a recent breakthrough of Holden and Sun [HS23].

Loops and CLE. The loops induced by the FK model (4.1) may be viewed as a decoration on the map. Indeed as we have already mentioned, given the map, they are the cluster boundaries of a self-dual FK percolation model on it with parameter q. It is therefore natural to wonder about their geometry in the scaling limit, after embeddings of the type discussed above. The widely shared belief is that they converge to so-called **conformal loop ensembles** $\text{CLE}_{\kappa'}$ where the parameter κ' is given by

$$\kappa' = \frac{16}{\gamma^2}; \text{ and thus } q = 2 + 2\cos\left(\frac{8\pi}{\kappa'}\right). \qquad (4.3)$$

In fact, one can also study the self-dual FK percolation model and its associated loops on a Euclidean lattice, and the same belief is held. That is, these loops are also conjectured to converge to $\text{CLE}_{\kappa'}$ in the scaling limit, where the relationship between q and κ' is the same as in (4.3). The fact that these two conjectures are the same should heuristically be considered as a consequence of conformal invariance. That is, if the scaling limit of FK loops is conformally invariant, it should be independent of the underlying metric: only their conformal type should matter.

For instance, we have already noticed that when $q = 1$, the associated FK model is just bond percolation. In this case, we already know (at least in the

4.3 Mullin–Bernardi–Sheffield's bijection for spanning trees

case of the triangular lattice) that the scaling limit of the associated loops is given by CLE with parameter $\kappa' = 6$ ([Smi01], [CN08]). This is consistent with the value $\gamma = \sqrt{8/3}$ being the Liouville quantum gravity parameter for the scaling limit of uniform planar maps, as described in Remark 4.1.

Likewise, for $q = 2$, the associated FK model is the FK representation of the critical Ising model. It was proven in [KS16] (see also [CDCH$^+$14] for interfaces and [BH19] for Ising loops) that the scaling limit of these loops is given by CLE$_{16/3}$. The associated parameter γ is thus $\gamma = \sqrt{3}$.

A short summary of these values is provided in Table 4.1.

Table 4.1 *Summary of the parameters associated with different FK models*

FK Model (4.1)	q	γ	κ'
General $q \in [0,4)$	$2 + 2\cos(\gamma^2 \pi/2)$	$\gamma \in [\sqrt{2}, 2)$	$16/\gamma^2 \in (4, 8]$
Uniform map + critical bond percolation	1	$\sqrt{8/3}$	6
Spanning tree decorated map	0	$\sqrt{2}$	8
Critical Ising decorated map	2	$\sqrt{3}$	16/3

4.3 Mullin–Bernardi–Sheffield's bijection in the case of spanning trees

We will now discuss the case where the map $M_n \in \mathcal{M}_n$ is chosen with probability proportional to the number of spanning trees it admits. Here a spanning tree is a collection of unoriented edges, covering every vertex, and which contains no cycle. (By contrast, in Section 4.4 we will also encounter spanning trees for which a designated vertex called the root has been singled out; in which case one may view the edges in the spanning tree as being oriented towards the root). In other words, for any (rooted) map $m_n \in \mathcal{M}_n$ with n edges and t_n a set of edges on it

$$\mathbb{P}(M_n = m_n, T_n = t_n) \propto \mathbf{1}_{\{t_n \text{ is a spanning tree on } m_n\}}. \tag{4.4}$$

This can be understood as the limit when $q \to 0^+$ of the Fortuin–Kasteleyn model discussed in (4.1), since in this limit the model concentrates on configurations where $\ell = 0$, equivalently, t_n is a tree. In fact, it is immediate in this case that given $M_n = m_n$, t_n is a uniform spanning tree (UST) on

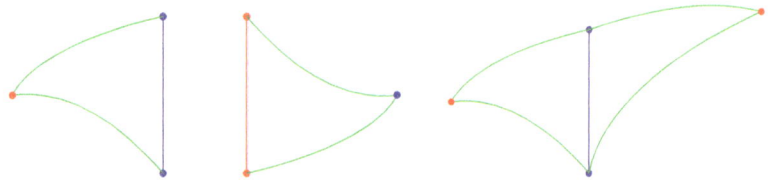

Figure 4.4 Refined or green edges split the map and its dual into primal and dual triangles. Each primal triangle sits opposite another primal triangle, resulting in a primal quadrangle as above.

m_n. We will discuss a powerful bijection due to Mullin [Mul67] and Bernardi [Ber07, Ber08] which is key to the study of such planar maps. This bijection is actually a particular case of a bijection due to Sheffield, which is sometimes called the "hamburger–cheeseburger" bijection. Sheffield's bijection can be used for arbitrary $q \geq 0$, however the case $q = 0$ of trees is considerably simpler and so we discuss it first. (We will use the language of Sheffield, in order to prepare for the more general case later.) Although the hamburger–cheeseburger bijection is the only example we will treat in detail here, we mention that there are other powerful bijections of a similar flavour that can be used to connect random planar map models to Liouville quantum gravity and SLE: see for example [BHS23, LSW24, GKMW18, KMSW19].

To describe the $q = 0$ hamburger–cheeseburger bijection, we first fix a deterministic pair (m_n, t_n) as above (with an oriented root edge chosen for m_n and t_n a spanning tree on m_n) – see Figure 4.6 – and describe how to associate with it a certain sequence of letters corresponding to "hamburgers" and "cheeseburgers". Recall that adding refinement edges to a map splits it into triangles of exactly two types: primal triangles (meaning two refined edges and one primal edge) or dual triangles (meaning two refined edges and one dual edge). For ease of reference, primal triangles will be associated with hamburgers, and dual triangles with cheeseburgers. Note that for a primal edge in a primal triangle, the triangle opposite that edge is obviously a primal triangle too. Hence it is better to think of the map as being split into quadrangles with either a primal or dual diagonal, as illustrated in Figure 4.4.

We will reveal the map, triangle by triangle, by exploring it along a space-filling (in the sense that it visits every triangle once) path. When we do this, we will keep track of the first time that the path enters a given quadrangle by saying that either a hamburger or a cheeseburger is produced, depending on whether the quadrangle is primal or dual. Later on, when the path comes back to the quadrangle for the second and final time, we will say that the burger has been

4.3 Mullin–Bernardi–Sheffield's bijection for spanning trees

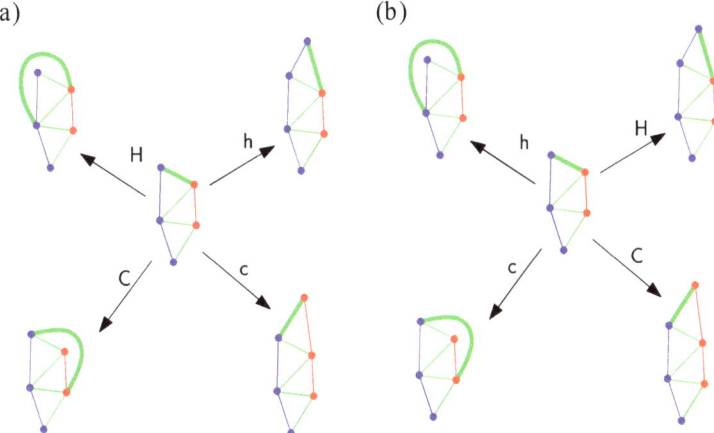

Figure 4.5 From symbols to map. The current position of the interface (or last discovered refined edge) is indicated with a bold line. (a) Reading the word sequence from left to right or *into the future*. The map in the centre is formed from the symbol sequence hch. (b) The corresponding operation when we go from right to left (or into the *past*); this is useful for instance when taking a local limit, see Section 4.6. The map in the centre now corresponds to the symbol sequence HCH.

eaten. We will use the letters h, c to indicate that a hamburger or cheeseburger has been produced and we will use the letters H, C to indicate that a burger has been eaten (or *ordered* and eaten immediately). So in this description, we will have one letter for every triangle.

It remains to specify in what order are the triangles visited; equivalently, to describe the space-filling path. In the case that we consider now, where the decoration t_n consists of a single spanning tree, the path is simply the contour path going around the tree (starting from the root); that is, the unique loop L_0 separating the primal and dual spanning trees, with its orientation inherited from that of the root edge of m_n. Hence in this case, we can associate to (m_n, t_n) a sequence w (or **word**) made up of M letters in the alphabet $\Theta = \{h, c, H, C\}$. We will see below that subject to certain natural conditions on the word w, this map is actually a bijection.

Observe that we always have $M = 2n$. To see why, recall that there is one letter for every triangle, so M is the total number of triangles. Moreover, each triangle can be identified with an edge (or in fact half an edge, because each edge is visited once when the burger is produced and once when it is eaten), and so

$$M = 2(E(t_n) + E(t_n^{\dagger})) = 2(V(t_n) - 1 + V(t_n^{\dagger}) - 1).$$

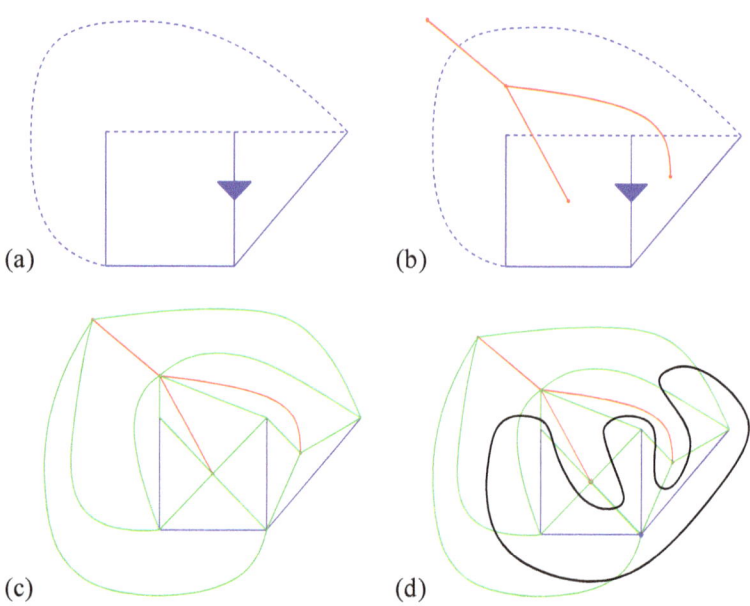

Figure 4.6 (a) a map with a spanning tree. (b) Spanning tree and dual tree. (c) Refinement edges. (d) Loop separating the primal and dual spanning trees, to which a root (refined) edge has been added in bold.

Now $V(t_n) = V(m_n)$, and $V(t_n^\dagger) = V(m_n^\dagger) = F(m_n)$. This gives that

$$M = 2(V(m_n) + F(m_n) - 2), \tag{4.5}$$

and applying Euler's formula together with the fact $E(m_n) = n$, we find that $M = 2n$. Alternatively note directly that $E(t_n) + E(t_n^\dagger) = E(m_n) = n$ since each edge of m_n corresponds to an edge that is either open in t_n or $t_n^\dagger$.

To summarise, given (m_n, t_n) a rooted, spanning tree decorated map with n edges, we can uniquely define a word w of length $2n$ in the letters $\{h, c, H, C\}$. Observe further that under the reduction rules

$$\overline{cC} = \overline{hH} = \emptyset, \ \overline{cH} = \overline{Hc} \text{ and } \overline{hC} = \overline{Ch},$$

we have $\overline{w} = \emptyset$ (here $\overline{w}$ denotes the reduction of the word w). This corresponds to the fact that every burger produced is eaten, and every food order corresponds to a burger that was produced before. Subject to the condition $\overline{w} = \emptyset$, it is easy to see that the map $(m_n, t_n) \mapsto w$ is a bijection. See, for example, Figure 4.5 for elements of a proof by picture.

Now we go a step further, and associate to this word w a pair $(X_k, Y_k)_{1 \leq k \leq 2n}$, which count the number of hamburgers and cheeseburgers, respectively, in

4.3 Mullin–Bernardi–Sheffield's bijection for spanning trees

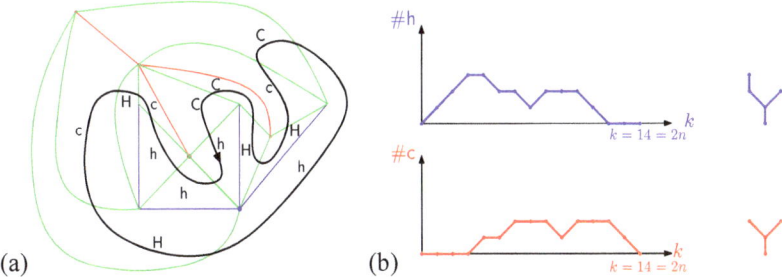

Figure 4.7 (a) The word associated to (m_n, t_n) is: $w = $ hhhcHcHhCcHHCC (b) The hamburger and cheeseburger counts, as well as the trees encoded by these excursions (which are identical to the primal and dual spanning trees, respectively).

the stack at any given time $1 \leq k \leq 2n$ (i.e. the number of hamburgers or cheeseburgers which have been produced prior to time k but get eaten after time k). Note that (X, Y) is a process which starts from the origin at time $k = 0$, and ends at the origin at time $k = 2n$. Moreover, by construction X and Y both stay non-negative throughout. We call a process $(X_k, Y_k)_{0 \leq k \leq 2n}$ satisfying these properties a **discrete excursion** (in the quarter plane). So at this point, we have associated with any (m_n, t_n) as above, a unique discrete excursion (X, Y) of length $2n$.

Conversely, given such a process (X, Y) we can associate to it a word w in the letters of Θ such that (X, Y) is the net burger count of w. Obviously w reduces to $\emptyset$ and so, as we have seen above, this word w specifies a unique pair (m_n, t_n).

Another property which is easy to check (and easily seen on Figure 4.7) is that the excursions X and Y encode the spanning tree t_n and dual spanning tree $t_n^\dagger$ in the sense that they are (after removing steps where X, respectively, Y, remain constant) the contour functions of these trees. More precisely, at a given time k, X_k denotes the height in the tree (distance to the root) of the last vertex discovered prior to time k.

Remark 4.2 It may be useful to recast the above connections in the language of **queues**, where customers are being served one at the time. More precisely, a queue (in discrete time) is a process where at each unit of time either a new customer arrives, or a customer at the front of the queue is being served and leaves the queue forever. Any queue can be equivalently described by a tree t or an excursion X. Indeed, a tree structure t can be defined from the queue, by declaring that any customer c arriving during the service of a customer c' is a child of c'. An excursion X can be defined by simply counting the queue length at each time. Note that X is nothing but the contour function of the tree t (meaning the discrete process which measures the height of the tree t as it goes around it in depth-first order; see [LG05] for much more about this). In our case, the tree t is simply either the primal spanning tree on the map or its dual.

When (M_n, T_n) are *random* and sampled according to (4.4), the corresponding random excursion (X, Y) is clearly chosen uniformly from the set of all possibilities. It therefore follows from classical results of Durrett, Iglehart and Miller [DIM77] that as $n \to \infty$,

$$\frac{1}{\sqrt{n}}(X_{\lfloor 2nt \rfloor}, Y_{\lfloor 2nt \rfloor})_{0 \leq t \leq 1} \to (e_t, e'_t)_{0 \leq t \leq 1},$$

where e, e' are independent Brownian (one-dimensional) excursions (i.e. the pair (e, e') is Brownian excursion in the quarter plane), for example, in the Skorokhod sense (alternatively for the topology of uniform convergence if the paths are linearly interpolated instead of piecewise constant as above). This property implies (see for example Lemma 2.4 in Le Gall's comprehensive survey [LG05]) that, in the Gromov–Hausdorff sense, the primal and dual spanning trees converge after rescaling the distances by a factor $n^{-1/2}$, to a pair of independent **Continuous Random Trees** (CRTs) [Ald93].

We summarise our findings, in the case of UST weighted random planar maps, in the following theorem.

Theorem 4.3 *The set of (rooted) spanning tree decorated maps (m_n, t_n) with n edges are in bijection with excursions $(X_k, Y_k)_{0 \leq k \leq 2n}$ in the quarter plane. When (M_n, T_n) is random and distributed according to (4.4), the pair of trees $(T_n, T_n^\dagger)$ converges, for the Gromov–Hausdorff topology and after scaling distances (in each tree) by a factor $n^{-1/2}$, to a pair of independent Continuous Random Trees (CRTs).*

Note that the map M_n itself can then be thought of as a glueing of two discrete trees (i.e. the primal and dual spanning trees, which are glued along the space-filling path). In the scaling limit, this pair of trees becomes a pair of independent CRTs. As it turns out, the procedure of glueing these two trees has a continuum analogue, which is described in the work of Duplantier, Miller and Sheffield [DMS21]. This is the **mating of trees** approach to LQG, and is an extremely powerful and fruitful point of view that we will describe in more detail later on.

4.4 The loop-erased random walk exponent

A loop-erased random walk (or LERW for short) is the process that one obtains when erasing the loops chronologically as they appear on a simple random walk trajectory. More precisely, fix a vertex x in a locally finite graph G and a subset U of vertices, and suppose that the hitting time $H_U < \infty$, $\mathbb{P}_x$-almost surely, where $\mathbb{P}_x$ denotes the law of simple random walk $(X_n)_{n \geq 0}$ on G starting from x, and $H_U = \inf\{n \geq 0 \colon X_n \in U\}$ is the hitting time of U for that walk.

Definition 4.4 A **loop-erased random walk** from x to U is the process obtained from $(X_n)_{0 \leq n \leq H_U}$ by chronologically erasing the loops from X. More precisely, the loop-erasure $\beta = (\beta_0, \ldots, \beta_\ell)$ is defined inductively as follows: $\beta_0 = x$. If $\beta_n \in U$ then $n = \ell$, else $\beta_{n+1} = X_L$, where $L = 1 + \max\{m \leq H_U \colon X_m = \beta_n\}$.

Somewhat remarkably, the loops can also be erased antichronologically, and this does not change the resulting distribution:

Lemma 4.5 *Let X be a random walk starting from x, stopped at the time $H = H_U$ when it hits U. Let β denote the loop-erasure of X, and let γ denote the loop-erasure of the time reversal $\hat{X} = (X_H, X_{H-1}, \ldots, X_0)$. Then γ has the same law as the time reversal of β.*

This allows us to speak unambiguously of the law of the loop-erasure of a (portion of) simple random walk.

There is a well-known and very deep connection between uniform spanning trees and loop-erased random walks, which was discovered by Wilson [Wil96] (see also Propp and Wilson [PW98]), and may be used to efficiently simulate such trees. This relation is known as **Wilson's algorithm**; see Chapter 4 of [LP16] for a thorough discussion. This relation extends to weighted graphs (provided that one replaces the uniform distribution on the set of spanning trees by a natural Gibbs distribution). However for simplicity, we will not discuss it here and continue to assume that our graphs are unweighted. (As we will see in the proof below, this connection is more naturally expressed when we think of our trees as being oriented towards a designated vertex called the root; beware however that this is in contrast with our definition of spanning trees in Sections 4.1–4.3 as being unoriented.)

Here we will only need the following result, which may be seen as a straightforward consequence of Wilson's algorithm, but which was first discovered by Pemantle [Pem91] (prior to [Wil96]). We state and prove it here, since the proof is short and rather beautiful.

Theorem 4.6 *Let G be an arbitrary finite (connected, unoriented) graph G,* and let T be a uniform spanning tree on G, that is, a uniformly chosen subset of unoriented edges which is acyclic and spanning. Let x, y be any fixed vertices in G. Then the unique branch of T between x and y has the same distribution as (the trace of) a loop-erased random walk from x run until it hits y.

Sketch of proof. A *rooted spanning tree* is just a pair (t, x), where t is an (unrooted) spanning tree and x a fixed vertex of G. Alternatively, we can view the rooted tree (t, x) as an oriented tree, where all the edges of t are oriented towards the root x. This is also known as an arborescence; it is a rooted, directed acyclic graph in which each vertex except the root has a unique oriented edge leading out of it. We will sketch the proof of the theorem for the measure on rooted trees given by

$$\mathbb{P}((T, X) = (t, x)) \propto \pi(x) = \deg(x) \qquad (4.6)$$

and by describing the law of the branch between y and X, conditional on $\{X = x\}$; the stated result then follows easily.

For a possibly infinite path $\gamma = (\gamma_0, \gamma_1, \ldots)$ on the vertices V of the graph, let $T(\gamma)$ be the set of oriented edges of the form $(\gamma_{H(w)-1}, w)$ where $w \neq \gamma_0$, and $H(w) = \inf\{n \geq 0 \colon \gamma_n = w\}$ is the first hitting time of w by the path γ. In other words, in $T(\gamma)$ we keep the edge (γ_n, γ_{n+1}) if and only if γ_{n+1} has not been previously visited. It is obvious that this generates an acyclic graph, and if the path visits every vertex (which will be almost surely the case when γ has the law of a random walk, by recurrence of G), then $T(\gamma)$ is a spanning tree rooted at $o = \gamma_0$. Note also that in $T(\gamma)$, the unique branch connecting the root o and a given vertex w can be described by chronologically erasing the loops of the time-reversed path $(\gamma_{H(w)}, \gamma_{H(w)-1}, \ldots, \gamma_0)$.

Appealing to Lemma 4.5, in order to conclude the proof of the theorem, it suffices to show that when γ_0 is chosen according to the stationary distribution of the graph G (i.e. proportionally to the degree of a vertex) and $\gamma = (\gamma_0, \gamma_1, \ldots,)$ is a random walk starting from γ_0, then the law of $(T(\gamma), \gamma_0)$ is the one in (4.6).

To do this, suppose that $(X_n)_{n \in \mathbb{Z}}$ is a bi-infinite stationary random walk on G (constructed, for example, using Kolmogorov's extension theorem), so that X_0 is distributed according to its equilibrium measure π, and let $\gamma^{(n)} = (X_{-n}, X_{-n+1}, \ldots)$ be the path started from $\hat{X}_n := X_{-n}$. Then the claim is that $(T(\gamma^{(n)}), \hat{X}_n)$ defines a certain Markov chain on rooted spanning trees. Indeed a straightforward computation using the definition of conditional probability and the fact that π is a reversible measure for X show that $\hat{X}_n$ is itself a Markov

4.4 The loop-erased random walk exponent

chain, with transition probabilities

$$\hat{p}(x,y) = p(y,x)\frac{\pi(y)}{\pi(x)},$$

where p are the original transitions of the simple random walk on G. In particular, given $(T(\gamma^{(n)}), \hat{X}_n) = (t,x)$, the probability that $\hat{X}_{n+1} = y$ is equal to $\hat{p}(x,y)$. Moreover, given $(T(\gamma^{(n)}), \hat{X}_n, \hat{X}_{n+1}) = (t,x,y)$, $T(\gamma^{(n+1)})$ is obtained deterministically from t by adding the edge $e = (x,y)$ (which creates a cycle) and removing from t the unique outgoing edge from y. The next state of the chain corresponds to $(T(\gamma^{(n+1)}), y)$. (This corresponds to what Lyons and Peres [LP16] describe as the forward procedure, but applied to the time-reversed chain $\hat{X}$.)

It is not hard to see that this Markov chain on the space of rooted trees is irreducible. A calculation (involving the so-called "backward procedure", see Section 4.4 in [LP16]) shows further that the unique stationary distribution of this chain is proportional to

$$\mathbb{P}((T,X) = (t,x)) \propto \psi((t,x)) := \prod_{\vec{e}} p(\vec{e}),$$

where the product is over all the oriented edges $\vec{e}$ of the rooted tree (t,x). (We warn the reader however that ψ is not in general a reversible measure for this chain.) Furthermore, it is a classical fact, known as the **Markov chain tree theorem**, that for general weighted graphs, ψ is in fact itself proportional to the invariant measure of the associated random walk. In the case which occupies us here where the graph is unweighted (i.e. all edges have unit weight), this is particularly easy to see: indeed, since t is spanning and every vertex except the root has exactly one outgoing oriented edge,

$$\prod_{\vec{e}} p(\vec{e}) = \prod_{v \neq x} \frac{1}{\deg(v)} = \frac{\deg(x)}{\prod_{v \in G} \pi(v)} \propto \pi(x);$$

that is, the invariant distribution is given exactly by (4.6). On the other hand, since $(\gamma^{(n)}, \hat{X}_n)_{n \in \mathbb{Z}}$ is stationary, so must be $(T(\gamma^{(n)}), \hat{X}_n)_{n \in \mathbb{Z}}$ (because $T(\gamma^{(n)})$ is a deterministic function of $\gamma^{(n)}$). In particular, $(T(\gamma^{(0)}), \hat{X}_0)$ has distribution (4.6), as required. □

Remark 4.7 The fact that $T(\gamma^{(0)})$ has the law (4.6) is closely related to the algorithm of Aldous [Ald90] and Broder [Bro89] for sampling a uniform spanning tree. As mentioned in [LP16], both authors credit Persi Diaconis for discussions. This algorithm was initially used in the study of the Uniform Spanning Tree (notably by Pemantle [Pem91]) before Wilson's algorithm ([Wil96, PW98]) became available.

However, we note that in fact, Wilson's algorithm to generate a full UST is a simple consequence of Theorem 4.6: indeed, to generate the tree T we can first sample the branch connecting a fixed vertex x to the boundary using a loop-erased random walk by Theorem 4.6. The conditional law of the rest of the tree is then a uniform spanning tree on a modified graph where this branch has been wired to make a single vertex and become part of the boundary. Applying Theorem 4.6 recursively in this manner then gives Wilson's algorithm.

Of course, direct (and relatively short) proofs of this algorithm exist. See, in particular, [LP16, Chapter 4.1] for a proof close to the original spirit of [Wil96], [LL10] for a proof using loop measures, and finally see [WP21, Chapter 2.1] for a proof based on the Green function and the discrete Laplacian.

We may deduce from Theorem 4.3 the following result about the loop-erased random walk.

Theorem 4.8 *Let* (M_n, T_n) *be chosen as in* (4.4) *and let* x, y *be two vertices chosen independently and uniformly on* M_n. *Let* $(\Lambda_k)_{0 \le k \le \xi_n}$ *be a LERW starting from x, run until the random time ξ_n when it first hits y. Then*

$$\frac{\xi_n}{\sqrt{n}} \to \xi_\infty$$

in distribution, where ξ_∞ is a random variable that has a non-degenerate distribution (in the sense that $\xi_\infty \in (0, \infty)$ almost surely).

Proof Let $(X_k, Y_k)_{1 \le k \le 2n}$ be the pair of excursions which describes the map (M_n, T_n). Then note that ξ_n may be identified with the "tree distance" $X(J_1) + X(J_2) - 2\min_{j \in [J_1, J_2]} X(j)$ where J_1, J_2 are uniformly (and independently) chosen between 1 and $2n$. As a consequence, Theorem 4.8 holds with $\xi_\infty = e(U_1) + e(U_2) - \inf_{u \in [U_1, U_2]} e(u)$, where e is a Brownian excursion and U_1, U_2 are chosen uniformly and independently from $(0, 1)$. □

Remark 4.9 In fact, as was already observed by Aldous [Ald93], the continuum random tree is invariant "under rerooting", that is, moving to the root to a uniformly chosen position. As a consequence, the law of the random variable ξ_∞ above may be more simply written as $e(U)$, where U is a uniform random variable on $(0, 1)$. In fact, as noted in [Ald93], this can be derived directly from a simple path transformation of the Brownian excursion. See also [DLG09] for a discussion in the more general context of Lévy trees.

Scaling exponent of LERW. We now explain how the above result can be used to compute an exponent for the loop-erased random walk. Let $\Lambda = \{\Lambda_0, \ldots, \Lambda_{\xi_n}\}$ denote the loop-erasure of a random walk on M_n, run from a uniformly chosen vertex x until the hitting time of another uniformly chosen vertex y, as above. Then Λ may be viewed as an independent random "fractal" set on M_n, whose size is $|\Lambda| = \xi_n = n^{1/2+o(1)}$ by Theorem 4.8. Since M_n has $n^{1+o(1)}$ vertices (indeed it has by definition n edges, and the degree distribution of a given vertex is known to be very concentrated), this means that Λ has a **quantum scaling exponent** given by

$$\Delta = 1/2$$

(recall our discussion from Section 3.12.4). We can therefore (at least informally) use the **KPZ relation** to compute the equivalent exponent for a loop-erased random walk on the square lattice. To do so, we must first find the correct value of γ: the constant in front of the GFF which describes the scaling limit of the conformally embedded planar map M_n. This is given by the relation (4.2) when $q = 0$ (which, as explained at the beginning of this section, indeed corresponds to the uniform spanning tree weighted map model of (4.4)). Plugging $q = 0$ yields

$$\gamma = \sqrt{2}.$$

Note that this is consistent with the conjecture (known to be true on the square lattice by results of [LSW04]) that the interface separating a uniform spanning tree from its dual, converges in the scaling limit to an SLE curve with parameter $\kappa' = 8$.

Therefore, the **Euclidean scaling exponent** x of the loop-erased random walk should satisfy

$$x = \frac{\gamma^2}{4}\Delta^2 + (1 - \frac{\gamma^2}{4})\Delta = 3/8.$$

In particular, we conclude that in the scaling limit, a loop-erased random walk on the square lattice has dimension

$$d_{\text{Hausdorff}} = 2 - 2x = 5/4.$$

This is in accordance with Beffara's formula [Bef08] for the dimension of SLE: indeed, in the scaling limit, LERW is known to converge to an SLE_κ curve with $\kappa = 2$. This is closely related to the above-mentioned scaling limit result for the UST, due to Lawler, Schramm and Werner [LSW04], and is also proved in [LSW04]. Beffara's result [Bef08] states that the Hausdorff dimension of SLE_κ is $(1 + \kappa/8) \wedge 2$. In the case $\kappa = 2$ this is exactly $5/4$, as above.

In fact, this exponent for LERW had earlier been derived by Kenyon in a remarkable paper [Ken00], building on his earlier work on the dimer model and the Gaussian free field [Ken01].

4.5 Sheffield's bijection in the general case

We now describe the situation when $\overline{m}_n \in \mathcal{M}_n$ but the collection of edges t_n is arbitrary (i.e. not necessarily a tree), which is more delicate. Note that in the case of spanning trees, there was only one loop present, but now there will generally be more than one. These loops are **densely packed** in the sense that every triangle is part of some loop, as illustrated in Figure 4.1. Indeed, each triangle consists of an edge of some type and a vertex of the opposite type, so must contain a loop separating the two associated clusters. In this case, we will see that we can still define a canonical space-filling interface (i.e. a curve which visits every single triangle exactly once). We will now describe this curve (see also Figure 4.8).

Recall that L_0 is the loop containing the root triangle of the map $\overline{m}_n$, oriented parallel to the orientation of the root edge of m_n. We view L_0 as an oriented collection of adjacent triangles (the triangles traversed by the loop). In general, L_0 does not cover every triangle of $\overline{m}_n$, and we may consider the connected components $C_1, \ldots, C_k$ which are obtained by removing all the triangles of L_0. Note that L_0 is adjacent to each of these components, in the sense that for each $1 \leq i \leq k$, it contains a triangle that is opposite a triangle in C_i. For each i, let T_i be the *last* (with respect to the orientation of the loop and its starting point) triangle that is adjacent to C_i. The triangle opposite T_i is in C_i and together they form a quadrangle. In order to explore all of the map and not just L_0, we will first modify the map by *flipping* the diagonal of this quadrangle, for every $1 \leq i \leq k$. It can be seen that having done so, we have reduced the number of loops on the map by exactly k (each such flipping has the effect of merging two loops). We may then iterate this procedure until there is only a single loop left, the loop L_0 (which now fills the whole map). This loop separates primal and dual clusters of the modified map, in the sense that it has only primal clusters on one side and dual clusters on the other (we will see below that these clusters are in fact spanning trees).

So we now have a canonical space-filling path which allows us to explore the map as in Section 4.3. As before, we can describe the type of triangles we see in this exploration using the symbols h, c, H, C. When we explore a triangle corresponding a flipped quadrangle for the first time, we record its type (either h, c) according to its type after having flipping the edge. However, when we visit its opposite triangle, we record the fact that this is a special edge (which must be flipped to recover the original map) by the symbol F. The letter F stands for "flexible" or "freshest" order. (We will see below a more precise interpretation in terms of queues, or hamburgers and cheeseburgers.) In this way, we may

4.5 Sheffield's bijection in the general case

associate to the decorated map (m_n, t_n) a list w of $2n$ symbols $w = (X_i)_{1 \le i \le 2n}$ taking values in the alphabet $\Theta = \{\mathsf{h}, \mathsf{c}, \mathsf{H}, \mathsf{C}, \mathsf{F}\}$.

We will see below the properties of this word (essentially, it reduces to $\emptyset$ with the appropriate definition of reduction when there is an F) and that the map from (m_n, t_n) to w, subject to this constraint, is a bijection. For now, we make the important observation that each loop corresponds to a unique symbol F, except for the loop through the root.

Inventory accumulation. Now, we can interpret an element in $\{\mathsf{h}, \mathsf{c}, \mathsf{H}, \mathsf{C}\}^{2n}$ as a last in, first out inventory accumulation process in a burger factory with two types of product: hamburgers and cheeseburgers. Think of a sequence of events, occurring once per unit time, in which either a burger is produced (either ham or cheese) or there is an order of a burger (either ham or cheese). The burgers are put in a single **stack** and every time there is an order of a certain type of burger, the freshest burger in the stack of the corresponding type is removed. The symbol h (resp. c) corresponds to a ham (resp. cheese) burger production and the symbol H (resp. C) corresponds to a ham (resp. cheese) burger order.

The inventory interpretation of the symbol F is the following: this corresponds to a customer demanding the freshest or the topmost burger in the stack, irrespective of the type. In particular, whether an F symbol corresponds to a hamburger or a cheeseburger order depends on the topmost burger type at the time of the order. Thus overall, we can think of the inventory process as a sequence of symbols in Θ with the following reduction rules

- $\overline{\mathsf{cC}} = \overline{\mathsf{cF}} = \overline{\mathsf{hH}} = \overline{\mathsf{hF}} = \emptyset$,
- $\overline{\mathsf{cH}} = \overline{\mathsf{Hc}}$ and $\overline{\mathsf{hC}} = \overline{\mathsf{Ch}}$.

Given a sequence of symbols w, we denote by $\overline{w}$ the reduced word formed via the above reduction rule.

Reversing the construction. Given a sequence w of symbols from Θ, such that $\overline{w} = \emptyset$, we can construct a decorated map (m_n, t_n) as follows. First, we convert all the F symbols to either an H or a C symbol depending on its order type. Then, we construct a spanning tree decorated map as described in Section 4.3 (see in particular Figure 4.5). The condition $\overline{w} = \emptyset$ ensures that we can do this. To obtain the original loop decorated map, we simply flip the type of every quadrangle which has one of the triangles corresponding to an F symbol. That is, if a quadrangle formed by primal triangles has one of its triangles coming from an F symbol, then we replace the primal map edge in that quadrangle by the corresponding dual edge and vice versa. The interface is now divided into

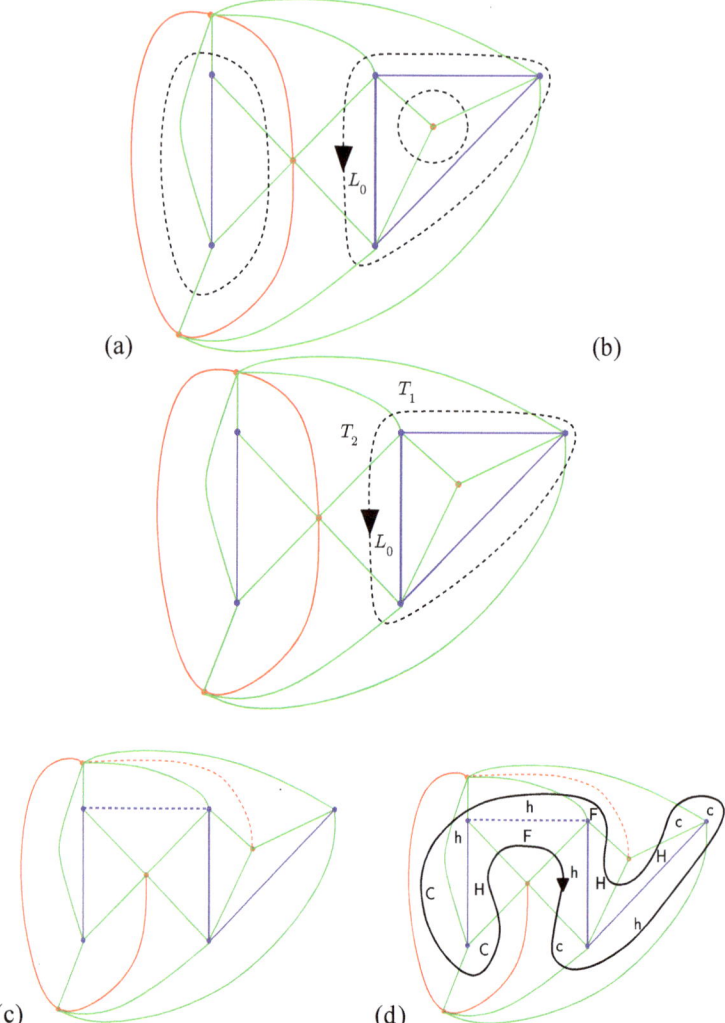

Figure 4.8 Generating a word from a decorated map in the general case. (a) The decorated map is as in Figure 4.1, with the (oriented) root loop L_0. (b) The complement of L_0 consists of two components, C_1 and C_2. T_1 and T_2 are the *last* triangles visited by the loop L_0 that share an edge with a triangle in C_1 and C_2 respectively. (c) We flip the diagonals of the quadrangles associated with T_1 and T_2. (d) We obtain a single space-filling loop (drawn in black). To this path, we can again associate a word in {h, c, H, C}. However, we also record the second visit to a flipped quadrangle by replacing the symbol C or H by the symbol F. The word here is thus hchccHHFhhCCHF. Note the non-obvious fact that after flipping, the primal and dual clusters have become trees.

4.5 Sheffield's bijection in the general case 155

several loops (and the number of loops is exactly one more than the number of F symbols). In particular:

Theorem 4.10 (Sheffield, [She16b]) *The map* $(\bm{m}_n, \bm{t}_n) \mapsto w$ *(subject to $\overline{w} = \emptyset$) is a bijection.*

Two canonical spanning trees. It is not obvious but true that after flipping, the corresponding primal and dual decorations of the map have become two mutually dual spanning trees: see Figure 4.9 for an illustration. One way to see this is as follows: observe that after flipping, we have (as already argued) a single space-filling loop which separates primal and dual clusters of the resulting modified map. These clusters are of course spanning, and they cannot contain non-trivial cycles, else the loop would either not be space-filling or consist of multiple loops. Therefore, we can again think of M_n as a glueing of two spanning trees, which are glued along the space-filling path (i.e. along their

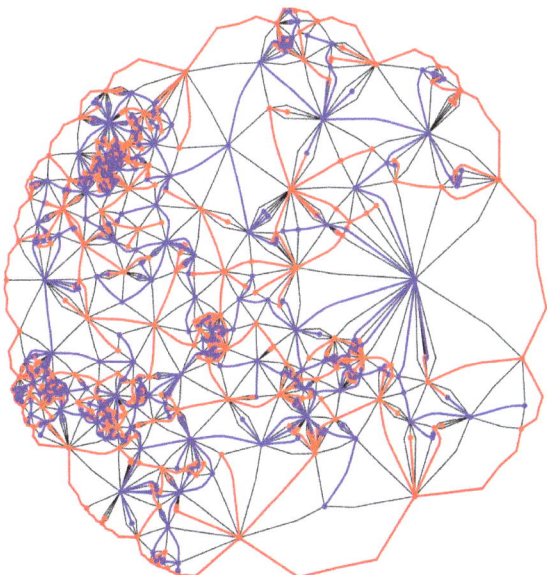

Figure 4.9 A random planar map with law (4.1) for $q = 1$ (uniform case), generated using Sheffield's bijection of Theorem 4.10. The map has been embedded using circle packing. Shown in blue and red are the primal and dual spanning trees. In the infinite volume limit and then in the scaling limit, Theorem 4.13 shows that these trees become correlated infinite CRTs. Simulation by Jason Miller.

contour functions). Again, this perspective is a crucial intuition which guides the **mating of trees approach** to Liouville quantum gravity [DMS21]. We will survey this later on (see Section 9.7.1).

Generating FK-weighted maps. A remarkable consequence of Theorem 4.10 is the following simple way of generating a random planar map from the FK model (4.1). Fix $p \in [0, 1/2)$, which will be suitably chosen (as a function of q) in (4.9). Let $(X_1, \ldots, X_{2n}) \in \Theta^{2n}$ be i.i.d. with the following law

$$\mathbb{P}(\mathsf{c}) = \mathbb{P}(\mathsf{h}) = \frac{1}{4}, \mathbb{P}(\mathsf{C}) = \mathbb{P}(\mathsf{H}) = \frac{1-p}{4}, \mathbb{P}(\mathsf{F}) = \frac{p}{2}, \quad (4.7)$$

conditioned on $\overline{X_1, \ldots, X_{2n}} = \emptyset$.

Let (M_n, T_n) be the random associated decorated map (via the bijection described above). Then observe that since n hamburgers and cheeseburgers must be produced, and since $\#\mathsf{H} + \#\mathsf{C} = n - \#\mathsf{F}$,

$$\mathbb{P}((M_n, T_n) = (\boldsymbol{m}_n, \boldsymbol{t}_n)) = \left(\frac{1}{4}\right)^n \left(\frac{1-p}{4}\right)^{\#\mathsf{H}+\#\mathsf{C}} \left(\frac{p}{2}\right)^{\#\mathsf{F}}$$

$$\propto \left(\frac{2p}{1-p}\right)^{\#\mathsf{F}} = \left(\frac{2p}{1-p}\right)^{\#\ell(\boldsymbol{m}_n, \boldsymbol{t}_n)-1}. \quad (4.8)$$

Thus we see that (M_n, T_n) is a realisation of the critical FK-weighted cluster random map model with

$$\sqrt{q} = \frac{2p}{(1-p)}. \quad (4.9)$$

Notice that $p \in [0, 1/2)$ corresponds to $q = [0, 4)$. From now on, we fix the value of p and q in this regime. Recall that $q = 4$ is believed to be a critical value for many properties of the map; indeed later on we will later show that a phase transition occurs at $p = 1/2$ ($q = 4$) for the geometry of the map. Intuitively, it is perhaps not surprising that the value $p = 1/2$ marks a distinction from the point of view of inventory accumulation.

4.6 Infinite volume limit

The following theorem due to Sheffield [She16b] and made more precise later by Chen [Che17], shows that the decorated map (M_n, T_n) has a local limit as $n \to \infty$ in the local topology. Roughly two (decorated) maps are close in the local topology if the finite maps (and their decorations) near a large neighbourhood of the root are isomorphic as decorated maps.

4.6 Infinite volume limit   157

Theorem 4.11 ([She16b, Che17])   *Fix $p \in [0, \frac{1}{2})$. We have*

$$(M_n, T_n) \xrightarrow[n \to \infty]{(d)} (M, T)$$

with respect to the local topology, where (M, T) is the unique infinite decorated map associated with a bi-infinite i.i.d. sequence of symbols $(X_n)_{n \in \mathbb{Z}}$ having law (4.7).

Sketch of proof   We now give the idea behind the proof of Theorem 4.11. Let $X_1, \ldots, X_{2n}$ be i.i.d. with law given by (4.7), and denote by E_{2n} the event that $\overline{X_1 \cdots X_{2n}} = \emptyset$.
A key step is to show the following.

Lemma 4.12 ([She16b, GS17])   *Let $X_1, \ldots, X_{2n}$ be i.i.d. with law (4.7). Then $\mathbb{P}(E_{2n})$ decays subexponentially in n, that is, $\log \mathbb{P}(E_{2n})/n \to 0$ as $n \to \infty$.*

We will not prove this statement (although we will later come back to it and explain it informally). Instead, we explain how Theorem 4.11 follows.
Notice that uniformly selecting a symbol $1 \leq I \leq 2n$ corresponds to selecting a uniform triangle in $(\overline{M}_n, T_n)$, which in turn corresponds to a unique oriented edge in M_n (recall that $\overline{M}_n$ denotes the refinement map associated to M_n). Because of invariance of the decorated map (M_n, T_n) under re-rooting, we claim that it suffices to check the convergence in distribution of a large neighbourhood of the triangle corresponding to X_I in $\overline{M}_n$.
Let $r > 0$. We will first show that for any fixed word w of length $2r + 1$ in the alphabet Θ,

$$\mathbb{P}(X_{I-r} \cdots X_{I+r} = w \mid E_{2n}) \to \mathbb{P}(w) := \mathbb{P}(X_{-r} \cdots X_r = w), \quad (4.10)$$

where on the left-hand side, the addition of indices has to be interpreted cyclically within $\{1, \ldots, 2n\}$, and on the right-hand side, $(X_n)_{n \in \mathbb{Z}}$ is the random bi-infinite word whose law is described in Theorem 4.11.
To see (4.10), observe that the conditional probability on the left-hand side (conditionally given the entire sequence $X = (X_1, \ldots, X_{2n})$ satisfying E_{2n}, and averaging just over I), is equal to $f + o(1)$ as $n \to \infty$, where f is the fraction of occurrences of w in X, that is, $f = (2n)^{-1} \sum_{i=r+1}^{2n-2r-1} 1_{\{X_{i-r}, \ldots, X_{i+r} = w\}}$, and the $o(1)$ term is uniform, accounting for boundary effects. Hence it suffices to check that $\mathbb{E}(f \mid E_{2n}) \to \mathbb{P}(w)$. To do this, for arbitrary $\varepsilon > 0$, we define $A_n = \{|f - \mathbb{P}(w)| \leq \varepsilon\}$, and write

$$\mathbb{E}(f \mid E_{2n}) = \mathbb{E}(f 1_{A_n} \mid E_{2n}) + \mathbb{E}(f 1_{A_n^c} \mid E_{2n}).$$

Now the first term $\mathbb{E}(f 1_{A_n} \mid E_{2n})$ is equal to $(\mathbb{P}(w) + O(\varepsilon))\mathbb{P}(A_n \mid E_{2n})$, while the second term satisfies

$$\mathbb{E}(f 1_{A_n^c} \mid E_{2n}) \leq \mathbb{P}(A_n^c \mid E_{2n}) \leq \frac{\mathbb{P}(A_n^c)}{\mathbb{P}(E_{2n})}.$$

However, $\mathbb{P}(A_n^c) \to 0$ *exponentially fast* as $n \to \infty$, by basic large deviation estimates (Cramer's theorem). This means that $\mathbb{E}(f 1_{A_n^c} | E_{2n})$ converges to zero by Lemma 4.12, and also that $\mathbb{P}(A_n \mid E_{2n}) \to 1$ as $n \to \infty$. We can conclude that $\mathbb{E}(f 1_{A_n})$ and therefore $\mathbb{E}(f \mid E_{2n})$ converges to $\mathbb{P}(w)$ as $n \to \infty$, which proves (4.10).

To conclude the theorem, it remains to show that convergence of the symbols locally around a letter implies local convergence of the maps. This is a consequence of Exercise 4.1; see also Figure 4.5. □

One important feature related to Theorem 4.11 is that every symbol in the i.i.d. sequence $\{X_i\}_{i \in \mathbb{Z}}$ has an almost sure unique **match**, meaning that every burger order is fulfilled (it corresponds to a burger that was produced at a finite time before), and every burger that is produced is consumed at some finite later time, both with probability 1; see [She16b, Proposition 3.2]. In the language of maps, this is equivalent to saying that the map M has no edge "to infinity". For future reference, let $\varphi(i)$ denote the match of the ith symbol. Notice that $\varphi \colon \mathbb{Z} \mapsto \mathbb{Z}$ defines an involution on the integers.

4.7 Scaling limit of the two canonical trees

We now state (without proof) one of the main results of Sheffield [She16b], which gives a scaling limit result for the geometry of the infinite volume map (M, T) defined in Theorem 4.11. Recall that (M, T) is completely described by a doubly infinite sequence $(X_n)_{n \in \mathbb{Z}}$ of i.i.d symbols in the alphabet Θ, having law (4.7). Associated to such a sequence, we can define two processes $(H_n)_{n \in \mathbb{Z}}$ and $(C_n)_{n \in \mathbb{Z}}$ which count the respective number of hamburgers and cheeseburgers present in the queue at time $n \in \mathbb{Z}$ (of course, we convert the flexible F orders into their appropriate values to count the numbers of hamburgers and cheeseburgers in the queue at time n). These numbers are defined relative to time 0, so $(H_0, C_0) = (0, 0)$. In other words, let $\tilde{w} = (\tilde{X}_n)_{n \in \mathbb{Z}}$ denote the infinite word obtained from $w = (X_n)_{n \in \mathbb{Z}}$ by transforming the F symbols into their actual values H and C, and let

$$H_n = \begin{cases} \sum_{i=1}^n 1_{\{\tilde{X}_i = \mathsf{h}\}} - 1_{\{\tilde{X}_i = \mathsf{H}\}} & \text{if } n > 0 \\ \sum_{i=n}^{-1} 1_{\{\tilde{X}_i = \mathsf{h}\}} - 1_{\{\tilde{X}_i = \mathsf{H}\}} & \text{if } n < 0, \end{cases}$$

similarly for C_n.

4.7 Scaling limit of the two canonical trees

This scaling limit is most conveniently phrased as a scaling limit for $H = (H_n)_{n \in \mathbb{Z}}$ and $C = (C_n)_{n \in \mathbb{Z}}$ (although the statement of Sheffield [She16b] concerns instead $H + C$ and the discrepancy $H - C$). We first state the result and then make some comments on its significance below.

Theorem 4.13 *Let $p \in [0, 1]$, and let C, H be as above. Then*

$$\left(\frac{H_{\lfloor nt \rfloor}}{\sqrt{n}}, \frac{C_{\lfloor nt \rfloor}}{\sqrt{n}} \right)_{-1 \le t \le 1} \to (L_t, R_t)_{-1 \le t \le 1}$$

in distribution as $n \to \infty$ for the topology of uniform convergence, where $(L_t, R_t)_{t \in \mathbb{R}}$ is a two-sided Brownian motion in $\mathbb{R}^2$, starting from 0 and having covariance matrix given by

$$\mathrm{Var}(L_t) = \mathrm{Var}(R_t) = \frac{1+\alpha}{4}|t| \quad ; \quad \mathrm{Cov}(L_t, R_t) = \frac{1-\alpha}{4}|t|$$

and

$$\alpha = \max(1 - 2p, 0).$$

See [She16b, Theorem 2.5] for a proof. We now make a few important remarks about this statement.

- This scaling limit result should be thought of as saying something about the large-scale geometry of the map (M, T) or, equivalently, what it looks like after scaling down by a large factor. However, what this actually means is not *a priori* obvious: really, the theorem only says that the pair of trees converges to correlated (infinite) CRTs. This is a (relatively weak) notion of convergence which has been called **peanosphere topology**; see more about this in Chapter 9. In particular, it does not say anything about convergence of the metric on M.
- Notice that when $p \ge 1/2$ (corresponding to $q \ge 4$ in terms of the FK model (4.1), see (4.9)) we have $\alpha = 0$, so $L_t = R_t$ for all $t \in \mathbb{R}$. This is because the proportion of F orders is large enough that there can be no discrepancy in the scaling limit between hamburgers and cheeseburgers.
- However, when $p \le 1/2$ (corresponding to $q \le 4$), the correlation between L and R is non-trivial. When $p = 0$ (corresponding to $q = 0$), they are actually independent. This last case should be compared with the case of spanning tree weighted maps (Theorem 4.3). In general, this suggests that the scaling limit of the map (M, T), if it exists, can be viewed as a gluing of two (possibly correlated) infinite CRTs; meaning that their contour (or alternatively their height) functions are described by a two-sided infinite Brownian motion (rather than a Brownian excursion of duration one). This fact is made rigorous (and will be discussed later on in Section 9.7.1) in the

mating of trees approach to LQG of [DMS21]. Note in particular that in the case $q \geq 4$, the two corresponding trees are identical, meaning that the map should degenerate to a CRT in the scaling limit. This is in contrast with the case $q < 4$, where the limit maps are expected to be homeomorphic to the sphere almost surely.

- H_n, C_n also have a geometric interpretation, as the boundary lengths at time n on the left- and right-hand sides of the space-filling interface (relative to time 0).

4.8 Exponents associated with FK-weighted random planar maps

Note: In this short section, some critical exponents of random planar maps are computed heuristically. This section can be skipped on a first reading, as none of those results are needed later on.

It is possible to use Theorem 4.13 to obtain very precise information on the geometry of loops on the map (M, T). In particular, it is possible to check that large loops have statistics that coincide with those of $\mathrm{CLE}_{\kappa'}$, where the value of κ' is related to $q \in (0, 4)$ via (4.3), thereby giving credence to the general conjectures formulated in Section 4.2. This line of reasoning has been pursued very successfully in a string of papers by Gwynne, Mao and Sun [GMS19, GS17, GS15]. We will present here a slightly less precise (but easier to state) result proved in [BLR17]. Let $(X_i)_{i \in \mathbb{Z}}$ denote the symbols encoding (M, T), and let us condition on the event $X_0 = \mathsf{F}$. This F symbol is associated to a loop in M (which by definition goes through the triangle encoded by X_0). Let L denote its length (the number of triangles through which the loop passes) and A its area (number of triangles surrounded by it).

Let

$$p_0 = \frac{\pi}{4 \arccos\left(\frac{\sqrt{2-\sqrt{q}}}{2}\right)} = \frac{\kappa'}{8} \in (1/2, 1), \tag{4.11}$$

where q and κ' are related as in (4.3). The following is the main result in [BLR17].

Theorem 4.14 *Let $0 < q < 4$. The random variables L and A satisfy*

$$\mathbb{P}(\mathsf{L} > k) = k^{-1/p_0 + o(1)}, \tag{4.12}$$

4.8 Exponents associated with FK-weighted random planar maps

and

$$\mathbb{P}(\mathsf{A} > k) = k^{-1+o(1)} \quad (4.13)$$

as $k \to \infty$.

As noted in [BLR17], the laws of L and A correspond, respectively, to the limits of the length and area of a *uniformly* chosen loop in the finite decorated planar map (M_n, T_n) as $n \to \infty$. (By contrast, if we consider without any conditioning the length and area of the loop going through the triangle encoded by X_0, this would lead to different exponents, due to a size-biasing effect.)

Results in [GMS19, GS17, GS15] are analogous and more precise, in particular showing regular variation of the tail at infinity. (As a consequence, the sum of loop lengths and areas, in the order that they are discovered by the space-filling path, can be shown to converge after rescaling to a stable Lévy process with appropriate exponent.)

A particular consequence of Theorem 4.14 is that we expect the longest loop in the map M_n to have size roughly $n^{p_0+o(1)}$; that is,

$$\max_{\ell \in (M_n, T_n)} |\ell| = n^{p_0+o(1)} \quad (4.14)$$

as $n \to \infty$. Heuristically, to derive (4.14), one then observes that M_n contains order n loops whose lengths are roughly i.i.d. with tail exponent $\alpha = 1/p_0$. The maximum value of this sequence of lengths is then easily shown to be of order $n^{1/\alpha+o(1)} = n^{p_0+o(1)}$.

We will not prove Theorem 4.14, but we will discuss in Exercise 4.4 an interesting application using the KPZ formula. These exponents are obtained (both in [BLR17] and [GMS19, GS17, GS15]) through a connection with a random walk in a cone. A simple setting, where it is easier to see this connection, is in the following result.

Proposition 4.15 ([GS17]) *Let $0 < q < 4$, and let E_{2n} be the event that the word $w = X_1 \cdots X_{2n}$ reduces to $\overline{w} = \emptyset$. Then*

$$\mathbb{P}(E_{2n}) = n^{-2p_0-1+o(1)} = n^{-1-\kappa'/4+o(1)},$$

as $n \to \infty$. In particular, $\mathbb{P}(E_{2n})$ decays subexponentially.

Sketch of proof We give a rough idea of where this exponent comes from, as it allows us to illustrate the connection to random walk in a cone, as mentioned above. A rigorous proof of this result may be found in [GS17].

The first step is to describe E_{2n} in terms of the burger count processes H and C of Theorem 4.13. In particular, we note that the event E_{2n} is equivalent to the conditions

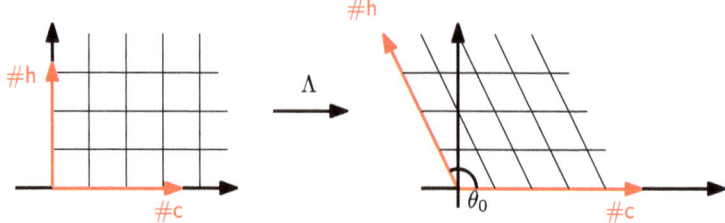

Figure 4.10 The coordinate transformation. In these new axes, the burger counts H and C become independent Brownian motions; the event E_{2n} then corresponds to $\Lambda(Z)$ making an excursion of duration $2n$ in the cone $C(\theta_0)$ of angle $\theta_0 = \pi/(2p_0) = 4\pi/\kappa'$, starting and ending at its apex.

- $C_i, H_i \geq 0$ for $0 \leq i \leq 2n$; and
- $C_{2n} = 0, H_{2n} = 0$

on H and C. Indeed, the first condition holds since if at some point $1 \leq k \leq 2n$ the burger count C or H becomes negative, this must be because of an order whose match in the bi-infinite sequence $(X_k)_{k\in\mathbb{Z}}$ was in the past, that is, $\varphi(k) < 0$. Therefore, the event E_{2n} is equivalent to the process $Z_k = (C_k, H_k)_{1 \leq k \leq 2n}$ being an excursion in the top right quadrant of the (C, H) plane, starting and ending at the origin.

This probability may be computed approximately (or rather, heuristically here) using Theorem 4.13. To do this, it is useful to apply first a linear map of the (C, H) plane so as to deal with independent Brownian coordinates in the limit. More precisely, we apply the linear map Λ defined by

$$\Lambda = (1/\sigma) \begin{pmatrix} 1 & \cos(\theta_0) \\ 0 & \sin(\theta_0) \end{pmatrix},$$

where $\theta_0 = \pi/(2p_0) = 4\pi/\kappa' = 2\arctan(\sqrt{1/(1-2p)})$ and $\sigma^2 = (1-p)/2$. A direct but tedious computation shows that $\Lambda(L_t, R_t)$ is indeed a standard planar Brownian motion. (The computation is easier to do by reverting to the original formulation, in [She16b], of Theorem 4.13 where it is shown that $C + H$ and $(C - H)/\sqrt{1-2p}$ converge to a standard planar Brownian motion.) Note that the top right quadrant transforms under Λ, see Figure 4.10, into the cone $C(\theta_0)$ of angle θ_0 with apex at zero.

We therefore consider an analogous question for two-dimensional Brownian motion. Namely, let B be a standard planar Brownian motion, starting from some point $z \in C(\theta_0)$ with $|z| = 1$. Let T be the first time that B leaves $C(\theta_0)$.

4.8 Exponents associated with FK-weighted random planar maps

Then from Theorem 4.13, it is reasonable to guess that

$$\mathbb{P}(E_{2n}) \approx \mathbb{P}_z(T > t; |B_t| \le 1), \text{ with } t = n^{1+o(1)}.$$

(Indeed, note that if $T > t$ and $|B_t| \le 1$, then the Brownian motion is likely to exit the cone soon after time t and not far from the apex. This intuition is for instance made rigorous in [BLR17] and [GMS19, GS17, GS15].)

For this, we first claim that

$$\mathbb{P}(T > t) = t^{-p_0 + o(1)} \tag{4.15}$$

as $t \to \infty$. To see why this is the case, consider the conformal isomorphism $z \mapsto z^{\pi/\theta_0}$, which sends the cone $C(\theta_0)$ to the upper half plane. In the upper half plane, the function $z \mapsto \Im(z)$ ($\Im(z)$ being the imaginary part of z) is harmonic with zero boundary condition, and so in the cone, the function

$$z \mapsto g(z) := r^{\pi/\theta_0} \sin\left(\frac{\pi\theta}{\theta_0}\right); \quad z \in C(\theta_0)$$

is also harmonic. Applying the optional stopping theorem at time t to the martingale $M_t := g(B_{t \wedge T})$, the only non-zero contribution to M_t comes from the event $T > t$. On the other, conditionally on $T > t$, B_t is likely to be at distance $\sqrt{t}$ from the origin, in which case $M_t \approx t^{\pi/(2\theta_0)} = t^{p_0}$. It is not hard to deduce (4.15).

We now claim that the desired probability satisfies

$$\mathbb{P}_z(T > t; |B_t| \le 1) = t^{-2p_0 - 1 + o(1)} \text{ as } t \to \infty. \tag{4.16}$$

To see this, we split the interval $[0, t]$ into three intervals of equal length $t/3$. In order for the event on the left-hand side to be satisfied, three things must happen during these three intervals.

- Over the interval $[0, t/3]$, B must not leave the cone. This has probability $t^{-p_0 + o(1)}$ by (4.15).
- At the other extreme, if we reverse the direction of time, we also have a Brownian motion started close to the tip of the cone that must not leave the cone for time $t/3$. Again, this has probability $t^{-p_0 + o(1)}$.
- Finally, given the behaviour of the process over $[0, t/3]$ and $[2t/3, t]$, the process must go from $B_{t/3}$ to $B_{2t/3}$ in the time interval $[t/3, 2t/3]$, and stay inside the cone. The latter requirement actually has probability bounded away from zero (because $B_{t/3}$ and $B_{2t/3}$ are typically far away from the boundary of the cone), so it remains to compute the probability to transition between these two endpoints. However this is roughly of order $t^{-1+o(1)}$, since we are dealing with a Brownian motion in dimension two.

Altogether, we obtain that $\mathbb{P}_z(T > t; |B_t| \le 1) = t^{-2p_0 - 1 + o(1)}$, as desired. □

4.9 Exercises

4.1 This exercise follows the arguments of Chen [Che17] and gives a very nice concrete construction of the planar map associated with a word.

Let $x_1, \ldots, x_{2n}$ be a sequence of $2n$ letters in the alphabet $\Theta = \{\mathsf{c}, \mathsf{h}, \mathsf{C}, \mathsf{H}, \mathsf{F}\}$ and suppose that the corresponding word $w = x_1 \cdots x_{2n}$ reduces to $\overline{w} = \emptyset$. For each $1 \le i \le 2n$, denote by $\varphi(i)$ the unique match of i: meaning that if i corresponds to production of a specific burger, then $\varphi(i)$ is the unique time at which this burger is consumed, and vice versa.

Let us draw a map as follows. Start with the line segment (drawn in the complex plane) having vertices $1, \ldots, 2n$ and horizontal nearest neighbour edges. Draw an arc between i and $\varphi(i)$ for each $1 \le i \le 2n$; this arc is drawn in the upper half plane for a hamburger, and in the lower half plane for a cheeseburger.

(a) Show that the arcs can be drawn in a planar way (so they don't cross one another); in other words, if $n_1 < n_2 < n_3 < n_4$ it is not possible that $\varphi(n_1) = n_3$ and $\varphi(n_2) = n_4$, unless $x_{n_1} \neq x_{n_2}$.

(b) Add an additional edge between 1 and $2n$ in the upper half plane, above every other edge, and call the resulting map **A**. Check that **A** has $2n$ vertices and $3n$ edges.

(c) Consider the planar dual **Δ** of **A**, and show this is a triangulation.

(d) Colour in blue the edges of **Δ** that stay in the upper half plane, and in red those that stay in the lower half plane. Show that the set of blue edges and the set of red edges define two trees. Let **Q** be the set of remaining edges in **Δ**, and colour them green. Show that **Q** is a quadrangulation.

(e) Explain how the map **Δ** is related to the triangulation $\mathbf{m}_n$ encoded by Sheffield's bijection and show that the straight line segment from 1 to $2n$ together with the additional edge linking the two extreme vertices in **A** corresponds to the space-filling loop in Sheffield's bijection.

(f) Deduce that local convergence of maps is equivalent to local convergence of the symbols encoding them via Sheffield's bijection, as claimed in Theorem 4.11.

4.2 **The reduced walk.** Consider the infinite decorated planar map (M, T) of Theorem 4.11, and let $(X_n)_{n \in \mathbb{Z}}$ denote the bi-infinite sequence of symbols encoding it via Sheffield's bijection. Let us assume that $q > 0$ or equivalently $p > 0$, where p and q are related via (4.9) and p is the proportion of F symbols. Define a *backward* exploration process $(c_n, h_n)_{n \ge 0}$ of the map, which keeps track of the number of C and H in the reduced word, as follows. Let $(c_0, h_0) = (0, 0)$. Suppose we have performed n steps of the

4.9 Exercises

exploration and defined c_n, h_n and in this process, we have revealed the symbols $(X_{-m}, \ldots, X_0)$. We inductively define the following.

- If X_{-m-1} is a C (resp. H), define $(c_{n+1}, h_{n+1}) = (c_n, h_n) + (1, 0)$ (resp. $(c_n, h_n) + (0, 1)$).
- If X_{-m-1} a c (resp. h), $(c_{n+1}, h_{n+1}) = (c_n, h_n) + (-1, 0)$ (resp. $(c_n, h_n) + (0, -1)$).
- If X_{-m-1} is F, then we explore $X_{-m-2}, X_{-m-3}, \ldots$ until we find the match of X_{-m-1}. Let $|\mathcal{R}_{n+1}|$ denote the number of symbols in the reduced word $\mathcal{R}_{n+1} = \overline{X_{\varphi(-m-1)} \ldots X_{-m-1}}$. Show that $\mathcal{R}_{n+1}$ contains only order symbols of one type. If $\mathcal{R}_{n+1}$ consists of H symbols, define $(c_{n+1}, h_{n+1}) = (c_n, h_n) + (0, |\mathcal{R}_{n+1}|)$. Otherwise, if $\mathcal{R}_{n+1}$ consists of C symbols define $(c_{n+1}, h_{n+1}) = (c_n, h_n) + (|\mathcal{R}_{n+1}|, 0)$.

Show that the walk $(c_n, h_n)_{n \geq 0}$ is a sum of *independent* and identically distributed random variables. Note that this is in contrast to Theorem 4.13. It can be shown that these random variables are in fact centred when $q \leq 4$ (see [She16b]).

4.3 **Bubbles.** Consider the infinite decorated planar map (M, T) of Theorem 4.11, and let $(X_n)_{n \in \mathbb{Z}}$ denote the bi-infinite sequence of symbols encoding it via Sheffield's bijection. Let us assume that $q > 0$ or equivalently $p > 0$, where p and q are related via (4.9) and p is the proportion of F symbols. Let us condition on the event $X_0 = $ F. Let $\varphi(0) \leq 0$ denote the match of this symbol. The word $w = X_{\varphi(0)} \cdots X_0$ encodes a finite planar map, called the **bubble** or envelope of the map at 0. This bubble corresponds to a finite number of loops of (M, T) (note that this is in general more than a single loop of (M, T) containing the root triangle, as there can be other F symbols in w). This notion was pivotal in [BLR17] where it was used to derive critical exponents of Theorem 4.14. This exercise gives one of the main steps in the derivation of this theorem.

(a) Assume without loss of generality that $X_{\varphi(0)} = $ h. Give a description of the reduced word $\overline{w}$. By considering the random length $N = |\varphi(0)|$ of w and the random length K of the reduced word $\overline{w}$, describe the event $\{N = n, K = k\}$ in terms of a certain cone excursion for the reverse two-dimensional walk $(C_{-i}, H_{-i})_{0 \leq i \leq n}$. Explain why N is the area of the bubble and K the length of its outer boundary.

(b) Arguing at the same level of rigour as in Proposition 4.15, show that there are exponents p_{area} and p_{boundary} such that

$$\mathbb{P}(N \geq n) = n^{-p_{\text{area}} + o(1)}; \quad \mathbb{P}(K \geq k) = k^{-p_{\text{boundary}} + o(1)},$$

where $p_{\text{boundary}}/2 = p_{\text{area}} = p_0$, and

$$p_0 = \frac{\pi}{4\arccos\left(\frac{\sqrt{2-\sqrt{q}}}{2}\right)} = \frac{\kappa'}{8} \in (1/2, 1),$$

was defined in (4.11).

The next three exercises use exponents derived in this chapter together with the KPZ formulas of the previous chapter to give predictions (in some cases proved through other methods) about the value of certain critical exponents associated with random fractals which can be defined without any reference to random planar maps.

4.4 Use (4.14), the KPZ relation, and the relation

$$q = 2 + 2\cos(8\pi/\kappa')$$

between q and κ' from (4.3), to recover (non-rigorously) that the dimension of $\text{SLE}_{\kappa'}$ is $1 + \kappa'/8$ for $\kappa' \in (4, 8)$.

4.5 Consider a simple random walk on the (infinite) uniform random planar map G, that is, take G to be the infinite volume of the FK-weighted maps for $q = 1$ defined in Theorem 4.11, and let $(X_n)_{n \geq 0}$ be a simple random walk on G starting from the root. If $n \geq 1$, a pioneer point for the walk $(X_1, \ldots, X_n)$ is a point x such that x is visited at some time $m \leq n$ and is on the boundary of the unbounded component of $G \setminus \{X_1, \ldots, X_m\}$. A beautiful theorem of Benjamini and Curien [BC13] shows that when such a simple random walk first exits a ball of radius R, it has had $\approx R^3$ pioneer points (technically this result is only proved when G is the so-called Uniform Infinite Planar Quadrangulation, although it is also believed to hold for infinite planar maps within the same universality class such as the one considered above).

Analogously, for $(B_s)_{s \geq 0}$ a planar Brownian motion, we define the set $\mathcal{P}_t$ for given $t > 0$ to be all points of the form B_s for some $0 \leq s \leq t$, such that B_s is on the "frontier" at time s (where by frontier we mean the boundary of the unbounded component of the complement of $B[0, s]$).

Using a (non-rigorous) KPZ-type argument, derive the dimension of the Brownian pioneer points $\mathcal{P}_t$ for any fixed $t \geq 0$. (The answer is 7/4, as rigorously proved in a famous paper of Lawler, Schramm and Werner [LSW01] using SLE techniques.)

4.6 Consider a simple random walk (X_n) on the infinite local limit of FK-weighted planar maps (as in Theorem 4.11), starting from the root. Try to argue using the KPZ relation (again without being fully rigorous), that the graph distance between X_n and X_0 must be approximately equal to $n^{1/D}$ where D is the dimension of the space. (Hint: the range of Brownian motion must satisfy $\Delta = 0$; more precisely, by the time a random walk leaves a ball of radius R, it has visited of order $R^2/\log R$ vertices with high probability.) In particular, on the Uniform Infinite Planar Triangulations (UIPT) one conjectures that this distance is $\approx n^{1/4}$. This has now been proven rigorously in [GH20] and [GM21e].

5
Introduction to Liouville Conformal Field Theory

In this chapter, we present a short introduction to the theory initiated in the pioneering paper of David, Kupiainen, Rhodes and Vargas [DKRV16], which we will refer to as Liouville conformal field theory, or Liouville CFT for short. We use this in order to avoid confusion with the SLE-based theory developed in Chapters 7–9, for which we choose to stick with the label of Liouville quantum gravity.

The main objectives of the two theories are similar (i.e. to say, making rigorous sense of Polyakov's conformal theory of quantum gravity), and indeed we will see concrete statements connecting these two approaches in Chapter 7. Nevertheless, they are entirely independent, and can be read in whichever order one chooses. In particular, the Liouville CFT we are about to present does not require knowledge of SLE (it depends only on Gaussian multiplicative chaos theory). It also presents the advantage of being closer to the original formulation of Polyakov.

We will start with some heuristics and then move on to a rigorous definition motivated by these heuristics, staying for simplicity in the context of the Riemann sphere. (See [GRV19] for an extension to more general Riemann surfaces; this extension is highly non-trivial due to the need to choose the conformal class at random with a suitable law, in contrast to the case of the sphere where this is not necessary.) We will then prove the existence of the theory (which is to say, the finiteness of some observables subject to the so-called **Seiberg bounds**). An absolutely remarkable feature of the theory is that it is in some sense integrable or exactly solvable. We will show a simple result which hints at this integrability: the k point function of the theory can be computed as a negative fractional moment of Gaussian multiplicative chaos. We conclude with a brief overview of some recent developments, including a short discussion of **conformal bootstrap** ([GKRV24]) and the proof by Kupiainen, Rhodes and Vargas [KRV20] of the celebrated **DOZZ formula**.

5.1 Preliminary background

5.1.1 Quantum and conformal field theory

It is helpful to begin with a brief and *very* informal overview of some underlying notions which help put Polyakov's proposal in context. A **statistical field theory** (also known as a **Euclidean field theory**) is, very roughly speaking, a random field $(\varphi(x))_{x \in \mathbb{R}^d}$, or collection of such fields, defined in the continuum space $\mathbb{R}^d$ (or some region $D \subset \mathbb{R}^d$). A probabilist might intuitively think about the scaling limits of discrete fields naturally arising in statistical mechanics; for example, the magnetisation field in the Ising model, at or away from the critical points (this field counts the sum of all Ising spins in a given region). As this example suggests, one should not expect the "statistical fields" to be defined pointwise; instead, like the GFF, they should be understood as random distributions. Physicists typically describe such fields via their k point **correlation functions**:

$$(x_1, \ldots, x_k) \mapsto \mathbb{E}[\varphi(x_1) \cdots \varphi(x_k)].$$

Although the field φ is typically not pointwise defined, such correlation functions *are* typically well defined. For instance, in the case of the Gaussian free field, they can be computed from the knowledge of the two-point function, a multiple of the Green function, and **Wick's rule** that expresses the k point functions of Gaussian fields in terms of their two-point functions. (Note however that the two-point function does not in general determine the k point function.)

Another subtlety is that, in many cases of interest, the underlying measure $\mathbb{P}$ with respect to which the above correlations are computed is in fact not a probability distribution but rather a positive measure which may well have infinite mass. For this reason, the correlations will usually be written as $\langle \varphi(x_1) \cdots \varphi(x_k) \rangle$ rather than as expectations. Furthermore, the quantities that are actually of interest to physicists are analytic continuations of these correlation functions in terms of the underlying parameters defining the model (for example, inverse temperature). Indeed the resulting quantities can be interpreted in terms of **quantum field theory**. Roughly speaking, the statistical field theory described above corresponds to a quantum field theory via what is known as a **Wick rotation**: essentially, multiplying one of the spatial coordinates (the 'time' coordinate) of a quantum field theory by i allows us to go from the quantum theory to a real valued, and indeed positive, measure, which describes the statistical field theory. See [Mus10] for an account of exactly solvable models of statistical field theory.

A (Euclidean) **Conformal Field Theory** is a particular case of statistical field theory, in which the theory is required to satisfy certain additional invariance properties under conformal mapping, often referred to as **conformal symmetries**. Note that this makes sense even in dimensions greater or equal to three, in which case conformal maps are simply diffeomorphisms that preserve angles locally. The central objects in conformal field theory are a family of **primary fields** denoted by $\{\psi_\alpha\}_{\alpha \in A}$. For instance, in the case of the Ising model, the primary fields are given by $\{1, \sigma, \mathcal{E}\}$ where σ is the spin field (the scaling limit of the sum of Ising spins in a given region) and $\mathcal{E}$ is the energy field (the energy of an edge $e = (x, y)$ is the contribution $\sigma_x \sigma_y$ to the total energy of the configuration, and $\mathcal{E}$ gives the scaling limit of the sum of these energies in a given region). In Liouville conformal field theory, the primary fields ψ_α will, roughly speaking, be given by $\psi_\alpha(z) = e^{\alpha h(z)}$ (suitably interpreted), and h will be "sampled" from an infinite measure which is related to the law of a Gaussian free field.

In a conformal field theory, these primary fields can be multiplied with one another in some formal sense, and this allows us to talk about correlation functions $\langle \psi_{\alpha_1}(z_1) \cdots \psi_{\alpha_k}(z_k) \rangle$, as above. The first assumption of conformal symmetry is that, whenever f is a Möbius map (i.e. a conformal isomorphism from the underlying domain D in which theory is defined to itself)

$$\langle \psi_{\alpha_1}(f(z_1)) \cdots \psi_{\alpha_k}(f(z_k)) \rangle = \left(\prod_{i=1}^{k} |f'(z_i)|^{-2\Delta_{\alpha_i}} \right) \langle \psi_{\alpha_1}(z_1) \ldots \psi_{\alpha_k}(z_k) \rangle, \tag{5.1}$$

for some numbers $\Delta_\alpha, \alpha \in A$ called the **conformal weights** associated to the primary fields. The assumption (5.1) describes a global symmetry condition as it imposes a constraint on how the correlation functions change under the application of a globally defined conformal isomorphism on D. Two-dimensional conformal field theories also satisfy a more local kind of symmetry condition. There are several viewpoints that may be used to formulate these more local symmetries. One way is via the so-called **Virasoro algebra**. This is beyond the scope of the present chapter, but roughly speaking, the Virasoro algebra is generated by a family of operators $(L_n)_{n \in \mathbb{Z}}$ together with a central element that commutes with every L_n and so is in the centre of the algebra. More generally, the L_n satisfy certain commutation relations involving the central element. Infinitesimal conformal symmetries are enforced by requiring that there is a representation of this algebra (often but not always unitary) on a vector space containing the primary fields (together with the so-called descendant fields). In this representation the central element can be identified with a number $c \in \mathbb{R}$ called **the central charge**. Moreover, it makes sense to "apply" L_n to a primary field ψ_α, and it is worth noting that the operators L_n associated to the levels

5.1 Preliminary background

$n = -1, 0, 1$ correspond in some informal sense to Möbius maps, so that this is indeed a generalisation of (5.1). While such a rigorous description has recently been announced for Liouville conformal field theory (see [BGK+24]), we will not pursue this here.

Another possibility (note that it is not *a priori* obvious whether the two descriptions are equivalent, and we do not claim this) is via the so-called **Weyl invariance** (or more precisely in our case **Weyl anomaly**) property. To state this, it is necessary to enrich the problem by considering the theory, with respect to which the correlation functions $\langle \psi_{\alpha_1}(z_1) \cdots \psi_{\alpha_k}(z_k) \rangle$ are computed, as being defined on a manifold M instead of a domain D and suppose that M is endowed with a background metric g. When we do so, for every metric g on M we should have an associated collection of correlation functions, which we denote $\langle \psi_{\alpha_1}(z_1) \cdots \psi_{\alpha_k}(z_k) \rangle_g$. To get a feeling for what this might correspond to in the case of the Ising model, say, consider the following toy example: let D be a domain and $U \subset D$ be a fixed subdomain, and take g to be twice the Euclidean metric in U, and the Euclidean metric in the complement $D \setminus U$. The corresponding correlation function should describe the scaling limits of Ising correlations for graphs in which the density of vertices in U is twice as large as that in $D \setminus U$.

With these notations, let us now describe the Weyl invariance property. If g is a metric, and $\rho \colon M \to \mathbb{R}$ is a smooth function, we obtain a conformally equivalent metric $\tilde{g}$ by setting $\tilde{g} = e^\rho g$: that is, the angle of the curves on M are locally the same under g and $\tilde{g}$, and the distances in $\tilde{g}$ are locally multiplied by e^ρ. Such a rescaling of the background metric is sometimes known as a **Weyl transformation**. Then, Weyl invariance would be the identity

$$\langle \psi_{\alpha_1}(z_1) \cdots \psi_{\alpha_k}(z_k) \rangle_{e^\rho g} = \langle \psi_{\alpha_1}(z_1) \cdots \psi_{\alpha_k}(z_k) \rangle_g. \tag{5.2}$$

However, while Weyl invariance is a natural requirement for conformal theories describing classical physics, in *quantum* conformal field theories, this is not the case; instead, one has the **Weyl anomaly**

$$\langle \psi_{\alpha_1}(z_1) \cdots \psi_{\alpha_k}(z_k) \rangle_{e^\rho g} = e^{\frac{c}{96\pi} A(\rho, g)} \langle \psi_{\alpha_1}(z_1) \cdots \psi_{\alpha_k}(z_k) \rangle_g, \tag{5.3}$$

where the anomaly term $A(\rho, g)$ is defined by

$$A(\rho, g) = \int_M (|\nabla^g \rho|^2 + 2R_g \rho) v_g.$$

(Sometimes, the Weyl anomaly formula is expressed slightly differently, see for instance Remark 5.26.) Here c is the **central charge** of the theory, v_g is the volume form on M associated to the metric g, R_g is the scalar curvature, and $\nabla^g \rho$ denotes the gradient of ρ computed in the metric g. The Weyl anomaly formula (5.3) replaces (5.2) and allows us to consider arbitrary rescalings of the metric; property (5.3) captures the desired "local" conformal transformations mentioned earlier. In the context of Liouville conformal field theory, we will be able to prove the Weyl anomaly formula (see Theorem 5.17). In particular, Theorem 5.17 identifies the central charge of the theory. (We also note that from this point of view, it is natural to consider infinitesimal deformations of the metric, that is, when $\rho = \rho_\delta = \delta\hat{\rho}$ for some fixed smooth function $\hat{\rho}\colon M \to \mathbb{R}$, so that $e^\rho g = (1 + \delta\hat{\rho} + o(\delta))g$; the corresponding change in the correlation functions would involve to the first order a quantity known as the **stress energy tensor** of the theory.)

Conformal field theory grew in the 1980s after it was observed (or rather predicted) by Polyakov that such conformal symmetries arise when the underlying statistical mechanics models are taken at their critical point [Pol70, BPZ84a, BPZ84b, FQS84]. Furthermore, it turns out that at least in two dimensions, adding this requirement of conformal symmetries to a natural list of axioms for quantum field theory (as introduced by Osterwalder and Schrader [OS75, OS73]), drastically impacts the space of solutions to these axioms. This leads to a classification of conformal field theories at least in the case of unitary representations.

5.1.2 Polyakov action

Having discussed the general context of quantum and conformal field theories, we now turn our attention to the specific case of Liouville conformal field theory, which will occupy us in this chapter. In order to assist the reader, we begin with rather general considerations on statistical mechanics. In physics, a **Hamiltonian** H is a function which assigns the energy $H(\sigma)$ to a configuration σ. In statistical physics, we are used to the idea of sampling a configuration σ according to the **Gibbs measure** (with respect to an underlying reference measure denoted by $d\sigma$), namely

$$\mathbb{P}(\sigma) \propto \exp(-\beta H(\sigma))d\sigma. \tag{5.4}$$

Here $\beta \geq 0$ is a parameter playing the role of the inverse temperature of the system.

An **action** is an energy integrated against time: it represents the amount of energy needed to bring the system from one configuration to another. For a two-dimensional field $\varphi\colon \mathbb{R}^2 \to \mathbb{R}$ (where as above we view the field as a random distribution, so that φ is not really pointwise defined), the **Polyakov action** $S(\varphi)$ associated to the field φ can be thought of directly as the energy

5.1 Preliminary background

of the configuration φ, so that for a probabilist used to statistical mechanics, there is no difference between the Hamiltonian of the system (the energy of the configuration φ) and the action $S(\varphi)$. The reason for this apparently confusing terminology is that in this 2d model of quantum gravity, one should remember that one of the two dimensions is space and the other is time. Thus by specifying the energy $S(\varphi)$, we have already integrated against time and are thus properly dealing with an action. We will keep this convention and refer to $S(\varphi)$ as the Polyakov action, but it should simply be thought of as the energy of the configuration φ. We are now ready to give an expression (which for the moment is purely formal) for this Polyakov action on the sphere.

To describe the action, we first need to fix a Riemannian metric g on the two-sphere $\mathbb{S} = \{x \in \mathbb{R}^3 : |x| \leq 1\}$. For computational purposes, it will often be easier to consider the pushforward of g under a conformal isomorphism

$$\psi : \mathbb{S} \to \hat{\mathbb{C}} \tag{5.5}$$

from $\mathbb{S}$ to the extended complex plane $\hat{\mathbb{C}} := \mathbb{C} \cup \{\infty\}$. From now on, we will assume that the map ψ in (5.5) has been fixed; for example, we could take it to be stereographic projection. We will write $\hat{g}(z)$ for the pushforward of the metric g on $\mathbb{S}$ to $\hat{\mathbb{C}}$, which we identify with a non-negative function on $\mathbb{C}$. So, a small region of area ε around the fixed point $z \in \mathbb{C}$ represents a region on $\mathbb{S}^2$ of area approximately $\hat{g}(z)\varepsilon$ as $\varepsilon \to 0$, while the distance between two points on the sphere is obtained by minimising the integral of $\sqrt{\hat{g}(z)}$ along paths between the two corresponding points on $\hat{\mathbb{C}}$.

We will be particularly interested in the spherical metric g_0 on $\mathbb{S}$, which corresponds on $\hat{\mathbb{C}}$ to the function

$$\hat{g}_0(z) = \frac{4}{(1 + |z|^2)^2}. \tag{5.6}$$

For instance one can check that $\int_{\mathbb{C}} \hat{g}_0(z)\, dz = 4\pi$, as required for the area of the unit sphere. In fact, without loss of generality, in what follows we will consider only metrics g on $\mathbb{S}$, conformally equivalent to g_0: this means that on $\hat{\mathbb{C}}$, $\hat{g}$ must take the form

$$\hat{g}(z) = e^{\rho(z)} \hat{g}_0(z); \quad z \in \mathbb{C}, \tag{5.7}$$

with ρ a twice differentiable function on $\mathbb{C}$ with finite limit at infinity such that $\int_{\mathbb{C}} |\nabla \rho|^2 < \infty$. We call v_g the associated volume form on $\mathbb{S}$ and $v_{\hat{g}}$ the associated volume form on $\hat{\mathbb{C}}$.

From now on, we also assume that the parameter $\gamma \in (0, 2)$ is fixed and let

$$Q = \frac{\gamma}{2} + \frac{2}{\gamma}, \tag{5.8}$$

which is the value first encountered in the change of coordinate formula for Liouville measure and the definition of random surfaces (see Theorem 2.8 and Definition 2.10, respectively). Finally, we let $\mu > 0$ denote a constant (the **cosmological constant**) whose value – apart from the important fact that it is positive – will not be of any relevance in the following.

With these notations, Polyakov's ansatz is to define the action as follows:

$$S(\varphi) = \frac{1}{4\pi} \int_{\mathbb{C}} \left[|\nabla^g \varphi(z)|^2 + R_g Q \varphi(z) + 4\pi \mu e^{\gamma \varphi(z)} \right] v_g(\mathrm{d}z), \quad (5.9)$$

where R_g is the scalar curvature associated to g. On $\hat{\mathbb{C}}$, the scalar curvature can be written explicitly as

$$R_{\hat{g}}(z) = -\frac{1}{\hat{g}(z)} \Delta \log \hat{g}(z); \; z \in \mathbb{C}. \quad (5.10)$$

The theory we are about to discuss is slightly simpler when the scalar curvature $R_g(z)$ is a constant. In particular, this includes the spherical metric $\hat{g}_0$, for which $R_{\hat{g}_0} \equiv 2$ (this can be seen by expressing the Laplacian in polar coordinates; we leave this as an exercise). We also note that in general, due to the Gauss–Bonnet theorem (see for example, [dC16]),

$$\int_{\mathbb{C}} R_{\hat{g}}(z) v_{\hat{g}}(\mathrm{d}z) = 8\pi.$$

Returning to (5.9), we call the reader's attention to the exponential term $e^{\gamma \varphi(z)}$ which is of course a priori not well defined for a generic distribution, but can be made sense of as in Chapter 3 via Gaussian multiplicative chaos provided that φ is a logarithmically correlated Gaussian field.

Given the action $S(\varphi)$, by analogy with (5.4), one is led to formally define the associated Gibbs measure

$$\mathbf{P}(\varphi) = \exp(-S(\varphi)) \mathrm{D}\varphi, \quad (5.11)$$

on a for now unspecified space of generalised functions defined on $\mathbb{S}$ (or $\hat{\mathbb{C}}$). The temperature has been set to 1 for simplicity; other choices lead to an equivalent theory since

$$\beta S(\varphi; \gamma, \mu) = S(\sqrt{\beta}\varphi, \frac{\gamma}{\sqrt{\beta}}, \beta\mu)$$

for $\beta > 0$. The crucial thing to notice in (5.11) is that the choice of the reference measure $\mathrm{D}\varphi$, which should be heuristically viewed as a kind of Lebesgue (uniform) measure over the space of fields on $\mathbb{S}$ (or $\hat{\mathbb{C}}$), is *not* specified precisely.

In this chapter, we will detail how [DKRV16] nevertheless succeeded in assigning a meaning to this Gibbs measure $\mathbf{P}$, which we will refer to in the rest of this chapter as the **Polyakov measure**. Note that this will in fact be an *infinite measure* (in particular, not a probability measure). When we integrate this against an observable F, we will more typically write $\int F(\varphi)\mathbf{P}(\mathrm{d}\varphi) = \langle F \rangle$ in agreement with the physics convention. We will compute these "expectations" for particular choices of F, and these will define the correlation functions of the theory. Informally, these F will be of the form $F(\varphi) = \exp(\alpha\varphi(z))$ and products thereof.

5.2 Spherical GFF

A key idea of [DKRV16] is to give meaning to the Polyakov measure in (5.11) by combining the term $\exp(-\int |\nabla^g \varphi(z)|^2 v_g(\mathrm{d}z))$ and the reference measure $\mathrm{D}\varphi$, and then suitably *reweighting* the resulting measure to account for the remaining terms on the right-hand side of (5.9). In view of Theorem 1.8, it is natural to want to interpret the product $\exp(-\int |\nabla^g \varphi(z)|^2 v_g(\mathrm{d}z))\mathrm{D}\varphi$ as the law of a Gaussian free field. However, in the absence of a boundary on which to impose boundary conditions, one has to choose a suitable version of the Gaussian free field, for which there is no obvious candidate at first sight. In [DKRV16], David, Kupiainen, Rhodes and Vargas made a simple but ingenious proposal, which consists of two steps. The first step requires us to define the Gaussian free field with zero average on the sphere $\mathbb{S}$ (or **spherical GFF** for short). In fact, one can do this on any compact surface (Σ, g), in which case we speak of the **zero average GFF on** Σ with respect to the metric g. We will explain the construction in this generality since it is not more difficult. The details are reminiscent of other Gaussian free fields discussed in the book (see Chapter 1 and the discussion of the **Neumann GFF** that will appear in Chapter 7). The reader who is keen to get on with the rigorous definition of Polyakov measure is encouraged to skip straight to Section 5.3, where the second step is described.

5.2.1 Laplacian on a compact manifold

Let (Σ, g) be a connected compact surface (i.e. a two-dimensional, closed, connected and bounded Riemannian manifold), where g refers to the Riemannian metric on Σ; in particular, Σ has no boundary. The Riemannian structure induces a Laplace operator which we will denote by $\Delta^{\Sigma,g}$. For instance, if $(\Sigma, g) = (\mathbb{S}, g)$ and $\hat{g}$ is the metric obtained after pushing forward g via the fixed conformal isomorphism $\psi: \mathbb{S} \to \hat{\mathbb{C}}$ of (5.5), then for smooth f defined

176   *Introduction to Liouville Conformal Field Theory*

on $\mathbb{S}$, we simply have

$$\Delta^{\mathbb{S},g} f(z) = \frac{1}{\hat{g}(\psi(z))} \Delta[f \circ \psi^{-1}](\psi(z)); \quad z \in \mathbb{S},$$

where $\Delta = \frac{\partial^2}{\partial x^2} + \frac{\partial^2}{\partial y^2}$ is the usual Laplacian on $\mathbb{C}$. In other words, the Brownian motion X on $\mathbb{S}$ with respect to g (i.e. the diffusion with infinitesimal generator $\Delta^{\mathbb{S},g}$) can be obtained by performing a time change to the standard Euclidean Brownian motion on $\mathbb{C}$

$$X_t = B_{F^{-1}(t)}; \quad F(t) = \int_0^t \hat{g}(B_s)\,ds, \tag{5.12}$$

and mapping back to $\mathbb{S}$ via the inverse of ψ. (A similar recipe, properly interpreted, may be used to define *Liouville Brownian motion*, see [Ber15] and [GRV16].)

It can be seen that on a compact connected surface, the negative Laplacian $-\Delta^{\Sigma,g}$ has a discrete spectrum, which we denote by

$$0 = \lambda_0 < \lambda_1 \leq \lambda_2 \cdots \uparrow +\infty,$$

with each distinct eigenvalue repeated according to its multiplicity. By the Sturm–Liouville decomposition (see, for example, [Cha84, VI.1]), we can assume that the corresponding eigenfunctions e_n form an orthonormal basis of $L^2(\Sigma; v_g)$. Note that this has bounded total mass, since Σ is bounded. The eigenvalue $\lambda_0 = 0$ is associated to the constant eigenfunction $e_0 = 1/\sqrt{v_g(\Sigma)}$, corresponding to the fact that the Brownian motion on Σ with respect to g converges to the uniform distribution.

We will give several equivalent definitions of the GFF with zero v_g average in Σ. The first one is as a random series. Before we give this definition, we briefly introduce the function space in which this series will converge, which is a variant of the Sobolev space $H^s(D)$ discussed in Chapter 1.

A distribution on (Σ, g) is simply a continuous linear functional on *test functions*, where test functions are simply smooth functions on (Σ, g) (since the space is bounded). Here continuity refers to the usual topology on test functions, meaning uniform convergence of derivatives of all orders. If f is a distribution on (Σ, g) and ϕ is a test function, we write

$$(f, \phi)_g = \text{``} \int_\Sigma f(x)\phi(x) v_g(dx) \text{''}$$

for the action of f on ϕ. This is exactly the $L^2(\Sigma, g)$ inner product when f is itself in $L^2(\Sigma, g)$ and in particular depends on the metric g. Note that the smooth function 1 is a valid test function, so that the total integral on Σ of a distribution f is well defined; we will write

$$(f, 1)_g =: v_g(f)$$

for this integral and refer to it as the average of f on (Σ, g). When this integral is equal to zero we say that the distribution has **zero average**. For $s \in \mathbb{R}$, we define $H^s(\Sigma, g)$ to be the space of zero average distributions f on Σ such that

$$\sum_{n \geq 1} (f, e_n)_g^2 \lambda_n^s < \infty.$$

Note that e_n is smooth for every $n \geq 1$ so is a valid test function. It is not hard to see that the left-hand side defines a Hilbert space norm $\|\cdot\|_{H^s(\Sigma,g)}$ on zero average distributions, and that convergence in that Hilbert space implies convergence in the sense of distributions. Note also that changing the metric g to a conformally equivalent one as in (5.7) leads to the same Sobolev spaces $\{H^s(\Sigma, g)\}_{s \in \mathbb{R}}$ in the sense that if $f \in H^s(\Sigma, \tilde{g})$, then $f - \bar{v}_g(f) \in H^s(\Sigma, g)$.

Let us also record that we have the Gauss–Green formula on (Σ, g): that is, for any twice continuously differentiable functions u, w on (Σ, g) with zero average

$$\int_\Sigma u(x) \Delta^{\Sigma,g} w(x) v_g(dx) = \int_\Sigma w(x) \Delta^{\Sigma,g} u(x) v_g(dx). \qquad (5.13)$$

See for example, [Aub98, (29); Chapter 1].

5.2.2 Definition of the zero average GFF on (Σ, g)

We can now introduce the GFF with zero v_g average on Σ:

Definition 5.1 Let $(X_n)_{n \geq 1}$ denote a sequence of i.i.d. standard normal random variables. The GFF with zero average on Σ is the random distribution $\mathbf{h}^{\Sigma,g}$ on Σ obtained from the series

$$\mathbf{h}^{\Sigma,g} = \sum_{n=1}^\infty \frac{X_n}{\sqrt{\lambda_n}} e_n,$$

which converges almost surely in any of the spaces $H^s(\Sigma, g)$ with $s < 0$, and hence in the space of (zero average) distributions. As usual, we will also write $h^{\Sigma,g} = \sqrt{2\pi} \mathbf{h}^{\Sigma,g}$.

The convergence of the series in this definition follows from Weyl's law (see [Cha84, VI.4, page 155]) in a manner similar to Lemma 1.46. An argument similar to (1.43) also shows that the convergence of this series would also hold if we replaced $\{\lambda_n^{-1/2} e_n\}_{n\geq 1}$ with any orthonormal basis of $H^1(\Sigma, g)$. Furthermore, if $f \in C^\infty(\Sigma, g)$ has zero v_g average, then

$$\text{Var}((\mathbf{h}^{\Sigma,g}, f)_g) = \sum_{n=1}^\infty \frac{(f, e_n)_{L^2(\Sigma,g)}^2}{\lambda_n} = \|f\|_{H^{-1}(\Sigma,g)}^2, \quad (5.14)$$

where the right-hand side indeed does not depend on the basis.

Observe that, as in Theorem 1.49, this allows us to alternatively define $\mathbf{h}^{\Sigma,g}$ to be a stochastic process indexed by $H^{-1}(\Sigma, g)$. More precisely,

$$\{(\mathbf{h}^{\Sigma,g}, f)_g\}_{f \in H^{-1}(\Sigma,g)}$$

defines a Gaussian stochastic process indexed by $H^{-1}(\Sigma, g)$, with $\mathbb{E}((\mathbf{h}^{\Sigma,g}, f)_g)$ $= 0$ for all f and

$$\text{cov}((\mathbf{h}^{\Sigma,g}, f)_g (\mathbf{h}^{\Sigma,g}, \tilde f)_g) = (f, \tilde f)_{H^{-1}(\Sigma,g)}$$
$$= \sum_{n=1}^\infty \frac{(f, e_n)_g (\tilde f, e_n)_g}{\lambda_n}; \quad f, \tilde f \in H^{-1}(\Sigma, g).$$

By (5.14) and the polarisation identity, the restriction of this stochastic process to $f \in C^\infty(\Sigma, g)$ agrees with the definition of $\mathbf{h}^{\Sigma,g}$ as a zero average distribution in Definition 5.1.

It will also be useful to have a description for the above covariance structure in terms of a covariance kernel. This will be the **centred Green function** of the g–Brownian motion on Σ; that is, the Brownian motion on Σ with respect to g (see Lemma 5.5). To define the centred Green function, let $(X_t, t \geq 0)$ denote this Brownian motion (so its generator is $\Delta^{\Sigma,g}$), and let $p_t^{\Sigma,g}(x, y)$ denote the **heat kernel**, which is characterised (for a fixed $x \in \Sigma$ and as a function of y), as the density of the law of X_t when started from x with respect to v_g. A standard fact (see [Cha84, VI.1, equation (13)]) is that we may decompose $p_t^\Sigma(x, y)$ according to the orthonormal basis of eigenfunctions e_n of $L^2(\Sigma; v_g)$ as

$$p_t^{\Sigma,g}(x, y) = \sum_{n=0}^\infty e^{-\lambda_n t} e_n(x) e_n(y). \quad (5.15)$$

This series converges absolutely and uniformly on $\Sigma \times \Sigma$ for any $t > 0$ (see again [Cha84, VI.1]). In fact, see [Gri09, Remark 10.15 and (10.51)], we have that for all $n \geq 0$,

$$\sup_{t > t_0} \sup_{x, y \in \Sigma} e^{\frac{\lambda_{n+1}}{2} t_0} \left| p_t^{\Sigma,g}(x, y) - \sum_{k=0}^n e^{-\lambda_k t} e_k(x) e_k(y) \right| \leq C(t_0), \quad (5.16)$$

5.2 Spherical GFF

where the constant $C(t_0)$ depends only on t_0 and in particular not on n. As a consequence, we can define the associated Green function with zero average

$$G^{\Sigma,g}(x,y) := \int_0^\infty \left[p_t^{\Sigma,g}(x,y) - \frac{1}{v_g(\Sigma)} \right] dt.$$

When $x \neq y$, $p_t^{\Sigma,g}(x,y)$ is bounded as $t \downarrow 0$, and so (5.16) and (5.15) imply that the above integral converges. In fact, they ensure that

$$G^{\Sigma,g}(x,y) = \sum_{n=1}^\infty \frac{1}{\lambda_n} e_n(x) e_n(y), \tag{5.17}$$

with the convergence of the series holding pointwise. Observe that if $x \neq y$ then by its definition as the transition density of the g-Brownian motion on Σ, $p_t^{\Sigma,g}(\cdot, y)$ is uniformly bounded as $t \to 0$ on a neighbourhood of x. This together with (5.16) implies that $G^{\Sigma,g}(\cdot, y)$ is continuous at x. In other words, $G^{\Sigma,g}$ is continuous away from the diagonal.

Note also that (5.16) implies the upper bound

$$|G^{\Sigma,g}(x,y)| \leq \int_0^1 p_t^{\Sigma,g}(x,y) \, dt + C,$$

where C is a constant not depending on $x, y \in \Sigma$. In particular, if f is uniformly bounded on Σ, then $\int_\Sigma G^{\Sigma,g}(x,y) f(y) v_g(dy)$ converges absolutely and

$$\int_\Sigma |G^{\Sigma,g}(x,y) f(y)| v_g(dy) \leq \int_0^1 \mathbb{E}(|f(B_t)|) \, dt + C \operatorname{vol}(G) \sup_\Sigma |f|$$
$$\leq C' \sup_\Sigma |f|. \tag{5.18}$$

This means that $G^{\Sigma,g}(x, \cdot)$ defines a distribution on (Σ, g) for every x. As should be expected from (5.17), we have

$$v_g(G^{\Sigma,g}(x, \cdot)) = \int_\Sigma G^{\Sigma,g}(x,y) v_g(dy) = 0$$

(i.e. $G^{\Sigma,g}(x, \cdot)$ does have zero average), and

$$\int_\Sigma G^{\Sigma,g}(x,y) e_n(x) v_g(dy) = \frac{e_n(x)}{\lambda_n} \text{ for all } n \geq 1,$$

which can both be justified rigorously using (5.16). As a result, $G^{\Sigma,g}$ is an "inverse" of (minus) the Laplacian in the following sense:

Let $\bar{v}_g(dx) = v_g(dx)/v_g(\Sigma)$ be the normalised volume measure associated to the metric g on Σ, and for $f: \Sigma \to \mathbb{R}$ a bounded measurable function, let

$$\bar{v}_g(f) := \frac{1}{v_g(\Sigma)} \int_\Sigma f(x) v_g(dx).$$

180 *Introduction to Liouville Conformal Field Theory*

Lemma 5.2 *If ϕ is a smooth function on (Σ, g), then for every $x \in \Sigma$,*

$$\int_\Sigma G^{\Sigma,g}(x,y) \Delta^{\Sigma,g} \phi(y) v_g(\mathrm{d}y) = -\phi(x) + \bar{v}_g(\phi). \tag{5.19}$$

Remark 5.3 This result should not be surprising, since with respect to the basis $\{e_n\}_{n\geq 1}$ of zero average functions, $-\Delta$ is a "diagonal" operator with diagonal entries $(\lambda_n)_{n\geq 1}$ and by (5.17), the Green function can also be viewed as a diagonal operator with diagonal entries $(\lambda_n^{-1})_{n\geq 1}$.

Proof We make use of the Sobolev embedding theorem for compact manifolds, [Aub98, Theorem 2.20], which implies in particular that for f smooth, the series $\sum_{n=0}^\infty (f, e_n)_{L^2(\Sigma,g)} e_n$ converges uniformly to f on Σ. This applies directly to ϕ as in the statement of the lemma, and also to $\Delta^{\Sigma,g}\phi$, where $(\Delta^{\Sigma,g}\phi, e_n)_{L^2(\Sigma,g)} = -\lambda_n(\phi, e_n)_{L^2(\Sigma,g)}$ for each $n \geq 0$ by (5.13). This means (using (5.18)) that the left-hand side of (5.19) is equal to

$$-\sum_{n=1}^\infty (G^{\Sigma,g}(x,\cdot), e_n)_{L^2(\Sigma,g)} \lambda_n (\phi, e_n)_{L^2(\Sigma,g)} = -\sum_{n=1}^\infty (\phi, e_n)_{L^2(\Sigma,g)} \phi(x)$$

$$= -\phi(x) + \bar{v}_g(\phi)$$

as required. $\square$

Remark 5.4 In fact, (5.19) can be extended, by approximation, to the case where ϕ is only assumed to be twice continuously differentiable on (Σ, g). See [Aub98, Proposition 4.14].

Let us now finally use the above to relate this Green function to the zero average GFF $\mathbf{h}^{\Sigma,g}$:

Lemma 5.5 *For a smooth function f on (Σ, g),*

$$\mathrm{Var}((\mathbf{h}^{\Sigma,g}, f)_g) = \int_\Sigma \int_\Sigma f(x) G^{\Sigma,g}(x,y) f(y) v_g(\mathrm{d}x) v_g(\mathrm{d}y). \tag{5.20}$$

In other words, the Green function is the covariance kernel of $\mathbf{h}^{\Sigma,g}$.

Proof By the Sobolev embedding theorem again, we have that for f smooth with zero v_g average, the series

$$\sum_{n=1}^\infty \lambda_n^{-1} e_n(x) (f, e_n)_{L^2(\Sigma,g)}$$

converges to a smooth ϕ with $\Delta^{\Sigma,g}\phi = f$. Hence $\int_\Sigma G^{\Sigma,g}(x,y) f(y) v_g(\mathrm{d}y) = \phi(x)$ and thus since $(\phi, e_n)_{L^2(\Sigma,g)} = \lambda_n^{-1}(f, e_n)_{L^2(v_g)}$ for each n:

$$\int_\Sigma \int_\Sigma f(x) G^{\Sigma,g}(x,y) f(y) v_g(\mathrm{d}x) v_g(\mathrm{d}y) = \|f\|_{H^{-1}(\Sigma,g)}^2.$$

5.2 Spherical GFF

Combining this equality with (5.14), we reach the conclusion. □

5.2.3 The spherical case

Let us now specialise to the case where $\Sigma = \mathbb{S}$ is the sphere, and the metric is still as in (5.7). In this case, we can easily observe how $G^{\mathbb{S},g}$ changes when we apply a Möbius transformation to $\mathbb{S}$.

Lemma 5.6 *Suppose that* $m \colon \mathbb{S} \to \mathbb{S}$ *is a Möbius transformation, and let* m_*g *be the pushforward of g under m. Then*

$$G^{\mathbb{S},m_*g}(x,y) = G^{\mathbb{S},g}(m^{-1}(x), m^{-1}(y))$$

for all $x \neq y \in \mathbb{S}$.

Proof It follows from a straightforward calculation that if $(e_n)_{n \geq 0}$ are an orthonormal basis of $L^2(\Sigma, \nu_g)$, such that $\Delta^{\Sigma,g} e_n = -\lambda_n e_n$ for $n \geq 1$, then $(e_n \circ m^{-1})_{n \geq 0}$ are an orthonormal basis of $L^2(\Sigma, \nu_{m_*g})$, such that $\Delta^{\Sigma,m_*g}(e_n \circ m^{-1}) = -\lambda_n(e_n \circ m^{-1})$ for $n \geq 1$. The result then follows from (5.17) (it is easy to check that the series definition (5.17) cannot depend on the choice of orthonormal basis of eigenfunctions, since each of the eigenspaces is finite dimensional). □

Using (5.20), this implies that $h^{\mathbb{S},g}$ transforms in the following way:

Corollary 5.7 *Suppose that* $m \colon \mathbb{S} \to \mathbb{S}$ *is a Möbius transformation. Then*

$$(\mathbf{h}^{\mathbb{S},g}, f \circ m)_g = (\mathbf{h}^{\mathbb{S},m_*g}, f)_{m_*g}$$

for $f \in C^\infty(\mathbb{S})$. In other words, if we define a zero average distribution $\mathbf{h}^{\mathbb{S},g} \circ m^{-1}$ on $(\mathbb{S}, m_*g)$ by setting $(\mathbf{h}^{\mathbb{S},g} \circ m^{-1}, f)_{m_*g} = (\mathbf{h}^{\mathbb{S},g}, f \circ m)_g$ for $f \in C^\infty(\mathbb{S})$. Then we have that

$$\mathbf{h}^{\mathbb{S},g} \circ m^{-1} \stackrel{(law)}{=} \mathbf{h}^{\mathbb{S},m_*g}$$

as zero average distributions on $(\mathbb{S}, m_*g)$.

As mentioned previously, in order to do explicit computations in this case, we will want to reparametrise $\mathbb{S}$ by the extended complex plane $\hat{\mathbb{C}}$. Recall from the conversation around (5.5) that $\psi \colon \mathbb{S} \to \hat{\mathbb{C}}$ is a fixed conformal isomorphism (for example, stereographic projection) and we write (with an abuse of notation) $\hat{g}(z)$ for the pushforward of a metric g on $\mathbb{S}$ under ψ. Then by the same proof as for Lemma 5.6, we have that

$$(\mathbf{h}^{\hat{\mathbb{C}},\hat{g}}, f)_{\hat{g}} = (\mathbf{h}^{\mathbb{S},g}, f \circ \psi)_g \qquad (5.21)$$

for all smooth functions f on $\hat{\mathbb{C}}$. In particular,

$$G^{\hat{\mathbb{C}},\hat{g}}(x,y) = G^{\mathbb{S},g}(\psi^{-1}(x),\psi^{-1}(y)) \text{ for } x \neq y \in \hat{\mathbb{C}}. \tag{5.22}$$

Remark 5.8 It follows that $G^{\hat{\mathbb{C}},\hat{g}}$ and $\mathbf{h}^{\hat{\mathbb{C}},\hat{g}}$ satisfy the same transformation rules under Möbius maps. If $m\colon \hat{\mathbb{C}} \to \hat{\mathbb{C}}$ is a Möbius transformation – that is, of the form $m(z) = (az+b)/(cz+d)$ with $ad - bc = 1$ – then

$$G^{\hat{\mathbb{C}},m_*\hat{g}}(x,y) = G^{\hat{\mathbb{C}},\hat{g}}(m^{-1}(x),m^{-1}(y)) \quad x \neq y \in \mathbb{C}$$

and

$$(\mathbf{h}^{\hat{\mathbb{C}},\hat{g}}, f \circ m)_{\hat{g}} = (\mathbf{h}^{\hat{\mathbb{C}},m_*\hat{g}}, f)_{m_*\hat{g}} \quad f \in C^\infty(\hat{\mathbb{C}}).$$

Lemma 5.9 *We have, for all $x \neq y \in \mathbb{C}$,*

$$G^{\hat{\mathbb{C}},\hat{g}}(x,y) = \frac{1}{2\pi}\left[-\log(|x-y|)+\bar{v}_{\hat{g}}\big(\log(|x-\cdot|)\big)+\bar{v}_{\hat{g}}\big(\log(|y-\cdot|)\big)-\theta_{\hat{g}}\right], \tag{5.23}$$

where

$$\theta_{\hat{g}} = \iint_{\mathbb{C}^2} \log|x-y|\bar{v}_{\hat{g}}(dx)\bar{v}_{\hat{g}}(dy). \tag{5.24}$$

In addition, we write

$$G^{\hat{\mathbb{C}},\hat{g}}(x,\infty) := \lim_{R\to\infty} G^{\hat{\mathbb{C}},\hat{g}}(x,R),$$

which is also well defined for $x \in \mathbb{C}$, by (5.22) and since $G^{\mathbb{S},g}(\psi^{-1}(x),y)$ is continuous at the pole $y = \psi^{-1}(\infty)$.

Remark 5.10 Note that $G^{\hat{\mathbb{C}},\hat{g}}$ is *not* harmonic away from the diagonal (in contrast to, say, the case of the Dirichlet Green function). Indeed, it follows from (5.19) that for fixed x, $\Delta G^{\hat{\mathbb{C}},\hat{g}}(x,\cdot) = -\delta_x + \bar{v}_{\hat{g}}$ as a distribution. This is necessary, due to the requirement that $G^{\hat{\mathbb{C}},\hat{g}}$ has zero average with respect to $v_{\hat{g}}$.

Proof Fix $x \in \mathbb{C}$ and define the function $y \mapsto F(y)$ as the difference between the left-hand side and the right-hand side of (5.23). Note that at the moment it is not clear whether F is defined at $x \neq y$, but we can still view it as a distribution on $\mathbb{C}$. We will show below that F is harmonic in the sense of distributions on $\mathbb{C}$; that is, for any compactly supported smooth test function f on $\mathbb{C}$:

$$(F,\Delta f) = \int_{\mathbb{C}} F(y)\Delta f(y)\,dy = 0. \tag{5.25}$$

Admitting (5.25), it then follows from elliptic regularity that F can be extended continuously to the point x and with this extension is a classically harmonic function on $\mathbb{C}$. It is also immediate that $\bar{v}_{\hat{g}}(F) = 0$, and F is bounded at infinity

5.2 Spherical GFF

(the boundedness holds since $\bar{v}_{\hat{g}}(\log|y-\cdot|) = \log|y| + O(1)$ as $y \to \infty$ and since $G^{\hat{\mathbb{C}},\hat{g}}(x,\cdot) = G^{\mathbb{S},g}(\psi^{-1}(x), \psi^{-1}(\cdot))$ is bounded at infinity by definition). This means that F must be identically zero, which completes the proof of the lemma.

So it remains to prove (5.25). For this, we first observe directly from (5.19) that

$$\int_{\mathbb{C}} G^{\hat{\mathbb{C}},\hat{g}}(x,y)\Delta f(y)\,dy = \int_{\mathbb{S}} G^{\mathbb{S},g}(x,y)\Delta^{\mathbb{S},g} f(y) v_g(dy) = -f(x) + \bar{v}_g(f). \tag{5.26}$$

On the other hand, we know that

$$-\frac{1}{2\pi}\int_{\mathbb{C}} \log|x-y|\Delta f(y)\,dy = -f(x) \tag{5.27}$$

since $\Delta \frac{1}{2\pi}\log|x-\cdot| = \delta_x(\cdot)$ in the distributional sense. We are left to compute the integral

$$\frac{1}{2\pi}\int_{\mathbb{C}} (\bar{v}_{\hat{g}}(\log(|x-\cdot|)) + \bar{v}_{\hat{g}}(\log(|y-\cdot|)) - \theta_{\hat{g}})\Delta f(y)\,dy$$

$$= \frac{1}{2\pi}\int_{\mathbb{C}} \bar{v}_g(\log|y-\cdot|)\Delta f(y)\,dy,$$

with the equality coming from the fact that two of the terms in the integral on the left-hand side do not depend on y, and the Gauss–Green formula. Here we can appeal to Fubini, since f is smooth and compactly supported, and write this as

$$\frac{1}{2\pi}\int_{\mathbb{C}}\int_{\mathbb{C}} \log(|y-w|)\Delta f(y)\,dy\,\bar{v}_{\hat{g}}(dw) = \int_{\mathbb{C}} f(w)\bar{v}_{\hat{g}}(dw) = \bar{v}_{\hat{g}}(f) \tag{5.28}$$

by (5.27). Combining (5.26) and (5.28), we conclude that

$$(\tilde{F}, \Delta f) = f(x) - \bar{v}_{\hat{g}}(f) - f(x) + \bar{v}_{\hat{g}}(f) = 0$$

as required. □

As a corollary, we obtain the following expression for the circle average variance. Recall that $h^{\mathbb{S},g} = \sqrt{2\pi}\mathbf{h}^{\mathbb{S},g}$ (similarly $h^{\hat{\mathbb{C}},\hat{g}} = \sqrt{2\pi}\mathbf{h}^{\hat{\mathbb{C}},\hat{g}}$) and that from (5.20) and the expression of the spherical Green function $G^{\hat{\mathbb{C}},g}$, circle averages of $h^{\hat{\mathbb{C}},\hat{g}}$ are well defined. Let $h_\varepsilon^{\hat{\mathbb{C}},\hat{g}}$ denote the circle average of $h^{\hat{\mathbb{C}},\hat{g}}$ at (Euclidean) distance ε.

Corollary 5.11 *As $\varepsilon \to 0$, we have*

$$\mathrm{Var}(h_\varepsilon^{\hat{\mathbb{C}},\hat{g}}(z)) = \log\frac{1}{\varepsilon} + v(z) + o(1), \tag{5.29}$$

where $v(z) = 2v_g(\log|z-\cdot|) - \theta_g$.

The following explicit formula for the spherical metric Green function, $G^{\hat{\mathbb{C}},\hat{g}_0}$ will also be useful.

Lemma 5.12 *We have*

$$G^{\hat{\mathbb{C}},\hat{g}_0}(x,y) = \frac{1}{2\pi}\left(-\log|x-y|-\frac{1}{4}(\log\hat{g}_0(x)+\log\hat{g}_0(y))+\log 2-1/2\right) \quad (5.30)$$

for $x \neq y \in \mathbb{C}$, *and* $G^{\hat{\mathbb{C}},\hat{g}_0}(x,\infty) = (1/2\pi)(-(1/4)\log\hat{g}_0(x)+(1/2)\log 2-1/2)$
for $x \in \mathbb{C}$. *Consequently, as* $\varepsilon \to 0$, *uniformly in* $z \in \mathbb{C}$,

$$\operatorname{Var}(h_\varepsilon^{\hat{\mathbb{C}},\hat{g}_0}(z)) = \log\frac{1}{\varepsilon} + \hat{v}(z) + o(1), \quad (5.31)$$

where

$$\hat{v}(z) = -\frac{1}{2}\log\hat{g}_0(z) + \log 2 - \frac{1}{2}. \quad (5.32)$$

Proof Recall that by (5.28), $\Delta\bar{v}_{\hat{g}_0}(\log|x-\cdot|) = 2\pi\bar{v}_{\hat{g}_0}(\cdot)$ in the sense of distributions on $\mathbb{C}$. On the other hand, since

$$2\hat{g}_0(z) = R_{\hat{g}_0}\hat{g}_0(z) = -\Delta\log\hat{g}_0(z),$$

by definition of the scalar curvature in (5.10) (and recalling that $R_{\hat{g}_0} = 2$), for any smooth compactly supported f on $\mathbb{C}$ we have

$$-\frac{1}{4}\int_\mathbb{C}\Delta f(y)\log\hat{g}_0(y)\,dy = -\frac{1}{4}\int_\mathbb{C}f(y)\Delta(\log\hat{g}_0(y))\,dy = \frac{1}{2}\int_\mathbb{C}f(y)\hat{g}_0(y)\,dy$$

$$= 2\pi\bar{v}_{\hat{g}_0}(f)$$

due to Gauss–Green. This implies that $\bar{v}_{\hat{g}_0}(\log|x-\cdot|) + \frac{1}{4}\log\hat{g}_0(x)$ is harmonic in $\mathbb{C}$, and the rest of the proof amounts to computing constants.

First, it is straightforward to check that $\bar{v}_{\hat{g}_0}(\log|y-\cdot|) + \frac{1}{4}\log(\hat{g}_0(y)) \to \frac{1}{2}\log 2$ as $x \to \infty$, so that by harmonicity $\bar{v}_{\hat{g}_0}(\log|y-\cdot|) + \frac{1}{4}\log(\hat{g}_0(y)) \equiv \frac{1}{2}\log 2$. Also note that $\bar{v}_{\hat{g}_0}(\log|x-\cdot|)$ has $\bar{v}_{\hat{g}_0}$ average $\theta_{\hat{g}_0}$ (by definition), and $-\frac{1}{4}\log\hat{g}_0$ has $\bar{v}_{\hat{g}_0}$ average $\frac{1}{2} - \frac{1}{2}\log 2$ (as can be shown by switching to polar coordinates and doing the integral explicitly). It therefore follows that $\theta_{\hat{g}_0} = 1/2$ and we obtain the result. □

Remark 5.13 More generally, if the curvature $R_{\hat{g}}$ is constant, then

$$\bar{v}_{\hat{g}}(\log|x-\cdot|) + \frac{1}{2R_g}\log\hat{g}(x) \equiv \theta_{\hat{g}} + \frac{1}{2R_{\hat{g}}}\bar{v}_{\hat{g}}(\log(\hat{g}))$$

so that

$$G^{\hat{\mathbb{C}},\hat{g}}(x,y) = \frac{1}{2\pi}\left(-\log|x-y| - \frac{1}{2R_{\hat{g}}}\log\hat{g}(x) - \frac{1}{2R_{\hat{g}}}\log\hat{g}(y) + c_{\hat{g}}\right)$$

5.2 Spherical GFF

and
$$v(z) = -\frac{1}{R_{\hat{g}}} \log \hat{g}(z) + c_{\hat{g}},$$
where
$$c_{\hat{g}} = \theta_g + \frac{1}{R_{\hat{g}}} \overline{v}_{\hat{g}}(\log(\hat{g})).$$

The following special case will come in useful later on, when studying the behaviour of Liouville theory under Möbius transformations. The proof uses similar reasoning to above, and we leave it as a guided exercise.

Lemma 5.14 *When $\hat{g} = m_*\hat{g}_0$, with m a Möbius transform of $\hat{\mathbb{C}}$, we have*

$$\theta_{m_*\hat{g}_0} = -\frac{1}{2}\overline{v}_{m_*\hat{g}_0}(\log(m_*(\hat{g}_0))) + \log(2) - \theta_{\hat{g}} \qquad (5.33)$$

and

$$R_{m_*\hat{g}_0} \equiv R_{\hat{g}_0} \equiv 2 \quad ; \quad c_{m_*\hat{g}_0} = c_{\hat{g}_0} = \log(2) - \frac{1}{2}. \qquad (5.34)$$

Proof See Exercise 5.5. □

Let us conclude this section by noting that $\mathbf{h}^{\mathbb{S},g}$ for different metrics g on $\mathbb{S}$ are simply recentrings of the same field. More precisely:

Lemma 5.15 *Suppose that g_1, g_2 are two metrics on $\mathbb{S}$ as in (5.7). Then*

$$\mathbf{h}^{\mathbb{S},g_1} - \overline{v}_{g_2}(\mathbf{h}^{\mathbb{S},g_1}) \stackrel{(\text{law})}{=} \mathbf{h}^{\mathbb{S},g_2}$$

as zero average distributions on $(\mathbb{S}, g_2)$. Equivalently,

$$\mathbf{h}^{\hat{\mathbb{C}},\hat{g}_1} - \overline{v}_{\hat{g}_2}(\mathbf{h}^{\hat{\mathbb{C}},\hat{g}_1}) \stackrel{(\text{law})}{=} \mathbf{h}^{\hat{\mathbb{C}},\hat{g}_2}.$$

Proof It is enough to prove the second statement. This can either be verified using the explicit expression for the covariance in Lemma 5.9, or using the fact that $\text{Var}(h^{\hat{\mathbb{C}},\hat{g}_2}, f)_{\hat{g}_2} = \|f\|^2_{H^{-1}(\hat{\mathbb{C}},\hat{g}_2)}$ for all smooth functions f on $\mathbb{C}$ (similarly with g_2 replaced by g_1). We leave this to the reader as Exercise 5.2. □

5.2.4 GMC on the Riemann sphere

Let $(\mathbb{S}, g)$ denote the sphere with a metric g assumed to be conformally equivalent to the standard round metric g_0, and let $h = \sqrt{2\pi}\mathbf{h}^{\mathbb{S},g}$ denote the (rescaled) Gaussian free field on $\mathbb{S}$ with zero average with respect to v_g. Associated with h there is a notion of Gaussian multiplicative chaos $\mathcal{M}_{h;g}$, formally described by

$$\mathcal{M}_{h;g}(\mathrm{d}x) = e^{\gamma h(x)} v_g(\mathrm{d}x),$$

186   *Introduction to Liouville Conformal Field Theory*

and understood rigorously as

$$\mathcal{M}_{h;g}(\mathrm{d}x) = \lim_{\varepsilon \to 0} \varepsilon^{\gamma^2/2} e^{\gamma h_\varepsilon(x)} v_g(\mathrm{d}x), \tag{5.35}$$

where $h_\varepsilon(x)$ denotes the mollification of h at scale ε with respect to the underlying metric g. Existence of the above limit requires a slight extension of the theory presented in Chapter 3 and more specifically Theorem 3.2, since the latter deals with only logarithmically correlated fields on $\mathbb{R}^d$. Rather than go through the necessary adjustments (which however do not present any difficulty, see Remark 3.12), it is equivalent and slightly simpler for our purposes to discuss the pushforward of $\mathcal{M}_{h;g}$ to the extended complex plane $\hat{\mathbb{C}}$ (see also [Cer22, DSHKS21] for a discussion of GMC measures arising naturally from higher-dimensional extensions of Liouville CFT).

Hence, we rigorously define $\mathcal{M}_{h;g}$ by conformally mapping h to $\hat{\mathbb{C}}$ using the fixed conformal isomorphism $\psi : \mathbb{S} \to \hat{\mathbb{C}}$ (see (5.5)), and then pulling back the resulting measure associated with $\hat{g}$. That is, if $\hat{h} = h \circ \psi^{-1}$, we define for Borel sets $A \subset \mathbb{S}$,

$$\mathcal{M}_{h;g}(A) = \lim_{\varepsilon \to 0} \mathcal{M}_{\hat{h}+(Q/2)\log \hat{g};\varepsilon}(\psi(A))$$

$$:= \lim_{\varepsilon \to 0} \int_{\psi(A)} \varepsilon^{\gamma^2/2} e^{\gamma(\hat{h}+\frac{Q}{2}\log \hat{g})_\varepsilon(z)} \, \mathrm{d}z, \tag{5.36}$$

where $\hat{g}$ denotes the pushforward of g by ψ. The subscript ε on the right-hand side here denotes a mollified version of the field $\hat{h} + (Q/2)\log \hat{g}$ at (Euclidean) radius ε. Notice that the field $\hat{h}$ needs to be shifted by $(Q/2)\log \hat{g}$, as specified by the change of coordinate formula: see Theorem 2.8, and note that $|(\psi^{-1})'(z)| = \sqrt{\hat{g}(z)}$ for $z \in \hat{\mathbb{C}}$.

For instance, when $g = g_0$, with this choice of normalisation, we have (using (5.32)),

$$\mathbb{E}(\mathcal{M}_{h;g_0}(\mathbb{S})) = \lim_{\varepsilon \to 0} \varepsilon^{\gamma^2/2} \int_{\hat{\mathbb{C}}} e^{\frac{\gamma^2}{2} \mathrm{Var}(\hat{h}_\varepsilon(z)) + \frac{\gamma Q}{2} \log \hat{g}_0(z)} \, \mathrm{d}z$$

$$= \int_{\hat{\mathbb{C}}} e^{\frac{\gamma^2}{2} \hat{v}(z) + (1+\frac{\gamma^2}{4})\log \hat{g}_0(z)} \, \mathrm{d}z$$

$$= \int_{\mathbb{C}} e^{\frac{\gamma^2}{2}(\log 2 - 1/2)} \hat{g}_0(z) \, \mathrm{d}z$$

$$= e^{\frac{\gamma^2}{2}(\log 2 - 1/2)} v_{g_0}(\mathbb{S}).$$

One can check (also using (5.32)) that this agrees with the limit of the expectation in the expression in (5.35).

The appearance of the (non-universal) constant $\log 2 - 1/2$ here is a consequence of our choice of normalisation for the GFF, which is required to have zero average. In the theory below, the choice of this additive constant does not play a role.

5.3 Defining the Polyakov measure

Step 1: GFF on the sphere with mean zero. We fix a metric g conformally equivalent to g_0 as in (5.7). We then consider the scaled version $h^{\mathbb{S},g} = \sqrt{2\pi}\mathbf{h}^{\mathbb{S},g}$ and, as usual, write $h^{\hat{\mathbb{C}},\hat{g}}$ for the same field parametrised by the extended complex plane, that is, $h^{\hat{\mathbb{C}},\hat{g}} = h^{\mathbb{S},g} \circ \psi^{-1}$ as in (5.21) where $\psi: \mathbb{S} \to \hat{\mathbb{C}}$ is the conformal isomorphism (for example, stereographic projection) that we fixed in (5.5).

Step 2: the Lebesgue shift. As observed earlier, there is nothing canonical about normalising the field to have zero average. In fact, the expression for $\exp(-\int_{\mathbb{C}} |\nabla^g \varphi(z)|^2 v_g(\mathrm{d}z))$ is clearly invariant under shifting the field φ by an additive constant. The heart of the construction of [DKRV16] is to start with $h^{\mathbb{S},g}$, and then add a constant c "distributed" according to Lebesgue measure on $\mathbb{R}$, the unique Radon measure (up to normalisation) on $\mathbb{R}$ invariant under additive shift. We then define the resulting "law" λ to be $\exp(-\int_{\mathbb{C}} |\nabla^g \varphi|^2 v_g(\mathrm{d}z))\mathrm{D}\varphi$. Before we explain precisely what this means, we mention that we write "law" in the above (and below) in quotation marks since it is in fact a measure of infinite total mass (because Lebesgue measure on $\mathbb{R}$ has infinite total mass). Let us be more specific. We define $H^{-1}(\mathbb{S})$ to be the subspace of distributions on $(\mathbb{S}, g)$ of the form $\{\varphi + c \,;\, \varphi \in H^{-1}(\mathbb{S}, g), c \in \mathbb{R}\}$ equipped with the topology induced by the natural product topology on $H^{-1}(\mathbb{S}, g) \times \mathbb{R}$. (It is easy to see that, unlike $H^{-1}(\mathbb{S}, g)$, this space is independent of the choice of $g = e^\rho g_0$ as in (5.7), hence we drop it from the notation.) Then we define a measure λ on $H^{-1}(\mathbb{S})$, by setting

$$\lambda(A) = \int_{c \in \mathbb{R}} \mathbb{P}(h^{\mathbb{S},g} + c \in A)\, \mathrm{d}c$$

for an arbitrary Borel set $A \subset H^{-1}(\mathbb{S})$.[1] Note that for a non-negative Borel functional F on $H^{-1}(\mathbb{S})$ (an "observable") we have

$$\int_{\varphi \in H^{-1}(\mathbb{S})} F(\varphi)\lambda(\mathrm{d}\varphi) = \int_{c \in \mathbb{R}} \mathbb{E}\Big[F(h^{\mathbb{S},g} + c)\Big]\mathrm{d}c. \qquad (5.37)$$

---

[1] In what follows, we use the generic notation $\mathbb{P}$ (and associated expectation $\mathbb{E}$) for the law of a field, for example $h^{\mathbb{S},g}$ or $h^{\hat{\mathbb{C}},\hat{g}}$, when the particular law in question is implicit from the notation.

This λ lets us assign a meaning to the term $\exp(-\frac{1}{4\pi} \int |\nabla^g \varphi|^2 v_g(\mathrm{d}z))\mathrm{D}\varphi$ in the Polyakov measure (5.11). Namely, we set

$$\exp\left(-\frac{1}{4\pi} \int |\nabla^g \varphi|^2 v_g(\mathrm{d}z)\right)\mathrm{D}\varphi := \lambda(\mathrm{d}\varphi).$$

The fact that the total mass of the measure λ is infinite is consistent with the fact that we expect the left-hand side to be invariant under shifting φ by an arbitrary constant c.

The measure λ can also be pushed forward by ψ to a measure $\hat{\lambda}$ on $H^{-1}(\hat{\mathbb{C}}) = \{\varphi + c; \varphi \in H^{-1}(\hat{\mathbb{C}}, \hat{g}_0), c \in \mathbb{R}\}$ (which is a subspace of $H^{-1}_{\mathrm{loc}}(\mathbb{C})$ as defined in Corollary 1.53). Similarly to λ, $\hat{\lambda}$ is then the "law" of $h^{\hat{\mathbb{C}},\hat{g}} + c$ with c distributed according to Lebesgue measure, and $h^{\hat{\mathbb{C}},\hat{g}} = h^{\mathbb{S},g} \circ \psi^{-1}$ as in (5.21).

Definition of the Polyakov measure(s). These two ingredients, the mean zero spherical GFF and its Lebesgue shift, allow us to give a rigorous definition of the Polyakov measure **P** of (5.11). As before, we will allow ourselves to consider two closely related versions **P** and $\widehat{\mathbf{P}}$, depending on whether we want to consider the fields on $\mathbb{S}$ or on $\hat{\mathbb{C}}$.

Simply put, the (spherical) Polyakov measure **P** corresponds to reweighting $\lambda(\mathrm{d}\varphi)$ by the remainder of the terms in $S(\varphi)$,

$$\mathbf{P}(\mathrm{d}\varphi) = \mathbf{P}_g(\mathrm{d}\varphi) := \exp\left(-\frac{Q}{4\pi}(\varphi, R_g)_g - \mu \mathcal{M}_{\varphi;g}(\mathbb{S})\right)\lambda(\mathrm{d}\varphi). \quad (5.38)$$

Here we recall that $\mathcal{M}_{\varphi;g}(\mathbb{S})$ is the total mass of the Gaussian multiplicative chaos measure with parameter γ associated to $\varphi = h^{\mathbb{S},g} + c$, as defined in (5.35). Thus for a non-negative functional F on $H^{-1}(\mathbb{S})$, we have

$$\mathbf{P}_g(F) = \int_{\varphi \in H^{-1}(\mathbb{S})} F(\varphi) \exp\left(-\frac{Q}{4\pi}(\varphi, R_g)_g - \mu \mathcal{M}_{\varphi;g}(\mathbb{S})\right)\lambda(\mathrm{d}\varphi)$$
$$= \int_{-\infty}^{\infty} \mathbb{E}\left[F(h^{\mathbb{S},g} + c) \exp\left(-\frac{Q}{4\pi}(h^{\mathbb{S},g} + c, R_g)_g - \mu e^{\gamma c} \mathcal{M}_{h^{\mathbb{S},g};g}(\mathbb{S})\right)\right] \mathrm{d}c,$$

where the expectation above is over the law of $h^{\mathbb{S},g}$.

Concretely, for computations, it is more convenient to work on the extended complex plane. If we want to express (5.38) in terms of an expectation over $h^{\hat{\mathbb{C}},\hat{g}}$, a straightforward rewriting gives the following expression:

$$\int_{c \in \mathbb{R}} \mathbb{E}\Big(F(h^{\hat{\mathbb{C}},\hat{g}} \circ \psi^{-1} + c) \times$$
$$\exp\left(-\frac{Q}{4\pi} \int_{\mathbb{C}} R_{\hat{g}}(h^{\hat{\mathbb{C}},\hat{g}} + c) v_{\hat{g}}(\mathrm{d}z) - \mu e^{\gamma c} \mathcal{M}_{h^{\hat{\mathbb{C}},\hat{g}} + \frac{Q}{2}\log \hat{g}}(\hat{\mathbb{C}})\right)\Big) \mathrm{d}c,$$
(5.39)

5.3 Defining the Polyakov measure

where the equality

$$\mu e^{\gamma c} M_{h^{\mathbb{S},g};g}(\mathbb{S}) = \mu e^{\gamma c} M_{h^{\hat{\mathbb{C}},\hat{g}} + \frac{Q}{2}\log \hat{g}}(\hat{\mathbb{C}})$$

is a consequence of (5.36). The above is, however, not quite the right definition for the law $\widehat{\mathbf{P}}$ of the Polyakov measure in $\hat{\mathbb{C}}$, since we have only partly taken into account the change of coordinates from $\mathbb{S}$ to $\hat{\mathbb{C}}$. Instead, we define $\widehat{\mathbf{P}}_{\hat{g}}$ to be the "law" of $\varphi \circ \psi^{-1} + (Q/2) \log \hat{g}$ under $\mathbf{P}_g$:

Definition 5.16 If F is now a non-negative Borel functional on $H^{-1}(\hat{\mathbb{C}})$, we set

$$\widehat{\mathbf{P}}_{\hat{g}}(F) = \mathbf{P}_g \left(F((\cdot + \tfrac{Q}{2}\log \hat{g}) \circ \psi) \right)$$

$$= \int_{\mathbb{R}} \mathbb{E}\left[F\left(h^{\hat{\mathbb{C}},\hat{g}} + \tfrac{Q}{2}\log \hat{g} + c\right) \times \right.$$
$$\left. \exp\left(-\tfrac{Q}{4\pi}(h^{\hat{\mathbb{C}},\hat{g}} + c, R_{\hat{g}})_{\hat{g}} - \mu e^{\gamma c} M_{h^{\hat{\mathbb{C}},\hat{g}} + \tfrac{Q}{2}\log \hat{g}}(\mathbb{C}) \right) \right] dc. \quad (5.40)$$

We will write $\langle F \rangle_{\hat{g}}$ for $\widehat{\mathbf{P}}_{\hat{g}}(F)$ in what follows, in agreement with physics conventions.

Note that $(c, R_{\hat{g}})_{\hat{g}} = 8\pi c$ by Gauss–Bonnet, hence

$$\exp\left(-\tfrac{Q}{4\pi}(c, R_{\hat{g}})_{\hat{g}} \right) = \exp(-2Qc).$$

Combining with (5.40), we reach the following explicit definition of the Polyakov measure $\widehat{\mathbf{P}}_{\hat{g}}$:

$$\widehat{\mathbf{P}}_{\hat{g}}(F) = \int_{\mathbb{R}} \mathbb{E}\left[F\left(h^{\hat{\mathbb{C}},\hat{g}} + \tfrac{Q}{2}\log \hat{g} + c\right) \times \right.$$
$$\left. \exp\left(-\tfrac{Q}{4\pi}(h^{\hat{\mathbb{C}},\hat{g}}, R_{\hat{g}})_{\hat{g}} - 2Qc - \mu e^{\gamma c} M_{h^{\hat{\mathbb{C}},\hat{g}} + \tfrac{Q}{2}\log \hat{g}}(\mathbb{C}) \right) \right] dc.$$

Observe that when g (or equivalently $\hat{g}$) has constant curvature, the term $\exp(-\tfrac{Q}{4\pi}(h^{\hat{\mathbb{C}},\hat{g}}, R_{\hat{g}})_{\hat{g}})$ also disappears, since $h^{\hat{\mathbb{C}},\hat{g}}$ has zero average with respect to $v_{\hat{g}}$ by definition. Thus in this case, we get a particularly simple expression, which we will use repeatedly below:

$$\langle F \rangle_{\hat{g}} = \int_{c \in \mathbb{R}} \mathbb{E}\left[F\left(h^{\hat{\mathbb{C}},\hat{g}} + \tfrac{Q}{2}\log \hat{g} + c\right) \times \right.$$
$$\left. \exp\left(-2Qc - \mu e^{\gamma c} M_{h^{\hat{\mathbb{C}},\hat{g}} + \tfrac{Q}{2}\log \hat{g}}(\mathbb{C}) \right) \right] dc. \quad (5.41)$$

5.4 Weyl anomaly formula

We have seen that the expression for $\langle F\rangle_{\hat{g}}$ simplifies considerably when $\hat{g}$ (or g on $\mathbb{S}$) has constant curvature. When $R_{\hat{g}}$ is not constant, we have the additional term $\exp(-\frac{Q}{4\pi}(h^{\mathbb{S},\hat{g}} + c, R_{\hat{g}})_{\hat{g}})$ in the expectation. The effect of this extra term is to further tilt the law of the field. It turns out the effect of this tilt can be described exactly, thanks to Girsanov's lemma.

Theorem 5.17 (Weyl anomaly) *Let g be a metric on $\mathbb{S}$, with pushforward $\hat{g}$ to $\hat{\mathbb{C}}$ by ψ of the form $\hat{g}(z) = e^{\rho(z)}\hat{g}_0(z)$ for ρ as in (5.7). Then for each non-negative Borel function F on $H^{-1}(\hat{\mathbb{C}})$, we have*

$$\langle F\rangle_{\hat{g}} = \exp\left(\frac{6Q^2}{96\pi}\int_{\mathbb{C}}[|\nabla^{\hat{g}_0}\rho(x)|^2 + 4\rho(x)]v_{\hat{g}_0}(dx)\right)\langle F\rangle_{\hat{g}_0}. \quad (5.42)$$

See Corollary 5.25 and Remark 5.26 for a Weyl anomaly formula valid for correlation functions.

In the above expression, $\nabla^{\hat{g}_0}$ is the gradient operator in the metric $\hat{g}_0$. The only thing that is needed about this operator is the fact that

$$\int_{\mathbb{C}}|\nabla^{\hat{g}_0}\rho(x)|^2 v_{\hat{g}_0}(dx) = -\int_{\mathbb{C}}\rho(x)\frac{1}{\hat{g}_0(x)}\Delta\rho(x)v_{\hat{g}_0}(dx)$$

(the Gauss–Green identity on $(\hat{\mathbb{C}}, \hat{g}_0)$).

Remark 5.18 In [DKRV16], their definition for $\langle F\rangle_{\hat{g}}$ includes an extra multiplicative factor $\exp(\frac{1}{96\pi}\int_{\mathbb{C}}[\nabla^{\hat{g}_0}\rho(x)|^2 + 4\rho(x)]v_{\hat{g}_0}(dx))$ which corresponds to the partition function of the Gaussian free field in (5.37). This leads to the anomaly

$$\langle F\rangle_{\hat{g}} = \exp\left(\frac{c_L}{96\pi}\int_{\mathbb{C}}[\nabla^{\hat{g}_0}\rho(x)|^2\rho(x) + 2R_{\hat{g}_0}\rho(x)]v_{\hat{g}_0}(dx)\right)\langle F\rangle_{\hat{g}_0} \quad (5.43)$$

with

$$c_L = 1 + 6Q^2.$$

This is the more classical Weyl anomaly formula for a conformal field theory with **central charge** c_L. Note that $c_L \in (25, \infty)$.

Proof Recalling Lemma 5.15 (applied to $h^{\hat{\mathbb{C}},\hat{g}}$), and applying the change of variables $c \mapsto c - \bar{v}_{\hat{g}}(h^{\hat{\mathbb{C}},\hat{g}_0})$ in the definition of the Polyakov measure, we first rewrite $\mathbf{P}_g(F) = \langle F\rangle_g$ as

$$\int \mathbb{E}\Bigg[\exp\Big(-\frac{Q}{4\pi}(h^{\hat{\mathbb{C}},\hat{g}_0}, R_{\hat{g}})_{\hat{g}}\Big)F\Big(h^{\hat{\mathbb{C}},\hat{g}_0} + \tfrac{Q}{2}\log\hat{g} + c\Big)\times$$
$$\exp\Big(-2Qc - \mu e^{\gamma c}M_{h^{\hat{\mathbb{C}},\hat{g}_0 + (Q/2)\log\hat{g}}}(\mathbb{C})\Big)\Bigg]dc.$$

5.4 Weyl anomaly formula

By Girsanov, the effect of the term $\exp(-\frac{Q}{4\pi}(h^{\hat{C},\hat{g}_0}, R_{\hat{g}})_{\hat{g}})$ is to shift the field $h^{\hat{C},\hat{g}_0}$ by

$$-\frac{Q}{4\pi}\int_{\mathbb{C}} 2\pi G^{\hat{C},\hat{g}_0}(x,y) R_{\hat{g}}(y)\hat{g}(y)\,dy \qquad (5.44)$$

and to multiply the whole expression by

$$\mathbb{E}\left[\exp\left(-\frac{Q}{4\pi}(h^{\hat{C},\hat{g}_0}, R_{\hat{g}})_{\hat{g}}\right)\right]. \qquad (5.45)$$

In fact, both of these expressions simplify quite nicely. Recalling that

$$R_{\hat{g}}\hat{g} = -\Delta\log\hat{g} = -\Delta\rho - \Delta\log\hat{g}_0 = -\Delta\rho + R_{\hat{g}_0}\hat{g}_0,$$

we have

$$-\frac{Q}{2}\int_{\mathbb{C}} G^{\hat{C},\hat{g}_0}(x,y) R_{\hat{g}}(y)\hat{g}(y)\,dy$$

$$= -\frac{Q}{2}\int_{\mathbb{C}} G^{\hat{C},\hat{g}_0}(x,y)\left(R_{\hat{g}_0}(y)\hat{g}_0(y) - \Delta\rho(y)\right)dy$$

$$= \frac{Q}{2}\int_{\mathbb{C}} G^{\hat{C},\hat{g}_0}(x,y)\Delta^{\hat{C},\hat{g}_0}\rho(y) v_{\hat{g}_0}(dy)$$

$$= -\frac{Q}{2}(\rho(x) - \overline{v}_{\hat{g}_0}(\rho)), \qquad (5.46)$$

where the third line follows because $R_{\hat{g}_0} = 2$ and $v_{\hat{g}_0}(G^{\hat{C},\hat{g}_0}(x,\cdot)) = 0$, while the final line follows from Remark 5.4. For this, we used the assumption that ρ is twice continuously differentiable. Similarly,

$$\mathbb{E}\left[\exp\left(-\frac{Q}{4\pi}(h^{\hat{C},\hat{g}_0}, R_{\hat{g}})_{\hat{g}}\right)\right]$$

$$= \exp\left(\frac{Q^2}{16\pi}\int_{\mathbb{C}} R_{\hat{g}}(x)\hat{g}(x)(\rho(x) - \overline{v}_{\hat{g}_0}(\rho))\,dx\right)$$

$$= \exp\left(\frac{Q^2}{16\pi}\int_{\mathbb{C}} (R_{\hat{g}_0}\hat{g}_0(x) - \Delta\rho(x))(\rho(x) - \overline{v}_{\hat{g}_0}(\rho))\,dx\right)$$

$$= \exp\left(-\frac{Q^2}{16\pi}\int_{\mathbb{C}} \rho(x)\Delta^{\hat{C},\hat{g}_0}\rho(x) v_{\hat{g}_0}(dx)\right), \qquad (5.47)$$

where the last line follows since

$$\int_{\mathbb{C}} R_{\hat{g}_0}\hat{g}_0(x)\rho(x)\,dx = 2v_{\hat{g}_0}(\rho) = 8\pi\overline{v}_{\hat{g}_0}(\rho) = \int_{\mathbb{C}} R_{\hat{g}_0}\hat{g}_0(x)\overline{v}_{\hat{g}_0}(\rho)\,dx$$

and $\int_{\mathbb{C}} \Delta\rho(x)\,dx = \int_{\mathbb{C}} \Delta^{\hat{C},\hat{g}_0}\rho(x) v_{\hat{g}_0}(dx) = 0$ by (5.13).

Notice that subtracting $-\frac{Q}{2}\rho$ from the field in our expression for $\mathbf{P}_g(F)$ has exactly the effect of turning $h^{\hat{\mathbb{C}},\hat{g}_0} + \frac{Q}{2}\log\hat{g}$ into $h^{\hat{\mathbb{C}},\hat{g}_0} - \frac{Q}{2}\log\hat{g}_0$. Combined with a further change of variables $c \mapsto c + \frac{Q}{2}\bar{v}_{\hat{g}_0}(\rho)$ in the integral we reach the conclusion:

$$\langle F \rangle_{\hat{g}} = \langle F \rangle_{\hat{g}_0} \exp\left(Q^2 \bar{v}_{\hat{g}_0}(\rho) - \frac{Q^2}{16\pi}\int_{\mathbb{C}} \rho(x)\Delta^{\hat{\mathbb{C}},\hat{g}_0}\rho(x) v_{\hat{g}_0}(dx)\right), \quad (5.48)$$

where the anomaly term can be rewritten as in the statement of the theorem. $\square$

Lemma 5.19 (Weyl Anomaly for Möbius transforms) *When $\hat{g} = m_*\hat{g}_0$, with m a Möbius transform of $\hat{\mathbb{C}}$, we have*

$$\langle F \rangle_{m_*\hat{g}_0} = \langle F \rangle_{\hat{g}_0}$$

for all non-negative Borel functions F.

Proof Recall from Lemma 5.14 that $R_{m_*\hat{g}_0} \equiv R_{\hat{g}_0} \equiv 2$, and by definition of ρ with $\hat{g} = m_*\hat{g}_0$,

$$-\Delta\rho(x) = R_{m_*\hat{g}_0} m_*\hat{g}_0(x) - R_{\hat{g}_0}\hat{g}_0(x) = 2m_*\hat{g}_0(x) - 2\hat{g}_0(x).$$

We therefore have

$$-\frac{Q^2}{16\pi}\int_{\mathbb{C}}\rho(x)\Delta\rho(x)\,dx = \frac{Q^2}{2}\left(\bar{v}_{m_*\hat{g}_0}(\rho) - \bar{v}_{\hat{g}_0}(\rho)\right) \quad (5.49)$$

while also

$$-\frac{Q^2}{16\pi}\int_{\mathbb{C}}\rho(x)\Delta\rho(x)\,dx = 2Q^2 \int_{\hat{\mathbb{C}}}\int_{\hat{\mathbb{C}}}(2\pi G^{\hat{\mathbb{C}},\hat{g}_0})(x,y)\bar{v}_{m_*\hat{g}_0}(dx)\bar{v}_{m_*\hat{g}_0}(dy) \quad (5.50)$$

by (5.47). Now, on the one hand, by (5.23), we have

$$2\pi G^{\hat{\mathbb{C}},m_*\hat{g}_0}(x,y) = -\log|x-y| + \bar{v}_{m_*\hat{g}_0}(\log|x-\cdot|) + \bar{v}_{m_*\hat{g}_0}(\log|x-\cdot|) - \theta_{m_*\hat{g}_0}.$$

On the other, since $2\pi G^{\hat{\mathbb{C}},m_*\hat{g}_0}$ and $2\pi G^{\hat{\mathbb{C}},\hat{g}_0}$ are the variances of $h^{\hat{\mathbb{C}},m_*\hat{g}_0}$ and $h^{\hat{\mathbb{C}},\hat{g}_0}$, respectively, and we know by Lemma 5.15 that $h^{\hat{\mathbb{C}},m_*\hat{g}_0}$ is equal in distribution to $h^{\hat{\mathbb{C}},\hat{g}_0} - \bar{v}_{m_*\hat{g}_0}(h^{\hat{\mathbb{C}},\hat{g}_0})$, we have

$$2\pi G^{\hat{\mathbb{C}},m_*\hat{g}_0}(x,y)$$
$$= 2\pi G^{\hat{\mathbb{C}},\hat{g}_0}(x,y) - \int_{\mathbb{C}} 2\pi G^{\hat{\mathbb{C}},\hat{g}_0}(x,y)\bar{v}_{m_*\hat{g}_0}(dx) - \int_{\mathbb{C}} 2\pi G^{\hat{\mathbb{C}},\hat{g}_0}(x,y)\bar{v}_{m_*\hat{g}_0}(dx)$$
$$+ \int_{\hat{\mathbb{C}}}\int_{\hat{\mathbb{C}}}(2\pi G^{\hat{\mathbb{C}},\hat{g}_0})(x,y)\bar{v}_{m_*\hat{g}_0}(dx)\bar{v}_{m_*\hat{g}_0}(dy).$$

Using that $2\pi G^{\hat{\mathbb{C}},\hat{g}_0} = -\log|x-y| - (1/4)(\log\hat{g}_0(x) + \log\hat{g}_0(y)) + \log(2) - \theta_{\hat{g}_0}$

5.5 Convergence of correlation functions and Seiberg bounds

and equating the two expressions for $2\pi G^{\hat{\mathbb{C}}, m_*\hat{g}_0}$ above, we are left with the equality

$$\int_{\hat{\mathbb{C}}}\int_{\hat{\mathbb{C}}}(2\pi G^{\hat{\mathbb{C}},\hat{g}_0})(x,y)\overline{v}_{m_*\hat{g}_0}(\mathrm{d}x)\overline{v}_{m_*\hat{g}_0}(\mathrm{d}y)$$
$$= -\theta_{m_*\hat{g}_0} + \log(2) - \theta_{\hat{g}_0} - \frac{1}{2}\overline{v}_{m_*\hat{g}_0}(\log \hat{g}_0)$$
$$= \frac{1}{2}\overline{v}_{m_*\hat{g}_0}(\log(m_*(\hat{g}_0))) - \frac{1}{2}\overline{v}_{m_*\hat{g}_0}(\log \hat{g}_0)$$
$$= \frac{1}{2}\overline{v}_{m^*\hat{g}_0}(\rho),$$

where the second equality follows from the expression (5.33) relating $\theta_{m_*\hat{g}_0}$ and $\theta_{\hat{g}_0}$. Combining this with (5.49) and (5.50), we deduce that

$$\frac{Q^2}{2}\left(\overline{v}_{m_*\hat{g}_0}(\rho) - \overline{v}_{\hat{g}_0}(\rho)\right) = Q^2 \overline{v}_{m_*\hat{g}_0}(\rho)$$

and so $\overline{v}_{m_*\hat{g}_0}(\rho) = -\overline{v}_{\hat{g}_0}(\rho)$. We conclude that the anomaly term

$$\exp\left(Q^2\overline{v}_{\hat{g}_0}(\rho) - \frac{Q^2}{16\pi}\int_{\mathbb{C}}\rho(x)\Delta^{\hat{\mathbb{C}},\hat{g}_0}\rho(x)v_{\hat{g}_0}(\mathrm{d}x)\right)$$
$$= \exp\left(Q^2\overline{v}_{\hat{g}_0}(\rho) - Q^2\overline{v}_{\hat{g}_0}(\rho)\right) = 1,$$

which completes the proof. □

5.5 Convergence of correlation functions within Seiberg bounds

Note. In this section, we drop the superscripts $\hat{\mathbb{C}}, \hat{g}$ for ease of notation. In particular, we write $h = h^{\hat{\mathbb{C}}, \hat{g}}$.

It is not immediately obvious for which observables F can we say that the associated expectation $\langle F \rangle_{\hat{g}}$ is finite. Let us consider the simplest case where $F = 1$ and $\hat{g} = \hat{g}_0$. Then recall that by Definition 5.16 of the Polyakov measure we have

$$\langle 1 \rangle_{\hat{g}_0} = \int_{c \in \mathbb{R}} \mathbb{E}\left[\exp\left(-2Qc - \mu e^{\gamma c}\mathcal{M}_{h+\frac{Q}{2}\log\hat{g}}(\mathbb{C})\right)\right]\mathrm{d}c. \quad (5.51)$$

The two possible divergences we need to worry about are at $c \to \infty$ and $c \to -\infty$. The first one (when $c \to \infty$) is not a problem since $Q > 0$ and the GMC mass is also strictly positive, so that overall the integrand decays (doubly) exponentially as $c \to \infty$, hence the integral converges for large c. The second

Introduction to Liouville Conformal Field Theory

limit however is divergent: indeed, when $c \to -\infty$, the exponential term is $\exp(-2Qc + o(1))$ which blows up exponentially. This implies $\langle 1 \rangle_{\hat{g}} = \infty$.

It turns out that we get a convergent expectation if we choose for our observable natural **correlation functions** of the model (denoted by $V = V_{\alpha_1,\ldots,\alpha_k}(\mathbf{z})$), that is, informally,

$$V(\varphi) = e^{\alpha_1 \varphi(z_1) + \cdots + \alpha_k \varphi(z_k)}, \tag{5.52}$$

where $\alpha_1, \ldots, \alpha_k$ are real numbers and $\mathbf{z} = (z_1, \ldots, z_k) \in \mathbb{C}^k$ with $z_i \neq z_j$ for $i \neq j$. In physics language, $e^{\alpha_i \varphi(z_i)}$ is a **vertex operator** and z_i is called an **insertion**.

As usual, since φ is a distribution, it is not entirely clear what one means by $e^{\alpha_i \varphi(z_i)}$, so in order to even speak about $\langle F \rangle_{\hat{g}}$ some regularisation is necessary. Let h_ε denote the circle average of h. Define

$$V_\varepsilon(\varphi) = \prod_{i=1}^{k} \varepsilon^{\alpha_i^2/2} e^{\alpha_i \varphi_\varepsilon(z_i)}, \tag{5.53}$$

so that

$$\langle V_\varepsilon \rangle_{\hat{g}} = \int_{c \in \mathbb{R}} \mathbb{E}\left[\prod_{i=1}^{k} \varepsilon^{\alpha_i^2/2} e^{\alpha_i \left(h_\varepsilon(z_i) + \frac{Q}{2} \log \hat{g}(z_i) + c\right)} \times \right.$$

$$\left. \exp\left(-2Qc - \mu e^{\gamma c} \mathcal{M}_{h + \frac{Q}{2} \log \hat{g}}(\mathbb{C})\right)\right] dc.$$

We will attempt to define $\langle V \rangle_{\hat{g}}$ by taking a limit of $\langle V_\varepsilon \rangle_{\hat{g}}$ as $\varepsilon \to 0$.

Note that if $\sum_{i=1}^{k} \alpha_i > 2Q$, the problem near $c = -\infty$ leading to the divergence of (5.51) should disappear. On the other hand, no new problem is created at $c = \infty$ because the decay of the integrand is doubly exponential in that region. This suggests that there is a chance that $\langle V \rangle_{\hat{g}} = \lim_{\varepsilon \to 0} \langle V_\varepsilon \rangle_{\hat{g}}$ may be finite if $\sum_{i=1}^{k} \alpha_i > 2Q$. On the other hand, if any of the α_i is too large, then the expectation can also explode as we are adding a logarithmic singularity of strength α_i to the field. One might naively guess that the maximal allowed value for α_i could be $\alpha_i = \gamma$ (corresponding to a Liouville typical point) or perhaps $\alpha_i = 2$ (corresponding to the maximal thickness of any point h). Surprisingly, the maximal allowed value is in fact strictly larger, namely it suffices to require $\alpha_i < Q$. That the expectation is convergent and non-zero under these two conditions, collectively known as the Seiberg bounds, is the content of the next theorem and one of the main results of [DKRV16]. With these notations, the main theorem of this section is the following:

5.5 Convergence of correlation functions and Seiberg bounds 195

Theorem 5.20 *Suppose* $\alpha_1, \ldots, \alpha_k \in \mathbb{R}$ *satisfy*

$$\sum_{i=1}^{k} \alpha_i > 2Q \qquad (5.54)$$

and $z_1, \ldots, z_k \in \mathbb{C}$ *are distinct. Then* $\langle V_\varepsilon \rangle_{\hat{g}} < \infty$. *Suppose furthermore*

$$\alpha_i < Q \quad \text{for } i = 1, \ldots, k. \qquad (5.55)$$

Then $\langle V_\varepsilon \rangle_{\hat{g}}$ *converges to a limit* $\langle V \rangle = \langle V \rangle_{\hat{g}}$ *in* $(0, \infty)$ *as* $\varepsilon \to 0$.

Before giving the proof of this theorem, we make a few comments. The two bounds (5.54) and (5.55) are known collectively as the Seiberg bounds. They are known to be optimal in the sense that if either of these bounds fail, then either $\langle V_\varepsilon \rangle_{\hat{g}} = \infty$ or it converges to zero as $\varepsilon \to 0$. Note that for these two bounds to be simultaneously satisfied, it is necessary that $k \geq 3$. The necessity of these **three insertions** will be discussed below both from a geometric and probabilistic perspective.

We now start the proof of this theorem.

Proof Without loss of generality, by the Weyl anomaly (Theorem 5.17), we take $\hat{g} = \hat{g}_0$. Our first task will be to re-express $\langle V_\varepsilon \rangle_{\hat{g}_0}$ via Girsanov's theorem (Lemma 2.5). To do this we will need to view each term $e^{\alpha h_\varepsilon(z_i)} \varepsilon^{\alpha_i^2/2}$ as an exponential biasing of the Gaussian field h. Notice however that the normalising factor, namely $\varepsilon^{\alpha_i^2/2}$, is not quite equal to $\mathbb{E}[e^{\alpha_i h_\varepsilon(z_i)}]^{-1}$, so we must account for this using Lemma 5.12.

The result is that we may write

$$\langle V_\varepsilon \rangle_{\hat{g}_0} = \int_{c \in \mathbb{R}} \mathbb{E}\Bigg[\prod_{i=1}^{k} \varepsilon^{\alpha_i^2/2} e^{\alpha_i (h_\varepsilon(z_i) + \frac{Q}{2}(\log \hat{g}_0)_\varepsilon(z_i) + c)} \times$$

$$\exp\Big(-2Qc - \mu e^{\gamma c} M_{h + \frac{Q}{2}\log \hat{g}_0}(\mathbb{C})\Big)\Bigg] dc$$

$$= e^{C_\varepsilon(z_1, \ldots, z_k)} \prod_i \hat{g}_0(z_i)^{\frac{\alpha_i}{2}(Q - \frac{\alpha_i}{2})} \int_{c \in \mathbb{R}} \mathbb{E}\Bigg[\exp\Big(\Big(\sum_{i=1}^{k} \alpha_i - 2Q\Big)c$$

$$- \mu e^{\gamma c} M_{\hat{h}^\varepsilon + \frac{Q}{2}\log \hat{g}_0}(\mathbb{C})\Big)\Bigg] dc, \qquad (5.56)$$

where the field $\hat{h}^\varepsilon$ is obtained from h by applying the Girsanov shift,

$$\hat{h}^\varepsilon(\cdot) = h(\cdot) + \sum_{i=1}^{k} \alpha_i \int_0^{2\pi} 2\pi G^{\hat{\mathbb{C}}, \hat{g}_0}(z_i + \varepsilon e^{i\theta}, \cdot) \frac{d\theta}{2\pi}. \qquad (5.57)$$

196   *Introduction to Liouville Conformal Field Theory*

(The factor 2π in front of the Green function is due to our normalisation: $h = \sqrt{2\pi}\mathbf{h}$.) The normalising constant C_ε above satisfies

$$C_\varepsilon(z_1, \ldots, z_k) := \sum_{i=1}^k \sum_{j>i} 2\pi \alpha_i \alpha_j G^{\hat{\mathbb{C}}, \hat{g}_0}(z_i, z_j) + \sum_{i=1}^k \frac{\alpha_i^2}{2}\left(\log(2) - \frac{1}{2}\right) + o(1).$$

Since this normalising factor has a well-behaved limit as $\varepsilon \to 0$, which we call $C(\mathbf{z})$, it suffices to consider the integral term in (5.56). Writing

$$Z_\varepsilon := \mathcal{M}_{\hat{h}^\varepsilon + \frac{Q}{2}\log \hat{g}_0}(\mathbb{C}),$$

applying Fubini's theorem, and changing variables $u = e^{\gamma c} Z_\varepsilon$, so that $du = \gamma e^{\gamma c} Z_\varepsilon \, dc = \gamma u \, dc$, we obtain that

$$\langle V_\varepsilon \rangle_{\hat{g}_0}$$

$$\sim e^{C(\mathbf{z})} \prod_i \hat{g}_0(z_i)^{\frac{\alpha_i}{2}(Q-\frac{\alpha_i}{2})} \mathbb{E}\left[\int_{c \in \mathbb{R}} \exp\left(\left(\sum_{i=1}^k \alpha_i - 2Q\right)c - \mu e^{\gamma c} Z_\varepsilon\right) dc\right]$$

$$\sim e^{C(\mathbf{z})} \prod_i \hat{g}_0(z_i)^{\frac{\alpha_i}{2}(Q-\frac{\alpha_i}{2})} \mathbb{E}\left[\int_{u>0} \left(\frac{u}{Z_\varepsilon}\right)^{\frac{\sum_i \alpha_i - 2Q}{\gamma}} e^{-\mu u} \frac{du}{\gamma u}\right]$$

$$\sim \frac{e^{C(\mathbf{z})}}{\gamma} \prod_i \hat{g}_0(z_i)^{\frac{\alpha_i}{2}(Q-\frac{\alpha_i}{2})} \int_{u>0} u^{\frac{\sum_i \alpha_i - 2Q}{\gamma} - 1} e^{-\mu u} du \cdot \mathbb{E}\left[Z_\varepsilon^{-\frac{\sum_i \alpha_i - 2Q}{\gamma}}\right].$$
(5.58)

Above we use the Landau notation $a_\varepsilon \sim b_\varepsilon$ to mean that the ratio $a_\varepsilon/b_\varepsilon \to 1$ as $\varepsilon \to 0$. The integral over u does not depend on ε (in fact, it is nothing but the Gamma function evaluated at $\frac{\sum_i \alpha_i - 2Q}{\gamma}$). Hence, the proof of the theorem eventually boils down to proving the following lemma:

Lemma 5.21 *Let* $s = \frac{\sum_i \alpha_i - 2Q}{\gamma} > 0$. *Then the limit*

$$\lim_{\varepsilon \to 0} \mathbb{E}(Z_\varepsilon^{-s}) =: \mathbb{E}(Z_0^{-s})$$

exists and lies in $(0, \infty)$.

Proof Recall that

$$Z_\varepsilon = \lim_{\delta \to 0} \int_\mathbb{C} \delta^{\gamma^2/2} e^{\gamma\left[\hat{h}^\varepsilon_\delta(z) + \frac{Q}{2}(\log \hat{g}_0)_\delta(z)\right]} dz$$

$$= \lim_{\delta \to 0} \int_\mathbb{C} e^{\gamma(H_\varepsilon)_\delta(z)} \delta^{\gamma^2/2} e^{\gamma\left[h_\delta(z) + \frac{Q}{2}(\log \hat{g}_0)_\delta(z)\right]} dz,$$

where the subscript δ is used everywhere to denote the circle average at radius

5.5 Convergence of correlation functions and Seiberg bounds

δ, and the convergence is in probability. Here

$$H_\varepsilon(z) = \sum_{i=1}^{k} \alpha_i \int_0^{2\pi} 2\pi G^{\hat{C},\hat{g}_0}(z_i + \varepsilon e^{i\theta}, z) \frac{d\theta}{2\pi}.$$

Since H_ε is a smooth function of z for fixed ε, $(H_\varepsilon)_\delta(z) \to H_\varepsilon(z)$ uniformly in z as $\delta \to 0$. Together with the fact that H_ε is uniformly bounded and that $\mathcal{M}_{h+(Q/2)\log\hat{g}_0}(\mathbb{C})$ is a limit in $L^1(\mathbb{P})$ of $\int_{\mathbb{C}} \delta^{\gamma^2/2} e^{\gamma[h_\delta(z)+(Q/2)(\log\hat{g}_0)_\delta(z)]} dz$, this implies that

$$Z_\varepsilon = \lim_{\delta \to 0} \int_{\mathbb{C}} e^{\gamma(H_\varepsilon)_\delta(z)} \delta^{\gamma^2/2} e^{\gamma\left[h_\delta(z)+\frac{Q}{2}(\log\hat{g}_0)_\delta(z)\right]} dz$$

$$= \int_{\mathbb{C}} e^{\gamma H_\varepsilon(z)} \mathcal{M}_{h+(Q/2)\log\hat{g}_0}(dz),$$

where the limit holds in $L^1(\mathbb{P})$ and in probability. In particular, Z_ε has finite expectation for each $\varepsilon > 0$.

Before proceeding with the proof, let us make a couple of remarks.

- It is clear that $H_\varepsilon(z)$ converges to a function $H(z) = \sum_{i=1}^{k} \alpha_i 2\pi G^{\hat{C},\hat{g}_0}(z_i, z)$ as $\varepsilon \to 0$. It is therefore natural to expect that Z_ε converges (in probability, say) to

$$Z_0 := \int_{\mathbb{C}} e^{\gamma H(z)} \mathcal{M}_{h+\frac{Q}{2}\log\hat{g}_0}(dz).$$

This will indeed follow from our proof.

- Most of the proof consists of checking that Z_0 is in fact finite almost surely under the second Seiberg bound (5.55) (i.e. $\alpha_i < Q$). This makes the negative moment $\mathbb{E}(Z_0^{-s})$ strictly positive. This is however far from obvious: for instance, the expectation of Z_0 actually blows up if one of the α_i satisfies $\gamma\alpha_i > 2$ (which is allowed since $Q > 2/\gamma$). Nevertheless, Z_0 remains finite almost surely, even though its expectation is infinite. The fact that Z_0 remains finite under the Seiberg condition (5.55) is instead a consequence of a scaling argument, as we will now see.

We divide the proof into several steps. The first step is the easy upper bound on $\mathbb{E}[(Z_\varepsilon)^{-s}]$ (corresponding to a lower bound on Z_ε).

Step 1. For any $q > 0$, there exists $C = C_s > 0$ such that $\mathbb{E}[(Z_\varepsilon)^{-q}] \leq C$ for all $\varepsilon > 0$. This is indeed easy to see, since if we consider any bounded set B (which does not even need to stay disjoint from the insertions z_i, $1 \leq i \leq k$), then

$$Z_\varepsilon \geq \int_B e^{\gamma H_\varepsilon(z)} \mathcal{M}_{h+(Q/2)\log\hat{g}_0}(dz) \geq C \mathcal{M}_{h+(Q/2)\log\hat{g}_0}(B),$$

where C is a uniform (in ε and $z \in B$) lower bound on $e^{\gamma H_\varepsilon(z)}$. Taking the negative moment of order $-q < 0$, the claimed upper bound therefore follows from Theorem 3.42.

Step 2. We fix $r > 0$ and let $A_r = \bigcup_{i=1}^{k} B(z_i, r)$. We decompose Z_ε according to whether $z \in A_r$ or not. Thus we write

$$Z_\varepsilon = \int_{A_r} e^{\gamma H_\varepsilon(z)} \mathcal{M}_{h+(Q/2)\log \hat{g}_0}(\mathrm{d}z) + \int_{A_r^c} e^{\gamma H_\varepsilon(z)} \mathcal{M}_{h+(Q/2)\log \hat{g}_0}(\mathrm{d}z)$$

$$=: Z_{r,\varepsilon} + Z_{r,\varepsilon}^c.$$

In this second step, we show that for fixed r, the term $Z_{r,\varepsilon}^c$ corresponding to points far away from the insertions is well behaved and has a limit in probability. Note that $H_\varepsilon \to H$ uniformly on A_r^c, and that $\mathcal{M}_{h+(Q/2)\log \hat{g}_0}(\mathrm{d}z)$ is a measure of uniformly bounded total expectation (as already observed, it is a limit in $L^1(\mathbb{P})$). Hence

$$\int_{A_r^c} |e^{\gamma H_\varepsilon(z)} - e^{\gamma H(z)}| \mathcal{M}_{h+(Q/2)\log \hat{g}_0}(\mathrm{d}z) \to 0$$

in $L^1(\mathbb{P})$ and in probability. We deduce that

$$Z_{r,\varepsilon}^c \to Z_{r,0}^c := \int_{A_r^c} e^{\gamma H(z)} \mathcal{M}_{h+(Q/2)\log \hat{g}}(\mathrm{d}z)$$

in probability as $\varepsilon \to 0$ for each fixed r.

Step 3. In this third step, we show that the term $Z_{r,\varepsilon}$ corresponding to the points close to the insertions does not blow up if $\alpha_i < Q$ for all $1 \le i \le k$. More precisely, we will show that there is a function $C(r) > 0$ such that $C(r) \to 0$ as $r \to 0$, and such that for sufficiently small $p > 0$ and all $\varepsilon > 0$,

$$\mathbb{E}[(Z_{r,\varepsilon})^p] \le C(r). \tag{5.59}$$

This is the most technical part of the proof. To begin with, we observe that by subadditivity of $x \mapsto x^p$ for $p < 1$, it suffices to prove the result for $k = 1$ insertions. Without loss of generality we take $z_1 = 0$, and we write $\alpha = \alpha_1$ for the power associated with the corresponding insertion. Recall that $\alpha < Q$ as per (5.55).

Since $Z_{r,\varepsilon}^p$ is decreasing with r for fixed ε, we may also assume without loss of generality that $r = e^{-k_0}$ with $k_0 \in \mathbb{N}$. We then decompose the ball $B(0, r)$ into the disjoint annuli $B_k = B(0, e^{-k}) \setminus B(0, e^{-k-1})$ so that $B(0, r) = \bigcup_{k \ge k_0} B_k$. Note that for any fixed $k \ge k_0$, one has

$$e^{\gamma H_\varepsilon(z)} \lesssim e^{k\alpha\gamma}$$

5.5 Convergence of correlation functions and Seiberg bounds

for $z \in B_k$, where the implied constant is uniform over $z \in B_k$ and $\varepsilon > 0$. It follows that

$$\mathbb{E}[(Z_{r,\varepsilon})^p] \lesssim \sum_{k=k_0}^{\lceil \log(1/\varepsilon) \rceil} e^{k\alpha\gamma p} \mathbb{E}[(\mathcal{M}_{h+(Q/2)\log \hat{g}_0}(B_k))^p] \lesssim \sum_{k \geq k_0} e^{k\alpha\gamma p} e^{-k\xi(p)},$$

where $\xi(p) = p(2 + \gamma^2/2) - p^2\gamma^2/2$ is the multifractal spectrum function. Here we have used Theorem 3.27 and more precisely (3.55), together with the obvious fact that for $p < 1$, by Jensen's inequality,

$$\mathbb{E}[(\mathcal{M}_{h+(Q/2)\log \hat{g}_0}(B(0,1)))^p] \leq \mathbb{E}[\mathcal{M}_{h+(Q/2)\log \hat{g}_0}(\mathbb{C})]^p \lesssim 1.$$

As a consequence, the claim (5.59) follows if we can find $0 < p < 1$ sufficiently small so that

$$\alpha\gamma p - \xi(p) < 0.$$

Linearising $\xi(p)$ around $p = 0$, it suffices that

$$\alpha\gamma - (2 + \gamma^2/2) < 0. \tag{5.60}$$

But from the Seiberg bound (5.55), since $\alpha < Q = (\gamma/2 + 2/\gamma)$, we see that $\alpha Q < 2 + \gamma^2/2$ so that (5.60) is fulfilled.

Step 4. We now conclude the proof of Lemma 5.21. Recall from Step 2 that the limit in probability $Z_{r,0}^c = \lim_{\varepsilon \to 0} Z_{r,\varepsilon}^c$ satisfies

$$Z_{r,0}^c = \int_{A_r^c} e^{\gamma H(z)} \mathcal{M}_{h+(Q/2)\log \hat{g}_0}(dz)$$

and is thus monotone increasing in r. We can therefore set

$$Z_0 = \lim_{r \to 0} Z_{r,0}^c,$$

where the above limit is in the almost sure sense. By Markov's inequality and (5.59) of Step 3, we also have that

$$\mathbb{P}(Z_{r,\varepsilon} > C(r)^{p/2}) \leq C(r)^{p/2} \to 0$$

as $r \to 0$, uniformly in ε. Using the triangle inequality and taking r small and then ε small, we deduce that

$$Z_\varepsilon \to Z_0 \text{ in probability as } \varepsilon \to 0.$$

Moreover, taking the difference, we see that $Z_{r,\varepsilon} = Z_\varepsilon - Z_{r,\varepsilon}^c \to Z_{r,0} := Z_0 - Z_{r,0}^c$ in probability for each $r > 0$.

From Step 1, we see that for our fixed $s := \gamma^{-1}(\sum \alpha_i - 2Q)$, $\mathbb{E}((Z_\varepsilon)^{-2s})$ is uniformly bounded in ε, so that $(Z_\varepsilon)^{-s}$ is uniformly integrable and hence

200   Introduction to Liouville Conformal Field Theory

$\mathbb{E}((Z_\varepsilon)^{-s}) \to \mathbb{E}((Z_0)^{-s}) < \infty$ as $\varepsilon \to 0$. To show that the limit is also non-zero, it suffices to show that $Z_0 < \infty$ (so that $(Z_0)^{-s} > 0$ and hence its expectation is also strictly positive). For this we take $r = 1$, and write $Z_0 = Z_{1,0} + Z_{1,0}^c$ in the notation of Step 2. The second term has finite expectation and is therefore finite almost surely. The first term has finite pth moment for sufficiently small $p > 0$ by (5.59) and Fatou's lemma. It is therefore also finite almost surely. We deduce that $Z_0 < \infty$, which concludes the proof of Lemma 5.21. □

Plugging Lemma 5.21 into (5.58), we conclude the proof of Theorem 5.20. □

Remark 5.22 It can be shown that the Seiberg bounds are sharp in the following sense. If $\sum_i \alpha_i \leq 2Q$, then $\langle V_\varepsilon \rangle_{\hat{g}_0} = \infty$, while if $\max_i \alpha_i \geq Q$, then the random variable Z_0 in Lemma 5.21 is almost surely infinite, so that its negative moment of order s is zero, hence $\langle V_\varepsilon \rangle_{\hat{g}_0} \to 0$. See (3.17) in [DKRV16]. Thus, the Seiberg bounds are a *necessary and sufficient condition* for the correlation functions to be well defined (at least without further normalisation).

In the course of the proof, we obtained a very important expression for the value of the correlation function $\langle V \rangle_{\hat{g}_0} = \lim_{\varepsilon \to 0} \langle V_\varepsilon \rangle_{\hat{g}_0}$. This shows that the correlation function of the model (which in a few moments we will view as the partition function of a random field) can be computed exactly as some fractional negative moment of a Gaussian multiplicative moment and the Gamma function, and is the first hint of the remarkable **integrability** of the model. It is worth restating this expression as a separate corollary.

Corollary 5.23 *Suppose that* $z_1, \ldots, z_k \in \mathbb{C}$ *are distinct, and* $\alpha_1, \ldots, \alpha_k$ *satisfy the Seiberg bounds (5.54) and (5.55). Write* $V = V_{\alpha_1, \ldots, \alpha_k}(\mathbf{z})$ *for the associated correlation function. Set*

$$\tilde{h}^{\hat{\mathbb{C}}} = h^{\hat{\mathbb{C}}, \hat{g}_0} + \tfrac{Q}{2} \log \hat{g}_0 + 2\pi \sum_{i=1}^{k} \alpha_i G^{\hat{\mathbb{C}}, \hat{g}_0}(\cdot, z_i); \tag{5.61}$$

$$C_\alpha(\mathbf{z}) = 2\pi \sum_{i=1}^{k} \sum_{j=i+1}^{k} \alpha_i \alpha_j G^{\hat{\mathbb{C}}, \hat{g}_0}(z_i, z_j) + \sum_{i=1}^{k} \alpha_i^2 c_{\hat{g}_0}; \tag{5.62}$$

where we recall that $c_{\hat{g}_0} = \log(2) - 1/2$; *and*

$$\Delta_\alpha = \tfrac{\alpha}{2}(Q - \tfrac{\alpha}{2}) \quad ; \quad s = \frac{\sum_{i=1}^{k} \alpha_i - 2Q}{\gamma}. \tag{5.63}$$

Then we have

$$\langle V \rangle_{\hat{g}_0} = \gamma^{-1} e^{C_\alpha(\mathbf{z})} \prod_i \hat{g}_0(z_i)^{\Delta_{\alpha_i}} \Gamma(s; \mu) \, \mathbb{E}\left(\mathcal{M}_{\tilde{h}^{\hat{\mathbb{C}}}}(\mathbb{C})^{-s} \right), \tag{5.64}$$

5.5 Convergence of correlation functions and Seiberg bounds

where $\Gamma(s;\mu) = \int_0^\infty u^{s-1} e^{-\mu u}\, du$ is the Gamma function as before. Here, following the general notation in the chapter, $\mathcal{M}_h(\mathbb{C}) = \lim_{\varepsilon \to 0} \int_{\mathbb{C}} \varepsilon^{\gamma^2/2} e^{\gamma h_\varepsilon(z)}\, dz$.

More generally, if F is a non-negative measurable functional on $H^{-1}(\hat{\mathbb{C}})$, we have

$$\langle VF \rangle_{\hat{g}_0} := \lim_{\varepsilon \to 0} \langle V_\varepsilon F \rangle_{\hat{g}_0}$$

$$= \gamma^{-1} e^{C_\alpha(\mathbf{z})} \prod_i \hat{g}_0(z_i)^{\Delta_{\alpha_i}} \times \int_{u>0} \mathbb{E}\bigg(F\Big(\tilde{h}^{\hat{\mathcal{C}}} + \frac{\log u}{\gamma} - \frac{\log \mathcal{M}_{\tilde{h}^{\hat{\mathcal{C}}}}(\mathbb{C})}{\gamma}\Big) \times$$

$$\mathcal{M}_{\tilde{h}^{\hat{\mathcal{C}}}}(\mathbb{C})^{-s}\bigg) u^{s-1} e^{-\mu u}\, du. \tag{5.65}$$

Remark 5.24 One can check that the proof of Theorem 5.20 goes through unchanged when $\hat{g}_0$ is replaced by a metric $\hat{g}$ as in (5.7) with constant scalar curvature. This results in analogous explicit expressions for $\langle V \rangle_{\hat{g}}$ and $\langle FV \rangle_{\hat{g}}$ (the constant $c_{\hat{g}_0}$ being replaced by its general expression $c_{\hat{g}}$ from Remark 5.13). This is in particular the case when $\hat{g}$ is the pushforward of $\hat{g}_0$ by a Möbius map m, and note furthermore that in this case $c_{\hat{g}} = c_{\hat{g}_0}$: see (5.14).

The Weyl anomaly formula of Theorem 5.17 was written for general "smooth" observable F on $H^{-1}(\hat{\mathbb{C}})$. But it is immediate to deduce from it a formula which includes the correlation functions.

Corollary 5.25 Let $V = V_{\alpha_1,\ldots,\alpha_k}(\mathbf{z})$ be the correlation functions as above; and let F be an arbitrary non-negative measurable functional on $H^{-1}(\hat{\mathbb{C}})$, and let $\hat{g} = e^\rho \hat{g}_0$ where ρ is a twice differentiable function on $\mathbb{C}$ with a finite limit at infinity and $\int_{\mathbb{C}} |\nabla \rho(z)|^2\, dz < \infty$, as in (5.7). Then

$$\langle VF \rangle_{\hat{g}} = \exp\bigg(\frac{6Q^2}{96\pi} \int_{\mathbb{C}} [|\nabla^{\hat{g}_0} \rho(x)|^2 + 2R_{\hat{g}_0} \rho(x)] v_{\hat{g}_0}(dx)\bigg) \langle VF \rangle_{\hat{g}_0}.$$

Proof By Theorem 5.17, we have

$$\langle V_\varepsilon F \rangle_{\hat{g}} = \exp\bigg(\frac{6Q^2}{96\pi} \int_{\mathbb{C}} [|\nabla^{\hat{g}_0} \rho(x)|^2 + 2R_{\hat{g}_0} \rho(x)] v_{\hat{g}_0}(dx)\bigg) \langle V_\varepsilon F \rangle_{\hat{g}_0}$$

for every $\varepsilon > 0$, so the result follows by taking $\varepsilon \to 0$. □

Remark 5.26 The above Weyl anomaly formula is partly a consequence of how we chose to define the correlation functions V in (5.52), since implicitly the chosen regularisation in (5.53) is given in terms of the Euclidean metric rather than the intrinsic metric g. If instead one thinks of choosing circle averages with respect to the metric g, a more intrinsic definition of $\langle V_\varepsilon \rangle_{\hat{g}}$ would be

$$\langle V_\varepsilon \rangle_{\hat{g}} = \int \mathbb{E}\left[\prod_{i=1}^{k} (\sqrt{\hat{g}(z_i)}\varepsilon)^{\alpha_i^2/2} e^{\alpha_i(h_\varepsilon(z_i)+c)} \times \right.$$
$$\left. \exp\left(-2Qc - \mu e^{\gamma c} M_{h+\frac{Q}{2}\log\hat{g}}(\hat{\mathbb{C}})\right) \right] dc.$$

Note that compared to (5.52), there is an extra factor $\sqrt{\hat{g}(z_i)}$ in front of the normalising factor ε and that the term $\frac{Q}{2}\log\hat{g}(z_i)$ is *not* included in the first exponential term. This would lead to a slightly different version of the Weyl anomaly formula: namely,

$$\langle VF \rangle_{\hat{g}} = \exp\left(\frac{6Q^2}{96\pi} A(\rho, \hat{g}_0) - \sum_{i=1}^{k} \Delta_{\alpha_i} \rho(z_i) \right) \langle VF \rangle_{\hat{g}_0},$$

where $A(\rho, \hat{g}_0) = \int_{\mathbb{C}}[|\nabla^{\hat{g}_0}\rho(x)|^2 + 2R_{\hat{g}_0}\rho(x)]v_{\hat{g}_0}(dx)$ is as in Corollary 5.25. This version of the Weyl anomaly formula is for instance the one that is used in [GKRV21, (1.3)].

As a consequence of Remark 5.24 and Corollary 5.25, together with Lemma 5.6 and Corollary 5.7, we obtain the following theorem describing how $\langle V \rangle_{\hat{g}_0}$ changes when the insertions $\{z_i\}_i$ are transformed using a Möbius map. This transform is described in [DKRV16] as a version of the KPZ formula, cf. Theorem 3.46.

Theorem 5.27 *Suppose that* $m: \hat{\mathbb{C}} \to \hat{\mathbb{C}}$ *is a Möbius transformation of the Riemann sphere, and* α_i, z_i *are as in Corollary 5.23. Then*

$$\langle V_{\alpha_1,\ldots,\alpha_k}(m(\mathbf{z})) \rangle_{\hat{g}_0} = \prod_i |m'(z_i)|^{-2\Delta_{\alpha_i}} \langle V_{\alpha_1,\ldots,\alpha_k}(\mathbf{z}) \rangle_{\hat{g}_0}, \tag{5.66}$$

where $m(\mathbf{z}) = (m(z_1), \ldots, m(z_k))$. *Moreover, if* F *is a non-negative functional on* $H^{-1}(\hat{\mathbb{C}})$,

$$\langle FV_{\alpha_1,\ldots,\alpha_k}(m(\mathbf{z})) \rangle_{\hat{g}_0} = \prod_i |m'(z_i)|^{-2\Delta_{\alpha_i}} \langle F_m V_{\alpha_1,\ldots,\alpha_k}(\mathbf{z}) \rangle_{\hat{g}_0}, \tag{5.67}$$

where $F_m(h) := F(h \circ m^{-1} + Q \log |(m^{-1})'|)$ *for* $h \in H^{-1}(\hat{\mathbb{C}})$.

Proof By setting $F = 1$, it suffices to prove the second statement. The Weyl anomaly formula, Lemma 5.19, gives that

$$\langle FV_{\alpha_1,\ldots,\alpha_k}(m(\mathbf{z})) \rangle_{\hat{g}_0} = \langle FV_{\alpha_1,\ldots,\alpha_k}(m(\mathbf{z})) \rangle_{m_*\hat{g}_0}.$$

On the other hand, using Remark 5.24 and using that $c_{m_*\hat{g}_0} = c_{\hat{g}_0} = \log(2) - 1/2$, we see that

5.5 Convergence of correlation functions and Seiberg bounds

$\langle FV_{\alpha_1,\ldots,\alpha_k}(m(\mathbf{z}))\rangle_{m_*\hat{g}_0}$
$= \gamma^{-1} e^{2\pi \sum_{i=1}^{k}\sum_{j=i+1}^{k} \alpha_i\alpha_j G^{\hat{\mathbb{C}},m_*\hat{g}_0}(m(z_i),m(z_j)) + \sum_i \alpha_i^2(\log(2)-1/2)} \times$

$$\prod_i (m_*\hat{g}_0)(m(z_i))^{\Delta_{\alpha_i}} \int_{u>0} \mathbb{E}\Big(F\big(\tilde{h} + \frac{\log u}{\gamma} - \frac{\log \mathcal{M}_{\tilde{h}}(\mathbb{C})}{\gamma}\big)\mathcal{M}_{\tilde{h}}(\mathbb{C})^{-s}\Big) u^{s-1} e^{-\mu u}\, du,$$

where

$$\tilde{h} = h^{\hat{\mathbb{C}},m_*\hat{g}_0} + \tfrac{Q}{2}\log m_*\hat{g}_0 + 2\pi \sum_{i=1}^{k}\alpha_i G^{\hat{\mathbb{C}},m_*\hat{g}_0}(\cdot, m(z_i)).$$

Recall Remark 5.8, which says that

$$h^{\hat{\mathbb{C}},m_*\hat{g}_0} \stackrel{d}{=} h^{\hat{\mathbb{C}},\hat{g}_0} \circ m^{-1}; \text{ and}$$

$$G^{\hat{\mathbb{C}},m_*\hat{g}_0}(m(z),m(w)) = G^{\hat{\mathbb{C}},\hat{g}_0}(z,w) \text{ for } z \neq w \text{ in } \hat{\mathbb{C}}.$$

Also using the explicit form $m_*\hat{g}_0(x) = \hat{g}_0(m^{-1}(x))|(m^{-1})'(x)|^2$, we therefore have that

$$\tilde{h} \stackrel{(d)}{=} \tilde{h}^{\hat{\mathbb{C}}} \circ m^{-1} + Q\log(|(m^{-1})'|)$$

and

$$2\pi \sum_{i=1}^{k}\sum_{j=i+1}^{k} \alpha_i\alpha_j G^{\hat{\mathbb{C}},m_*\hat{g}_0}(m(z_i),m(z_j)) + \sum_i \alpha_i^2(\log(2)-1/2) = C_\alpha(\mathbf{z}),$$

where $\tilde{h}^{\hat{\mathbb{C}}}$ and $C_\alpha(\mathbf{z})$ are as in Corollary 5.23. Finally, observe that

$$\prod_i (m_*\hat{g}_0)(m(z_i))^{\Delta_{\alpha_i}} = \prod_i (\hat{g}_0(z_i))^{\Delta_{\alpha_i}} \prod_i |(m^{-1})'(m(z_i))|^{2\Delta_{\alpha_i}}$$
$$= \prod_i (\hat{g}_0(z_i))^{\Delta_{\alpha_i}} \prod_i |m'(z_i)|^{-2\Delta_{\alpha_i}},$$

which yields

$$\langle FV_{\alpha_1,\ldots,\alpha_k}(m(\mathbf{z}))\rangle_{\hat{g}_0} = \langle FV_{\alpha_1,\ldots,\alpha_k}(m(\mathbf{z}))\rangle_{m_*\hat{g}_0}$$
$$= \prod_i |m'(z_i)|^{-2\Delta_{\alpha_i}} \langle F_m V_{\alpha_1,\ldots,\alpha_k}(\mathbf{z})\rangle_{\hat{g}_0},$$

as desired. □

Corollary 5.28 *Set $k=3$ and suppose α_1,α_2 and α_3 satisfy the Seiberg bounds. Then there exist a constant $C(\alpha_1,\alpha_2,\alpha_3) > 0$ called the **structure constant** such that for any $z_1,\ldots,z_3 \in \hat{\mathbb{C}}$*

$$\langle V_{\alpha_1,\alpha_2,\alpha_3}(\mathbf{z})\rangle_{\hat{g}_0} = C(\alpha_1,\alpha_2,\alpha_3)|z_1-z_2|^{2\Delta_{1,2}}|z_2-z_3|^{2\Delta_{2,3}}|z_3-z_1|^{2\Delta_{1,3}}$$

where $\Delta_{1,2} = \Delta_{\alpha_3} - \Delta_{\alpha_1} - \Delta_{\alpha_2}$, $\Delta_{2,3} = \Delta_{\alpha_1} - \Delta_{\alpha_2} - \Delta_{\alpha_3}$, and $\Delta_{1,3} = \Delta_{\alpha_2} - \Delta_{\alpha_1} - \Delta_{\alpha_3}$.

Introduction to Liouville Conformal Field Theory

Proof Since $k = 3$ we can find a Möbius map m sending $z_1 \mapsto 0$, $z_2 \mapsto 1$, $z_3 \mapsto \infty$. The map m has an explicit form, namely

$$m(z) = \frac{z - z_1}{z - z_3} \frac{z_2 - z_3}{z_2 - z_1}.$$

Note that then

$$m'(z) = \frac{z_1 - z_3}{(z - z_3)^2} \frac{z_2 - z_3}{z_2 - z_1}.$$

Hence

$$m'(z_1) = \frac{z_2 - z_3}{(z_2 - z_1)(z_1 - z_3)}; m'(z_2) = \frac{z_1 - z_3}{(z_2 - z_1)(z_2 - z_3)}.$$

while

$$m'(z_3) = \infty; \text{ and } m'(z) \sim \frac{(z_1 - z_3)(z_2 - z_3)}{z_2 - z_1} \times \frac{1}{(z - z_3)^2}$$

as $z \to z_3$. Let us evaluate (5.66) at $\mathbf{z} = z_1, z_2, z$ with $z \to z_3$. Then writing $\Delta_i = \Delta_{\alpha_i}$, we have

$$\frac{\langle V_{\alpha_1, \alpha_2, \alpha_3}(0, 1, m(z)) \rangle_{\hat{g}_0}}{\langle V_{\alpha_1, \alpha_2, \alpha_3}(z_1, z_2, z) \rangle_{\hat{g}_0}} \sim$$
$$|z - z_3|^{4\Delta_3} |z_1 - z_2|^{2(\Delta_1 + \Delta_2 + \Delta_3)} |z_2 - z_3|^{2(-\Delta_1 - \Delta_3 + \Delta_2)} |z_1 - z_3|^{2(\Delta_1 - \Delta_2 - \Delta_3)}$$

as $z \to z_3$. From this we learn that, writing $y = m(z)$,

$$\lim_{y \to \infty} |y|^{4\Delta_3} \langle V_{\alpha_1, \alpha_2, \alpha_3}(0, 1, y) \rangle_{\hat{g}_0} = C(\alpha_1, \alpha_2, \alpha_3)$$

exists, and equals

$$\langle V_{\alpha_1, \alpha_2, \alpha_3}(\mathbf{z}) \rangle_{\hat{g}_0} |z_1 - z_2|^{-2\Delta_{1,2}} |z_2 - z_3|^{-2\Delta_{2,3}} |z_3 - z_1|^{-2\Delta_{1,3}}.$$

This concludes the proof. □

Definition 5.29 (Correlation functions with an insertion at ∞) Generalising the argument in the proof above we see that for $\alpha_1, \ldots, \alpha_k$ satisfying (5.54) and (5.55),

$$\lim_{y \to \infty} |y|^{4\alpha_k} \langle V_\alpha(z_1, \ldots, z_{k-1}, y) \rangle_{\hat{g}_0} =: \langle V_\alpha(z_1, \ldots, z_{k-1}, \infty) \rangle_{\hat{g}_0}$$

exists.

It then follows from taking a limit as $y \to \infty$ in Corollary 5.23, with $z_k = y$, that for any non-negative Borel function on $H^{-1}(\hat{\mathbb{C}})$:

$$\langle V_\alpha(z_1,\ldots,z_{k-1},\infty)F\rangle_{\hat{g}_0}$$
$$= c\gamma^{-1}e^{C_{\alpha_1,\ldots,\alpha_{k-1}}(z_1,\ldots,z_k)}\prod_{i=1}^{k-1}\hat{g}_0(z_i)^{\Delta_{\alpha_i}-\frac{\alpha_i\alpha_k}{4}}\times$$
$$\int_{u>0}\mathbb{E}\big(F(\tilde{h}^{\hat{\mathbb{C}}}+\tfrac{\log u}{\gamma}-\tfrac{\log\mathcal{M}_{\tilde{h}^{\hat{\mathbb{C}}}}(\mathbb{C})}{\gamma})\mathcal{M}_{\tilde{h}^{\hat{\mathbb{C}}}}(\mathbb{C})^{-s}\big)u^{s-1}e^{-\mu u}\,du, \quad (5.68)$$

with $c = c(\alpha)$ depending only on α (and not on F nor $z_1,\ldots,z_{k-1}$), $s = \gamma^{-1}(\sum_{i=1}^{k}\alpha_i - 2Q)$, and

$$\tilde{h}^{\hat{\mathbb{C}}} = h^{\hat{\mathbb{C}},\hat{g}_0} + (\tfrac{Q}{2}-\tfrac{\alpha_k}{4})\log\hat{g}_0 + 2\pi\sum_{i=1}^{k-1}\alpha_i G^{\hat{\mathbb{C}},\hat{g}_0}(\cdot,z_i). \quad (5.69)$$

Moreover, if $m(z) = az + b$ is a Möbius transformation fixing ∞, then

$$\langle V_\alpha(m(z_1)),\ldots,m(z_{k-1}),\infty)\rangle_{\hat{g}_0}$$
$$= \prod_{i=1}^{k-1}|m'(z_i)|^{-2\Delta_{\alpha_i}}a^{2\Delta_{\alpha_k}}\langle V_\alpha(z_1,\ldots,z_{k-1},\infty)\rangle_{\hat{g}_0}. \quad (5.70)$$

5.6 An alternative choice of background metric

It is common in the literature on spherical Liouville CFT (for example in [RV19, RV23, KRV20]) to define the correlations starting with the field $h^{\mathfrak{c}}$, the whole plane GFF with zero average on the unit circle, in place of $\hat{h}^{\hat{\mathbb{C}},\hat{g}_0}$. One can, for instance, define $h^{\mathfrak{c}}$ by setting $h^{\mathfrak{c}} := h^{\hat{\mathbb{C}},\hat{g}_0} - (h^{\hat{\mathbb{C}},\hat{g}_0},\rho_1)$ where ρ_1 is uniform measure on the unit circle.

Let us begin by computing the covariance $G^{\mathfrak{c}}$ of $h^{\mathfrak{c}}$.

Lemma 5.30

$$2\pi G^{\mathfrak{c}}(x,y) = -\log|x-y| + \log(|x|\vee 1) + \log(|y|\vee 1) \quad (5.71)$$

for $x \neq y \in \mathbb{C}$.

Proof From the definition of $h^{\mathfrak{c}}$, we have

$$G^{\mathfrak{c}}(x,y) = G^{\hat{\mathbb{C}},\hat{g}_0}(x,y) - \int G^{\hat{\mathbb{C}},\hat{g}_0}(x,z)\rho_1(dz) - \int G^{\hat{\mathbb{C}},\hat{g}_0}(y,z)\rho_1(dz)$$
$$+ \iint G^{\hat{\mathbb{C}},\hat{g}_0}(z,w)\rho_1(dz)\rho_1(dw).$$

Recall that a formula for $G^{\hat{\mathbb{C}},\hat{g}_0}(x,z)$ is provided in (5.30). Furthermore, we claim that $\log \hat{g}_0(z) = 1$ if $|z| = 1$, and for $x \in \mathbb{C}$,

$$\int \log|x-z|\rho_1(\mathrm{d}z) = \log(|x| \vee 1). \tag{5.72}$$

To justify (5.72) we consider the cases $|x| > 1$ and $|x| < 1$ separately (the case $|x| = 1$ follows by dominated convergence and continuity). When $|x| > 1$, (5.72) is straightforward by harmonicity of $\log|x - \cdot|$ in $B(0,1)$. When $|x| < 1$, we note that for $z \in \partial B(0,1)$,

$$|x-z| = |x-z||\bar{z}| = |1-\bar{z}x| = |1-\bar{x}z|,$$

so that

$$\int \log|x-z|\rho_1(\mathrm{d}z) = \int \log|1-\bar{x}z|\rho_1(\mathrm{d}z).$$

As z varies across the unit circle, $1 - \bar{x}z$ varies across a circle centred at 1 of radius $|x| < 1$. Using harmonicity of the log function, we deduce that the right-hand side is 0, which proves (5.72).

Together with (5.30) this immediately implies that

$$2\pi \int G^{\hat{\mathbb{C}},\hat{g}_0}(x,y)\rho_1(\mathrm{d}y) = -\log(|x| \vee 1) - \tfrac{1}{4}\log \hat{g}_0(x) + \log(2) - 1/2, \tag{5.73}$$

for $x \in \mathbb{C}$. The lemma follows. $\square$

Recalling (5.23), h^c formally corresponds to the GFF with zero average for the metric $\hat{g} = \hat{g}_c$, $\hat{g}_c(z) = (|z| \vee 1)^{-4}$. (This metric corresponds to gluing two copies of a unit Euclidean disc along their boundaries, but this fact will not be used below.) As this metric is not of the form $e^\rho \hat{g}_0$ with ρ twice differentiable, it does not quite fit into the earlier framework.

Nonetheless if we set

$$\langle F \rangle_c = \int_{c \in \mathbb{R}} \mathbb{E}\Big[F\big(h^c - 2Q\log(|\cdot| \vee 1) + c\big) \times$$

$$\exp\big(-2Qc - \mu e^{\gamma c} M_{h^c - 2Q\log(|\cdot| \vee 1)}(\mathbb{C})\big)\Big]\,\mathrm{d}c, \tag{5.74}$$

analogously to (5.41), then we can prove the following.

Lemma 5.31

$$\langle F \rangle_c = e^{-2Q^2(\log(2) - 1/2)} \langle F \rangle_{\hat{g}_0} \tag{5.75}$$

for all non-negative Borel functions F on $H^{-1}(\hat{\mathbb{C}})$. Moreover, for $V = V_\alpha(\mathbf{z})$ as in Theorem 5.20,

$$\langle VF \rangle_c := \lim_{\varepsilon \to 0} \langle V_\varepsilon F \rangle_c$$

5.6 An alternative choice of background metric

exists and can be explicitly expressed as

$$\langle VF\rangle_c$$
$$= \gamma^{-1}\prod_{i=1}^{k}(|z_i|\vee 1)^{-4\Delta\alpha_i+\alpha_i\sum_{l\neq i}\alpha_l}\prod_{j=i+1}^{k}|z_i-z_j|^{-\alpha_i\alpha_j}$$
$$\times \int_{u>0}\mathbb{E}(F(\tilde{h}^{\mathfrak{c}}+\gamma^{-1}(\log u-\log \mathcal{M}_{\tilde{h}^{\mathfrak{c}}}(\mathbb{C})))\mathcal{M}_{\tilde{h}^{\mathfrak{c}}}(\mathbb{C})^{-s})u^{s-1}e^{-\mu u}\,du,$$
(5.76)

with $s = \gamma^{-1}(\sum \alpha_j - 2Q)$ and $\tilde{h}^{\mathfrak{c}} = h^{\mathfrak{c}} - 2Q\log(|\cdot|\vee 1) + \sum\alpha_i 2\pi G^{\mathfrak{c}}(\cdot,z_i)$.
Equivalently,

$$\tilde{h}^{\mathfrak{c}} = h^{\mathfrak{c}} + (\sum\alpha_i - 2Q)\log(|\cdot|\vee 1) - \sum\alpha_i\log|\cdot - z_i| + \sum\alpha_i\log(|z_i|\vee 1).$$

Proof We start by proving (5.75) using Girsanov's theorem. Sine $h^{\mathfrak{c}}$ is equal in law to $h^{\hat{\mathbb{C}},\hat{g}_0} - (h^{\hat{\mathbb{C}},\hat{g}_0},\rho_1)$ we can write

$$\langle F\rangle_c = \int_{c\in\mathbb{R}}\mathbb{E}\Big[F\big(h^{\hat{\mathbb{C}},\hat{g}_0} - (h^{\hat{\mathbb{C}},\hat{g}_0},\rho_1) - 2Q\log(|\cdot|\vee 1) + c\big)\times$$
$$\exp\big(-2Qc - \mu e^{\gamma c}\mathcal{M}_{h^{\hat{\mathbb{C}},\hat{g}_0}-(h^{\hat{\mathbb{C}},\hat{g}_0},\rho_1)-2Q\log(|\cdot|\vee 1)}(\mathbb{C})\big)\Big]\,dc$$

which by applying the change of variables $\hat{c} = c - (h^{\hat{\mathbb{C}},\hat{g}_0},\rho_1)$, is equal to

$$\int_{\hat{c}\in\mathbb{R}}\mathbb{E}\Big[F\big(h^{\hat{\mathbb{C}},\hat{g}_0} - 2Q\log(|\cdot|\vee 1) + \hat{c}\big)$$
$$\exp\big(-2Q(h^{\hat{\mathbb{C}},\hat{g}_0},\rho_1) - 2Q\hat{c} - \mu e^{\gamma\hat{c}}\mathcal{M}_{h^{\hat{\mathbb{C}},\hat{g}_0}-2Q\log(|\cdot|\vee 1)}(\mathbb{C})\big)\Big]\,d\hat{c}.$$

Defining $\tilde{\mathbb{P}}$ by

$$\frac{d\tilde{\mathbb{P}}}{d\mathbb{P}} := \frac{\exp(-2Q(h^{\hat{\mathbb{C}},\hat{g}_0},\rho_1))}{\mathbb{E}(\exp(-2Q(h^{\hat{\mathbb{C}},\hat{g}_0},\rho_1)))} = \frac{\exp(-2Q(h^{\hat{\mathbb{C}},\hat{g}_0},\rho_1))}{\exp(2Q^2(\log(2)-1/2)}$$

(recall (5.73)) we obtain that

$$\langle F\rangle_c = e^{2Q^2(\log(2)-1/2)}\int_{\hat{c}\in\mathbb{R}}\tilde{\mathbb{E}}\Big[F\big(h^{\hat{\mathbb{C}},\hat{g}_0} - 2Q\log(|\cdot|\vee 1) + \hat{c}\big)\times$$
$$\exp\big(-2Q\hat{c} - \mu e^{\gamma\hat{c}}\mathcal{M}_{h^{\hat{\mathbb{C}},\hat{g}_0}-2Q\log(|\cdot|\vee 1)}(\mathbb{C})\big)\Big]\,d\hat{c}.$$

By Girsanov's theorem, the law of $h^{\hat{\mathbb{C}},\hat{g}_0}$ under $\tilde{\mathbb{P}}$ is the same as under $\mathbb{P}$, except with mean shifted by

$$-2Q\int 2\pi G^{\hat{\mathbb{C}},\hat{g}_0}(\cdot,y)\rho_1(dy) = 2Q\log(|\cdot|\vee 1) + (Q/2)\log\hat{g}_0 - 2Q(\log(2)-\tfrac{1}{2}),$$

again using (5.73). This yields that

$$\langle F\rangle_{\mathfrak{c}} = e^{2Q^2\left(\log(2)-\frac{1}{2}\right)}\int_{\hat{c}\in\mathbb{R}}\mathbb{E}\bigg[F\big(h^{\hat{\mathbb{C}},\hat{g}_0} + \tfrac{Q}{2}\log\hat{g}_0 + (\hat{c} - 2Q(\log(2) - \tfrac{1}{2}))\big)\times$$
$$\exp\big(-2Q(\hat{c} - 2Q(\log(2) - \tfrac{1}{2})) + 4Q^2(\log(2) - \tfrac{1}{2})$$
$$- \mu e^{\gamma(\hat{c}-2Q(\log(2)-\frac{1}{2}))}M_{h^{\hat{\mathbb{C}},\hat{g}_0}+\frac{Q}{2}\log\hat{g}_0}(\mathbb{C})\big)\bigg]\,\mathrm{d}\hat{c}.$$

Performing one final change of variables $c = \hat{c} - 2Q(\log(2) - \tfrac{1}{2})$, we obtain that

$$\langle F\rangle_{\mathfrak{c}} = e^{-2Q^2\left(\log(2)-\frac{1}{2}\right)}\int_{c\in\mathbb{R}}\mathbb{E}\bigg[F\big(h^{\hat{\mathbb{C}},\hat{g}_0} + \tfrac{Q}{2}\log\hat{g}_0 + c\big)\times$$
$$\exp\big(-2Qc - \mu e^{\gamma c}M_{h^{\hat{\mathbb{C}},\hat{g}_0}+\frac{Q}{2}\log\hat{g}_0}(\mathbb{C})\big)\bigg]\,\mathrm{d}c,$$

which by (5.41) is exactly (5.75).

The explicit expression for $\langle VF\rangle_{\mathfrak{c}} := \lim_{\varepsilon\to 0}\langle V_\varepsilon F\rangle_{\mathfrak{c}}$ follows from exactly the exact same argument as in the $\hat{g}_0$ case (Theorem 5.20 and Corollary 5.23). In summary:

- The field $\tilde{h}^{\mathfrak{c}}$ is obtained by shifting $h^{\mathfrak{c}} - 2Q\log(|\cdot|\vee 1)$ by $\sum_i \alpha_i(2\pi G^{\mathfrak{c}}(\cdot, z_i))$; this shift arises from Girsanov's theorem exactly as in (5.56). $\tilde{h}^{\mathfrak{c}}$ is analogous to $\tilde{h}^{\hat{\mathbb{C}}}$ in Corollary 5.23.
- There is a compensation term $\lim_{\varepsilon\to 0}\varepsilon^{\alpha_i^2/2}\exp(\tfrac{1}{2}\operatorname{Var}(\sum_i \alpha_i h_\varepsilon^{\mathfrak{c}}(z_i)))$ coming from Girsanov's theorem, as in (5.56). Using the expression for $2\pi G^{\mathfrak{c}}$, this is equal to

$$\exp\bigg(\frac{1}{2}\sum_i \alpha_i^2(2\log(|z_i|\vee 1)) - \sum_{i=1}^{k}\sum_{j=i+1}^{k}\alpha_i\alpha_j\log|z_i - z_j|$$
$$+ \sum_{i=1}^{k}\sum_{j\neq i}\alpha_i\alpha_k\log(|z_i|\vee 1)\bigg),$$

which can be rewritten as

$$\prod_i (|z_i|\vee 1)^{\alpha_i^2 - \alpha_i\sum_{j\neq i}\alpha_j}\prod_i\prod_{j=i+1}^{k}|z_i - z_j|^{-\alpha_i\alpha_j}.$$

There is also a term $e^{-\sum 2\alpha_i Q\log(|z_i|\vee 1)}$ arising from the insertions, which is equal to $\prod_i(|z_i|\vee 1)^{-2Q\alpha_i}$. Putting these together gives the factor

$$\prod_{i=1}^{k}(|z_i|\vee 1)^{-4\Delta_{\alpha_i}+\alpha_i\sum_{l\neq i}\alpha_l}\prod_{j=i+1}^{k}|z_i - z_j|^{-\alpha_i\alpha_j}$$

(analogous to $e^{C_\alpha(\mathbf{z})}\prod\hat{g}_0(z_i)^{\Delta_{\alpha_i}}$ in Corollary 5.23). □

5.7 Interpretation of Seiberg bounds   209

We also have a similar expression when one of the insertions is at ∞ (we assume this is the kth and final insertion for simplicity). Recall Definition 5.29.

Remark 5.32 (Correlations with an insertion at ∞) Suppose that $z_1, \ldots, z_{k-1}$ are distinct and $\alpha_1, \ldots, \alpha_k$ are as in Theorem 5.20. Then

$$\langle V_\alpha(z_1, \ldots, z_{k-1}, \infty) F \rangle_{\hat{g}_0} = e^{2Q^2(\log(2) - 1/2)} \langle V_\alpha(z_1, \ldots, z_{k-1}, \infty) F \rangle_{\mathfrak{c}}$$

is equal to

$$c' \gamma^{-1} \prod_{i=1}^{k-1} (|z_i| \vee 1)^{-4\Delta_{\alpha_i} + \alpha_i \sum_{l \neq i} \alpha_l} \prod_{j=i+1}^{k-1} |z_i - z_j|^{-\alpha_i \alpha_j}$$

$$\times \int_{u > 0} \mathbb{E}(F(\tilde{h}^{\mathfrak{c}} + \gamma^{-1}(\log u - \log \mathcal{M}_{\tilde{h}^{\mathfrak{c}}}(\mathbb{C}))) \mathcal{M}_{\tilde{h}^{\mathfrak{c}}}(\mathbb{C})^{-s}) u^{s-1} e^{-\mu u} \, du,$$

(5.77)

with $c' = c'(\alpha)$ depending only on α and

$$\tilde{h}^{\mathfrak{c}} = h^{\mathfrak{c}} + \left(\sum_{i=1}^{k} \alpha_i - 2Q \right) \log(|\cdot| \vee 1) - \sum_{i \neq k} \alpha_i \log |\cdot - z_i| + \sum_{i \neq k} \alpha_i \log(|z_i| \vee 1) \right).$$

5.7 Geometric and probabilistic interpretation of Seiberg bounds

The following discussion is intended to guide intuition but is not meant to be fully rigorous.

Assuming that the Seiberg bounds hold, the finite partition function $\langle V \rangle_{\hat{g}}$ in Theorem 5.20 implicitly defines a random field (i.e. sampled from a probability distribution) that we will soon call the **Liouville field**, see Definition 5.33. It is believed, as will be detailed more precisely in Remark 5.37, that this Liouville field and its multiplicative chaos describe the scaling limit of suitably (i.e. conformally) embedded random planar maps; the precise formulation of this conjecture goes back to [DKRV16]. This gives a **probabilistic justification** as to why insertions are necessary to define correlation functions, and why we need at least three of them. Indeed, conformal embeddings into the sphere are typically only unique up to Möbius transforms, so that in order to get a well defined, unique embedding, it is necessary to choose the embedded position of three vertices of the map in advance. It is natural to choose these three vertices uniformly at random on the planar map; in that case, note that (because of Girsanov's theorem) their associated insertion weights should correspond to

210 Introduction to Liouville Conformal Field Theory

$\alpha_i = \gamma$. Conversely, if we take $k = 3$ and $\alpha_i = \gamma$, the Seiberg bounds can only be satisfied when

$$3\gamma > 2Q$$

or equivalently

$$\gamma > \sqrt{2}.$$

We remind the reader that the range $\gamma \in [\sqrt{2}, 2)$ is exactly the range of values that one should obtain in the scaling limit of FK-decorated planar maps, see Section 4.2.

To appreciate the necessity of the insertions in order to get a finite partition function from a **geometric** point of view, it is useful to pause the exposition of the theory and to make a few heuristic considerations. Since the probability distribution associated to $\langle V \rangle_{\hat{g}}$ should formally be a Gibbs measure as in (5.4), it is intuitively useful to view this field as a random perturbation around the **ground state** of the theory, that is, the state φ of minimal energy, particularly when $\gamma \to 0$ and the field is essentially deterministic. To begin with, one might wonder what the ground state corresponding to the Polyakov action (5.9) looks like without any insertion. Let us simply study the associated variational problem; that is, let φ be a minimiser of $S(\varphi)$, let f be an arbitrary test function, let $\varepsilon > 0$ and consider the action $S(\varphi + \varepsilon f)$. Then

$$S(\varphi + \varepsilon f) = \frac{1}{4\pi} \int \left[|\nabla^{\hat{g}} (\varphi + \varepsilon f)|^2 + R_{\hat{g}} Q(\varphi + \varepsilon f) + 4\pi\mu e^{\gamma(\varphi + \varepsilon f)} \right] v_{\hat{g}}(\mathrm{d}z)$$

$$= \frac{1}{4\pi} \int \left[|\nabla^{\hat{g}} \varphi|^2 + R_{\hat{g}} Q\varphi + 4\pi\mu e^{\gamma\varphi} \right] v_{\hat{g}}(\mathrm{d}z) +$$

$$+ \frac{\varepsilon}{4\pi} \int \left[2\langle \nabla^{\hat{g}} \varphi, \nabla^{\hat{g}} f \rangle + Q R_{\hat{g}} f + 4\pi\mu\gamma e^{\gamma\varphi} f \right] v_{\hat{g}}(\mathrm{d}z) + o(\varepsilon)$$

so that, by the Gauss–Green formula, since φ is a minimiser,

$$\frac{1}{4\pi} \int \left[-2\Delta^{\hat{g}} \varphi + Q R_{\hat{g}} + 4\pi\mu\gamma e^{\gamma\varphi} \right] f v_{\hat{g}}(\mathrm{d}z) = 0.$$

Because f is arbitrary, we deduce (first in the sense of distributions and then in the pointwise sense using elliptic arguments, which are not required here since this is anyway entirely heuristic), that

$$\Delta^{\hat{g}} \varphi = \frac{Q}{2} R_{\hat{g}} + 2\pi\mu\gamma e^{\gamma\varphi}. \tag{5.78}$$

Now let $u = \gamma\varphi$, and consider the situation as $\gamma \to 0$ and $\mu\gamma^2$ is kept constant (in particular, the cosmological constant μ tends to infinity, this is the so-called semiclassical limit). Then we get an equation of the form

$$\Delta^{\hat{g}} u = R_{\hat{g}} - K e^u; \quad \text{where } K = -2\pi\mu\gamma^2 < 0. \tag{5.79}$$

5.7 Interpretation of Seiberg bounds

This equation is called **Liouville's equation**, and arises when searching for metrics $\tilde{g}$ conformally equivalent to $\hat{g}$ and of constant curvature K. Indeed, let us write $\tilde{g} = e^{\rho}\hat{g}$ and suppose $R_{\tilde{g}} = K$. Since $R_{\tilde{g}} = -\Delta^{\tilde{g}}\log\tilde{g}$, this becomes

$$-\Delta^{\tilde{g}}\log\tilde{g} = K$$
$$-e^{-\rho}\Delta^{\hat{g}}\log(\hat{g}e^{\rho}) = K$$
$$-\Delta^{\hat{g}}\log\hat{g} - \Delta^{\hat{g}}\rho = Ke^{\rho}$$
$$R_{\hat{g}} - \Delta^{\hat{g}}\rho = Ke^{\rho},$$

which is the same as (5.79) with $\rho = u$ and $K = -2\pi\mu\gamma^2$ (recall that $\mu\gamma^2$ is chosen to be a constant). Thus the Polyakov action is formally minimised by a function ρ corresponding to a metric $\tilde{g} = e^{\rho}\hat{g}$ which has constant negative curvature K. But of course this is impossible on the sphere, in view of the Gauss–Bonnet theorem, which implies that the integral of the curvature should be 8π.

Formally, adding insertions in the computation of $\langle V \rangle$ can be thought of as changing the Polyakov action; the minimiser then satisfies

$$\frac{1}{4\pi}\int\left[-2\Delta^{\hat{g}}\varphi + QR_{\hat{g}} + 4\pi\mu\gamma e^{\gamma\varphi}\right]fv_{\hat{g}}(\mathrm{d}z) - \sum_{i=1}^{k}\alpha_i f(z_i) = 0,$$

or in other words,

$$\Delta^{\hat{g}}\varphi = \frac{Q}{2}R_{\hat{g}} + 2\pi\mu\gamma e^{\gamma\varphi} - 2\pi\sum_{i=1}^{k}\alpha_i\delta_{\{z_i\}}. \tag{5.80}$$

instead of (5.78), where $\delta_{\{z_i\}}$ is the Dirac mass on $\mathbb{S}$ (with respect to the underlying metric $\hat{g}$). Scaling the weights α_i by defining new weights $\tilde{\alpha}_i = \gamma\alpha_i$, we get that the new weights satisfy the rescaled Seiberg bounds:

$$\sum_{i=1}^{k}\tilde{\alpha}_i > 2\gamma Q, \tilde{\alpha}_i < \gamma Q$$

which as $\gamma \to 0$ becomes

$$\sum_{i=1}^{k}\tilde{\alpha}_i > 4, \tilde{\alpha}_i < 2. \tag{5.81}$$

Then with these rescaled weights, setting $u = \gamma\varphi$ in (5.80) and letting $\gamma \to 0$ with $\mu\gamma^2$ constant as above (and $\tilde{\alpha}_i = \gamma\alpha_i$ fixed), the equation satisfied by u becomes

$$\Delta^{\hat{g}}u = R_{\hat{g}} + Ke^u - 2\pi\sum_{i=1}^{k}\tilde{\alpha}_i\delta_{\{z_i\}}. \tag{5.82}$$

This modified form of Liouville's equation describes a metric $\tilde{g} = e^u \hat{g}$ such that $\tilde{g}$ has constant curvature $K = -2\pi\mu\gamma^2$ away from the points $\{z_i\}_i$, but conical singularities at each of the z_i. That is, the metric is locally of the form $1/|z_i - z|^{\tilde{\alpha}_i}$ as $z \to z_i$. In this setting, there is no obstruction from Gauss–Bonnet to the existence of such metrics.

This connection is made precise by Lacoin, Rhodes and Vargas [LRV17, LRV22]. Indeed, the authors prove that in the limit $\gamma \to 0$ (with $\mu\gamma^2$ and $\gamma\alpha_i$ kept fixed as above) the associated normalised Liouville field concentrates near the solution of the modified Liouville equation (5.82). Furthermore, they show that the fluctuations are asymptotically given by a massive Gaussian free field, and obtain a large deviation theorem where the rate function is given by the Polyakov action (shifted by the insertions); as expected, this rate function is thus zero at the minimiser.

5.8 Liouville fields

As mentioned above, the finiteness of the partition function $\langle V \rangle_{\hat{g}}$ when the Seiberg bounds are satisfied, allows us not only to define "expectations" of observables such as $\langle VF \rangle_{\hat{g}}$ in Corollary 5.23, but actually random *fields* sampled from the associated probability distribution.

Definition 5.33 We define the Liouville field (associated to insertions $\mathbf{z} = (z_1, \ldots, z_k)$ and parameters $\alpha = (\alpha_1, \ldots, \alpha_k)$ satisfying the Seiberg bounds (5.54) and (5.55)) to be the random field $h^L_{\alpha,\mathbf{z}}$ in $H^{-1}(\hat{\mathbb{C}})$, such that for any observable F,

$$\mathbb{E}[F(h^L_{\alpha,\mathbf{z}})] := \frac{\langle FV_\alpha(\mathbf{z}) \rangle_{\hat{g}}}{\langle V_\alpha(\mathbf{z}) \rangle_{\hat{g}}}.$$

This law does not depend on the choice of the metric $\hat{g}$.

The remarkable independence of this law from the metric $\hat{g}$ is a direct consequence of the Weyl anomaly formula (see Theorem 5.17 and Remark 5.24). In what follows, we will always work with the definition starting from the spherical metric $\hat{g}_0$ for simplicity.

Remark 5.34 By (5.74) we also have $\mathbb{E}[F(h^L_{\alpha,\mathbf{z}})] = \frac{\langle FV_\alpha(\mathbf{z}) \rangle_c}{\langle V_\alpha(\mathbf{z}) \rangle_c}$, where $\langle \cdot \rangle_c$ is as defined in Section 5.6.

Theorem 5.27 also implies that the field transforms in the following way under Möbius transformations of the sphere.

5.8 Liouville fields

Corollary 5.35 *Suppose that* $m\colon \hat{\mathbb{C}} \to \hat{\mathbb{C}}$ *is a Möbius transform of the Riemann sphere, and* $\{\alpha_i, z_i\}$ *are as in Definition 5.33. Then*

$$h^L_{\alpha,\mathbf{z}} \stackrel{\text{(law)}}{=} h^L_{\alpha,m(\mathbf{z})} \circ m + Q \log |m'|,$$

where $m(\mathbf{z}) = (m(z_1), \ldots, m(z_k))$.

Note that the law of $h^L_{\alpha,\mathbf{z}}$ also depends on γ, but we omit this from the notation (as with everything previously in this chapter). A natural next question is to identify the law of the Liouville field in a way that is more explicit than the definition. We will not do this right away, but in the end we will get a very nice description by conditioning on the area; this will give us the *unit volume Liouville sphere* in Section 5.9. For now, we will first make a simple (but surprising) observation about the law of the total mass of the multiplicative chaos measure associated to the Liouville field $h^L_{\alpha,\mathbf{z}}$.

Lemma 5.36 *Suppose that* $\{\alpha_i, z_i\}$ *are as in Definition 5.33. Then*

$$\mathcal{M}_{h^L_{\alpha,\mathbf{z}}}(\mathbb{C}) \sim \Gamma(s, \mu); \quad s = \tfrac{\sum_i \alpha_i - 2Q}{\gamma} > 0.$$

That is, $\mathcal{M}_{h^L_{\alpha,\mathbf{z}}}(\mathbb{C})$ *has density proportional to* $u^{s-1}e^{-\mu u}\mathbf{1}_{\{u>0\}}$ *with respect to Lebesgue measure on* $\mathbb{R}$.

Proof Recall the definition $\tilde{h}^{\hat{\mathbb{C}}} = h^{\hat{\mathbb{C}},\hat{g}_0} + \tfrac{Q}{2}\log \hat{g}_0 + \sum_{i=1}^k \alpha_i G^{\hat{\mathbb{C}},\hat{g}_0}(\cdot, z_i)$. For $A \subset \mathbb{R}_+$, we have that

$$\mathbb{P}(\mathcal{M}_{h^L_{\alpha,\mathbf{z}}}(\mathbb{C}) \in A) = \frac{\langle V_{\alpha,\mathbf{z}} \mathbf{1}_{\{\mathcal{M}_{h^L_{\alpha,\mathbf{z}}}(\mathbb{C}) \in A\}} \rangle_{\hat{g}_0}}{\langle V_{\alpha,\mathbf{z}} \rangle_{\hat{g}_0}}$$

$$= \frac{\int_{u>0} \mathbb{E}\big(\mathbf{1}_{\{\mathcal{M}_h(\mathbb{C}) \in A\}} (\mathcal{M}_{\tilde{h}^{\hat{\mathbb{C}}}}(\mathbb{C}))^{-s}\big) u^{s-1} e^{-\mu u}\, du}{\int_{u>0} \mathbb{E}\big((\mathcal{M}_{\tilde{h}^{\hat{\mathbb{C}}}}(\mathbb{C}))^{-s}\big) u^{s-1} e^{-\mu u}\, du},$$

where the second line follows from (5.65) with

$$h = \tilde{h}^{\hat{\mathbb{C}}} + \gamma^{-1}\log u - \gamma^{-1}\mathcal{M}_{\tilde{h}^{\hat{\mathbb{C}}} + (Q/2)\log \hat{g}_0}(\mathbb{C}).$$

Notice however that $\mathcal{M}_h(\mathbb{C}) = u$ by definition, so that the above becomes

$$\frac{\int_{u \in A} \mathbb{E}\big((\mathcal{M})_{\tilde{h}^{\hat{\mathbb{C}}}}(\mathbb{C})\big)^{-s} u^{s-1} e^{-\mu u}\, du}{\int_{u>0} \mathbb{E}\big((\mathcal{M}_{\tilde{h}^{\hat{\mathbb{C}}}}(\mathbb{C}))^{-s}\big) u^{s-1} e^{-\mu u}\, du} = \frac{\int_{u \in A} u^{s-1} e^{-\mu u}\, du}{\int_{u>0} u^{s-1} e^{-\mu u}\, du},$$

as required. □

Remark 5.37 This Gamma law is precisely what one expects to get for the limiting distribution of the total area of a random planar map when it is chosen according to an appropriate Boltzmann–Gibbs measure (i.e. with a random number of vertices and edges). Let us explain more precisely what we mean by this. In a celebrated work, Tutte showed that the number of planar maps with n edges and k designated roots grows like $Ce^{n\beta}n^{-7/2+k}$, where $C > 0$ and $\beta > 0$ are two (essentially) unimportant constants; here $\beta = \log 12$. If we want to embed this conformally using for example circle packing, it is natural to take $k = 3$ (so we have k designated points that can be mapped to three fixed locations on the Riemann sphere), so we can rewrite this as $Ce^{n\beta}n^{1/2-1}$. If we assign each map with n edges a weight equal to $e^{-n\beta(1+\varepsilon\mu)}$ (this is the slightly subcritical Boltzmann–Gibbs law, which samples very large maps when ε is small) then we see from Tutte's formula that we should have in the limit as $\varepsilon \to 0$, after suitable rescaling, a distribution for the area which is proportional to $e^{-\mu u}u^{1/2-1}$, that is, a Gamma(s, μ) law with the parameter $s = 1/2$. This matches the value that one obtains for $\gamma = \sqrt{8/3}$, $k = 3$, $\alpha_i = \gamma$. Indeed, in that case the formula in Lemma 5.36 gives

$$s = \frac{3\gamma - 2Q}{\gamma} = 3 - (1 + 4/\gamma^2) = 3 - 5/2 = 1/2,$$

as desired. This is no mere coincidence, and analogous results should hold for more general planar maps weighted by the $O(n)$ model or FK-weighted planar maps as considered in Chapter 4. This led [DKRV16] to formulate the precise conjecture that after conformally embedding these maps with the three roots sent to some fixed points $\mathbf{z} = (z_1, z_2, z_3)$ of the Riemann sphere, the uniform measure on vertices of the map converges to the Gaussian multiplicative chaos measure associated to the Liouville field $h^L_{\alpha,\mathbf{z}}$ with $\alpha = (\gamma, \gamma, \gamma)$.

5.9 Unit volume Liouville sphere

In order to describe the Liouville field $h^L_{\alpha,\mathbf{z}}$ defined in Section 5.8, the next step is to identify the law of $h^L_{\alpha,\mathbf{z}}$ conditional on the total area. Remarkably, the result does **not** depend on the actual area except for a (conditionally) deterministic shift corresponding to the area itself. The resulting field will be called the **unit volume Liouville sphere**.

Proposition 5.38 *Let $\{\alpha_i, z_i\}$ be as in Definition 5.33. Then $\mathcal{M}_{h^L_{\alpha,\mathbf{z}}}(\mathbb{C})$ and $h^L_{\alpha,\mathbf{z}} - \mathcal{M}_{h^L_{\alpha,\mathbf{z}}}(\mathbb{C})$ are independent, and the law of $h^L_{\alpha,\mathbf{z}} - \mathcal{M}_{h^L_{\alpha,\mathbf{z}}}(\mathbb{C})$ is equal to that of*

$$\tilde{h}^{\hat{\mathbb{C}}} - \gamma^{-1}\log(\mathcal{M}_{\tilde{h}^{\hat{\mathbb{C}}}}(\mathbb{C}))$$

weighted by $\left(\mathcal{M}_{\tilde{h}^{\hat{\mathbb{C}}}}(\mathbb{C})\right)^{-s}$, where $\tilde{h}^{\hat{\mathbb{C}}} = h^{\hat{\mathbb{C}},\hat{g}_0} + \frac{Q}{2}\log \hat{g}_0 + \sum_{i=1}^{k}\alpha_i G^{\hat{\mathbb{C}},\hat{g}_0}(z_i, \cdot)$.

5.9 Unit volume Liouville sphere

Remark 5.39 Note that this law *does* depend on $(\alpha_i, z_i)_i$ because $s = \gamma^{-1}(\sum_i \alpha_i - 2Q) > 0$, and because the field $\tilde{h}^{\hat{\mathbb{C}}}$ also depends on them.

Proof Let F be a non-negative Borel measurable function on $H^{-1}(\hat{\mathbb{C}})$ and let A be a Borel subset of $[0, \infty)$. Then just as in the proof of Lemma 5.36 (using the fact that $M_h(\mathbb{C}) = u$ when $h = \tilde{h}^{\hat{\mathbb{C}}} - \gamma^{-1} M_{\tilde{h}^{\hat{\mathbb{C}}}}(\mathbb{C}) + \gamma^{-1} \log u$), we have

$$\mathbb{E}^L_{\alpha, \mathbf{z}}(F(h^L_{\alpha, \mathbf{z}} - \gamma^{-1} \log(M_{h^L_{\alpha, \mathbf{z}}}(\mathbb{C}))) \mathbf{1}_{\{M_{h^L_{\alpha, \mathbf{z}}}(\mathbb{C}) \in A\}})$$

$$= \frac{\int_{u \in A} \mathbb{E}\big(F(\tilde{h}^{\hat{\mathbb{C}}} - \gamma^{-1} \log M_{\tilde{h}^{\hat{\mathbb{C}}}}(\mathbb{C}))(M_{\tilde{h}^{\hat{\mathbb{C}}}}(\mathbb{C}))^{-s}\big) u^{s-1} e^{-\mu u}\, du}{\int_{u > 0} \mathbb{E}\big((M_{\tilde{h}^{\hat{\mathbb{C}}}}(\mathbb{C}))^{-s}\big) u^{s-1} e^{-\mu u}\, du}$$

$$= \frac{\mathbb{E}\big(F(\tilde{h}^{\hat{\mathbb{C}}} - \gamma^{-1} \log M_{\tilde{h}^{\hat{\mathbb{C}}}}(\mathbb{C}))(M_{\tilde{h}^{\hat{\mathbb{C}}}}(\mathbb{C}))^{-s}\big)}{\mathbb{E}\big((M_{\tilde{h}^{\hat{\mathbb{C}}}}(\mathbb{C}))^{-s}\big)} \cdot \frac{\int_{u \in A} u^{s-1} e^{-\mu u}\, du}{\int_{u > 0} u^{s-1} e^{-\mu u}\, du}$$

$$= \frac{\mathbb{E}\big(F(\tilde{h}^{\hat{\mathbb{C}}} - \gamma^{-1} \log M_{\tilde{h}^{\hat{\mathbb{C}}}}(\mathbb{C}))(M_{\tilde{h}^{\hat{\mathbb{C}}}}(\mathbb{C}))^{-s}\big)}{\mathbb{E}\big((M_{\tilde{h}^{\hat{\mathbb{C}}}}(\mathbb{C}))^{-s}\big)} \mathbb{P}^L_{\alpha, \mathbf{z}}(M_{h^L_{\alpha, \mathbf{z}}}(\mathbb{C}) \in A).$$

This immediately yields the statement of the proposition. □

Definition 5.40 (Unit volume Liouville sphere) The unit volume Liouville sphere $h^{L,1}_{\alpha, \mathbf{z}}$ is the random field in $H^{-1}(\hat{\mathbb{C}})$ whose law is that of

$$\tilde{h}^{\hat{\mathbb{C}}} - \gamma^{-1} \log(M_{\tilde{h}^{\hat{\mathbb{C}}}}(\mathbb{C})) \text{ weighted by } (M_{\tilde{h}^{\hat{\mathbb{C}}}}(\mathbb{C}))^{-s},$$

where $\tilde{h}^{\hat{\mathbb{C}}} = h^{\hat{\mathbb{C}}, \hat{g}_0} + \frac{Q}{2} \log \hat{g}_0 + \sum_{i=1}^k \alpha_i G^{\hat{\mathbb{C}}, \hat{g}_0}(z_i, \cdot)$.

Remark 5.41 (Extended Seiberg bounds) It can be shown that $\mathbb{E}(M_{\tilde{h}^{\hat{\mathbb{C}}}}(\mathbb{C})^{-s}) < \infty$ if (and only if) $\alpha_i < Q$ for all i and $Q - \frac{1}{2} \sum \alpha_i < \frac{2}{\gamma} \wedge \min_i(Q - \alpha_i)$. See [DKRV16, Lemma 3.10]. Therefore, a "unit volume Liouville field" with the law described in Definition 5.40 can still be defined in this extended setting.

Note that this law does not depend on the cosmological constant μ. Furthermore, by Proposition 5.38, the law is unaffected if we replace $\hat{g}_0$ with any metric $\hat{g} = e^\rho \hat{g}_0$ as in (5.7).

Remark 5.42 Recalling Lemma 5.36, we also obtain the following decomposition of the Liouville field $h^L_{\alpha, \mathbf{z}}$:

$$h^L_{\alpha, \mathbf{z}} = h^{L,1}_{\alpha, \mathbf{z}} + \frac{1}{\gamma} \log X,$$

where $h^{L,1}_{\alpha, \mathbf{z}}$ is the unit volume Liouville sphere, and X is an independent random variable with the Gamma$(s; \mu)$ distribution.

Remark 5.43 (The case $\mu = 0$)   Define the infinite measure

$$m_{\alpha,\mathbf{z}}^L(F) := \lim_{\mu \to 0} \langle V_\alpha(\mathbf{z}) F \rangle_{\hat{g}_0}$$

$$= \gamma^{-1} e^{C_\alpha(\mathbf{z})} \prod_i \hat{g}_0(z_i)^{\Delta_{\alpha_i}} \int_{u>0} \mathbb{E}\big(F(\tilde{h}^{\hat{C}} + \tfrac{\log u}{\gamma} - \tfrac{\log \mathcal{M}_{\tilde{h}^{\hat{C}}}(\mathbb{C})}{\gamma}) \mathcal{M}_{\tilde{h}^{\hat{C}}}(\mathbb{C})^{-s}\big) u^{s-1} \, du.$$

This infinite measure will play a role in the identification of the unit volume Liouville field with the "unit volume quantum sphere" which will be introduced in Chapter 7. Then we can write

$$m_{\alpha,\mathbf{z}}^L(F) = \gamma^{-1} e^{C_\alpha(\mathbf{z})} \prod_i \hat{g}_0(z_i)^{\Delta_{\alpha_i}} \int_{u>0} u^{s-1} \mathbb{E}(F(h_{\alpha,\mathbf{z}}^{L,1} + \gamma^{-1} \log u)) \, du$$

by Definition 5.40 (of $h_{\alpha,\mathbf{z}}^{L,1}$). In other words, as in the $\mu > 0$ case, we can disintegrate the infinite measure $m_{\alpha,\mathbf{z}}^L$ on $H^{-1}(\hat{\mathbb{C}})$ with respect to the total GMC mass of the field. The marginal of the mass is proportional to $u^{s-1} \, du$ and the law of the field conditioned to have mass u is simply that of the unit volume Liouville sphere $h_{\alpha,\mathbf{z}}^{L,1}$ plus the constant $\gamma^{-1} \log(u)$.

Remark 5.44   It will be useful later on to express $m_{\alpha,\mathbf{z}}^L$ in terms of the field $h^{\mathfrak{c}}$ with average zero on the unit circle, defined in Section 5.6, and $\mathbf{z} = (0, \infty, z)$ for $z \neq 0$, $z \in \mathbb{C}$. Namely, using (5.75), we obtain that

$$m_{\alpha,\mathbf{z}}^L(F) = C(|z| \vee 1)^{-4\Delta_{\alpha_3} + \alpha_3(\alpha_1 + \alpha_2)} |z|^{-\alpha_1 \alpha_3} \times$$
$$\int_{u>0} \mathbb{E}(F(\tilde{h}^{\mathfrak{c}} + \gamma^{-1}(\log u - \log \mathcal{M}_{\tilde{h}^{\mathfrak{c}}}(\mathbb{C}))) \mathcal{M}_{\tilde{h}^{\mathfrak{c}}}(\mathbb{C})^{-s}) u^{s-1} \, du$$
(5.83)

with C depending only on $\alpha_1, \alpha_2, \alpha_3$ and

$$\tilde{h}^{\mathfrak{c}} = h^{\mathfrak{c}} + (\alpha_1 + \alpha_2 - 2Q) \log(|\cdot| \vee 1) - \alpha_1 \log|\cdot| + 2\pi \alpha_3 G^{\mathfrak{c}}(z, \cdot).$$

5.10 Some integrability results

We have so far discussed the way the Liouville correlation functions evolve under global geometric deformations. However, a key step in the development of conformal field theory was accomplished in a celebrated paper of Beliavin, Polyakov and Zamolodchikov [BPZ84a] in which "infinitesimal" geometric deformations were considered and shown to lead to differential identities for the correlation functions. Unlike the identities such as the KPZ identity of Theorem 5.27, which expresses global invariance of the correlations under Möbius maps and thus with three degrees of freedom, we get as a result of

5.10 Some integrability results

these considerations an infinite hierarchy of equations (one for each number of insertions, that is, number of points where the correlation functions are being evaluated) and hence infinitely many degrees of freedom for the parameters of these equations. Compared to the Virasoro point of view on CFT which was briefly alluded to at the start of this chapter, this is analogous to the fact that the Virasoro algebra contains not only the operators L_{-1}, L_0, and L_1 but more generally the infinite-dimensional family of operators $\{L_n\}_{n \in \mathbb{Z}}$.

A first set of identities is obtained by considering the behaviour of the correlation functions $\langle V_{\alpha_1, \ldots, \alpha_k}(\mathbf{z}) \rangle_g$ under infinitesimal deformations of the metric $g \to g + \varepsilon f$. Taking a derivative in the Weyl anomaly formula (Theorem 5.17) would lead to a field called the **stress energy tensor** $T(z)$. Correlations between $T(z)$ and the insertion operators are shown to satisfy, as a function of z, two families of differential equations known as the **Ward identities**. We will not enter into details here except to refer the interested reader to [KRV19] where this is discussed in detail and furthermore rigorously. Instead we state here the so-called **BPZ equations**.

Recall from (5.63) that for $\alpha > 0$, the conformal weight Δ_α of the operator V_α is given by $\Delta_\alpha = \frac{\alpha}{2}(Q - \frac{\alpha}{2})$.

Theorem 5.45 (Theorem 2.2 in [KRV19]) *Fix $\alpha \in \{-\frac{\gamma}{2}, -\frac{2}{\gamma}\}$, and suppose $k \geq 2$ and $\alpha_1, \ldots, \alpha_k$ satisfy $\sum_{i=1}^k \alpha_i + \alpha > 2Q$, $\alpha_i < Q$ for $1 \leq i \leq k$. Then*

$$\left(\frac{1}{\alpha^2} \partial_z^2 + \sum_{i=1}^k \frac{\Delta_{\alpha_i}}{(z-z_i)^2} + \sum_{i=1}^k \frac{1}{z-z_i} \partial_{z_i} \right) \langle V_\alpha(z) \prod_{i=1}^k V_{\alpha_i}(z_i) \rangle_{\hat{g}} = 0. \quad (5.84)$$

Outline of proof The proof is very technical, and we will only give an extremely rough summary here; of course, readers are once again referred to [KRV19] for details. A key step in the proof of (5.84) is the following identity. Write

$$G(x; \mathbf{z}) = \langle V_\gamma(x) V_{\alpha_1}(z_1) \cdots V_{\alpha_k}(z_k) \rangle_{\hat{g}}.$$

Note that this is not quite the correlation function appearing on the left-hand side of (5.84) as here the "weight" of the insertion is γ, whereas in (5.84) it is $\alpha \in \{-\frac{2}{\gamma}, -\frac{\gamma}{2}\}$. Write also $G(\mathbf{z})$ for $\langle V_{\alpha_1, \ldots, \alpha_k}(\mathbf{z}) \rangle_{\hat{g}}$.

Then using Gaussian integration by parts (already mentioned in the proof of Kahane's convexity inequality in Theorem 3.19) and plenty of careful estimates (which require ingenious tricks) one can check that for every fixed $1 \leq i \leq k$,

$$\partial_{z_i} G(\mathbf{z}) = -\frac{1}{2} \sum_{j \neq i} \frac{\alpha_i \alpha_j}{z_i - z_j} G(\mathbf{z}) + \frac{\alpha \mu \gamma}{2} \int_{\mathbb{C}} G(x; \mathbf{z}) \, dx. \quad (5.85)$$

218   *Introduction to Liouville Conformal Field Theory*

This corresponds to (3.27) in [KRV19]. (A priori it is not even clear that the integral on the right-hand side is finite, but this could be deduced with some work from Corollary 5.23, cf. the proof of Proposition 5.5 in Section 6.8 of [KRV19].)

In a second step, we may apply the formula with $\mathbf{z}$ replaced by $(z, \mathbf{z})$ and $\alpha_1, \ldots, \alpha_k$ replaced by $\alpha, \alpha_1, \ldots, \alpha_k$. We can then differentiate this identity with respect to z a second time using (5.85) itself to identify the derivatives on the right-hand side. After a long calculation and some remarkable cancellations when $\alpha \in \{-\frac{2}{\gamma}, \frac{\gamma}{2}\}$, the authors of [KRV19] end up with the identity (5.84). □

The proof of the BPZ equations in [KRV19] is a major step in the proof of the celebrated **DOZZ formula** (named after Dorn, Otto, Zamolodchikov and Zamolodchikov) which gives an explicit formula for the **structure constant** $C(\alpha_1, \alpha_2, \alpha_3)$ determining the three-point correlation function, and its proof in [KRV20] is a landmark of Liouville conformal field theory. Write

$$\ell(z) = \frac{\Gamma(z)}{\Gamma(1-z)},$$

and furthermore define the special Upsilon function by

$$\log \Upsilon_{\frac{\gamma}{2}}(z) = \int_0^\infty \left((\tfrac{Q}{2} - z)^2 e^{-t} - \frac{\sinh^2((\tfrac{Q}{2}-z)\tfrac{t}{2})}{\sinh(\tfrac{t\gamma}{4})\sinh(\tfrac{t}{\gamma})} \right) \frac{dt}{t}; \quad 0 < \Re(z) < Q,$$

which can be analytically continued to $\mathbb{C}$ (this is by no means obvious and in fact follows from functional identities satisfied by the function).

Theorem 5.46 *For any $\alpha_1, \alpha_2, \alpha_3$ satisfying the Seiberg bounds, setting $\overline{\alpha} = \alpha_1 + \cdots + \alpha_3$*

$$C_\gamma(\alpha_1, \alpha_2, \alpha_3) = \left(\pi \mu \ell(\tfrac{\gamma^2}{4})(\tfrac{\gamma}{2})^{2-\tfrac{\gamma^2}{2}} \right)^{\frac{2Q-\overline{\alpha}}{\gamma}} \frac{\Upsilon'_{\frac{\gamma}{2}}(0)\Upsilon_{\frac{\gamma}{2}}(\alpha_1)\Upsilon_{\frac{\gamma}{2}}(\alpha_2)\Upsilon_{\frac{\gamma}{2}}(\alpha_3)}{\Upsilon_{\frac{\gamma}{2}}(\tfrac{\overline{\alpha}-2Q}{2})\Upsilon_{\frac{\gamma}{2}}(\tfrac{\overline{\alpha}}{2}-\alpha_1)\Upsilon_{\frac{\gamma}{2}}(\tfrac{\overline{\alpha}}{2}-\alpha_2)\Upsilon_{\frac{\gamma}{2}}(\tfrac{\overline{\alpha}}{2}-\alpha_3)}$$

Given the BPZ equations, a relatively short sketch of the main arguments can be found in section 5 of the lecture notes [RV23].

The **conformal bootstrap**, recently proved by Guillarmou, Kupiainen, Rhodes and Vargas [GKRV24], allows one to express correlation functions of order $n+1$ in terms of those of order n. In combination with the above DOZZ formula (Theorem 5.46), this gives exact formulae for correlation functions of *all* orders.

5.11 Exercises

5.1 By using spherical polar coordinates, show that the spherical metric $\hat{g}$ satisfies

$$\int_{\mathbb{C}} \hat{g}_0(z)\,dz = 4\pi \text{ and } R_{\hat{g}_0}(z) = -\frac{\Delta \log \hat{g}_0(z)}{\hat{g}_0(z)} \equiv 2.$$

5.2 Prove Lemma 5.15, using the fact that for general g as in (5.7),

$$\text{Var}((h^{\hat{\mathbb{C}},g}, f)_g) = \|f\|^2_{H^{-1}(\hat{\mathbb{C}},g)}$$

for all $f \in H^{-1}(\hat{\mathbb{C}}, g)$ with v_g average zero.

5.3 Let $\{\alpha_i\}_{i=1}^{k}$ satisfy the Seiberg bounds and $\mathbf{z} = (z_1, \ldots, z_k)$ be fixed. Denote $V_\mu := V_{\alpha_1,\ldots,\alpha_k,\mathbf{z}}$ when the cosmological constant is equal to $\mu > 0$. Using Corollary 5.23, show that

$$\langle V_\mu \rangle_{\hat{g}} = \mu^{\frac{2Q - \sum_i \alpha_i}{\gamma}} \langle V_1 \rangle_{\hat{g}}.$$

5.4 Give a proof of (5.65).

5.5 Suppose that $m \colon \hat{\mathbb{C}} \to \hat{\mathbb{C}}$ is a Möbius transformation of the Riemann sphere, $z_1, \ldots, z_k$ are points on $\mathbb{C}$ and $\alpha_1, \ldots, \alpha_k$ satisfy the Seiberg bounds (5.54) and (5.55). Suppose that $m \colon \hat{\mathbb{C}} \to \hat{\mathbb{C}}$ is a Möbius transformation, and

$$m_* \hat{g}_0(z) = \hat{g}_0(m^{-1}(z))|(m^{-1})'(z)|^2,$$

that is, viewed as metrics, $m_* \hat{g}_0$ is the pushforward of $\hat{g}_0$ by the map m.
 (a) Recalling the definition $R_{\hat{g}} = -(1/\hat{g})\Delta \log(\hat{g})$, show that $R_{m_* \hat{g}_0} \equiv 2$.
 (b) Using that

$$G^{\hat{\mathbb{C}},\hat{g}}(x,y) = \frac{1}{2\pi}\left(-\log|x-y| - \frac{1}{2R_{\hat{g}}}\log\hat{g}(x) - \frac{1}{2R_{\hat{g}}}\log\hat{g}(y) + c_{\hat{g}}\right)$$

for $\hat{g}$ with constant curvature, and Möbius invariance of the Green function, that is $G^{\hat{\mathbb{C}},m_* \hat{g}_0}(x,y) = G^{\hat{\mathbb{C}},\hat{g}_0}(m^{-1}(x), m^{-1}(y))$ for $x \neq y \in \mathbb{C}$, deduce that $c_{m_* \hat{g}_0} = c_{\hat{g}_0}$. Hint: it may be helpful to write $m(z)$ in the explicit form $(az+b)/(cz+d)$.
 (c) Finally, using Remark 5.13, show that

$$\theta_{m_* \hat{g}_0} = -\frac{1}{2}\bar{v}_{m_* \hat{g}_0}(\log(m_*(\hat{g}_0))) + \log(2) - \theta_{\hat{g}}.$$

5.6 Write down a general formula for $\langle V \rangle_g$ when g does not have constant scalar curvature but is in the same conformal class as $\hat{g}_0$. Deduce that the law of the Liouville field does *not* depend on the choice of g.

6
Gaussian Free Field with Neumann Boundary Conditions

So far in this book, we have encountered:

- Gaussian free fields on graphs (Section 1.1);
- Gaussian free fields with zero boundary conditions on proper, regular domains of $\mathbb{R}^d$ (Chapter 1); and
- Gaussian free fields on compact surfaces (Section 5.2.2).

The purpose of this chapter is to introduce a different version of the GFF on simply connected domains of $\mathbb{C}$, but now with non-zero boundary conditions. This object will be the so-called **Neumann** or **free boundary** GFF. It is the basic building block for constructing the special "scale-invariant quantum surfaces" that will be the focus of Chapter 7.

In general if we wish to add boundary data to a GFF, it is natural to simply add a function that is harmonic in the domain (though it can have relatively wild behaviour on the boundary). We will seek to impose **Neumann boundary conditions**. Recall that for a smooth function, this means that the normal derivative of the function vanishes along the boundary (if the domain is smooth). Of course for an object as rough as the GFF, it is a priori unclear what this condition should mean. Indeed, we will see that the resulting object is actually the same as when we don't impose any conditions at all (which is why the field can also be called a free boundary GFF, as is done for example in the papers [She16a] and [DMS21]). Indeed, note that in the discrete, a random walk on a graph with Neumann/"reflecting" boundary conditions or no/"free" boundary conditions are by definition the same thing (and both converge to reflecting Brownian motion, whose generator is $\frac{1}{2}$ the Laplace operator with Neumann boundary conditions).

Outlook Let D be a proper, simply connected domain of $\mathbb{C}$. We will first show how to define the Neumann GFF as a random distribution on D, just as in

Gaussian Free Field with Neumann Boundary Conditions

Section 1.7 for the Dirichlet GFF and Section 5.2.2 for the GFF on a Riemann surface. This allows for a straightforward deduction of several nice properties, which is why we present this point of view first. In Section 6.3, we will then go on to show that the Neumann GFF can be defined as a stochastic process (as in the Dirichlet case), and that this object coincides with the random distribution defined here when its index set is restricted appropriately. In the penultimate section of this chapter, we will discuss some further variants of the Gaussian free field, and how they relate to one another, and conclude in the final section with an analysis of *boundary* Gaussian multiplicative chaos.

Warning One technical complication when working with the Neumann GFF, compared to the Dirichlet case, is that it is really only defined up to a global additive constant. This corresponds to the fact that if one tries to extend the Dirichlet inner product $(\cdot, \cdot)_\nabla$ to test functions that are not necessarily compactly supported in D, it is no longer an inner product. Indeed, functions that are constant on the domain will have zero Dirichlet norm. Alternatively (as we will see later), one can think of the additive constant as arising from the fact that the Green function with Neumann boundary conditions is *not* canonically defined (or equivalently, that Brownian motion reflected on the boundary of D is recurrent).

Note that this complication was already present for the GFF on a Riemann surface: see Section 5.2.2. In that setting, we fixed the additive constant by requiring the field to have zero average with respect to the Riemannian volume form. In this chapter, it will be useful to have access to both of the following viewpoints.

1. We can view the Neumann GFF as a **distribution modulo constants** (two distributions are equivalent if their difference is a constant function). Equivalently, a distribution modulo constants can be defined as a continuous linear functional on test functions whose integral is required to be zero.
2. We can specify a **particular representative** of the Neumann GFF's **equivalence class modulo constants** (for example by requiring that the average of the field over a specific region is zero). We will then speak of "fixing the additive constant". Note that while this point of view may appear to be more concrete, fixing the additive constant for the free field in this way actually causes it to lose some useful properties, such as conformal invariance.

When using the Neumann GFF, we will therefore always need to be careful to say whether we consider the modulo constants version, or a version that has had the constant fixed in a particular way.

6.1 The Neumann GFF as a random distribution

Let $\overline{\mathcal{D}}(D)$ be the space of C^∞ functions in D with $(f, f)_\nabla < \infty$ ("finite Dirichlet energy"), defined **modulo constants**. That is, two functions are equivalent if their difference is a constant function. Note that these functions are *not* assumed to have compact support in D. It is clear that on this space, $(\cdot, \cdot)_\nabla$ really is an inner product. Hence we can define $\overline{H}^1(D)$ to be the Hilbert space closure of $\overline{\mathcal{D}}(D)$ with respect to $(\cdot, \cdot)_\nabla$.

We define a distribution modulo constants to be a continuous linear functional on the space of test functions $f \in \mathcal{D}_0(D)$ such that $\int_D f(x)\,dx = 0$, and denote the set of such test functions by $\tilde{\mathcal{D}}_0(D)$. We write $\overline{\mathcal{D}}'_0(D)$ for the space of distributions modulo constants, and equip it with the topology of weak-$\star$ convergence. That is, a sequence T_n of distributions modulo constants converges to a distribution T if and only if $(T_n, f) \to (T, f)$ for any test function $f \in \tilde{\mathcal{D}}_0(D)$.

Remark 6.1 (Notation)   In this section, we will use the general notation $\overline{\cdot}$ to refer to spaces of objects or objects defined modulo constants, and the notation $\tilde{\cdot}$ for spaces of objects or objects with zero average over D.

As in Section 1.4, a random variable X defined on a probability space $(\Omega, \mathcal{F}, \mathbb{P})$ and taking values in the space of distributions modulo constants, is simply a function $X \colon \Omega \to \overline{\mathcal{D}}'_0(D)$ which is measurable with respect to the Borel σ-field on $\overline{\mathcal{D}}'_0(D)$ induced by the weak-$\ast$ topology. Arguing as in Lemma 1.34, we see that convergence of a sequence of random variables $X_n \in \overline{\mathcal{D}}_0(D)$ is a measurable event. Thus, it makes sense to ask about almost sure convergence of such sequences.

We now give the definition of the Neumann GFF as a random element of $\overline{\mathcal{D}}'_0(D)$.

Theorem 6.2   *Let $\{\overline{f}_j\}_{j \geq 1}$ be any orthonormal basis of $\overline{H}^1(D)$, and $\{X_j\}_{j \geq 1}$ be a sequence of independent $\mathcal{N}(0, 1)$ random variables. Then the random series*

$$\overline{\mathbf{h}}_n := \sum_1^n X_j \overline{f}_j \qquad (6.1)$$

converges almost surely in the space of distributions modulo constants. Moreover, the law of the limit $\overline{\mathbf{h}} = \overline{\mathbf{h}}^D$ does not depend on the choice of orthonormal basis $\{\overline{f}_j\}_j$ and can be written as the sum of a Dirichlet boundary condition GFF on D and an independent harmonic function modulo constants.

Definition 6.3 (Neumann GFF as a distribution modulo constants)   We define

6.1 The Neumann GFF as a random distribution

the Neumann GFF $\overline{\mathbf{h}}$ to be the random distribution modulo constants constructed in Theorem 6.2.

Remark 6.4 (Neumann boundary conditions) Suppose that $D = \mathbb{D}$. In defining $\overline{H}^1(\mathbb{D})$, we started from the space $\overline{\mathcal{D}}(\mathbb{D})$ of smooth functions (modulo constants) on $\mathbb{D}$ with no restriction on their boundary conditions. However, we could equally have started with the space of smooth functions (modulo constants) with Neumann boundary conditions and ended up with the same space $\overline{H}^1(\mathbb{D})$ after taking the closure with respect to $(\cdot, \cdot)_\nabla$. Indeed, there exists an orthonormal basis of $L^2(\mathbb{D})$ made up of eigenfunctions of the Laplacian with Neumann boundary conditions (see for example [Jos02, Theorem 8.5.2]). Then omitting the first eigenfunction (which has eigenvalue 0) and dividing the rest by the square roots of their respective eigenvalues and considering them modulo constants provides an orthonormal basis of $\overline{H}^1(\mathbb{D})$. Thus, one can think of the Neumann GFF as either having no imposed ("free") boundary conditions, or as having Neumann boundary conditions.

The connection with Neumann boundary conditions will also become more apparent when we define the Neumann GFF as a stochastic process. Indeed, we will see that its covariance function is given by a Green function in the domain, with Neumann instead of Dirichlet boundary conditions. As already mentioned, in the discrete, a random walk on a graph with Neumann/"reflecting" boundary conditions or no/"free" boundary conditions are really one and the same thing. So the discrete Green's function will be the same if either free or Neumann boundary conditions are imposed.

Proof of Theorem 6.2 We will carry out the proof in two steps: first assuming that D is the unit disc $\mathbb{D}$; and then extending to general D by conformal invariance.

Step 1 ($D = \mathbb{D}$). Write $\overline{\text{Harm}}(\mathbb{D})$ for the space of harmonic functions on $\mathbb{D}$ with finite Dirichlet energy, viewed modulo constants. By the same reasoning as in Lemma 1.54, we can decompose

$$\overline{H}^1(\mathbb{D}) = H_0^1(\mathbb{D}) \oplus \overline{\text{Harm}}(\mathbb{D}) \qquad (6.2)$$

as a direct orthogonal sum with respect to the Dirichlet inner product.

We can therefore define f_j^0 and $\overline{f}_j^H$ to be the projections onto $H_0^1(\mathbb{D})$ and $\overline{\text{Harm}}(\mathbb{D})$, respectively, of each $\overline{f}_j$ in our orthonormal basis of $\overline{H}^1(\mathbb{D})$. Accordingly, we set $\mathbf{h}_n^0 := \sum_{j=1}^n X_j f_j^0$ and $\overline{\mathbf{h}}_n^H = \sum_{j=1}^n X_j \overline{f}_j^H$, so that $\overline{\mathbf{h}}_n = \mathbf{h}_n^0 + \overline{\mathbf{h}}_n^H$ for each n.

First, we claim that $\mathbf{h}_n^0$ converges almost surely in the space $H_0^s(\mathbb{D})$ (for any

$s < 0$) to a limit $\mathbf{h}$ with the law of a zero boundary condition GFF in $\mathbb{D}$. The proof is very similar to that of Theorem 1.45, but we reproduce it here, since there are some technical differences. We let $(e_m)_{m \geq 1}$ be an orthonormal basis of $L^2(\mathbb{D})$ made up of eigenfunctions of $-\Delta$, with corresponding eigenvalues $(\lambda_m)_{m \geq 0}$, and recall that $\lambda_m \asymp m$ as $m \to \infty$ by Weyl's law. Then we have, for $n \geq 1$,

$$\mathbb{E}\big((\mathbf{h}_n^0, \mathbf{h}_n^0)_{H_0^s}\big) = \sum_{j=1}^{n}(f_j^0, f_j^0)_{H_0^s}, \qquad (6.3)$$

where, using the Gauss–Green formula, the fact that $e_m \in H_0^1(\mathbb{D})$ for each m, and Fubini:

$$\sum_{j \geq 1}(f_j^0, f_j^0)_{H_0^s} = \sum_{j \geq 1}\sum_{m \geq 1} \lambda_m^s (f_j^0, e_m)_{L^2}^2 = \sum_{j \geq 1}\sum_{m \geq 1} \lambda_m^{-1+s}(f_j^0, \tfrac{e_m}{\sqrt{\lambda_m}})_\nabla^2$$

$$= \sum_{j \geq 1}\sum_{m \geq 1} \lambda_m^{-1+s}(\overline{f}_j, \tfrac{e_m}{\sqrt{\lambda_m}})_\nabla^2 = \sum_{m \geq 1} \lambda_m^{-1+s}\sum_{j \geq 1}(\overline{f}_j, \tfrac{e_m}{\sqrt{\lambda_m}})_\nabla^2$$

$$= \sum_{m \geq 1} \lambda_m^{-1+s} < \infty;$$

the finiteness holding as long as $s < 0$. We deduce that $\mathbf{h}_n^0$ converges almost surely to a limit $\mathbf{h}^0$ in H_0^s, by exactly the same reasoning as in the proof of Theorem 1.45. Moreover, as a limit of centred and jointly Gaussian random variables, $((\mathbf{h}^0, e_j))_{j \geq 1}$ are centred and jointly Gaussian, with

$$\mathbb{E}((\mathbf{h}^0, e_j)(\mathbf{h}^0, e_k))$$
$$= \lim_{N \to \infty}\sum_{n=1}^{N}(f_n^0, e_j)(f_n^0, e_k) = \lim_{N \to \infty}\sum_{n=1}^{N}(\overline{f}_n, \lambda_j^{-1}e_j)_\nabla(\overline{f}_n, \lambda_k^{-1}e_k)_\nabla$$
$$= (\lambda_j^{-1}e_j, \lambda_k^{-1}e_k)_\nabla = \lambda_j^{-1}\mathbf{1}_{j=k}, \qquad (6.4)$$

for each $j, k \geq 1$. This implies that $\mathbf{h}^0$ is equal in law (as a random element of H_0^s and therefore as a distribution) to a zero boundary GFF in $\mathbb{D}$ (indeed, it is immediate from the definition as a Fourier series that the above holds for such a GFF).

Next, we will show that $\overline{\mathbf{h}}_n^H = \sum_{j=1}^n X_j \overline{f}_j^H$ converges almost surely in $\overline{\mathcal{D}}_0'(\mathbb{D})$, to a random element $\overline{\mathbf{h}}^H$ of $\overline{\mathrm{Harm}}(\mathbb{D})$. For this, we will make use of a specific orthonormal basis of $\overline{\mathrm{Harm}}(\mathbb{D})$, which will play a similar role to the basis $(e_m)_m$ of eigenfunctions of the Laplacian used above. This basis is given by

$$\overline{u}_j(z) = \frac{1}{\sqrt{\pi j}}\Re(z^j) \text{ and } \overline{v}_j(z) = \frac{1}{\sqrt{\pi j}}\Re(iz^j) \text{ for } j \geq 1$$

(viewed modulo constants), which are easily checked to be orthonormal with

6.1 The Neumann GFF as a random distribution   225

respect to $(\cdot,\cdot)_\nabla$. They also span the space $\overline{\mathrm{Harm}}(\mathbb{D})$, because any harmonic function on $\mathbb{D}$ is the real part of an analytic function, and therefore admits a Taylor series expansion of the form $a + \sum_{j=1}^\infty b_j \Re(z^j) + \sum_{j=1}^\infty c_j \Re(iz^j)$ with $a, \{b_j, c_j\}_j \in \mathbb{R}$.

For each $j \geq 1$, let us denote by f_j^H the representative of $\overline{f}_j^H$ with $f_j^H(0) = 0$. Similarly, we set $u_m = (\pi m)^{-1/2} \Re(z^m)$, $v_m = (\pi m)^{-1/2} \Re(iz^m)$ for $m \geq 1$. Another simple calculation verifies that $((u_m, v_m))_{m \geq 1}$ are orthogonal with respect to the L^2 inner product on $\mathbb{D}$, and $(u_m,, u_m)_{L^2} = (v_m, v_m)_{L^2} = 1/(2m(m+1))$ for each $m \geq 1$. We write

$$\mathbb{E}\left(\Big\|\sum_{j=1}^n X_j f_j^H\Big\|_{L^2(\mathbb{D})^2}\right) = \sum_{j=1}^n (f_j^H, f_j^H)_{L^2(\mathbb{D})} \qquad (6.5)$$

and by Parseval's identity, can express

$$\sum_{j \geq 1}(f_j^H, f_j^H)_{L^2(\mathbb{D})} = \sum_{j \geq 1}\sum_{m \geq 1} 2m(m+1)\big((f_j^H, u_m)_{L^2}^2 + (f_j^H, v_m)_{L^2}^2\big). \qquad (6.6)$$

Now, for each $j \geq 1$, since $\overline{f}_j^H \in \overline{\mathrm{Harm}}(\mathbb{H})$, it has an expansion

$$\overline{f}_j^H = \sum_{m \geq 1} (\overline{f}_j^H, \overline{u}_m)_\nabla \overline{u}_m + (\overline{f}_j^H, \overline{v}_m)_\nabla \overline{v}_m$$

which converges in $\overline{H}^1(\mathbb{D})$, and therefore (since the (u_m, v_m) are also orthogonal for the L^2 inner product and since $(\overline{f}_j^0, \overline{u}_m)_\nabla = 0$)

$$(f_j^H, u_m)_{L^2} = (\overline{f}_j^H, \overline{u}_m)_\nabla (u_m, u_m)_{L^2} = \frac{1}{2m(m+1)}(\overline{f}_j^H, \overline{u}_m)_\nabla$$

$$= \frac{1}{2m(m+1)}(\overline{f}_j, \overline{u}_m)_\nabla$$

for each $j, m \geq 1$. The analogous equation holds for v_m. Thus the right-hand side of (6.6) becomes

$$\sum_{j \geq 1}\sum_{m \geq 1} \frac{1}{2m(m+1)}\big((\overline{f}_j, \overline{u}_m)_\nabla^2 + (\overline{f}_j, \overline{v}_m)_\nabla^2\big)$$

$$= \sum_{m \geq 1} \frac{1}{2m(m+1)} \sum_{j \geq 1}\big((\overline{f}_j, \overline{u}_m)_\nabla^2 + (\overline{f}_j, \overline{v}_m)_\nabla^2\big)$$

$$= \sum_{m \geq 1} \frac{1}{2m(m+1)}(\|\overline{u}_m\|_\nabla^2 + \|\overline{v}_m\|_\nabla^2) < \infty,$$

where we applied Fubini, and the fact that $(\overline{f}_j)_{j \geq 1}$ are an orthonormal basis of $\overline{H}^1(\mathbb{D})$. By the same argument used in the case of $(\mathbf{h}_n^0)_n$, this implies

226   *Gaussian Free Field with Neumann Boundary Conditions*

that $\sum_j X_j f_j^H$ converges almost surely in $L^2(\mathbb{D})$, and in particular, $\overline{\mathbf{h}}_n^{H} = \sum_{j=1}^n X_j \overline{f}_j^{H}$ converges almost surely in $\overline{\mathcal{D}}'_0(\mathbb{D})$.

By analogous reasoning to (6.4), it holds that the almost sure $L^2(\mathbb{D})$ limit of $\sum_j X_j \overline{f}_j^{H}$ has to be independent of the choice of $\{\overline{f}_j\}_j$. It remains to justify that the almost sure limit of

$$\sum_1^n \sqrt{\frac{1}{\pi j}} \Re((\alpha_j + i\beta_j) z^j) \; ; \; \alpha_j, \beta_j \overset{\text{i.i.d}}{\sim} \mathcal{N}(0,1) \qquad (6.7)$$

is harmonic. This simply follows from the fact limits in $\mathcal{D}'(\mathbb{D})$ of harmonic functions are harmonic (by definition, distributional limits of weakly harmonic functions are weakly harmonic, and then true harmonicity follows by elliptic regularity).

Finally, we claim that $\overline{\mathbf{h}}^{H} = \lim_{n\to\infty} \overline{\mathbf{h}}_n^{H}$ and $\mathbf{h}^0 = \lim_{n\to\infty} \mathbf{h}_n^0$ are independent. In other words, that for any $\rho, \eta \in \tilde{\mathcal{D}}_0(\mathbb{D})$, $(\mathbf{h}^0, \rho)$ and $(\overline{\mathbf{h}}^{H}, \eta)$ are independent. Since $(\mathbf{h}^0, \rho), (\overline{\mathbf{h}}^{H}, \eta)$ are the almost sure limits of $\sum_{j=1}^n (X_j f_j^0, \rho)$ and $\sum_{j=1}^n (X_j \overline{f}_j^{H}, \eta)$, respectively, they are centred and jointly Gaussian. Hence it suffices to check that

$$\lim_{n\to\infty} \mathbb{E}\left[\left(\sum_{j=1}^n X_j f_j^0, \rho\right)\left(\sum_{k=1}^n X_k \overline{f}_k^{H}, \eta\right)\right] = \lim_{n\to\infty} \sum_{j=1}^n (f_j^0, \rho)(\overline{f}_j^{H}, \eta) = 0.$$

For this, recall the definitions of $(e_m, \overline{u}_m, \overline{v}_m)_{m\geq 1}$ appearing previously in the proof. Then for each j, we can write

$$f_j^0 = \sum_{m\geq 1} (f_j^0, (\lambda_m)^{-1/2} e_m)_\nabla (\lambda_m)^{-1/2} e_m, \text{ and}$$

$$\overline{f}_j^{H} = \sum_{m\geq 1} ((\overline{f}_j^{H}, \overline{u}_m)_\nabla \overline{u}_m + (\overline{f}_j^{H}, \overline{v}_m)_\nabla \overline{v}_m,$$

with both sums converging in $\overline{H}^1(\mathbb{D})$. By Parseval, we can therefore write

$$\sum_{j\geq 1} (f_j^0, \rho)(\overline{f}_j^{H}, \eta) = \sum_{j\geq 1} \sum_{m,n\geq 1} \frac{(f_j^0, e_m)_\nabla}{\sqrt{\lambda_m}} (\overline{f}_j^{H}, \overline{u}_n)_\nabla (e_m, \rho)(\overline{u}_n, \eta)$$

$$+ \frac{(f_j^0, e_m)_\nabla}{\sqrt{\lambda_m}} (\overline{f}_j^{H}, \overline{v}_n)_\nabla (f_j^0, \rho)(\overline{v}_m, \eta)$$

$$= \sum_{m,n\geq 1} (e_m, \rho)(\overline{u}_n, \eta) \sum_{j\geq 1} \frac{(f_j^0, e_m)_\nabla}{\sqrt{\lambda_m}} (\overline{f}_j^{H}, \overline{u}_n)_\nabla$$

$$+ \sum_{m,n\geq 1} (e_m, \rho)(\overline{v}_n, \eta) \sum_{j\geq 1} \frac{(f_j^0, e_m)_\nabla}{\sqrt{\lambda_m}} (\overline{f}_j^{H}, \overline{v}_n)_\nabla,$$

6.1 The Neumann GFF as a random distribution

where in the second line we also applied Fubini. Note that this is justified since (restricting to the first of the two double sums by symmetry)

$$\sum_{m,n\geq 1}\sum_{j\geq 1}\left|(e_m,\rho)(\bar{u}_n,\eta)\frac{(f_j^0,e_m)_\nabla}{\sqrt{\lambda_m}}(\bar{f}_j^H,\bar{u}_n)_\nabla\right|$$

$$\leq \sum_{m,n\geq 1}|(e_m,\rho)(\bar{u}_n,\eta)|\sqrt{\sum_{j\geq 1}\frac{(f_j^0,e_m)_\nabla^2}{\lambda_m}}\sqrt{\sum_{j\geq 1}(\bar{f}_j^H,\bar{u}_n)_\nabla^2},$$

where the two sums over j are bounded by $(e_m,e_m)_\nabla/\lambda_m = 1$ and $(\bar{u}_n,\bar{u}_n)_\nabla = 1$ (using Parseval) and $\sum_{m,n\geq 1}|(e_m,\rho)(\bar{u}_n,\eta)| < \infty$ since $\rho,\eta \in \mathcal{D}_0(\mathbb{D})$.

Noticing that

$$(\bar{f}_j,\bar{u}_n)_\nabla = (\bar{f}_j^H,\bar{u}_n)_\nabla, (\bar{f}_j,\bar{v}_n) = (\bar{f}_j^H,\bar{v}_n) \text{ and}$$
$$(\bar{f}_j,\lambda_m^{-1/2}e_m)_\nabla = (f_j^0,\lambda_m^{-1/2}e_m)_\nabla,$$

for each j,m,n by orthogonality of $H_0^1(\mathbb{D})$ and $\overline{\text{Harm}}(\mathbb{D})$, we conclude that

$$\sum_{j\geq 1}(f_j^0,\rho)(\bar{f}_j^H,\eta)$$

$$= \sum_{m,n\geq 1}(e_m,\rho)(\bar{u}_n,\eta)\sum_{j\geq 1}\left(\bar{f}_j,\frac{e_m}{\sqrt{\lambda_m}}\right)_\nabla(\bar{f}_j,\bar{u}_n)_\nabla$$

$$+ \sum_{m,n\geq 1}(e_m,\rho)(\bar{v}_n,\eta)\sum_{j\geq 1}\left(\bar{f}_j,\frac{e_m}{\sqrt{\lambda_m}}\right)_\nabla(\bar{f}_j,\bar{v}_n)_\nabla$$

$$= \sum_{m,n\geq 1}(e_m,\rho)(\bar{u}_n,\eta)(e_m,\bar{u}_n)_\nabla + \sum_{m,n\geq 1}(e_m,\rho)(\bar{v}_n,\eta)(e_m,\bar{v}_n)_\nabla = 0,$$

as required.

Step 2 (general D). Suppose that $D \subsetneq \mathbb{C}$ is simply connected, and let $\{\bar{f}_j\}_j$ be an orthonormal basis for $\overline{H}^1(D)$. We would like to show that $\bar{\mathbf{h}}_n = \sum_1^n X_j f_j$ converges almost surely in $\tilde{\mathcal{D}}_0'(D)$ (when the X_j are i.i.d. $\mathcal{N}(0,1)$).

For this, we are going to use Step 1 and conformal invariance. Let $T: \mathbb{D} \to D$ be a conformal isomorphism (which exists by the Riemann mapping theorem). Then by conformal invariance of the Dirichlet inner product, $\{\bar{f}_j \circ T\}_j$ forms an orthonormal basis of $\overline{H}^1(\mathbb{D})$. We therefore know, by Step 1, that $\bar{\mathbf{h}}_n \circ T := \sum_1^n X_j(\bar{f}_j \circ T)$ converges almost surely in $\tilde{\mathcal{D}}_0'(\mathbb{D})$. That is, with probability 1, there exists $\bar{\mathbf{h}}^\mathbb{D} \in \tilde{\mathcal{D}}_0'(\mathbb{D})$ such that $(\bar{\mathbf{h}}_n \circ T, g) \to (\bar{\mathbf{h}}^\mathbb{D}, g)$ as $n \to \infty$ for all $g \in \tilde{\mathcal{D}}_0(\mathbb{D})$.

Since for $f \in \tilde{\mathcal{D}}_0(D)$ the function $g(z) = |T'(z)|^2 (f \circ T)(z)$ is in $\tilde{\mathcal{D}}_0(\mathbb{D})$, this tells us – in particular – that with probability 1:

$$(\overline{\mathbf{h}}_n \circ T, |T'|^2 (f \circ T)) \to (\overline{\mathbf{h}}^{\mathbb{D}}, |T'|^2 (f \circ T)) \text{ as } n \to \infty, \text{ for all } f \in \tilde{\mathcal{D}}_0(D).$$

Defining $\overline{\mathbf{h}} \in \overline{\mathcal{D}}_0'(D)$ by $(\overline{\mathbf{h}}, f) = (\overline{\mathbf{h}}^{\mathbb{D}}, |T'|^2 (f \circ T))$ for all $f \in \overline{\mathcal{D}}_0(D)$, this is exactly saying that with probability 1, $(\overline{\mathbf{h}}_n, f) \to (\overline{\mathbf{h}}, f)$ as $n \to \infty$ for all $f \in \tilde{\mathcal{D}}_0(D)$. That is, $\overline{\mathbf{h}}_n \to \overline{\mathbf{h}}$ in $\overline{\mathcal{D}}_0'(D)$, almost surely as $n \to \infty$.

Finally, by the same argument, if $T \colon \mathbb{D} \to D$ is conformal, then the law of $\overline{\mathbf{h}}$ must be given by the law of $\overline{\mathbf{h}}^{\mathbb{D}} \circ T^{-1}$, where $\overline{\mathbf{h}}^{\mathbb{D}}$ is the (unique in law) limit of (6.3) when $D = \mathbb{D}$. Note that this does not depend on the choice of T, since the law of $\overline{\mathbf{h}}^{\mathbb{D}}$ is conformally invariant (we can see this by applying the reasoning of the previous sentence with $D = \mathbb{D}$, together with the uniqueness in Step 1). Thus, the law of $\overline{\mathbf{h}}$ is unique for general D.

Using the description of this law when $D = \mathbb{D}$ from Step 1, plus conformal invariance of the Dirichlet GFF (Theorem 1.57) and the fact that conformal isomorphisms preserve harmonicity, we see that in general the law of $\overline{\mathbf{h}}$ satisfies the description in Definition 6.3. □

By conformal invariance of the Dirichlet inner product, we obtain the following (the details are spelled out in the proof above):

Corollary 6.5 *Let $\overline{\mathbf{h}}^D$ be the Neumann GFF (viewed modulo constants) in D, as in Definition 6.3. Then the law of $\overline{\mathbf{h}}^D$ is conformally invariant. That is, if $T \colon D \to D'$ is a conformal isomorphism between simply connected domains, then*

$$\overline{\mathbf{h}}^{D'} \stackrel{(d)}{=} \overline{\mathbf{h}}^D \circ T^{-1},$$

where $(\overline{\mathbf{h}}^D \circ T^{-1}, f) := (\overline{\mathbf{h}}^D, |T'|^2 (f \circ T))$ for all $f \in \tilde{\mathcal{D}}_0(D')$.

Straight from the definition, we also know that if $\overline{\mathbf{h}}$ is the Neumann GFF (viewed as a distribution modulo constants) in D, then $\overline{\mathbf{h}}$ can be written as the sum $\mathbf{h}_0 + u$, where $\mathbf{h}_0$ has the law of a zero boundary GFF in D, and u is an independent harmonic function modulo constants in D. By applying the Markov property of the Dirichlet GFF (Theorem 1.52) to $\overline{\mathbf{h}}$, we get an analogous decomposition for the Neumann GFF.

Theorem 6.6 (Markov property) *Fix $U \subset D$, open. Let $\overline{\mathbf{h}}$ be a Neumann GFF viewed as a distribution modulo constants in D, as in Definition 6.3. Then we may write*

$$\overline{\mathbf{h}} = \mathbf{h}_0 + \varphi,$$

where:

6.1 The Neumann GFF as a random distribution

1. $\mathbf{h}_0$ *is a zero boundary condition GFF in U, and is zero outside of U;*
2. φ *is a harmonic function viewed modulo constants in U;*
3. $\mathbf{h}_0$ *and* φ *are independent.*

For a more explicit Markov decomposition in the case $D = \mathbb{H}$ and U a semidisc centred on the real line, see Proposition 6.33.

Recall that we defined a distribution modulo constants to be a continuous linear functional on the space $\tilde{\mathcal{D}}_0(D)$ of test functions with average 0. Equivalently, we could define it to be an equivalence class of distributions (elements of $\mathcal{D}'_0(D)$), under the equivalence relation identifying distributions ϕ_1 and ϕ_2 whenever $\phi_1 - \phi_2 \equiv C$ for $C \in \mathbb{R}$.

Remark 6.7 (Fixing the additive constant, see also Definition 6.21) With the latter perspective, it is quite natural (and will sometimes be useful) to fix the additive constant for the GFF in some way (i.e. to pick an equivalence class representative). For example, we could define the Neumann GFF $\mathbf{h}$ with average zero when tested against some fixed test function $\rho_0 \in \mathcal{D}_0(D)$, by setting

$$(\mathbf{h}, \rho) = (\overline{\mathbf{h}}, \rho - \frac{\int_D \rho(\mathrm{d}x)}{\int_D \rho_0(\mathrm{d}x)} \rho_0) \quad \text{for } \rho \in \mathcal{D}_0(D),$$

where $\overline{\mathbf{h}}$ is as in Defintion 6.3. Since $\overline{\mathbf{h}}$ is almost surely a random distribution modulo constants, the above can be defined simultaneously for all $\rho \in \mathcal{D}_0(D)$, and almost surely defines an element of $\mathcal{D}'_0(D)$, that is, a distribution on D. In fact, by Corollary 1.53, it almost surely defines an element of $H^{-1}_{\mathrm{loc}}(D)$: the **local Sobolev space** of distributions whose restriction to any $U \Subset D$ (i.e. such that $\overline{U}$ is a compact subset of D) is an element of $H^{-1}_0(U)$.

Note that the choice of constant, or equivalence class representative, changes the resulting element of $\mathcal{D}'_0(D)$, but not how it acts when tested against functions (with average zero) in $\tilde{\mathcal{D}}_0(D)$.

Remark 6.8 Although it is sometimes helpful to fix the additive constant for the Neumann GFF, one should take care with the conformal invariance and Markovian properties discussed above. In particular:

- if $\overline{\mathbf{h}}$ is a Neumann GFF in D with additive constant fixed in some way, then it is *no longer* conformally invariant;
- in this case one can still write $\overline{\mathbf{h}} = \mathbf{h}_0 + u$ with $\mathbf{h}_0$ a Dirichlet GFF in D and u a harmonic function, but $\mathbf{h}$ and u *need not be independent*;
- on the other hand, if one starts with a Neumann GFF modulo constants, decomposes it as a Dirichlet GFF plus a harmonic function modulo constants, and then fixes the constant for the GFF in a way that only depends on the

harmonic function (for example, by specifying the value of the harmonic function at a point), then the two summands *will be* independent.

6.2 Covariance formula: the Neumann Green function

Recalling the definition of the Dirichlet GFF in a domain D, it is quite natural to guess that the Neumann GFF will have "covariance" given by a version of the Green function with Neumann boundary conditions in D. This is indeed the case, but with the caveat that the Green function with Neumann boundary conditions is not *uniquely* defined (see discussion below).

We say that a function G on $D \times D$ is **a covariance function for the Neumann GFF** in D if

$$\mathbb{E}((\overline{\mathbf{h}}, \rho_1)(\overline{\mathbf{h}}, \rho_2)) = \int_{D \times D} \rho_1(x) G(x, y) \rho_2(y) \, dx \, dy \qquad (6.8)$$

for every $\rho_1, \rho_2 \in \tilde{\mathcal{D}}_0(D)$, where $\overline{\mathbf{h}}$ is a Neumann GFF (viewed as a distribution modulo constants in D) as in Definition 6.3.

Let us immediately make a couple of remarks.

- This need not uniquely define G, as the equality is only required to hold for test functions with average 0. For example, adding any nice enough functions $v(x)$ and $w(y)$ to G will not affect the value of $\int_{D \times D} \rho_1(x) G(x, y) \rho_2(y) \, dx \, dy$. This ill definition is also an inherent property of the Neumann Green function (see below).
- As a consequence of Corollary 6.5, if $G(x, y)$ is a covariance function for the Neumann GFF in D, and $T: D' \to D$ is conformal, then $G'(x, y) = G(T(x), T(y))$ is a covariance function for the Neumann GFF in D'.

We will now show that any choice of **Neumann Green function** in D (if it exists), will be a valid covariance function for the GFF in D. To explain this, we first need to introduce the **Neumann problem** in D. This is the problem, given $\{\psi, \nu\}$, of

$$\text{finding } f \text{ such that: } \begin{cases} \Delta f = \psi & \text{in } D \\ \partial f / \partial n = \nu & \text{on } \partial D, \end{cases} \qquad (6.9)$$

subject to suitable regularity conditions on D. A requirement for the existence of a (weak) solution is that ψ, ν satisfy the Stokes condition:

$$\int_D \psi = \int_{\partial D} \nu. \qquad (6.10)$$

6.2 Covariance formula: the Neumann Green function

This condition comes from the divergence theorem; the integral of v along the boundary measures the total flux of ∇f across the boundary, while the integral of $\Delta f = \operatorname{div}(\nabla f)$ inside the domain measures the total divergence of ∇f. This solution is then (subject to appropriate conditions on the regularity of D) unique, up to a global additive constant. That is, this solution is unique in $\overline{H}^1(D)$. Existence of a solution is also known, for example, when D is smooth and bounded and $v = 0, \psi \in L^2(D)$, [Eva10, Section 6] (but we will not use any of these facts).

Definition 6.9 (Neumann Green function) We say that G is a **(choice of) Neumann Green function** in D, if for every $\rho \in \tilde{\mathcal{D}}_0(D)$:

$$f(x) := \int_D G(x, y)\rho(y)\,dy \tag{6.11}$$

is a solution of the Neumann problem (6.9) in D, with $\psi = -\rho$ and $v = 0$.

Proposition 6.10 *Suppose that $D \subset \mathbb{C}$ is simply connected and has C^1 smooth boundary. Then if G is a choice of Neumann Green function, it is a valid choice of covariance for the Neumann GFF $\overline{\mathbf{h}}$ in D. That is, for every $\rho \in \tilde{\mathcal{D}}_0(D)$,*

$$\mathbb{E}((\overline{\mathbf{h}}, \rho)^2) = \int_{D \times D} \rho(x) G(x, y) \rho(y)\,dx\,dy. \tag{6.12}$$

Remark 6.11 Note that adding an arbitrary function of x to G will not affect whether f defined in (6.11) is a solution to the Neumann problem. In other words, we have the same lack of uniqueness for G as for the covariance of the Neumann GFF.

Proof of Proposition 6.10 We need to check that if $\rho \in \tilde{\mathcal{D}}_0(D)$ and $\overline{\mathbf{h}}$ is a Neumann GFF in D, then

$$\mathbb{E}((\overline{\mathbf{h}}, \rho)^2) = \int_{D \times D} \rho(x) G(x, y) \rho(y)\,dx\,dy. \tag{6.13}$$

Defining $f(x) := \int_D G(x, y)\rho(y)\,dy$, we will show that both sides are equal to $\|f\|_\nabla^2$.

Note that by assumption, the right-hand side of (6.13) is equal to

$$\int_D -\Delta f(x) f(x)\,dx,$$

which by applying the Gauss–Green formula and the Neumann boundary condition for f is equal to

$$\int_D \nabla f(x) \cdot \nabla f(x)\,dx = \|f\|_\nabla^2.$$

232 Gaussian Free Field with Neumann Boundary Conditions

For the left-hand side, we use the construction of $\overline{\mathbf{h}}$ as the limit as $n \to \infty$ of $\sum_1^n X_j \overline{f}_j$ where the X_js are i.i.d. $\mathcal{N}(0, 1)$ and the $\overline{f}_j$s are an orthonormal basis of $\overline{H}^1(D)$. Since this is an almost sure limit in the space of distributions modulo constants, we have that

$$(\overline{\mathbf{h}}, \rho) = \lim_{n \to \infty} \sum_{j=1}^n X_j(\overline{f}_j, \rho) \text{ almost surely.}$$

Furthermore, by the Gauss–Green formula again, we have that $(\overline{f}_j, \rho) = (\overline{f}_j, f)_\nabla$ for each j, and so

$$\mathbb{E}\left(\left(\sum_{j=1}^n X_j(\overline{f}_j, \rho)\right)^2\right) = \sum_{j=1}^n (\overline{f}_j, f)_\nabla^2.$$

Note that this is bounded above by $\|f\|_\nabla^2$ for every n. Hence, $\sum_{j=1}^n X_j(\overline{f}_j, \rho)$ defines a martingale that is bounded in L^2, and so

$$\mathbb{E}((\overline{\mathbf{h}}, \rho)^2) = \mathbb{E}\left(\lim_{n \to \infty}\left(\sum_{j=1}^n X_j(\overline{f}_j, \rho)\right)^2\right) = \lim_{n \to \infty}\mathbb{E}\left(\left(\sum_{j=1}^n X_j(\overline{f}_j, \rho)\right)^2\right) = \|f\|_\nabla^2,$$

as desired. □

Example 6.12 We can define a choice of Neumann Green function in the unit disc $\mathbb{D}$ by

$$G_N^\mathbb{D}(x, y) = -(2\pi)^{-1} \log|(x-y)(1-x\overline{y})|; \quad x \neq y \in \mathbb{D}.$$

Indeed, a tedious but straightforward calculation can be used to verify that if $g_y(x) := G_N^\mathbb{D}(x, y)$ for fixed $y \in \mathbb{D}$, then (in the sense of distributions on $\mathbb{D}$)

$$\begin{cases} \Delta g_y &= -\delta_y \\ \partial g_y/\partial n &= -1/(2\pi) \text{ on } \partial \mathbb{D}. \end{cases}$$

This implies that if $\rho \in \tilde{D}_0(\mathbb{D})$ then $f(x) = \int_\mathbb{D} G_N^\mathbb{D}(x, y)\rho(y)\, dy$ as in (6.11) is a solution of the Neumann problem with $\psi = -\rho$ and $\nu = 0$. Indeed,

- $\Delta f(x) = \int_\mathbb{D} \Delta g_y(x)\rho(y)\, dy = -\int_\mathbb{D} \delta_y(x)\rho(y)\, dy = -\int_\mathbb{D} \delta_x(y)\rho(y)\, dy = -\rho(x)$;
- and for $x \in \partial \mathbb{D}$,

$$(\partial f/\partial n)(x) = \int_\mathbb{D} (\partial g_y/\partial n)(x)\rho(y)\, dy = -(2\pi)^{-1}\int_\mathbb{D} \rho(y)\, dy = 0.$$

Hence $G_N^\mathbb{D}$ is a choice of Neumann Green function in $\mathbb{D}$, and so also a valid choice of covariance for the Neumann GFF in $\mathbb{D}$.

6.2 Covariance formula: the Neumann Green function

Example 6.13 Define

$$G_N^{\mathbb{H}}(x,y) = -\frac{1}{2\pi} \log |(x-y)| - \frac{1}{2\pi} \log |(x-\bar{y})|; \quad x \ne y \in \mathbb{H}. \quad (6.14)$$

In this case, defining the conformal isomorphism $T \colon \mathbb{H} \to \mathbb{D}$ by $T(z) = (i-z)(i+z)^{-1}$, we have that if $g_y(x) := G^{\mathbb{H}}(T^{-1}(x), T^{-1}(y))$, then $\Delta g_y = -\delta_y$ and $\partial g_y/\partial n = -\delta_{-1}$ on $\partial \mathbb{D}$. Similarly to in the previous example, this implies that $G^{\mathbb{H}}(T^{-1}(\cdot), T^{-1}(\cdot))$ is a valid choice of Neumann Green function on $\mathbb{D}$. Hence by Proposition 6.10, it defines a valid choice of covariance function for the Neumann GFF on $\mathbb{D}$. Finally, by conformal invariance of the Neumann GFF (Corollary 6.5), we see that $G^{\mathbb{H}}$ is a valid choice of covariance function for the Neumann GFF in $\mathbb{H}$.

Note: it may seem that we have taken a rather long-winded approach in this example. Indeed, one can easily verify that if $g_y^{\mathbb{H}}(x) = G^{\mathbb{H}}(x,y)$ then $\Delta g_y^{\mathbb{H}} = -\delta_y$ on $\mathbb{H}$ and $\partial g_y^{\mathbb{H}}/\partial n = 0$ on $\mathbb{R}$. It is tempting to say that $G^{\mathbb{H}}$ therefore defines a choice of Green function on $\mathbb{H}$, and so by Proposition 6.10, is a valid covariance for the Neumann GFF on $\mathbb{H}$. However, one needs to take care that there is an extra "point at ∞" on the boundary of $\mathbb{H}$ (where, as one can see from the calculations in Example 6.13, we actually have a Dirac mass for $\partial g_y^{\mathbb{H}}/\partial n$). To make this example rigorous, it is therefore necessary to map to the unit disc and appeal to conformal invariance – as carried out above.

Remark 6.14 Recall that the Green's function G_0^D for a GFF with zero boundary conditions on a domain D could be defined in terms of the expected occupation time of ($\sqrt{2}$ times a) Brownian motion killed when leaving D:

$$G_0^D(x,y) = \int_0^\infty p_t^D(x,y) \, dt \quad (x \ne y).$$

There is a similar relationship between the Neumann Green's function and Brownian motion reflected on the boundary of D. The fact that the Neumann Green's function is not uniquely defined is related to the fact that reflected Brownian motion is recurrent. This means if $\tilde{p}_t^D(x,y)$ is the transition density for this reflected Brownian motion, then $\int_0^\infty \tilde{p}_t^D(x,y) \, dt$ does not actually converge, so one needs to normalize in some way to obtain a finite quantity. There are many possible ways to do this – hence the non-uniqueness.

Let us describe this more precisely in the case where $D = \mathbb{H}$. Denoting by $p_t(x,y)$, the transition density of Brownian motion in $\mathbb{C}$, it is easy to see that

$$\tilde{p}_t^{\mathbb{H}}(x,y) = p_t(x,y) + p_t(x,\bar{y})$$
$$= (4\pi t)^{-1} \left(\exp(-|x-y|^2/4t) + \exp(-|x-\bar{y}|^2/4t) \right),$$

which does not have finite integral over $t \in [0, \infty)$. However, if we look at

$\tilde{p}_t^{\mathbb{H}}(x,y) - \tilde{p}_t^{\mathbb{H}}(x_0,y)$ for some fixed $x_0 \in \mathbb{H}$ (for instance) then the corresponding integral does converge: to $G^{\mathbb{H}}(x,y)$ as defined in (6.14) plus the function $\log|x_0 - y|$. It is straightforward to check that this integral, for any choice of x_0, does define a valid choice of Neumann Green function on $\mathbb{H}$.

Remark 6.15 (A choice of covariance for general D)   Let us remark again that by Corollary 6.5, if G_N^D is a valid choice of covariance function for the Neumann GFF on some domain D, and $T: D' \to D$ is conformal, then $G_N^D(T(\cdot), T(\cdot))$ is a valid choice of covariance function for the Neumann GFF on D'.

From this observation and the above examples, we obtain a recipe to define a valid covariance function for the Neumann GFF in any simply connected domain D. This works even when the boundary of D is too rough to make sense of the Neumann problem.

We emphasise that any valid choice gives the same value for $\mathbb{E}((\bar{\mathbf{h}}, \rho_1)(\bar{\mathbf{h}}, \rho_2))$ when $\bar{\mathbf{h}}$ is a Neumann GFF (viewed as a distribution modulo constants) in D and $\rho_1, \rho_2 \in \tilde{\mathcal{D}}_0(D)$.

6.3 Neumann GFF as a stochastic process

In this section, we will define the Neumann GFF as a stochastic process, similarly to the definition of the Dirichlet GFF in Section 1.3. As with the Dirichlet GFF, this will allow us to "test" the Neumann GFF against a wider range of functions: in particular, they need not be smooth or compactly supported inside the domain D and can be non-zero near the boundary. However they can of course not be too singular near the boundary either. We formulate below a condition which, although not optimal, is easy to check in many examples and hence very practical.

For $x \in \overline{\mathbb{D}}$ and $A \subset \partial \mathbb{D}$ a Borel set let $q_x(A)$ denote the harmonic measure viewed from x, that is, the law of Brownian motion at the first time it leaves $\mathbb{D}$ (if $x \in \partial \mathbb{D}$ then q_x is just a Dirac mass at x). Given a Radon measure m on $\overline{\mathbb{D}}$, let $\nu_m(A) = \int_{x \in \overline{\mathbb{D}}} q_x(A) m(\mathrm{d}x)$. (If m is a probability measure, then ν_m is simply the law of a Brownian motion at its first exit of $\mathbb{D}$, starting from a point distributed according to m; if m is supported on $\partial \mathbb{D}$ then note that $\nu_m = m$).

Definition 6.16   Let m be a non-negative Radon measure on $\overline{\mathbb{D}}$ (i.e. a finite non-negative measure on $\overline{\mathbb{D}}$). We say that $m \in \mathfrak{M}_N^+(\mathbb{D})$ if:

- $m|_{\mathbb{D}} \in \mathfrak{M}_0^{\mathbb{D}}$, that is, $\iint_{\mathbb{D}^2} m(\mathrm{d}x) m(\mathrm{d}y) G_0^{\mathbb{D}}(x,y) < \infty$;
- and $\nu_m \in H^{-1/2}(\partial \mathbb{D})$.

6.3 Neumann GFF as a stochastic process

If $m = m^+ - m^-$ is a signed Radon measure on $\overline{D}$, let us say that $m \in \mathfrak{M}_N(\mathbb{D})$ if $m^\pm \in \mathfrak{M}_N^+(\mathbb{D})$.

Finally, let D be a simply connected domain with a locally connected boundary ∂D. Fix T a conformal isomorphism $T \colon \mathbb{D} \to D$. By definition we say that $\rho \in \mathfrak{M}_N(D)$ if $\rho = T_*m$ for some $m \in \mathfrak{M}_N(\mathbb{D})$, where T_*m denotes the pushforward of m by T, that is, $T_*m(T(A)) = m(A)$ for $A \subset \overline{\mathbb{D}}$.

For convenience, we note that the condition $v_m \in H^{-1/2}(\partial \mathbb{D})$ is implied by the more concrete condition $v_m \in L^2(\partial \mathbb{D})$.

This definition calls for a few comments. First of all, when D is simply connected with a locally connected boundary, a conformal isomorphism from $\mathbb{D}$ to D extends to a continuous map from $\overline{\mathbb{D}}$ to $\overline{D}$ ([Pom92]). In fact, in terms of the so-called "prime ends of D" (equivalently, the Martin boundary of D, [BN11]) the extended map is a homeomorphism. The pushforward T_*m should therefore be viewed as a measure on D and its boundary in this sense. Second, given such a measure ρ, to check if $\rho \in \mathfrak{M}_N(D)$, we therefore need to check if $m = T_*^{-1}\rho \in \mathfrak{M}_N(\mathbb{D})$. Notice that this does not depend on the choice of the conformal isomorphism T: indeed, any two such conformal isomorphisms differ by a conformal automorphism of $\mathbb{D}$ which is a Möbius map and therefore extends analytically to a neighbourhood of $\mathbb{D}$.

Example 6.17 As an example, any $\rho \in \mathfrak{M}_0$ compactly supported in D is clearly in $\mathfrak{M}_N$. As another example, suppose $D = \mathbb{D}$ and m is the uniform measure on a circular arc of $\partial \mathbb{D}$ of positive length. Then $m \in \mathfrak{M}_N(\mathbb{D})$. Indeed, in this case, clearly $v_m = m \in L^2(\partial \mathbb{D}) \subset H^{-1/2}(\partial \mathbb{D})$. As a final example, the measure ρ on the intersection $\gamma \cap D$ of a smooth curve γ and a Jordan domain D satisfies $\rho \in \mathfrak{M}_N(D)$, even if γ is not fully in D.

We can now define the index set of the Neumann GFF, which we denote by $\widetilde{\mathfrak{M}}_N(D) \subset \mathfrak{M}_N(D)$. By definition, this consists of those measures $\rho = \rho^+ - \rho^-$ with $\rho^+(\overline{D}) = \rho^-(\overline{D})$ (which corresponds to requiring that the total mass of ρ is zero). More precisely, $\rho \in \widetilde{\mathfrak{M}}_N(D)$ if $\rho^\pm = T_*m^\pm$ with T a conformal isomorphism from $\mathbb{D}$ to D, $m^\pm \in \mathfrak{M}_N(\mathbb{D})$, and $m^+(\overline{\mathbb{D}}) = m^-(\overline{\mathbb{D}})$.

Theorem 6.18 *Let D be simply connected with a locally connected boundary. Let $\rho \in \widetilde{\mathfrak{M}}_N(D)$. Then if $\overline{\mathbf{h}}_n = \sum_{j=1}^n X_j \overline{f}_j$ is as in (6.1),*

$$\lim_{n \to \infty} (\overline{\mathbf{h}}_n, \rho) =: (\overline{\mathbf{h}}, \rho)$$

exists almost surely and in $L^2(\mathbb{P})$.

Proof of Theorem 6.18 By conformal invariance of $H^1(D)$ and the definition of $\mathfrak{M}_N(D)$, we assume without loss of generality that $D = \mathbb{D}$. The first potential issue to address in the above theorem is whether $(\bar{\mathbf{h}}_n, \rho)$ makes sense for each fixed n. We will check that $\rho(\bar{g})$ makes sense for general $\bar{g} \in \overline{H}^1(D)$. (In fact, we will check that our definition of $\mathfrak{M}_N(\mathbb{D})$ implies that a measure $\rho \in \mathfrak{M}_N(\mathbb{D})$ defines a continuous linear functional on $\overline{H}^1(\mathbb{D})$ and thus is an element in the dual space of $\overline{H}^1(\mathbb{D})$; this will imply the result.)

By (6.2), $\bar{g}$ can be decomposed as $\bar{g} = g^0 + \bar{g}^H$ with $g^0 \in H_0^1(\mathbb{D})$ and $\bar{g}^H \in \overline{\text{Harm}}(\mathbb{D})$. The assumptions on ρ in Definition 6.16 mean that $\rho|_\mathbb{D} \in \mathfrak{M}_0 \subset H_0^{-1}(\mathbb{D}) = (H_0^1(\mathbb{D}))'$ and therefore $\rho(g^0) = \rho|_\mathbb{D}(g^0)$ is well defined, see for example Remark 1.41. In fact,

$$|\rho(g^0)| \leq \|g^0\|_\nabla \|\rho\|_{H_0^{-1}}. \tag{6.15}$$

Let g^H denote the representative of $\bar{g}^H$ which has mean zero over $\mathbb{D}$ (since $\rho^+(\overline{D}) = \rho^-(\overline{D})$, the choice of representative does not affect the value of $\rho(g^H)$). Then by the trace theorem, see for example [AF03, Theorem 5.36], g^H is the harmonic extension of a function g_∂ on the boundary with $g_\partial \in H^{1/2}(\partial \mathbb{D})$. Moreover,

$$\|g_\partial\|_{H^{1/2}(\partial \mathbb{D})} \leq C \|g^H\|_\nabla. \tag{6.16}$$

Since g^H is harmonic and $g_\partial \in H^{1/2}(\partial \mathbb{D})$, we have for a *fixed* $x \in \mathbb{D}$, $g^H(x) = \mathbb{E}_x(g_\partial(B_{\tau_\mathbb{D}}))$ (with B a Brownian motion started from x under $\mathbb{E}_x$ and τ_D its hitting time of $\partial \mathbb{D}$). (Here the regularity of g_∂ on $\partial \mathbb{D}$ is not important, it would suffice that g_∂ is for example an L^1 function on $\partial \mathbb{D}$.)

By Fubini's theorem and the Cauchy–Schwarz inequality, we have

$$\begin{aligned}|\rho(\bar{g}^H)| = |\rho(g^H)| &= \left|\int_{\overline{\mathbb{D}}} \mathbb{E}_x(g_\partial(B_{\tau_\mathbb{D}}))\rho^+(dx) - \int_{\overline{\mathbb{D}}} \mathbb{E}_x(g_\partial(B_{\tau_\mathbb{D}}))\rho^-(dx)\right| \\ &= |\nu_{\rho^+}(g_\partial) - \nu_{\rho^-}(g_\partial)| \\ &\leq (\|\nu_{\rho^+}\|_{H^{-1/2}(\partial \mathbb{D})} + \|\nu_{\rho^-}\|_{H^{-1/2}(\partial \mathbb{D})}) \|g_\partial\|_{H^{1/2}(\partial \mathbb{D})} \\ &\leq C_\rho \|g^H\|_\nabla = C_\rho \|\bar{g}^H\|_\nabla, \tag{6.17}\end{aligned}$$

for some $C_\rho < \infty$. In the last line, we also used (6.16) and the assumption that $\nu_{\rho^\pm} \in H^{-1/2}(\partial \mathbb{D})$ in Definition 6.16. Hence $\rho(\bar{g})$ is well defined, and combining with (6.15), ρ defines a continuous linear functional on $\overline{H}^1(\mathbb{D})$.

By the Riesz representation theorem, there exists $\bar{g}_\rho \in \overline{H}^1(\mathbb{D})$ with $\rho(\bar{f}) = (\bar{g}_\rho, \bar{f})_\nabla$ for all $f \in \overline{H}^1(\mathbb{D})$. This means that

$$(\bar{\mathbf{h}}_n, \rho) = \sum_{j=1}^n X_j (\bar{f}_j, \bar{g}_\rho)_\nabla$$

6.3 Neumann GFF as a stochastic process

is a martingale with mean zero, and uniformly bounded variance:

$$\mathrm{Var}(\overline{\mathbf{h}}_n, \rho) = \sum_{j=1}^{n} (\overline{f}_j, \overline{g}_\rho)_\nabla^2 \leq (\overline{g}_\rho, \overline{g}_\rho)_\nabla^2 \quad \text{for all } n \geq 1.$$

The martingale convergence theorem yields the result. □

For $\rho_1, \rho_2 \in \widetilde{\mathfrak{M}}_N(D)$, we denote

$$\Gamma_N(\rho_1, \rho_2) = \Gamma_N^D(\rho_1, \rho_2) := \mathrm{Cov}((\overline{\mathbf{h}}, \rho_1), (\overline{\mathbf{h}}, \rho_2)), \tag{6.18}$$

where $(\overline{\mathbf{h}}, \rho_1), (\overline{\mathbf{h}}, \rho_2)$ are the almost sure limits from Theorem 6.18. This brings us to the following definition.

Definition 6.19 (Neumann GFF modulo constants as a stochastic process) Let D be a simply connected domain with locally connected boundary. There exists a unique stochastic process

$$(\overline{\mathbf{h}}_\rho)_{\rho \in \widetilde{\mathfrak{M}}_N} = ((\overline{\mathbf{h}}, \rho))_{\rho \in \widetilde{\mathfrak{M}}_N},$$

indexed by $\widetilde{\mathfrak{M}}_N$, such that for every choice of $\rho_1, \ldots, \rho_n \in \widetilde{\mathfrak{M}}_N$, $(\overline{\mathbf{h}}_{\rho_1}, \ldots, \overline{\mathbf{h}}_{\rho_n})$ is a centred Gaussian vector with covariance

$$\mathrm{Cov}(\overline{\mathbf{h}}_{\rho_i}, \overline{\mathbf{h}}_{\rho_j}) = \Gamma_N^D(\rho_1, \rho_2).$$

By construction, if we restrict the process in Definition 6.19 to

$$(\overline{\mathbf{h}}, \rho)_{\rho \in \tilde{\mathcal{D}}_0(D)},$$

then there exists a version of this process defining a random distribution modulo constants, with the same law as the Neumann GFF in Definition 6.3.

Remark 6.20 (Conformal invariance) We can also talk about conformal invariance of the Neumann GFF viewed as a stochastic process. Indeed, suppose that $T \colon D \to \mathbb{D}$ is conformal. Then, as discussed in the proof of Theorem 6.18, $\rho \in \widetilde{\mathfrak{M}}_N(D)$ if and only if the pushforward measure $T_*\rho \in \widetilde{\mathfrak{M}}_N(\mathbb{D})$, and by conformal invariance of the Dirichlet inner product, we have $\Gamma_N^D(\rho_1, \rho_2) = \Gamma_N^{\mathbb{D}}(T_*\rho_1, T_*\rho_2)$ for all $\rho_1, \rho_2 \in \widetilde{\mathfrak{M}}_N(D)$. It follows that if $\overline{\mathbf{h}}^D$ and $\overline{\mathbf{h}}^{\mathbb{D}}$ are the stochastic processes from Definition 6.19, corresponding to the domains D and $\mathbb{D}$, then

$$((\overline{\mathbf{h}}^D, \rho))_{\rho \in \widetilde{\mathfrak{M}}_N^D} \stackrel{\text{(law)}}{=} ((\overline{\mathbf{h}}^{\mathbb{D}}, T_*\rho))_{\rho \in \widetilde{\mathfrak{M}}_N^D}.$$

With this definition of the Neumann GFF as a stochastic process, it still makes sense to speak of *fixing the additive constant* for the field. In fact, let us now make this notion more precise.

238 Gaussian Free Field with Neumann Boundary Conditions

Definition 6.21 (Neumann GFF with fixed additive constant) Let D be simply connected with locally connected boundary. Suppose that $\rho_0 \in \mathfrak{M}_N(D) \setminus \widetilde{\mathfrak{M}}_N(D)$. The Neumann GFF $\mathbf{h}$ with additive constant fixed so that $(\mathbf{h}, \rho_0) = 0$ is the stochastic process defined from $\overline{\mathbf{h}}$ in Definition 6.19 by setting

$$(\mathbf{h}, \rho) = \left(\overline{\mathbf{h}}, \rho - \frac{\fint_{\overline{D}} \rho(\mathrm{d}x)}{\fint_{\overline{D}} \rho_0(\mathrm{d}x)} \rho_0\right)$$

for each $\rho \in \mathfrak{M}_N(D)$, where, with an abuse of notation, we write $\fint_{\overline{D}} \rho(\mathrm{d}x)$ for $\fint_{\overline{\mathbb{D}}} T_*\rho(\mathrm{d}x)$, and $T: D \to \mathbb{D}$ is a conformal isomorphism.

Remark 6.22 For any $\rho_0 \in \mathfrak{M}_N(D) \setminus \widetilde{\mathfrak{M}}_N(D)$, the Neumann GFF with additive constant fixed so that $(\mathbf{h}, \rho_0) = \mathbf{0}$ has a version which almost surely defined a random distribution, that is, an element of $\mathcal{D}'_0(D)$. Indeed, suppose without loss of generality that $I_{\rho_0} := \int \rho_0 = 1$, and fix an arbitrary $\rho' \in \mathcal{D}_0(D)$ with $I_{\rho'} = 1$. Then, by Definition 6.3 and Theorem 6.18, there exists a probability space and a version of $\overline{\mathbf{h}}$ defined on this probability space such that $\overline{\mathbf{h}}$ defines a distribution modulo constants and $(\overline{\mathbf{h}}, \rho' - \rho_0)$ is also defined. Then

$$(\mathbf{h}, \rho) := (\overline{\mathbf{h}}, \rho - I_\rho \rho') + I_\rho(\overline{\mathbf{h}}, \rho' - \rho_0)$$

is defined simultaneously for all $\rho \in \mathcal{D}_0(D)$, and defines a version of the Neumann GFF with fixed additive constant from Definition 6.21. Moreover, $\rho \mapsto (\mathbf{h}, \rho)$ is clearly linear in ρ and $(\mathbf{h}, \rho_n) \to 0$ for any sequence ρ_n converging to 0 in $\mathcal{D}_0(D)$. Thus $\mathbf{h}$ defines a random element of $\mathcal{D}'_0(D)$.

In the following, whenever we talk of a Neumann GFF with *arbitrary* fixed additive constant, we mean a Neumann GFF with additive constant fixed – as defined above – for some *arbitrary, deterministic* $\rho_0 \in \mathfrak{M}_N(D) \setminus \widetilde{\mathfrak{M}}_N(D)$.

Example 6.23 (Semicircle averages) Suppose that $D = \mathbb{H}$ and for $x \in \mathbb{R}$ and $\varepsilon > 0$, let $\rho_{x,\varepsilon}$ be the uniform probability distribution on $\partial B(x, \varepsilon) \cap \mathbb{H}$ of radius ε about x. Then it is straightforward to check that $\rho_{x,\varepsilon} \in \mathfrak{M}_N(\mathbb{H})$. Therefore if $\mathbf{h}$ is a Neumann GFF with a fixed additive constant, we can define the ε-semicircle average $(\mathbf{h}, \rho_{x,\varepsilon})$ of $\mathbf{h}$ about x.

Remark 6.24 Notice that if $\rho_1, \rho_2 \in \mathfrak{M}_N(D)$ with $\fint_{\overline{D}} \rho_1 = \fint_{\overline{D}} \rho_2$, then $\rho_1 - \rho_2 \in \widetilde{\mathfrak{M}}_N(D)$. Hence we can define $(\overline{\mathbf{h}}, \rho_1 - \rho_2)$ when $\overline{\mathbf{h}}$ is a Neumann GFF modulo constants. We can also define $(\mathbf{h}, \rho_1 - \rho_2)$ whenever $\mathbf{h}$ is a Neumann GFF with fixed additive constant, and its law will not depend on the choice of additive constant: it will be exactly that of $(\overline{\mathbf{h}}, \rho_1 - \rho_2)$.

6.4 Other boundary conditions

6.4.1 Whole plane GFF

In this section, we will discuss the **whole plane GFF**, which we will define as:

- a distribution modulo constants on the whole complex plane $\mathbb{C}$ whose *odd* and *even* parts are given by reflecting the Dirichlet GFF and Neumann GFF, respectively, in the x-axis.

Equivalently, we will see that the whole plane GFF coincides with:

- a stochastic process with covariance $-(2\pi)^{-1}\log|x-y|$ in a suitable sense,
- a local limit of the Dirichlet GFF on large disks,
- the spherical GFF constructed in Chapter 5 (when the latter is viewed modulo constants and the sphere is identified with the extended complex plane $\hat{\mathbb{C}} = \mathbb{C} \cup \{\infty\}$).

Just as before, but now with $D = \mathbb{C}$, we define the space of distributions modulo constants on $\mathbb{C}$, $\overline{\mathcal{D}}_0'(\mathbb{C})$, to be the space of continuous linear functionals on $\tilde{\mathcal{D}}_0(\mathbb{C}) = \{f \in C^\infty(\mathbb{C})$ with compact support and $\int_\mathbb{C} f = 0\}$, equipped with the weak-* topology.

Definition 6.25 The whole plane GFF, $\overline{\mathbf{h}}^\infty$, is the random distribution modulo constants on $\mathbb{C}$ defined by

$$\overline{\mathbf{h}}^\infty = \overline{\mathbf{h}}^\infty_{\text{even}} + \overline{\mathbf{h}}^\infty_{\text{odd}}, \qquad (6.19)$$

where for every $f \in \tilde{\mathcal{D}}_0(\mathbb{C})$ with conjugate $f^* : z \mapsto f(\overline{z})$,

$$(\overline{\mathbf{h}}^\infty_{\text{even}}, f) = \frac{(\overline{\mathbf{h}}^\mathbb{H}, f|_\mathbb{H} + f^*|_\mathbb{H})}{\sqrt{2}} \quad ; \quad (\overline{\mathbf{h}}^\infty_{\text{odd}}, f) = \frac{(\mathbf{h}^\mathbb{H}_0, f|_\mathbb{H} - f^*|_\mathbb{H})}{\sqrt{2}}.$$

Here $\overline{\mathbf{h}}^\mathbb{H}$, $\mathbf{h}^\mathbb{H}_0$ are independent Neumann (modulo constants) and Dirichlet GFFs in $\mathbb{H}$, respectively.

The definition (6.19) is natural and should be compared with the fact that any function on $\mathbb{C}$ can be written as the sum of an even and an odd function, respectively (where even and odd refer to reflection with respect to the real axis).

Recalling that the Dirichlet Green function in $\mathbb{H}$ is given by $G^\mathbb{H}_0(x,y) = \frac{1}{2\pi}(-\log|x-y| + \log|x-\overline{y}|)$ and a valid covariance for the Neumann GFF in

240 *Gaussian Free Field with Neumann Boundary Conditions*

$\mathbb{H}$ is given by $G_N^{\mathbb{H}}(x,y) = \frac{1}{2\pi}(-\log|x-y| - \log|x-\bar{y}|)$, a simple calculation (which we leave as an exercise) gives that

$$\text{Cov}((\overline{\mathbf{h}}^\infty, f_1)(\overline{\mathbf{h}}^\infty, f_2)) = \frac{1}{2\pi}\iint_{\mathbb{C}\times\mathbb{C}} \log(\tfrac{1}{|x-y|}) f_1(x) f_2(y)\,dx\,dy \quad (6.20)$$

for each $f_1, f_2 \in \tilde{\mathcal{D}}_0(\mathbb{C})$. In other words, the covariance of the whole plane GFF (modulo constants) is equal to $-\frac{1}{2\pi}\log(|x-y|)$.

Recall also from Lemma 5.9 that the zero average GFF with respect to a Riemannian metric g on the sphere $\hat{\mathbb{C}}$, had covariance function

$$G^{\hat{\mathbb{C}},g}(x,y) = \frac{1}{2\pi}\Big[-\log(|x-y|) + \overline{v}_g\big(\log(|x-\cdot|)\big) + \overline{v}_g\big(\log(|y-\cdot|)\big) - \theta_g\Big],$$

where $\overline{v}_g, \theta_g$ were defined in that lemma. In particular, for any f_1, f_2 such that $\int_{\mathbb{C}} f_1(x)\,dx = \int_{\mathbb{C}} f_2(x)\,dx = 0$, we will have

$$\iint_{\mathbb{C}\times\mathbb{C}} G^{\hat{\mathbb{C}},g}(x,y) f_1(x) f_2(y)\,dx\,dy = \iint_{\mathbb{C}\times\mathbb{C}} \frac{1}{2\pi}\log(\tfrac{1}{|x-y|}) f_1(x) f_2(y)\,dx\,dy. \quad (6.21)$$

This implies the following:

Lemma 6.26 *Let g be a Riemannian metric on the sphere and $\overline{\mathbf{h}}^{\hat{\mathbb{C}},g}$ be $\mathbf{h}^{\hat{\mathbb{C}},g}$ viewed as a distribution modulo constants. Then*

$$\overline{\mathbf{h}}^\infty \stackrel{\text{(law)}}{=} \overline{\mathbf{h}}^{\hat{\mathbb{C}},g}.$$

In fact, just as with the Neumann GFF, we can define the whole plane GFF as a distribution on $\hat{\mathbb{C}}$ (not modulo constants) by fixing the additive constant in some way. For example, we can take the equivalence class representative of $\overline{\mathbf{h}}^\infty$ which has average 0 with respect to $g(z)\,dz$. We leave it as an exercise to check that this has precisely the same law as $\mathbf{h}^{\hat{\mathbb{C}},g}$.

As mentioned at the start of this subsection, there is another natural way to describe the whole plane GFF, and that is as the local limit of Dirichlet GFFs in large domains. This limit actually exists in a strong sense, and for this, we need to recall the definition and some basic properties of the **total variation distance**. For two random variables X, Y taking values in the same measurable space $(E, \mathcal{E})$, with respective laws μ and ν, we define the total variation distance between them by

$$d_{TV}(\mu, \nu) = \sup_{A\in\mathcal{E}} |\mu(A) - \nu(A)|.$$

With an abuse of notation we also write $d_{TV}(X, Y)$ for $d_{TV}(\mu, \nu)$. Suppose that

6.4 Other boundary conditions

ν is absolutely continuous with respect to μ, with Radon–Nikodym derivative $Z = d\nu/d\mu$. Then for any $A \in \mathcal{E}$,

$$|\nu(A) - \mu(A)| = \mathbb{E}_\mu[1_A(Z-1)] \leq \mathbb{E}_\mu[|Z-1|].$$

Since $A \in \mathcal{E}$ is arbitrary, we deduce that

$$d_{TV}(\mu, \nu) \leq \mathbb{E}_\mu[|Z-1|]. \tag{6.22}$$

Now suppose that E is a metric space and $\mathcal{E}$ is the associated Borel σ-algebra. Then it is well known (and easy to check) that given two laws μ and ν on $(E, \mathcal{E})$ and a coupling (X, Y) (measurable with respect to the product Borel σ-algebra) of these two laws, then $d_{TV}(\mu, \nu) \leq \mathbb{P}(X \neq Y)$. Conversely, if $(E, \mathcal{E})$ is a separable metric measure space, then there necessarily exists a *maximal coupling* of μ and ν, that is, a coupling (X, Y) measurable with respect to the product σ-algebra, such that $\mathbb{P}(X \neq Y) = d_{TV}(X, Y)$. See, for example, [Che04, Section 5.1].

It is straightforward to check that the set of measures on $(E, \mathcal{E})$, equipped with the total variation distance, is a complete metric space. The main point is to verify that if μ_n forms a Cauchy sequence with respect to the total variation distance then $\mu_n(A)$ converges to a limit $\ell(A)$ for any fixed set $A \in \mathcal{E}$ (and in fact the convergence is uniform). As a consequence, the limits $\ell(A)$ necessarily satisfy σ additivity (with respect to A), and thus define a probability measure on $(E, \mathcal{E})$.

Finally, if $(E, \mathcal{E})$ is a metric measure space, then convergence of a sequence of measures μ_n on $(E, \mathcal{E})$ to μ in the sense of total variation distance implies weak convergence: indeed, by the portmanteau theorem, the latter is equivalent to convergence of $\mu_n(A)$ to $\mu(A)$ for every μ-continuity set $A \in \mathcal{E}$, whereas convergence in the total variance is equivalent to the uniform (in $A \in \mathcal{E}$) convergence of $\mu_n(A)$.

We can now state the result.

Theorem 6.27 *Fix $a > 0$ and let $R > a$. Let $\mathbf{h}_0^R$ be a Dirichlet (zero) boundary condition GFF on $R\mathbb{D}$. Then as $R \to \infty$,*

$$d_{TV}(\mathbf{h}_0^R|_{a\mathbb{D}}, \overline{\mathbf{h}}^\infty|_{a\mathbb{D}}) \to 0,$$

when $\mathbf{h}_0^R|_{a\mathbb{D}}$ and $\overline{\mathbf{h}}^\infty|_{a\mathbb{D}}$ are considered as distributions modulo constants *in $a\mathbb{D}$. In fact the same statement holds when both of these are considered as elements of $H_{loc}^{-1}(a\mathbb{D})$ modulo constants.*

Remark 6.28 The fact that $\overline{\mathbf{h}}^\infty|_{a\mathbb{D}}$ may be viewed as an element of $H_{loc}^{-1}(a\mathbb{D})$ modulo constants follows from the definition of the whole plane GFF in (6.19) and Remark 6.7.

242   *Gaussian Free Field with Neumann Boundary Conditions*

Proof of Theorem 6.27   We will first show that

$$\sup_{R_1,R_2 \geq R} d_{TV}(\mathbf{h}_0^{R_1}|_{a\mathbb{D}}, \mathbf{h}_0^{R_2}|_{a\mathbb{D}}) \to 0, \tag{6.23}$$

when $\mathbf{h}_0^{R_1}|_{a\mathbb{D}}$ and $\mathbf{h}_0^{R_2}|_{a\mathbb{D}}$ are considered as distributions *modulo constants* in $a\mathbb{D}$. Without loss of generality, suppose that $R_2 \geq R_1$. Then by the Markov property of the Dirichlet GFF (Theorem 1.52), we can write $\mathbf{h}^{R_2} = \tilde{\mathbf{h}}^{R_1} + \varphi$, where $\tilde{\mathbf{h}}^{R_1}$ has the law of $\mathbf{h}^{R_1}$, and φ is independent of $\tilde{\mathbf{h}}^{R_1}$ and almost surely harmonic in $R_1\mathbb{D}$. The proof of this lemma will essentially follow from the fact that, when viewed modulo constants and restricted to $a\mathbb{D}$, φ is very small.

Indeed, if we define $\varphi_0 = \varphi - \varphi(0)$ then by independence and harmonicity, $\text{Var}(\mathbf{h}_1^{R_2}(w) - \mathbf{h}_1^{R_2}(0)) = \text{Var}(\mathbf{h}_1^{R_1}(w) - \mathbf{h}_1^{R_1}(0)) + \text{Var}(\varphi_0(w))$ for any $w \in \partial(8a\mathbb{D})$ (say). Since we have the explicit expressions

$$2\pi G_0^{R_i\mathbb{D}}(x, y) = \log R_i + \log|1 - (\overline{x}y/R_i^2)| - \log(|x - y|) \tag{6.24}$$

for $x \neq y \in R_i\mathbb{D}$ ($i = 1, 2$), it follows easily that

$$\sup_{R_1,R_2 \geq R} \sup_{w \in \partial(8a\mathbb{D})} \text{Var}(\varphi_0(w)) \to 0 \text{ as } R \to \infty. \tag{6.25}$$

Now, note that $\mathbf{h}^{R_2}$ and $\tilde{\mathbf{h}}^{R_1} + \varphi_0$ differ by exactly a constant in $R_1\mathbb{D}$. So we would be done with the proof of (6.23) if we could show that the laws of

$$\tilde{\mathbf{h}}^{R_1} + \varphi_0 \text{ and } \tilde{\mathbf{h}}^{R_1}$$

are close in total variation distance when restricted to $a\mathbb{D}$ (uniformly in $R_2 \geq R_1 \geq R$ as $R \to \infty$). The idea for this is to use the explicit expression for the Radon–Nikodym derivative between a zero boundary GFF and a zero boundary GFF plus an H_0^1 function; see Proposition 1.51.

The first obstacle here is that φ_0 is not actually $H_0^1(R_1\mathbb{D})$. To get around this, we introduce $\tilde{\varphi}(z) = \psi(|z|)\varphi_0(z)$ for $z \in R_1\mathbb{D}$, where $\psi \colon [0, R_1] \to [0, 1]$ is smooth, equal to 1 on $[0, a]$, and equal to 0 on $[2a, R_1]$. Note that $\tilde{\varphi} \in H_0^1(R_1\mathbb{D})$ and that $\tilde{\varphi} = \varphi_0$ in $a\mathbb{D}$. Moreover, *conditionally* on $\tilde{\varphi}$, the Radon–Nikodym derivative between the laws of $\tilde{\mathbf{h}}^{R_1}$ and $\tilde{\mathbf{h}}^{R_1} + \tilde{\varphi}$ is given by

$$Z := \frac{\exp((\tilde{\mathbf{h}}^{R_1}, \tilde{\varphi})_\nabla)}{\exp((\tilde{\varphi}, \tilde{\varphi})_\nabla)}, \tag{6.26}$$

see Proposition 1.51. To complete the proof of (6.23) it suffices (by the definition of total variation distance) to show that (6.26) tends to 1 in $L^1(\mathbb{P})$, uniformly over $R_2 \geq R_1 \geq R$ as $R \to \infty$.

To show this, we will first prove that

$$\sup_{R_1,R_2 \geq R} \mathbb{E}(e^{(\tilde{\varphi},\tilde{\varphi})_\nabla} - 1) \to 0 \tag{6.27}$$

6.4 Other boundary conditions

as $R \to \infty$. To see this, note that $\nabla \tilde{\varphi} = 0$ outside $2a\mathbb{D}$ and, for $x \in 2a\mathbb{D}$, $|\nabla \tilde{\varphi}| \leq c_1(|\nabla \varphi_0| + \sup_{x \in 2a\mathbb{D}} |\varphi_0|)$, where the constant c_1 depends only on ψ.

We now make use of the fact that φ_0 is harmonic in $2a\mathbb{D}$ and of two well-known inequalities for harmonic functions:

Lemma 6.29 *Let u be a harmonic function in $4a\mathbb{D}$. Then there exists a universal constant $C > 0$ such that*

$$\sup_{x \in 2a\mathbb{D}} |\nabla u| \leq C \sup_{x \in \partial(4a\mathbb{D})} |u|.$$

This follows for example from Theorem 7 in [Eva10, Section 2.2] and the maximum principle for harmonic functions. The second inequality we use is a consequence of Harnack's inequality:

Lemma 6.30 *Let u be a harmonic function in $8a\mathbb{D}$. Then there exists a universal constant $C > 0$ such that for any $x \in 4a\mathbb{D}$,*

$$|u(x)| \leq C|u(0)|.$$

See, for example, Theorem 11 in [Eva10, Section 2.2] for a proof when u is assumed to be non-negative; the general case follows by considering $u - \inf_{8a\mathbb{D}} u$.

Combining these two estimates, we deduce

$$\sup_{x \in 2a\mathbb{D}} |\nabla \tilde{\varphi}(x)| \leq c_2 |\varphi_0(0)| = c_2 \left| \int_{\partial(8a\mathbb{D})} \varphi_0(x) \rho(dx) \right|,$$

where ρ is the uniform measure on the circle $\partial(8a\mathbb{D})$. Therefore applying Cauchy–Schwarz,

$$\mathbb{E}(e^{(\tilde{\varphi}, \tilde{\varphi})_\nabla} - 1) \leq \mathbb{E}\left(e^{c_2 \int_{\partial(8a\mathbb{D})} |\varphi_0(w)|^2 \rho(dw)} - 1\right)$$

which by Jensen's inequality is less than

$$\mathbb{E}\left(\int_{\partial(8a\mathbb{D})} \left(e^{c_2 |\varphi_0(w)|^2} - 1\right) \rho(dw)\right) \leq \int_{\partial(8a\mathbb{D})} \mathbb{E}\left(e^{c_2 |\varphi_0(w)|^2} - 1\right) \rho(dw).$$

Note that since c_2 is a fixed constant and $\varphi_0(w)$ is a centred Gaussian random variable with arbitrarily small variance (uniformly over $\partial(8a\mathbb{D})$) as $R \to \infty$, these expectations will all be finite for $R_2 \geq R_1 \geq R$ large enough. Moreover, the right-hand side of the above expression will go to 0 uniformly in $R_2 \geq R_1 \geq R$ as $R \to \infty$. To conclude, we simply observe that conditionally on $\tilde{\varphi}$, the random variable Z from (6.26) is log normal with parameters $(-(\tilde{\varphi}, \tilde{\varphi})_\nabla/2, (\tilde{\varphi}, \tilde{\varphi})_\nabla)$. Hence

$$\mathbb{E}(|Z - 1|^2) = \mathbb{E}\Big(\mathbb{E}(|Z - 1|^2 \mid \tilde{\varphi})\Big) = \mathbb{E}(e^{(\tilde{\varphi}, \tilde{\varphi})_\nabla} - 1).$$

By (6.27), this shows that Z converges to 1 in $L^2(\mathbb{P})$ (hence in $L^1(\mathbb{P})$ and thus completes the proof of (6.23).

With (6.23) in hand, we know that $\mathbf{h}_0^R|_{a\mathbb{D}}$ (viewed as an element of $H_{\text{loc}}^{-1}(a\mathbb{D})$, modulo constants) is a Cauchy sequence with respect to total variation distance, and so its law has a limit (say μ) in total variation distance as $R \to \infty$. It remains to identify μ with the law of $\overline{\mathbf{h}}^\infty|_{a\mathbb{D}}$.

Fix a test function $\varphi \in \tilde{\mathcal{D}}_0(a\mathbb{D})$. Then

$$(h_0^R, \varphi) \sim \mathcal{N}(0, \sigma_\varphi^2) \text{ where } \sigma_\varphi^2 = \iint_{(R\mathbb{D})^2} G_0^{R\mathbb{D}}(x,y)\varphi(x)\varphi(y)\,dx\,dy.$$

Using the expression in (6.24) for $G_0^{R\mathbb{D}}$ and the fact that $\int_{a\mathbb{D}} \varphi(x)\,dx = 0$, we see that

$$\sigma_\varphi^2 \to \frac{-1}{2\pi} \iint_{\mathbb{C}^2} \log|x-y|\varphi(x)\varphi(y)\,dx\,dy = \text{Var}((\overline{\mathbf{h}}^\infty|_{a\mathbb{D}}, \varphi)). \quad (6.28)$$

Now let $\varphi_1, \ldots, \varphi_k$ be arbitrary test functions in $\tilde{\mathcal{D}}_0(a\mathbb{D})$ and fix $x_1, \ldots, x_k \in \mathbb{R}$. Consider the event $A = \{h \in H_{\text{loc}}^{-1}(a\mathbb{D}): (\mathbf{h}, \varphi_1) < x_1, \ldots, (h, \varphi_k) < x_k\}$ and let $\mathcal{A}$ denote the set of events of this form. Since the law of $\mathbf{h}_0^R$ converges to μ in total variation, we immediately deduce that for all $A \in \mathcal{A}$,

$$\mu(A) = \lim_{R \to \infty} \mathbb{P}(\mathbf{h}_0^R \in A),$$

but this also agrees with $\mathbb{P}(\overline{\mathbf{h}}^\infty|_{a\mathbb{D}} \in A)$ by (6.28) and properties of Gaussian random variables. Thus μ agrees with the law of $\overline{\mathbf{h}}^\infty|_{a\mathbb{D}}$ on $\mathcal{A}$. However the latter is a π-system which clearly generates the Borel σ-field on $H_{\text{loc}}^{-1}(a\mathbb{D})$ modulo constants, hence we conclude by Dynkin's lemma. □

As a corollary, we deduce that the whole plane GFF restricted to $\mathbb{D}$ inherits from the Dirichlet GFF the same Markov property:

Corollary 6.31 (Markov property for the whole plane GFF)

$$\mathbf{h}^\infty|_\mathbb{D} = \mathbf{h}^\mathbb{D} + \varphi,$$

where $\mathbf{h}^\mathbb{D}$ *has the law of a Dirichlet boundary condition GFF in* $\mathbb{D}$, *and* φ *is a harmonic function modulo constants that is independent of* $\mathbf{h}^\mathbb{D}$.

6.4.2 Dirichlet–Neumann GFF

Another variant of the GFF that is important, because it appears in a natural Markov property for the Neumann GFF, is the GFF with "mixed" boundary conditions. Here we will discuss one specific version, which is a distribution defined in $\mathbb{D}_+ = \mathbb{D} \cap \mathbb{H}$ and (heuristically speaking) has free/Neumann boundary conditions on $[0, 1]$ and zero/Dirichlet boundary conditions on $\partial \mathbb{D} \cap \mathbb{H}$.

6.4 Other boundary conditions 245

Definition 6.32 (Dirichlet–Neumann GFF) Suppose that $\mathbf{h}_0^{\mathbb{D}}$ is a Dirichlet GFF in $\mathbb{D}$. Then the Dirichlet–Neumann GFF, $\mathbf{h}^{DN}$, is defined to be $\sqrt{2}$ times its even part

$$\mathbf{h}^{DN} := \sqrt{2}(\mathbf{h}_0^{\mathbb{D}})_{even},$$

where

$$((\mathbf{h}_0^{\mathbb{D}})_{even}, \rho) := \frac{(\mathbf{h}_0^{\mathbb{D}}, \rho) + (\mathbf{h}_0^{\mathbb{D}}, \rho^*)}{2} \text{ for } \rho \in \mathcal{D}_0(\mathbb{D}_+)$$

which is a random distribution on $\mathbb{D}_+$.

Putting this together with Theorem 6.27 and Definition 6.25, we obtain a useful boundary Markov property for the Neumann GFF. Indeed recall that by definition of the whole plane GFF,

$$\overline{\mathbf{h}}^{\mathbb{H}}|_{\mathbb{D}_+} = \sqrt{2}\overline{\mathbf{h}}_{even}^{\infty}|_{\mathbb{D}_+},$$

where we recall that $\overline{\mathbf{h}}^{\mathbb{H}}$ is a Neumann boundary condition GFF in $\mathbb{H}$, modulo constants. On the other hand, by the Markov property of the whole plane GFF (Corollary 6.31), we also know that

$$\sqrt{2}\overline{\mathbf{h}}_{even}^{\infty}|_{\mathbb{D}_+} = \sqrt{2}(\mathbf{h}_0^{\mathbb{D}})_{even}|_{\mathbb{D}_+} + \sqrt{2}\varphi_{even}|_{\mathbb{D}_+},$$

where φ_{even} is the even part of the harmonic function φ appearing in Corollary 6.31 and is thus also harmonic over all of $\mathbb{D}$, and $(\mathbf{h}_{\mathbb{D}}^0)_{even}$ is the even part of a Dirichlet GFF in $\mathbb{D}$. By definition, the first term on the right-hand side is the Dirichlet–Neumann GFF on $\mathbb{D}_+$. We thus obtain the following (since $\sqrt{2}\varphi_{even}$ is also a harmonic function in $\mathbb{D}$, and changing notations slightly for later convenience).

Proposition 6.33 (Boundary Markov property) *Let $\mathbf{h}^{\mathbb{H}}$ be a Neumann GFF on $\mathbb{H}$ (considered modulo constants). Then we can write*

$$\mathbf{h}^{\mathbb{H}}|_{\mathbb{D}_+} = \mathbf{h}^{DN} + \varphi_{even},$$

where the two summands are independent, $\mathbf{h}^{DN}$ has the law of a Dirichlet–Neumann GFF in $\mathbb{D}_+$, and φ_{even} is a harmonic function modulo constants in $\mathbb{D}_+$, smooth up to and including $(-1, 1)$ and satisfying Neumann boundary conditions along $(-1, 1)$.

We conclude this section with one further comment that will be useful at the end of this chapter and in Chapter 8. It can be used to say, roughly speaking, that any (nice enough) way of fixing the additive constant for a Neumann GFF in $\mathbb{H}$ will produce a field with the same behaviour when looking very close to

246   *Gaussian Free Field with Neumann Boundary Conditions*

the origin. Moreover, this will still be true if we condition on the realisation of the field far away from the origin.

Lemma 6.34 *Suppose that* $\mathbf{h}$ *is a Neumann GFF in* $\mathbb{H}$, *with additive constant fixed so that it has average 0 on the upper unit semicircle (this makes sense by Example 6.23). Let* $\mathbf{h}^{\mathrm{DN}}$ *be an independent Dirichlet–Neumann GFF in* $\mathbb{D}_+$. *Then, for any* $K > 1$, *the total variation distance between*

- *the joint law of* $(\mathbf{h}|_{K\mathbb{D}_+\backslash \mathbb{D}_+}, \mathbf{h}|_{\delta \mathbb{D}_+})$ *and*
- *the (independent product) law* $(\mathbf{h}|_{K\mathbb{D}_+\backslash \mathbb{D}_+}, \mathbf{h}^{\mathrm{DN}}|_{\delta \mathbb{D}_+})$,

tends to 0 as $\delta \to 0$. *Note that the fields can be viewed as distributions here, rather than just distributions modulo constants.*

The same thing is true if $\mathbf{h}$ *is replaced by* $\mathbf{h} + \mathfrak{h}$, *where* $\mathfrak{h}$ *is harmonic in a neighbourhood of 0 and has* $\mathfrak{h}(0) = 0$.

Remark 6.35 Note that the lemma in particular implies that

$$d_{\mathrm{TV}}(\mathbf{h}|_{\delta \mathbb{D}_+}, \mathbf{h}^{\mathrm{DN}}|_{\delta \mathbb{D}_+}) \to 0 \quad \text{as } \delta \to 0.$$

So the lemma also holds if we replace the second pair $(\mathbf{h}|_{K\mathbb{D}_+\backslash \mathbb{D}_+}, \mathbf{h}^{\mathrm{DN}}|_{\delta \mathbb{D}_+})$ by $(\mathbf{h}|_{K\mathbb{D}_+\backslash \mathbb{D}_+}, \tilde{\mathbf{h}}|_{\delta \mathbb{D}_+})$, where $\tilde{\mathbf{h}}$ is independent of $\mathbf{h}$ but with the same marginal law.

Proof The proof essentially follows along the same lines as Theorem 6.27 and taking even parts. More precisely, let $R \gg K$ be large, and write

$$\tilde{\mathbf{h}}^{R\mathbb{D}} = \mathbf{h}^{R\mathbb{D}} - \mathbf{h}_1^{R\mathbb{D}}(0),$$

for $\mathbf{h}^{R\mathbb{D}}$ a Dirichlet GFF in $R\mathbb{D}$ and $\mathbf{h}_1^{R\mathbb{D}}(0)$ its unit circle average around 0. By Proposition 6.33 and considering even parts (and multiplying by a factor $\sqrt{2}$), it suffices to prove that as $R \to \infty$, and for $\mathbf{h}^{\mathbb{D}}$ a Dirichlet GFF in $\mathbb{D}$ that is independent of $\tilde{\mathbf{h}}^{R\mathbb{D}}$

$$\lim_{\delta \to 0} \lim_{R \to \infty} d_{\mathrm{TV}}\Big((\tilde{\mathbf{h}}^{R\mathbb{D}}|_{K\mathbb{D}\backslash\mathbb{D}}, \tilde{\mathbf{h}}^{R\mathbb{D}}|_{\delta \mathbb{D}}), (\tilde{\mathbf{h}}^{R\mathbb{D}}|_{K\mathbb{D}\backslash\mathbb{D}}, \mathbf{h}^{\mathbb{D}}|_{\delta \mathbb{D}}) \Big) \to 0. \quad (6.29)$$

This follows from the same argument as Theorem 6.27. That is, we use the Markov property of the GFF to write $\tilde{\mathbf{h}}^{R\mathbb{D}}|_{\mathbb{D}} = \mathbf{h} + \varphi$ where φ is measurable with respect to $\tilde{\mathbf{h}}^{R\mathbb{D}}|_{R\mathbb{D}\backslash\mathbb{D}}$ and harmonic in $\mathbb{D}$ with $\varphi(0) = 0$, while $\mathbf{h}$ is independent of $\tilde{\mathbf{h}}^{R\mathbb{D}}|_{R\mathbb{D}\backslash\mathbb{D}}$ with the law of $\mathbf{h}^{\mathbb{D}}$. To show (6.29), as in the proof of Theorem 6.27, it suffices to prove that

$$\limsup_{R \to \infty} \sup_{z \in \partial(8\delta\mathbb{D})} \mathrm{Var}(\varphi(z)) \to 0 \quad (6.30)$$

as $\delta \to 0$ (analogous to (6.24)). Indeed, once (6.30) is shown, we can use the same considerations as in Theorem 6.27 on harmonic functions to control the

Radon–Nikodym deriatve, so that (6.29) follows. On the other hand, (6.30) follows from a straightforward Green's function calculation as in the proof of Theorem 6.27.

If a deterministic function $\mathfrak{h}$ that is harmonic in a neighbourhood of the origin with $\mathfrak{h}(0) = 0$ is added to the field, one must replace φ with $\varphi + \mathfrak{h}$ in the above argument. However, since $\mathfrak{h}|_{\partial(8\delta\mathbb{D})} \to 0$ as $\delta \to 0$, the same argument goes through. □

6.5 Semicircle averages and boundary Liouville measure

Let
$$\overline{h} = \sqrt{2\pi}\overline{\mathbf{h}}, \quad (6.31)$$

where $\overline{\mathbf{h}}$ is a Neumann GFF on $\mathbb{H}$ modulo constants (recall that we use a bar in order to distinguish statements concerning the Neumann GFF modulo constants and Neumann GFFs with fixed additive constants). We will refer to both $\overline{h}$ and $\overline{\mathbf{h}}$ as "a Neumann GFF" in what follows: the use of bold font distinguishing between the different multiples as in the Dirichlet GFF setting. An immediate consequence of our previous considerations is the following fact. Recall our notation from Example 6.23 that for $x \in \mathbb{R}$ and $\varepsilon > 0$, $\rho_{x,\varepsilon}$ denotes the uniform distribution on the upper semicircle of radius ε around x (and recall also that $\rho_{x,\varepsilon} \in \mathfrak{M}_N(\mathbb{H})$).

Theorem 6.36 *For any $x \in \mathbb{R}$, the finite-dimensional distributions of the process*
$$(X_t)_{t \in \mathbb{R}} := ((\overline{h}, \rho_{x,e^{-t}} - \rho_{x,1}))_{t \in \mathbb{R}}$$
are those of a two-sided Brownian motion with variance 2 (so $\mathrm{Var}(X_t) = 2|t|$).

Note that the statement of the theorem makes sense, since for any $\varepsilon > 0$, $\rho_{x,\varepsilon} - \rho_{x,1} \in \widetilde{\mathfrak{M}}_N^{\mathbb{H}}$. By Remark 6.24, this also means that if h is a Neumann GFF in $\mathbb{H}$ with additive constant fixed *in any way*, and $h_\varepsilon(x) := (h, \rho_{x,\varepsilon})$, then
$$(h_{e^{-t}}(x) - h_1(x))_{t \in \mathbb{R}}$$
is a two-sided Brownian motion with variance 2.

Proof of Theorem 6.36 Without loss of generality, we may take $x = 0$. Then by conformal invariance (actually just scale invariance) of $\overline{h}$, it follows that X has stationary increments. Moreover, by applying the Markov property (a scaled version of Proposition 6.33) in the semidisc of radius e^{-t} about 0 for

248 Gaussian Free Field with Neumann Boundary Conditions

any t, we see that $(X_r)_{r \le t}$ and $(X_s - X_t)_{s \ge t}$ are independent. Hence, X has stationary and independent increments.

Since the increments are also Gaussian with mean zero and finite variance, it must be that $X_t = B_{\kappa t}$ for some $\kappa > 0$, where B is a standard Brownian motion. It remains to check that $\kappa = 2$, but this follows from the fact that a choice of Neumann GFF covariance in the upper half plane is given by $G^{\mathbb{H}}(0, y) = (2\pi)^{-1} \times 2\log(1/\varepsilon)$ if $|y| = \varepsilon$: see (6.14). □

Having identified the "boundary behaviour" of the Neumann GFF, we can now construct a random measure supported on the boundary of $\mathbb{H}$. As it turns out, the measure of interest to us is again given by an "exponential of the Neumann GFF", but the multiplicative factor in the exponential is $\gamma/2$ rather than γ. It might initially seem that the factor $(1/2)$ appearing in this definition comes from the fact that we are measuring lengths rather than areas. We want to emphasise that this is however *not* the real reason: instead, it is more related to the fact that the variance of the Brownian motion describing circle averages on the boundary has variance two (see Theorem 6.36). This will guarantee that the boundary measure enjoys the same change of coordinate formula as the bulk measure, as should be the case. Alternatively, this can be seen as a consequence of the fact that the so-called "quantum length of SLE" can be measured via this boundary length, and an application of the KPZ formula shows that the corresponding quantum scaling exponent is, as it turns out, always $\Delta = 1/2$.

Theorem 6.37 (Boundary Liouville measure for the Neumann GFF on $\mathbb{H}$) *Let h a Neumann GFF in $\mathbb{H}$ as in (6.31) but with additive constant fixed in an arbitrary way. Define a measure $\mathcal{V}_\varepsilon$ on $\mathbb{R}$ by setting $\mathcal{V}_\varepsilon(\mathrm{d}x) = \varepsilon^{\gamma^2/4} e^{(\gamma/2) h_\varepsilon(x)} \, \mathrm{d}x$. Then for $\gamma < 2$, the measure $\mathcal{V}_\varepsilon$ converges almost surely along the dyadic subsequence $\varepsilon = 2^{-k}$ to a non-trivial, non-atomic measure $\mathcal{V}$ called the boundary Liouville measure.*

Proof This can be proved as in Chapter 2, proof of Theorem 2.1, using the Markov property and Theorem 6.36. We leave the details as Exercise 6.6. □

Note the scaling in $\mathcal{V}_\varepsilon$, which is by $\varepsilon^{\gamma^2/4}$. This is because, as proved in Theorem 6.36 (also see the discussion below), when $x \in \mathbb{R}$ and $h = \sqrt{2\pi}\mathbf{h}$ for $\mathbf{h}$ a Neumann GFF with arbitary fixed additive constant, we have $\operatorname{Var} h_\varepsilon(x) = 2\log(1/\varepsilon) + O(1)$.

Remark 6.38 The law of $\mathcal{V}$ above *does* depend on the choice of additive constant for h. If one starts with a Neumann GFF modulo constants, then the boundary Liouville measure can be defined as a measure *up to a multiplicative constant*.

6.5 Semicircle averages and boundary Liouville measure

Note: As with $\mathcal{M}$, we will sometimes also use the notation $\mathcal{V}_h$ or $\mathcal{V}_h^\gamma$ to indicate the dependence of $\mathcal{V}$ on the underlying field h or the field h and the parameter γ.

For general D, $h = \sqrt{2\pi}\mathbf{h}$ a Neumann GFF in D with arbitrary fixed additive constant, and z, ε such that $B(z, \varepsilon) \subset D$, we can also define the circle average $(h, \rho_{z,\varepsilon}) =: h_\varepsilon(z)$. Although we use the same notation $h_\varepsilon(\cdot)$ for circle averages and semicircle averages, it should always be clear which one we refer to, depending whether the argument lies, respectively, in the bulk or on the boundary of D.

Definition 6.39 (Bulk Liouville measure for the Neumann GFF) When $h = \sqrt{2\pi}\mathbf{h}$ is a Neumann GFF with some arbitrary fixed additive constant and $\gamma < 2$, we can also define the bulk Liouville measure

$$\mathcal{M}(dz) := \lim_{\varepsilon \to 0} \varepsilon^{\gamma^2/2} e^{\gamma h_\varepsilon(z)} \, dz,$$

exactly as for the Dirichlet GFF.

The existence of this limit follows from the construction of GMC measures for general log-correlated Gaussian processes in Chapter 3. The analogue of Remark 6.38 also applies in this case.

Remark 6.40 Adapting the results of Chapter 3, it is not hard to see that for any fixed compact set of $\mathbb{R}$ (respectively D) the boundary (respectively bulk) Liouville measure of a Neumann GFF on $\mathbb{H}$ (respectively, D) will assign finite and strictly positive mass to that set with probability 1. On the other hand, note that for the Neumann GFF on D, since there is no bounded function g such that (3.1) holds uniformly on $\overline{D}$, it is not obvious that the bulk Liouville measure of D is finite. This will be addressed in Section 6.6, more specifically in Theorem 6.42.

The conformal covariance properties of the boundary and bulk Liouville measures are not quite as straightforward as for the Dirichlet GFF. The first problem is that conformal invariance of the Neumann GFF only holds when we view it as a distribution modulo constants. The second is that we have only defined the boundary measure on the domain $\mathbb{H}$, where semicircles centred on the boundary can be defined. We could extend this definition to linear boundary segments of other domains, but it is unclear what to do when the boundary of the domain is very wild.

Let us start with the bulk measure, where we only need to deal with the first problem. In this case, the statement

$$\mathcal{M}_h \circ T^{-1} = \mathcal{M}_{h \circ T^{-1} + Q \log |(T^{-1})'|}$$

250 *Gaussian Free Field with Neumann Boundary Conditions*

of Theorem 2.8 still holds (by absolute continuity with respect to the Dirichlet GFF) when $T: D \to D'$ is a deterministic, conformal isomorphism and we replace the Dirichlet GFF with $\sqrt{2\pi}$ times a Neumann GFF h in D with some arbitrary fixed additive constant. However, now $h \circ T^{-1}$ is a Neumann GFF in D' with a *different* additive constant. The exact analogue of Theorem 2.8 only holds if we consider Neumann GFFs modulo additive constants, and their associated bulk Liouville measures modulo multiplicative constants (see exercises).

Now for the boundary measure, suppose that h is a Neumann GFF on $\mathbb{H}$ with some arbitrary fixed additive constant, and $T: \mathbb{H} \to D$ is a conformal isomorphism. Then $h' := h \circ T^{-1}$ is a Neumann GFF on D with another additive constant. Moreover, if ∂D contains a linear boundary segment $L \subset \partial D \cap \mathbb{R}$, the measure $\mathcal{V}_{h'}(dx) = \lim_{\varepsilon \to 0} e^{(\gamma/2) h'_\varepsilon(x)} \varepsilon^{\gamma^2/4} \, dx$ is well defined and

$$\mathcal{V}_h \circ T^{-1} = \mathcal{V}_{h \circ T^{-1} + Q \log |(T^{-1})'|} = e^{\gamma Q \log |(T^{-1})'|} \mathcal{V}_{h'}. \qquad (6.32)$$

on L with probability 1. In fact, by [SW16, Theorem 4.3], the measure is well defined and the above formula holds with probability 1 for all conformal $T: \mathbb{H} \to D$ with $\partial D \cap \mathbb{R} \neq \emptyset$ simultaneously.

We will use this formula to *define* the boundary Liouville measure for GFF-like fields on the conformal boundary[1] of an arbitrary simply connected domain.

Definition 6.41 (Boundary Liouville measure for the GFF on D) Suppose that h is a random variable in $\mathcal{D}'_0(D)$, and that for some conformal isomorphism $T: \mathbb{H} \to D$ the field $h \circ T + Q \log |T'|$ has the law of a Neumann GFF (with some fixed additive constant) plus an almost surely continuous function on some neighbourhood in $\mathbb{H}$ of $L \subset \mathbb{R}$. Then the measure $\mathcal{V}_{h \circ T + Q \log |T'|}$ is almost surely well defined on L, and we may define

$$\mathcal{V}_h := \mathcal{V}_{h \circ T + Q \log |T'|} \circ T^{-1} \qquad (6.33)$$

to be the Liouville measure for h, on the part of the conformal boundary of D corresponding to the image of L under T. With probability one, this defines the same measure simultaneously for all choices of T.

Note that the behaviour of conformal isomorphisms near the boundary of a domain can be very wild. For instance, if D is a domain whose boundary is only Hölder with a certain exponent, then the boundary Liouville measure defined as above may not be easy to construct directly by approximation.

---

[1] The conformal boundary of a simply connected domain D, equivalent to the Martin boundary (see [BN11, Section 1.3]), is the set of limit points of D with respect to the metric $d(x, y) = d(\phi(x), \phi(y))$ for $\phi: D \to \mathbb{D}$ a conformal isomorphism.

6.6 Finiteness of the GMC on a disc with Neumann boundary conditions

When studying quantum surfaces and the mating of trees approach to Liouville quantum gravity in the following chapters, the following question comes up naturally. Let **h** be a Gaussian free field with Neumann boundary conditions on the unit disc $\mathbb{D}$, normalised to have zero unit circle average (say) and let $h = \sqrt{2\pi}\mathbf{h}$. For $0 \leq \gamma < 2$ let $\mathcal{M}(\mathrm{d}x) = \lim_{\varepsilon \to 0} \varepsilon^{\gamma^2/2} e^{\gamma h_\varepsilon(x)} \, \mathrm{d}x$ be the bulk Gaussian multiplicative chaos measure (with circle average normalisation) associated with h. This limit exists in probability as $\varepsilon \to 0$ for the topology of vague (as opposed to weak) convergence of measures, as a consequence of the standard GMC theory. Indeed, when we restrict h to an open subset A such that $\overline{A} \subset \mathbb{D}$, $h|_A$ is a logarithmically correlated Gaussian field in the sense of Section 3.2. Theorem 3.2 thus shows that the above limit exists on any such subset A of $\mathbb{D}$, from which vague convergence in probability easily follows. This defines unambiguously a GMC measure (which we denote by $\mathcal{M}$) on $\mathbb{D}$.

From the above, it is clear that $\mathbb{E}[\mathcal{M}(A)] < \infty$ for any compact subset $A \subset \mathbb{D}$. It is however not obvious if $\mathcal{M}(\mathbb{D}) < \infty$ almost surely. An easy to way to be convinced that there is something non-trivial to show is to observe that $\mathbb{E}[\mathcal{M}(\mathbb{D})] = \infty$ as soon as $\gamma \geq \sqrt{2}$. The almost sure finiteness of this measure is the content of the next theorem (shown by Huang, Rhodes and Vargas in [HRV18]).

Theorem 6.42 *Let $0 \leq \gamma < 2$ and let $\mathcal{M}$ be the above GMC measure on the disc. Then $\mathcal{M}(\mathbb{D}) < \infty$ almost surely, and $\mathbb{E}[\mathcal{M}(\mathbb{D})^\alpha] < \infty$ for $\alpha = \min(1, 1/\gamma)$.*

Proof The proof below is partly inspired by [HRV18] (but we use a different strategy for the key estimate (6.36) below: indeed, we appeal to Theorem 3.35, based on a rough scaling argument, rather than through a comparison to branching random walk and the delicate work of Madaule [Mad17]). We will show that there is some $0 < \alpha < 1$ such that $\mathbb{E}[\mathcal{M}(\mathbb{D})^\alpha] < \infty$, which clearly implies the result. For $n \geq 1$, let A_n be the dyadic annulus $A_n = \{z \in \mathbb{D}: 2^{-n-1} < 1 - |z| < 2^{-n}\}$. Since the boundary of A_n has zero Lebesgue measure, it is easy to see that $\mathcal{M}(\mathbb{D}) = \sum_{n \geq 1} \mathcal{M}(A_n)$. Moreover, since $0 \leq \alpha < 1$, the function $x > 0 \mapsto x^\alpha$ is subadditive, hence

$$\mathbb{E}[\mathcal{M}(\mathbb{D})^\alpha] \leq \sum_{n=0}^{\infty} \mathbb{E}[\mathcal{M}(A_n)^\alpha].$$

To estimate $\mathbb{E}[\mathcal{M}(A_n)^\alpha]$, we will make a comparison via Kahane's inequality (Theorem 3.19) to a field defined on the unit circle and which is scale invariant.

252   Gaussian Free Field with Neumann Boundary Conditions

However, in order to do this, it is important to first rewrite $\mathcal{M}$ in terms of the variance normalisation of h, rather than in terms of the global factor $\varepsilon^{\gamma^2/2}$ used in the statement of the theorem. For this, we observe from Example 6.12 that the covariance of h at $x \neq y$ is equal to

$$2\pi G_N^{\mathbb{D}}(x,y) = -\log(|x-y||1-\bar{x}y|).$$

Thus if $x \in \mathbb{D}$ and $B(x,\varepsilon) \subset \mathbb{D}$ we see that

$$\operatorname{Var} h_\varepsilon(x) = \log(1/\varepsilon) - \int_{\partial B(x,\varepsilon)} \log(|1-\bar{x}y|) \rho_{x,\varepsilon}(\mathrm{d}y).$$

The above integral can be computed explicitly, since the function $z \mapsto \log|1-\bar{x}z|$ is harmonic in $\mathbb{D}$, and we find that

$$\operatorname{Var} h_\varepsilon(x) = \log(1/\varepsilon) - \log(1-|x|^2).$$

Consequently, for any open subset A such that $\bar{A} \subset \mathbb{D}$,

$$\mathcal{M}(A) = \lim_{\varepsilon \to 0} \int_A e^{\gamma h_\varepsilon(x) - \frac{\gamma^2}{2}\operatorname{Var}(h_\varepsilon(x))} \frac{1}{(1-|x|^2)^{\gamma^2/2}} \, \mathrm{d}x. \quad (6.34)$$

Now, we fix $n \geq 1$ and consider $\mathcal{M}(A_n)$. We let $(Y(x))_{x \in \mathbb{U}}$ denote a rotationally invariant centered Gaussian field on the unit circle $\mathbb{U}$, with covariance K_Y satisfying

$$K_Y(x,x') = -\log|x-x'| + O(1),$$

where $|x-x'|$ denotes the distance between x and x' in $\mathbb{R}^2$ or equivalently (possibly up to a change of the implied constant $O(1)$ on the right-hand side) their distance viewed as points on the unit circle. (An example of such a field is for instance provided by an appropriate multiple of the Gaussian free field with Neumann boundary conditions on $\mathbb{D}$, normalised to have zero unit circle average, and restricted to $\mathbb{U}$, see Example 6.17.) Then an approximation to Y at scale 2^{-n} can be compared to the field h on A_n in the following sense.

Lemma 6.43   *There exists a constant $C > 0$ independent of n and $x, y \in A_n$ such that for $x = re^{i\alpha}, y = se^{i\beta} \in A_n$, if $Y_{2^{-n}}(z)$ denotes the average of Y over an arc of circle of length 2^{-n} whose midpoint is $z \in \mathbb{U}$,*

$$2\pi G_N^{\mathbb{D}}(x,y) = -\log(|x-y||1-\bar{x}y|) \geq 2\mathbb{E}[Y_{2^{-n}}(e^{i\alpha})Y_{2^{-n}}(e^{i\beta})] - C. \quad (6.35)$$

Proof   Note that $\mathbb{E}[Y_{2^{-n}}(e^{i\alpha})Y_{2^{-n}}(e^{i\beta})] = -\log(|x-y| \vee 2^{-n}) + O(1) \leq -\log|x-y| + O(1)$ so to show (6.35) it suffices to show the inequality

$$-\log|1-\bar{x}y| \geq -\log(|e^{i\alpha} - e^{i\beta}| \vee 2^{-n}) - C$$

for some $C > 0$ large enough, uniformly over $x, y \in A_n$, and over $n \geq 0$.

6.6 Finiteness of GMC with Neumann boundary conditions

For this, we observe that

$$|1 - \bar{x}y| = |1 - rse^{i(\beta-\alpha)}|$$
$$\leq |1 - e^{i(\beta-\alpha)}| + |e^{i(\beta-\alpha)} - rse^{i(\beta-\alpha)}|$$
$$= |e^{i\alpha} - e^{i\beta}| + |(1-r)(1-s) + r(1-s) + s(1-r)|$$
$$\leq |e^{i\alpha} - e^{i\beta}| + 5 \cdot 2^{-n}$$
$$\leq 6(|e^{i\alpha} - e^{i\beta}| \vee 2^{-n}),$$

from which we conclude by taking logarithms. □

From (6.35) we deduce the following. For $x \in A_n$ and $\varepsilon < 2^{-n-1}$, let $\rho_{\varepsilon,x}$ be the uniform distribution on the circle of radius ε around $x \in A_n$. Define the field $\tilde{Y}_{\varepsilon,2^{-n}}$ on A_n by setting for $x \in A_n$, $\tilde{Y}_{\varepsilon,2^{-n}}(x) = \int Y_{2^{-n}}(e^{i \arg(x')}) \rho_{\varepsilon,x}(dx')$, and note that if n is fixed but $\varepsilon \to 0$, $\tilde{Y}_{\varepsilon,2^{-n}}(x) \to Y_{2^{-n}}(e^{i \arg(x)})$ almost surely uniformly over A_n, and this limit is a field which is purely angular (i.e. depends only on the argument). Furthermore, by Lemma 6.43 there exists $C > 0$ such that for all $n \geq 0$ and $\varepsilon < 2^{-n-1}$, and for all $x, y \in A_n$,

$$\mathbb{E}[h_\varepsilon(x) h_\varepsilon(y)] \geq 2 \mathbb{E}[\tilde{Y}_{\varepsilon,2^{-n}}(x) \tilde{Y}_{\varepsilon,2^{-n}}(y)] - C.$$

Thus, applying Kahane's inequality (Theorem 3.19) to the field $h_\varepsilon(x) + \mathcal{N}$ (where $\mathcal{N}$ is an independent normal random variable) and the field $\sqrt{2}\tilde{Y}_{\varepsilon,2^{-n}}(x)$, we deduce that there exists $C > 0$ such that for all $0 < \alpha < 1$ and for all $n \geq 1$, and for all $\varepsilon < 2^{-n-1}$,

$$\mathbb{E}\left[\left(\int_{A_n} e^{\gamma h_\varepsilon(x) - \frac{\gamma^2}{2}\mathbb{E}[h_\varepsilon(x)^2]} dx\right)^\alpha\right]$$
$$\leq C \mathbb{E}\left[\left(\int_{A_n} e^{\sqrt{2}\gamma \tilde{Y}_{\varepsilon,2^{-n}}(x) - \gamma^2 \mathbb{E}[\tilde{Y}_{\varepsilon,2^{-n}}(x)^2]} dx\right)^\alpha\right].$$

Letting $\varepsilon \to 0$ in this inequality, and recalling (see (6.34)) that $\sigma(dx) = dx/(1 - |x|^2)^{\gamma^2/2} \leq C 2^{n\gamma^2/2} dx$ on A_n, we deduce that

$$\mathbb{E}[\mathcal{M}(A_n)^\alpha] \leq C 2^{n\alpha\gamma^2/2} \mathbb{E}\left[\left(\int_{A_n} e^{\sqrt{2}\gamma Y_{2^{-n}}(e^{i \arg(x)}) - \gamma^2 \mathbb{E}[Y_{2^{-n}}(e^{i \arg(x)})^2]} dx\right)^\alpha\right].$$

The integral inside the expectation on the right-hand side is the mass on A_n of a multiplicative chaos of a field which is purely angular. As a result, and by Fubini's theorem, we obtain

$$\mathbb{E}[\mathcal{M}(A_n)^\alpha] \leq C 2^{n\alpha\gamma^2/2} 2^{-n\alpha} \mathbb{E}\left[\left(\int_{\mathbb{U}} e^{\sqrt{2}\gamma Y_{2^{-n}}(z) - \gamma^2 \mathbb{E}[Y_{2^{-n}}(z)^2]} dz\right)^\alpha\right],$$

where dz denotes (with a small abuse of notation) Lebesgue measure on $\mathbb{U}$. When $\gamma < \sqrt{2}$ then we may simply take $\alpha = 1$ in the above; then the expectation on the right-hand side is constant and $\sum_{n\geq 1} \mathbb{E}[\mathcal{M}(A_n)] < \infty$ so $\mathcal{M}(\mathbb{D}) < \infty$ almost surely. This no longer applies when $\gamma \in [\sqrt{2}, 2)$. Instead, we use the fact that when $\gamma > 1$, $\sqrt{2}\gamma$ is greater than the critical value for the multiplicative chaos associated to the one-dimensional field Y, which is $\sqrt{2d} = \sqrt{2}$. As a consequence[2] of Theorem 3.35 (with $d = 1$ and γ replaced by $\sqrt{2}\gamma$), taking

$$\alpha = \sqrt{2d}/(\sqrt{2}\gamma) = 1/\gamma < 1,$$

we have that

$$\mathbb{E}\left[\left(\int_U e^{\sqrt{2}\gamma Y_{2^{-n}}(z) - \gamma^2 \mathbb{E}[Y_{2^{-n}}(z)^2]} dz\right)^\alpha\right] \leq C 2^{-n\alpha(\gamma-1)^2}. \quad (6.36)$$

for C not depending on n. Thus we obtain that

$$\mathbb{E}[\mathcal{M}(A_n)^\alpha] \leq C 2^{n\alpha\left(\frac{\gamma^2}{2} - 1 - (\gamma-1)^2\right)}.$$

Since $\frac{\gamma^2}{2} - 1 - (\gamma - 1)^2 < 0$ for all $0 \leq \gamma < 2$, this shows that $\mathbb{E}[\mathcal{M}(\mathbb{D})^\alpha] \leq \sum_n \mathbb{E}[\mathcal{M}(A_n)^\alpha] < \infty$ and hence $\mathcal{M}(D) < \infty$ almost surely. This concludes the proof of Theorem 6.42. □

6.7 Exercises

6.1 Let $D = (0, 1)^2$ be the unit square. Find an orthonormal basis of $L^2(D)$ consisting of eigenfunctions of $-\Delta$ in D, with *Neumann* boundary conditions, and write down their eigenvalues. Now consider the setting of Proposition 1.63, and set $V_N = D \cap (\mathbb{Z}^2/N)$ for $N \geq 1$. Come up with a definition of the discrete Neumann GFF in V_N with Neumann boundary conditions, and prove that it converges as $N \to \infty$ to a continuum GFF in D with Neumann boundary conditions in a suitable sense.

6.2 Consider the Hilbert space completion $(\overline{H}_\mathbb{C}, (\cdot, \cdot)_\nabla)$ of the set of smooth functions modulo constants in $\mathbb{C}$ with finite Dirichlet norm. Let

$$\overline{H}_{\text{even}} = \{h \in \overline{H}_\mathbb{C} : h(z) - h(0) = h(\overline{z}) - h(0), z \in \mathbb{C}\}$$

and likewise let

$$\overline{H}_{\text{odd}} = \{h \in \overline{H}_\mathbb{C} : h(z) - h(0) = -(h(\overline{z}) - h(0)), z \in \mathbb{C}\}$$

(note that $h(z) - h(0)$ is well defined for a function modulo constants).

---

[2] We note that this theorem was stated for Gaussian multiplicative chaos in $\mathbb{R}^d$ with respect to Lebesgue measure. However the reader can verify readily that it applies without any changes to Gaussian multiplicative chaos on the unit circle of $\mathbb{R}^2$ with respect to uniform measure, and more generally to smooth manifolds.

6.7 Exercises   255

Show that $\overline{H}_{\mathbb{C}} = \overline{H}_{\text{even}} \oplus \overline{H}_{\text{odd}}$. (Hint: orthogonality follows from the change of variables $z \mapsto \overline{z}$.)

Show that the series
$$\sum_n X_n \overline{f}_n,$$
where X_n are i.i.d. standard normal random variables, and $\overline{f}_n$ is an orthonormal basis of $\overline{H}_{\mathbb{C}}$, converges almost surely in the space $\overline{\mathcal{D}}'_0(\mathbb{C})$ and the limiting distribution modulo constants agrees in law with the whole plane GFF, $\overline{\mathbf{h}}^\infty$.

6.3 Prove (6.20) using Definition 6.25 of the whole plane GFF, and the explicit expressions for the Neumann and Dirichlet Green functions in $\mathbb{H}$.

6.4 Give a rigorous definition (i.e. as a random distribution) of the whole plane GFF with additive constant fixed to have average 0 with respect to a Riemannian volume form $g(z)\,dz$ on the Riemann sphere $\mathbb{C} \cup \{\infty\}$, in a manner analogous to Definition 6.21. Show that this is equal in law to the spherical GFF $\mathbf{h}^{\mathbb{S},g}$.

6.5 Write down a definition of the Dirichlet–Neumann GFF in the upper unit semidisc $\mathbb{D}_+$ as a stochastic process, giving an explicit expression for its covariance function in the upper unit semidisc.

6.6 Give a complete proof of Theorem 6.37, using the same strategy as in Chapter 2. Explain briefly why Theorem 3.2 does not apply directly to this setting.

6.7 Prove (6.32) (see the proof of Theorem 2.8). Check that the boundary Liouville measure ν satisfies the same KPZ relation as the bulk Liouville measure.

6.8 Let $\mathcal{V}^\sharp$ be boundary Liouville measure for a Neumann GFF on $\mathbb{H}$ with some fixed choice of additive constant, restricted to $(0,1)$, and renormalised so that it is a probability distribution. Sample x from $\mathcal{V}^\sharp$. Is the point x thick for the field (in terms of semi-circle averages)? If so, how thick?

7
Quantum Wedges and Scale-Invariant Random Surfaces

7.1 Convergence of random surfaces

Note. from this point onwards, we will almost exclusively work with the multiplicative normalisation $h = \sqrt{2\pi}\mathbf{h}$ as in (6.31) for the Neumann GFF and its variants.

Recall that we defined a *random surface* to be an equivalence class of pairs (D, h) where D is a simply connected domain and h is a distribution on D, under the relation identifying (D_1, h_1) and (D_2, h_2) if for some $f: D_1 \to D_2$ conformal,

$$h_2 = h_1 \circ f^{-1} + Q \log |(f^{-1})'|.$$

The reason for this was that if h_1 is a Dirichlet Gaussian free field in D_1, then all members of the equivalence class of (D_1, h_1) describe the same Liouville measure up to taking conformal images.

Now, we have seen that the same thing is true when h_1 is a Neumann GFF with an arbitrary fixed additive constant. And indeed if we want to view the Neumann GFF as a quantum surface, then we have to fix the additive constant, since the definition of quantum surface involves distributions and not distributions modulo constants. But the Neumann GFF is only really *uniquely* defined as a distribution modulo constants. This manifests itself in the following problem: different ways of fixing the additive constant do not yield the same quantum surface in law (see Example 7.4). So if we want to view the Neumann GFF as a quantum surface, which way of fixing the additive constant should we pick? The lack of a canonical answer to this suggests that, at least when working with quantum surfaces, it is perhaps more natural to look at a slightly different object.

Another point of view is the following: if we consider a Neumann GFF h with some arbitrary fixed additive constant, and also the field $h + C$ for some

7.1 Convergence of random surfaces 257

C, then the Liouville measure for $h + C$ is just $e^{\gamma C}$ times the Liouville measure for h. So we can think that the quantum surface described by $h + C$ represents "zooming in" on the quantum surface defined by h. (Note that this is distinct from rescaling space by a fixed factor and applying the change of coordinate formula, since by definition this does not change the quantum surface.) For some purposes, it will be natural to work with quantum surfaces that are invariant (in law) under such a zooming operation. Such a property can be thought of as a type of **scale invariance for quantum surfaces**.

In order to construct a surface $(\mathbb{H}, h)$ which does have this invariance property (once again, by Example 7.4, this is not true when h is a Neumann GFF with some arbitrary fixed additive constant), Sheffield [She16a] introduced the notion of **quantum wedge**. This will play an important role in our study of the *quantum gravity zipper* in Chapter 8. Roughly speaking, a quantum wedge is the limiting surface that one obtains by "zooming in" to a Neumann GFF close to a point on the boundary. Since this surface is obtained as a scaling limit, it automatically satisfies the desired scale invariance. Later on, we will also study scale-invariant quantum surfaces without boundaries (quantum cones) and finite volume versions of both wedges and cones (namely, so-called quantum discs and spheres).

In order to make proper sense of the above discussion, we first need to provide a notion of convergence for random surfaces – and more precisely, for surfaces with marked points.

Definition 7.1 (Quantum surface with k marked points) A quantum surface that has k marked boundary points is an equivalence class of tuples $(D, h, x_1, \ldots, x_k)$ where $D \subset \mathbb{C}$ is a domain, $h \in \mathcal{D}'_0(D)$, and $x_1, \ldots, x_k$ are points on the (conformal) boundary or in the interior of D, under the equivalence relation $(D, h, x_1, \ldots, x_k) \sim (D', h', x'_1, \ldots, x'_k)$ if and only if for some $T \colon D \to D'$ conformal with $T(x_i) = x'_i$ for $1 \le i \le k$ (note that T extends to a to map between conformal boundaries by definition):

$$h' = h \circ T^{-1} + Q \log |(T^{-1})'|. \tag{7.1}$$

We recall that $Q = Q_\gamma = 2/\gamma + \gamma/2$ depends on the LQG parameter γ, and therefore so does the notion of *quantum surface*, but we drop this from the notation for simplicity. Note that since h is assumed to be in the space of distributions $\mathcal{D}'_0(D)$, this definition may be applied to a Neumann GFF with an arbitrary fixed additive constant.

In order to define a quantum surface $\mathcal{S}$ with k marked points, we need only to specify a single equivalence class representative $(D, h, x_1, \ldots, x_k)$. We will

258 *Quantum Wedges and Scale-Invariant Random Surfaces*

call such a representative an **embedding** or **parametrisation** of the quantum surface.

This means that our usual topology on the space of distributions induces a topology on the space of quantum surfaces (with k marked points).

Definition 7.2 (Quantum surface convergence) A sequence of quantum surfaces S^n converges to a quantum surface S as $n \to \infty$ if there exist representatives $(D, h^n, x_1, \ldots, x_k)$ of S^n and $(D, h, x_1, \ldots, x_k)$ of S, such that $h_n \to h$ in the space of distributions as $n \to \infty$.

(We note that this notion of convergence is somewhat different from the notions used in [She16a] or [DMS21], but this definition has the advantage that it makes sense for all deterministic distributions viewed as quantum surfaces rather than a special class of random ones. It is also, in any case, the one that actually used to verify convergence statements for quantum surfaces, as will be discussed below.)

Now, when we are actually working with quantum surfaces, it will often be very useful to specify a surface by describing a particular *canonically chosen* embedding. Of particular interest are *random* surfaces (like the Neumann GFF or the quantum wedges defined below), and this allows for certain special choices of embedding (we will see several in the rest of this chapter and the next).

Example 7.3 Suppose that h is equal to a continuous function plus a Neumann GFF (with some fixed additive constant) in D simply connected, and $z_0, z_1 \in \partial D$ are such that the *bulk* Liouville measure $\mathcal{M}_h$ for h assigns finite mass to any finite neighbourhood of z_0, and infinite mass to any neighbourhood of z_1.[1] Then the doubly marked quantum surface (D, h, z_0, z_1) has a unique representative $(\mathbb{H}, \tilde{h}, 0, \infty)$ such that $\mathcal{M}_{\tilde{h}}(\mathbb{D} \cap \mathbb{H}) = 1$. The distribution $\tilde{h}$ is called the **canonical description** of the quantum surface in [She16a].[2] In fact, in practice, this is a difficult embedding to work with and we usually prefer others; this will be discussed further in the following section.

Example 7.4 (Zooming in – important!) Let h be a Neumann GFF in $\mathbb{H}$, for concreteness, normalised to have average zero in $\mathbb{D} \cap \mathbb{H}$. Then the canonical descriptions of h and of $h + 100$ (say), viewed as quantum surfaces in $\mathbb{H}$ with marked points at 0 and ∞, are *very different*. This can be confusing at first, since

---

[1] If h is just a Neumann GFF with arbitrary fixed additive constant in an unbounded domain D, then this will be the case by Theorem 6.42 whenever $z_1 = \infty$ and z_0 is another ($\neq \infty$) boundary point where the boundary is smooth (say).

[2] But bear in mind that it is only well defined when h is in a particular class of distributions for which the Liouville measure makes sense.

7.1 Convergence of random surfaces

h is in some sense defined "up to a constant", but the point is that "equivalence as quantum surfaces" and "equivalence as distributions modulo constants" *are not the same*.

Indeed to find the canonical description of h, we just need to find the (random) r such that $\mathcal{M}_h(B(0,r) \cap \mathbb{H}) = 1$, and apply the conformal isomorphism $z \mapsto z/r$; the resulting field

$$\tilde{h}(z) = h(rz) + Q\log(r)$$

defines the canonical description $\tilde{h}$ of the surface $(\mathbb{H}, h, 0, \infty)$. On the other hand, in order to find the canonical description of $h + 100$, we need to find $s > 0$ such that $\mathcal{M}_{h+100}(B(0,s) \cap \mathbb{H}) = 1$. That is, we need to find $s > 0$ such that $\mathcal{M}_h(B(0,s) \cap \mathbb{H}) = e^{-100\gamma}$. The resulting field,

$$h^*(z) = h(sz) + Q\log(s) + 100,$$

defines the canonical description of $(\mathbb{H}, h + 100, 0, \infty)$.

Note that in this example, the ball of radius s is much smaller than the ball of radius r. Yet in $\tilde{h}$, the ball of radius r has been scaled to become the unit disc, while in h^* it is the ball of radius s which has been scaled to become the unit disc. In other words, and since s is much smaller than r, the surface $(\mathbb{H}, h + 100, 0, \infty)$ is obtained by taking the surface $(\mathbb{H}, h, 0, \infty)$ and **zooming in** at 0.

7.1.1 Thick quantum wedges

As we will see very soon, a (thick) quantum wedge is the abstract random surface that arises as the $C \to \infty$ limit of the doubly marked surface $(h + C, \mathbb{H}, 0, \infty)$, when h is a Neumann GFF in $\mathbb{H}$ with some fixed additive constant plus certain logarithmic singularity at the origin. Thus, as explained in the example above, it corresponds to zooming in near the origin of $(h, \mathbb{H}, 0, \infty)$.

In practice however, we prefer to work with a concrete definition of the quantum wedge and then prove that it can indeed be seen as a scaling limit. It turns out to be most convenient to define it in the infinite strip $S = \mathbb{R} \times (0, \pi)$ rather than the upper half plane, with the two marked boundary points being $+\infty(=: \infty)$ and $-\infty$, respectively. A conformal isomorphism transforming $(S, \infty, -\infty)$ into $(\mathbb{H}, 0, \infty)$ is given by $z \mapsto -e^{-z}$, and under this conformal isomorphism, vertical line segments are mapped to semicircles. To be precise, the segment $\{z \colon \mathfrak{R}(z) = s\}$ is mapped to $\partial B(0, e^{-s}) \cap \overline{\mathbb{H}}$ for every $s \in \mathbb{R}$.

The following lemma will be used repeatedly in the rest of this chapter and the next.

260 *Quantum Wedges and Scale-Invariant Random Surfaces*

Lemma 7.5 (Radial decomposition) *Let S be the infinite strip $S = \{z = x + iy \in \mathbb{C}: y \in (0, \pi)\}$. Let $\overline{\mathcal{H}}_{\mathrm{rad}}$ be the subspace of $\overline{H}^1(S)$ obtained as the closure of smooth functions which are constant on each vertical segment, viewed modulo constants. Let $\mathcal{H}_{\mathrm{circ}}$ be the subspace obtained as the closure of smooth functions which have mean zero on all vertical segments. Then*

$$\overline{H}^1(S) = \overline{\mathcal{H}}_{\mathrm{rad}} \oplus \mathcal{H}_{\mathrm{circ}}.$$

Proof Suppose that g_1 is a smooth function modulo constants in S, that is constant on vertical lines, and that g_2 is a smooth function in S that has mean zero on every vertical line. Then it is straightforward to check that $\iint_S \nabla g_1 \cdot \nabla g_2 = 0$. Indeed $\nabla g_1 = (\partial_x g_1, 0)$ and $\nabla g_2 = (\partial_x g_2, \partial_y g_2)$ where the partial derivative $\partial_x g_1$ is constant on vertical lines and $\partial_x g_2$ has average 0 on vertical lines. This means that $\nabla g_1 \cdot \nabla g_2$ has average 0 on every vertical line, and consequently has average 0 over S. By definition of $\overline{\mathcal{H}}_{\mathrm{rad}}$ and $\mathcal{H}_{\mathrm{circ}}$ (as closures with respect to $(\cdot, \cdot)_\nabla$) the two spaces are therefore orthogonal with respect to $(\cdot, \cdot)_\nabla$.

To check that they span $\overline{H}^1(S)$, note that if we consider smooth $f \in \mathcal{D}(S)$ and we set $f_{\mathrm{rad}}(z)$ to be the average of f on the line $\Re z + i[0, 2\pi]$, then $f_{\mathrm{rad}} \in \overline{\mathcal{H}}_{\mathrm{rad}}$. Moreover, defining $f_{\mathrm{circ}} = f - f_{\mathrm{rad}}$, it is clear that $f_{\mathrm{circ}} \in \mathcal{H}_{\mathrm{circ}}$. From this it follows that if $f \in \overline{H}^1(S)$, then we can write $f = \lim_n f_n$ for a sequence $(f_n)_n \in \mathcal{D}(S)$, and by decomposing each f_n we have $\lim_n f_n = \lim_n ((f_n)_{\mathrm{rad}} + (f_n)_{\mathrm{circ}})$. By orthogonality, the sequences $(f_n)_{\mathrm{rad}}$ and $(f_n)_{\mathrm{circ}}$ are each Cauchy and have individual limits $f_{\mathrm{rad}} \in \overline{\mathcal{H}}_{\mathrm{rad}}$ and $f_{\mathrm{circ}} \in \mathcal{H}_{\mathrm{circ}}$. Hence $f = f_{\mathrm{rad}} + f_{\mathrm{circ}}$, and the two spaces do indeed span $\overline{H}^1(S)$. □

Similarly to the domain Markov property for the Neumann GFF (that we saw arises from the orthogonal decomposition $\overline{H}^1(D) = H_0^1(D) \oplus \overline{\mathrm{Harm}}(D)$), this results in another representation of the Neumann GFF on S modulo constants. Namely, as a stochastic process indexed by $\widetilde{\mathfrak{M}}_N^S$, it can be written as $\overline{h} = \overline{h}_{\mathrm{rad}}^S + h_{\mathrm{circ}}^S$, where

- $\overline{h}_{\mathrm{rad}}^S, h_{\mathrm{circ}}^S$ are independent;
- $\overline{h}_{\mathrm{rad}}^S(z) = \overline{B}_{2\Re(z)}$, where $\overline{B}$ is a standard Brownian motion modulo constants (by Theorem 6.36 and conformal invariance);
- $h_{\mathrm{circ}}^S(z)$ has mean zero on each vertical segment.

To justify the above, notice that given $\overline{h}$, we can define $\overline{h}_{\mathrm{rad}}^S$ to be constant on each vertical segment with value equal to the average of $\overline{h}$ on that segment. Then we know by Theorem 6.36 and conformal invariance that $\overline{h}_{\mathrm{rad}}^S$ has the law described in the second bullet point. Thus it remains to justify is that

7.1 Convergence of random surfaces

$(\bar{h}-\bar{h}_{\text{rad}}^S,\rho)$ and $(\bar{h}_{\text{rad}}^S,\rho)$ are independent for any $\rho \in \tilde{\mathfrak{M}}_N^S$. For this, observe that if $(\bar{h}_n)_n$ are as in Theorem 6.18 (but multiplied by $\sqrt{2\pi}$) then $(\bar{h}_n,\rho)$ converges in $L^2(\mathbb{P})$ and in probability, to a random variable with the law of $(\bar{h},\rho)$. Moreover, $((\bar{h}_n)_{\text{rad}},\rho)$ and $(\bar{h}_n - (\bar{h}_n)_{\text{rad}},\rho)$ are independent for every n, with $\text{Var}((\bar{h}_n)_{\text{rad}},\rho) \le \text{Var}((\bar{h}_{\text{rad}}^S,\rho))$ and $\text{Var}(\bar{h}_n - (\bar{h}_n)_{\text{rad}},\rho) \le \text{Var}(\bar{h} - \bar{h}_{\text{rad}}^S,\rho)$. This implies that $(\bar{h} - \bar{h}_{\text{rad}}^S,\rho)$ and $(\bar{h}_{\text{rad}}^S,\rho)$ are uncorrelated and hence, by Gaussianity, independent.

Note that the $\bar{h}_{\text{rad}}^S$ part is defined modulo constants, while the h_{circ}^S part really has additive constant fixed. As such, we can actually define h_{circ}^S to be a stochastic process indexed by $\mathfrak{M}_N^S$ rather than just $\tilde{\mathfrak{M}}_N^S$.[3]

Also observe that all the roughness of h is contained in the h_{circ}^S part, as $\bar{h}_{\text{rad}}^S$ is a nice continuous function modulo constants. On the upper half plane, this would correspond to a decomposition of $\bar{h}$ into a part which is a radially symmetric continuous function (modulo constants), and one which has zero average on every semicircle (hence the notation).

Remark 7.6 (Translation invariance of h_{circ}^S) Note that the Neumann GFF h on S is invariant under horizontal translations (modulo constants), as it is conformally invariant (modulo constants). Since the radial part is simply a two-sided Brownian motion, the translation invariance of this part modulo constants is also clear. Thus, we may deduce that the circular part h_{circ}^S is translation invariant as well. (Note that the additive constant here is specified).

Let $0 \le \alpha \le Q = 2/\gamma + \gamma/2$. We will define an α-**(thick) quantum wedge** to be a quantum surface $(S, h, +\infty, -\infty)$, where the law of the representative field h on S will be defined by specifying, separately, its averages on vertical line segments, and what is left when we subtract these. The second of these components will be an element of $\mathcal{H}_{\text{circ}}$, having exactly the same law as the corresponding projection h_{circ}^S of the standard Neumann GFF. It is only the "radially symmetric part" which is different.

The bound $\alpha \le Q$ corresponds to the fact that we are defining a so-called "thick" quantum wedge. When $\alpha > Q$ it is possible to define something called a "thin" quantum wedge, as introduced in [DMS21], but we will discuss this separately later on.

---

[3] Concretely, we can set h_{circ}^S to be $h - h_{\text{rad}}$, where h is $\bar{h}$ with additive constant fixed so that its average on $(0, i\pi)$ is zero, as in Definition 6.21, and h_{rad} is constant on each vertical segment with value equal to the average of h on that segment.

262 Quantum Wedges and Scale-Invariant Random Surfaces

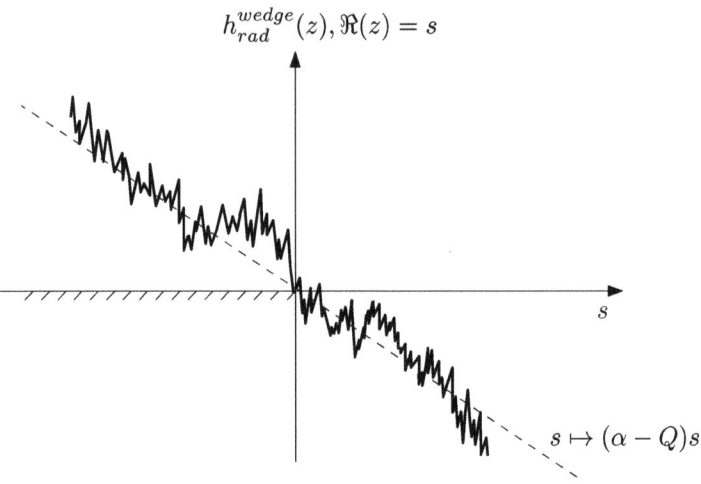

Figure 7.1 Schematic representation of the radially symmetric part of a quantum wedge in a strip. When $s < 0$, the function is conditioned to be positive.

Definition 7.7 Let

$$h_{\text{rad}}^{\text{wedge}}(z) = \begin{cases} B_{2s} + (\alpha - Q)s & \text{if } \Re(z) = s \text{ and } s \geq 0 \\ \widehat{B}_{-2s} + (\alpha - Q)s & \text{if } \Re(z) = s \text{ and } s < 0, \end{cases} \quad (7.2)$$

where $B = (B_t)_{t \geq 0}$ is a standard Brownian motion, and $\widehat{B} = (\widehat{B}_t)_{t \geq 0}$ is an independent Brownian motion conditioned so that $\widehat{B}_{2t} + (Q - \alpha)t > 0$ for all $t > 0$ (see below for what this means precisely, and Figure 7.1 for an illustration).

Let $h_{\text{circ}}^{\text{wedge}}$ be a stochastic process indexed by $\mathfrak{M}_N^S$, that is independent of $h_{\text{rad}}^{\text{wedge}}$ and has the same law as h_{circ}^S. Finally, set $h^{\text{wedge}} = h_{\text{rad}}^{\text{wedge}} + h_{\text{circ}}^{\text{wedge}}$ (which, since $h_{\text{rad}}^{\text{wedge}}$ is just a continuous function, can again be defined as a stochastic process indexed by $\mathfrak{M}_N^S$). We call $h^{\text{wedge}} = h_{\text{rad}}^{\text{wedge}} + h_{\text{circ}}^{\text{wedge}}$ the α**-quantum wedge** field in $(S, +\infty, -\infty)$.

The α-quantum wedge itself is defined to be the doubly marked quantum surface represented by $(S, h^{\text{wedge}}, +\infty, -\infty)$ (see Remark 7.8).

The conditioning defining the process $\widehat{B}$ above bears on an event of probability zero and thus requires a careful definition. For instance, when $\alpha = Q$, this is the limit (as $C \to \infty$ and then $\varepsilon \to 0$) of a (speed two) Brownian motion starting from ε condition to hit C before zero. This turns out to be identical to a (speed two) Bessel process of dimension 3 (this classical fact can be proved using standard techniques in stochastic calculus and in particular Doob's h-

transform; for a proof see, for example, Lemma 8.2 in [BN11]). On the other hand, if $\alpha < Q$, then the process $\hat{B}$ can be defined as the limit, as $\varepsilon \to 0$, of a (speed two) Brownian motion with drift $(Q - \alpha)$, started from $\varepsilon > 0$ and conditioned to stay positive for all time. As observed in particular by Williams [Wil74], under this conditioning, $(\hat{B}_t + (Q - \alpha)t)_{t \geq 0}$ (equivalently, $h_{\text{rad}}(z)$) for $\mathfrak{R}(z) = -t$, viewed as a function of $t \geq 0$), is a strong Markov process with generator $\frac{d^2}{dx^2} + b \coth(bx)\frac{d}{dx}$, where $b = Q - \alpha > 0$. This again follows from Doob's h-transform after observing that $x \mapsto \phi(x) = e^{-2bx}$ is harmonic for Brownian motion with drift $-b$.

However, a simpler definition of $h_{\text{rad}}^{\text{wedge}}(z)$ for $\mathfrak{R}(z) = t$ which works for all $t \in \mathbb{R}$ whether positive or negative, (and works for all $\alpha \leq Q$, including $\alpha = Q$), is the following. Fix $C > 0$, and let Y_t be a Brownian motion with speed two, starting from C at time 0, and with drift $-b = -(Q - \alpha) \leq 0$. That is, $Y_t = B_{2t} - bt$. Let $\tau = \inf\{t \geq 0 \colon Y_t = 0\}$. Then we set (for $\mathfrak{R}(z) = t$)

$$h_{\text{rad}}^{\text{wedge}}(z) := \tilde{Y}_t; \text{ where } \tilde{Y}_t = \lim_{C \to \infty} Y_{\tau+t}. \qquad (7.3)$$

As this procedure provides a consistent definition of $(Y_{\tau+t})_{t \in \mathbb{R}}$ as C increases (thanks to the strong Markov property of Y), it is obvious that this limit exists. Note that this provides a definition which works for all $t \in \mathbb{R}$ (i.e. there is no need to distinguish between $t \geq 0$ and $t \leq 0$). We will refer to $\tilde{Y}$ as Brownian motion with speed two and drift $-b = -(Q - \alpha)$, from $+\infty$ to $-\infty$ (although this terminology is a slight abuse of language when $\alpha = Q$). It turns out to give the same process as the one described above by the conditioning using Doob's h-transform; however this equivalence is not obvious. Instead this is a consequence of William's path reversal theorem [Wil74, Theorem 2.5], which implies that the time-reversal of a Brownian motion with drift $-b$, started from a value $C > 0$, considered until its first hitting time of 0, has the same law as the process with generator $\frac{1}{2}\frac{d^2}{dx^2} + b\coth(bx)\frac{d}{dx}$, considered until its last hitting time of C (the extra factor 1/2 in the generator comes from the speed of Brownian motion). In later arguments, we will see that using the definition (7.3) leads to simple proofs, and these do not rely on the above-mentioned equivalence, so we bypass in this manner the use of results requiring delicate stochastic calculus techniques. This is therefore the definition we adopt.

We warn the reader that the process $\tilde{Y}$ defined in (7.3) is somewhat different (even modulo horizontal translation) from a two-sided Brownian motion with drift $-b$ (i.e. $(B_{2t} - bt)_{t \in \mathbb{R}}$, where $(B_t)_{t \in \mathbb{R}}$ is a two-sided Brownian motion). Indeed, the latter has a local time at zero given by the sum of *two* independent exponential random variables (corresponding to the local time accumulated in the negative and positive times, respectively), whereas $\tilde{Y}$ has a local time at the origin which is a single exponential random variable. Essentially, the difference comes from the fact that the latter can be viewed as the process $\tilde{Y}$ biased by its local time at the origin.

To emphasise once more, our definition of quantum wedge fields is such that they come with a specific way of fixing the additive constant; in other words, they are stochastic processes indexed by $\mathfrak{M}_N^S$ rather than just $\tilde{\mathfrak{M}}_N^S$. We will *not* want to consider these wedge fields modulo constants Figure 7.1.

Remark 7.8 Observe that by the corresponding property of the Neumann GFF, if we restrict the index set of h defined above to $\mathcal{D}_0(S)$, it gives rise to a stochastic process having a version that almost surely defines a distribution in S, that is, an element of $\mathcal{D}'_0(S)$. In fact, by Remark 6.7, it is almost surely an element of $H^{-1}_{\text{loc}}(S)$.

We can then define the α-quantum wedge as a doubly marked random surface, by letting it be the equivalence class of $(S, h^{\text{wedge}}, +\infty, -\infty)$, in the sense of Definition 7.1.[4] Using the change of coordinate formula, we could thus also view it as being parametrised by the upper half plane, and we would obtain a distribution $\hat{h}$ defined on $\mathbb{H}$. However, the expression for $\hat{h}$ is not particularly nice, and makes the following proofs more difficult to follow, which is why we usually take the strip S as our domain of reference.

Note also that, when we parametrise this quantum surface by the infinite strip $S = \{x + iy : x \in \mathbb{R}, y \in (0, \pi)\}$ with the two marked points at $+\infty$ and $-\infty$, that is, restrict to equivalence class representatives of the form $(S, h^{\text{wedge}}, +\infty, -\infty)$, where h^{wedge} is a field on S, then there is still one degree of freedom in the choice of the field h^{wedge}, given by translations. Namely, because the term $Q \log |(T^{-1})'|$ in the change of coordinates formula (7.1) disappears when T is a translation,

$$(S, h^{\text{wedge}}, +\infty, -\infty) \text{ and } (S, h^{\text{wedge}}(\cdot + a), +\infty, -\infty)$$

are equivalent as doubly marked quantum surfaces, for any $a \in \mathbb{R}$. In other words, we should not distinguish between h^{wedge} and $h^{\text{wedge}}(+ \cdot a)$. In Definition 7.7 of the thick quantum wedge, we chose a specific representative field h^{wedge} by fixing the horizontal translation so that the radial part of the field hit 0 for the first time at 0. But, in light of the discussion above, we could alternatively think of the wedge as being the doubly marked quantum surface, which when parametrised by the strip S with marked points at $+\infty, -\infty$, is represented by $h^{\text{wedge}}_{\text{circ}} + \tilde{Y}_{\Re(z)}$ *modulo translation*, where $h^{\text{wedge}}_{\text{circ}}$ has the law of $h^{\text{GFF}}_{\text{circ}}$, and $\tilde{Y}$

---

[4] Note that this is the same definition as in [She07, DMS21] for the thick quantum wedge as a doubly marked quantum surface, but in [She07, DMS21], it is represented by $(S, \bar{h}, -\infty, +\infty)$ instead, where $\bar{h}(\cdot) = h(-\cdot)$.

7.1 Convergence of random surfaces

is a Brownian motion with speed two and drift $-b = -(Q - \alpha)$, from $+\infty$ to $-\infty$, that is has the law (7.3), as per Definition 7.7.

In fact later, it will be slightly more convenient to (equivalently) parametrise the wedge by the strip S with marked points $-\infty, +\infty$ (i.e. switched). In this case, the thick quantum wedge is represented by the field

$$h_{\text{circ}} + (B_{2\Re(\cdot)} + (Q - \alpha)\Re(\cdot)) \text{ modulo horizontal translation}, \quad (7.4)$$

where the radial part $(B_{2\Re(\cdot)} + (Q - \alpha)\Re(\cdot))$ is a Brownian motion with speed two and drift $b = (Q - \alpha)$ from $-\infty$ to $+\infty$, (i.e. with drift of the opposite sign), using the terminology (7.3).

We note finally that there is an embedding of a thick quantum wedge in the upper half plane for which the associated field has a nice description in $\mathbb{D}_+$:

Remark 7.9 Note that when $s \geq 0$, $h_{\text{rad}}^{\text{wedge}}(s)$ is simply a Brownian motion with a drift of coefficient $\alpha - Q \leq 0$. This means that, embedding in the upper half plane using $z \mapsto -e^{-z}$ and taking into account the conformal change of variables formula, the obtained representative $(\mathbb{H}, \hat{h}^{\text{wedge}}, 0, \infty)$ of the quantum wedge has a logarithmic singularity of coefficient α near zero. In fact,

$$\hat{h}^{\text{wedge}}(z)\big|_{\mathbb{D}_+} \stackrel{(\text{law})}{=} (h + \alpha \log 1/|z|)\big|_{\mathbb{D}_+},$$

where h has the law of a Neumann GFF in $\mathbb{H}$, normalised so that it has zero average on the semicircle of radius 1.

Remark 7.10 (Unit circle embedding) Suppose that h is a distribution on S of the form $h_{\text{circ}}^S + h_{\text{rad}}$ where h_{rad} is constant on each vertical segment $\{\Re(z) = s\}$, and these constant values define a continuous function $h_{\text{rad}}(s)$ that is positive for all s less than some $s_0 \in \mathbb{R}$ but touches zero at some $s_1 \geq s_0$. Consider the unique translation of the strip so that the image of h_{rad} under this translation hits 0 for the first time at $s = 0$, and let $\tilde{h}$ be the image of h after applying this translation, mapping to $\mathbb{H}$ using the map $z \mapsto -e^{-z}$ and applying the change of coordinates formula.

If a quantum surface has a representative of the form $(S, h, \infty, -\infty)$ with h as above, then we call $(\mathbb{H}, \tilde{h}, 0, \infty)$ the **unit circle embedding** of this quantum surface.

The unit (semi)circle clearly plays a special role in this embedding since it is the image of the vertical segment with $\Re(z) = 0$ on the strip. Note that if $\hat{h}^{\text{wedge}}$ is defined as in Remark 7.9, then $(\mathbb{H}, \hat{h}^{\text{wedge}}, 0, \infty)$ is the unit circle embedding of the α-quantum wedge.

We can now state the result about the quantum wedge being the scaling limit of a Neumann GFF with a logarithmic singularity near the origin.

266 Quantum Wedges and Scale-Invariant Random Surfaces

Theorem 7.11 *Fix $0 \le \alpha < Q$. Then the following hold:*

(i) Let $\tilde{h}$ be a Neumann GFF in $\mathbb{H}$ with additive constant fixed so that its upper unit semicircle average is equal to 0, and set $h(z) = \tilde{h}(z) + \alpha \log(1/|z|)$. Let h^C be such that $(\mathbb{H}, h^C, 0, \infty)$ is the unit circle embedding of $(\mathbb{H}, h+C, 0, \infty)$, and let $(\mathbb{H}, \hat{h}^{\text{wedge}}, 0, \infty)$ be the unit circle embedding of a quantum wedge. Then for any $R > 0$, $h^C|_{RD_+}$ converges in total variation distance to $\hat{h}^{\text{wedge}}|_{RD_+}$ as $C \to \infty$.

(ii) If $(\mathbb{H}, h^{\text{wedge}}, 0, \infty)$ is an α-quantum wedge, then $(\mathbb{H}, h^{\text{wedge}}, 0, \infty)$ and $(\mathbb{H}, h^{\text{wedge}} + C, 0, \infty)$ have the same law as quantum surfaces.

To summarise in the language introduced at the start of this chapter: (ii) says that a quantum wedge is invariant under rescaling, while (i) says that a quantum wedge is the limit, zooming in near zero, of the surface described by $\tilde{h}(z) + \alpha \log 1/|z|$. The fact that the convergence holds in the strong sense of total variation is very useful (as we shall see in Chapter 8). Note that for this theorem, we have to restrict to the case $\alpha < Q$.

Proof Note that point (ii) of the theorem follows from point (i), since the limit in law of $(\mathbb{H}, h + C, 0, \infty)$ and $(\mathbb{H}, h + C + C', 0, \infty)$ as $C \to \infty$, must coincide (in other words, scaling limits are by definition invariant under scaling). It thus suffices to prove point (i).

For this, we embed the field h into the strip S using the conformal isomorphism $z \in S \mapsto \phi(z) = -e^{-z} \in \mathbb{H}$, and apply the change of coordinates formula (7.1). Then the law of the resulting field can be written as

$$h = h^S_{\text{circ}} + h_{\text{rad}}, \tag{7.5}$$

where h^S_{circ} has the law specified by the radial decomposition of the Neumann GFF (see after Lemma 7.5); h_{rad} is independent of h^S_{circ}; and finally h_{rad} is constant equal to $B_{2s} + (\alpha - Q)s$ on each vertical segment $\{\Re(z) = s\}$, with B a standard two-sided Brownian motion equal to 0 at time 0. Here the $+\alpha$ comes from the logarithmic singularity of h, and the $-Q$ from the change of coordinates formula.

Observe that, by definition, the unit circle embedded field of $h + C$ is the image (after mapping back to $\mathbb{H}$ using the change of coordinates formula with ϕ) of $h^S_{\text{circ}}(\cdot + s_C) + h_{\text{rad}}(\cdot + s_C) + C$, where

$$s_C = \inf\{s \in \mathbb{R}\colon B_{2s} + (\alpha - Q)s = -C\}$$

is the first hitting time of $-C$ by the radial part of h, and is independent of h^S_{circ}.

Since h^S_{circ} is independent of s_C and is translation invariant, and since h^{wedge} has an angular part also independent of its radial part and with the same law, it

7.1 Convergence of random surfaces

suffices to show that as $C \to \infty$,

$$C + h_{\text{rad}}(\cdot + s_C) \to h_{\text{rad}}^{\text{wedge}}(\cdot), \tag{7.6}$$

in total variation, uniformly over intervals of $\mathbb{R}$ of the form $[-\log R, \infty)$. However, from our definition of $h_{\text{rad}}^{\text{wedge}}(\cdot)$ in (7.3), the two objects can be exactly coupled on $[-\log R, \infty)$ provided that $s_C \geq \log R$. As this is an event of probability tending to 1 as $C \to \infty$ and R is fixed, we deduce the claim. □

In fact, part (i) of Theorem 7.11 still holds if we take an arbitrary choice of additive constant for the Neumann GFF $\tilde{h}$. We state something slightly stronger in the next lemma: not only does this still hold if we choose the additive constant of the field slightly differently, but also the way the scaling needs to be done when we zoom in is asymptotically independent of the initial choice of this additive constant.

Lemma 7.12 *Let $\tilde{h}, h$ be as in Theorem 7.11. Also let $\tilde{k}$ be a Neumann GFF in $\mathbb{H}$ with additive constant fixed so that $(\tilde{k}, \rho_0)$ is equal to 0 for some $\rho_0 \in \mathfrak{M}_N^\mathbb{H} \setminus \mathfrak{M}_N^\mathbb{H}$ (i.e. with non-zero average) that is compactly supported away from the origin, and let $\mathfrak{h}$ be a deterministic function, that is harmonic in a neighbourhood of the origin. Set $k(z) = \tilde{k}(z) + \mathfrak{h}(z) + \alpha \log(1/|z|)$. Let φ_C (resp. ψ_C), be the scaling maps such that $h^C := h \circ \varphi_C^{-1} + Q \log |(\varphi_C^{-1})'| + C$ (resp. $k^C := k \circ \psi_C^{-1} + Q \log |(\psi_C^{-1})'| + C)$ is the unit circle embedding of $h + C$ (resp. $k + C$). Then for any bounded set $K \subset \mathbb{H}$,*

$$d_{\text{TV}}\left((\varphi_C, h^C|_K), (\psi_C, k^C|_K)\right) \to 0$$

as $C \to \infty$.

Proof We use the same notations as in Theorem 7.11. We first check that for any fixed $a \in \mathbb{R}$,

$$d_{\text{TV}}\left((\varphi_C, h^C|_K), (\varphi_{C+a}, h^{C+a}|_K)\right) \to 0 \tag{7.7}$$

as $C \to \infty$. This will in fact follow from the observation that

$$d_{\text{TV}}(s_C, s_{C+a}) \to 0 \tag{7.8}$$

as $C \to \infty$, for any fixed $a \in \mathbb{R}$. Indeed, once (7.8) is known, let us fix $\varepsilon > 0$, and suppose that $K \subset B(0, R)$ for some $R > 0$. (7.8) means that for large enough C we can choose a coupling of $s_{C/2}$ and $s_{C/2+a}$ which succeeds with probability at least $1 - \varepsilon$ (and we may further assume that $s_{C/2} = s_{C/2+a} > 0$). Then, conditionally on this event and using the Markov property of Brownian motion, we can further couple the two processes $(h_{\text{rad}}(s_{C/2} + \cdot))$ and $(h_{\text{rad}}(s_{C/2+a} + \cdot))$ so that they differ exactly by the constant a on $\{\Re(\cdot) \geq 0\}$. In particular, on this event of probability $> 1 - \varepsilon$, the radial parts hit $-C$ and $-(C+a)$, respectively, at the same time, say s, and unless $|s_{C/2} - s| < \log R$ (which has probability tending to zero as $C \to \infty$) the radial parts agree up to the additive constant a on $[s - \log R, \infty)$. The expression (7.7) then follows by definition of h^C and h^{C+a} (since the scaling maps φ_C and φ_{C+a} depend only on s_C and s_{C+a}, respectively).

Let us therefore check (7.8). First, we may instead work with τ_C and τ_{C+a} which are the first hitting times of $-C$ and $-(C+a)$ *after time 0*, for a Brownian motion equal to 0 at 0 and with drift $-b := -(Q - \alpha) < 0$ (so $b > 0$). (This suffices since the probability that s_C or s_{C+a} is negative converges to 0 as $C \to \infty$.) Then we observe, by an elementary calculation using the reflection principle and Girsanov's transform that the density f_{τ_C} of τ_C with respect to Lebesgue measure on $[0, \infty)$ is given by

$$f_{\tau_C}(t) = \frac{C}{\sqrt{4\pi t^3}} e^{-\frac{(C-bt)^2}{4t}}.$$

It is easy to check from this expression that

$$\int_{C/b+C^{2/3}}^{\infty} f_{\tau_C}(t)\, dt \leq \int_{C/b+C^{2/3}}^{\infty} f_{\tau_{C+a}}(t)\, dt \to 0$$

as $C \to \infty$. Moreover,

$$\int_0^{C/b-C^{2/3}} f_{\tau_{C+a}}(t)\, dt \leq \int_0^{C/b-C^{2/3}} f_{\tau_C}(t)\, dt \to 0$$

as $C \to \infty$, since the right-hand side can be bounded above by the probability that the maximum of Brownian motion at time less than a constant times C is bigger than a constant times $C^{2/3}$. Thus, to prove (7.8) it is enough to show that

$$\sup_{t \in [C/b - C^{2/3}, C/b + C^{2/3}]} \left| \frac{f_{\tau_{C+a}}(t)}{f_{\tau_C}(t)} - 1 \right| \to 0$$

as $C \to \infty$. For this, we use that

$$\left| \frac{f_{\tau_{C+a}}(t)}{f_{\tau_C}(t)} - 1 \right| = \left| 1 - (1 + \frac{a}{C}) e^{a(-\frac{a}{2t} - \frac{C}{t} + b)} \right|$$

and that $|(-\frac{a}{2t} - \frac{C}{t} + b)| = O(C^{-1/3})$ when $t \in [C/b - C^{2/3}, C/b + C^{2/3}]$.

This concludes the proof of (7.7). It is easy to see that this implies that if X is an independent finite random variable,

$$d_{TV}\left((\varphi_C, h^C|_K), (\varphi_{C+X}, h^{C+X}|_K)\right) \to 0 \qquad (7.9)$$

as $C \to \infty$. To see this, suppose that X_n, Y_n are two random variables on a

7.1 Convergence of random surfaces 269

Polish space such that $d_{TV}(X_n + a, Y_n) \to 0$ as $n \to \infty$. Let μ_n^a denote the law of $X_n + a$ and let ν_n denote the law of Y_n. Let $\lambda(da)$ denote the law of X. Then

$$\sup_{A \in \mathcal{F}} |\mathbb{P}(X_n + X \in A) - \mathbb{P}(Y_n \in A)| = \sup_{A \in \mathcal{F}} \left| \int \mu_n^a(A) \lambda(da) - \nu_n(A) \right|$$

$$\leq \sup_{A \in \mathcal{F}} \left| \int [\mu_n^a(A) - \nu_n(A)] \lambda(da) \right|$$

$$\leq \int \sup_{A \in \mathcal{F}} |\mu_n^a(A) - \nu_n(A)| \lambda(da)$$

$$= \int d_{TV}(X_n + a, Y_n) \lambda(da).$$

The total variation distance inside the integral tends to zero pointwise by assumption, and is bounded by one. The right-hand side thus tends to zero by dominated convergence, as desired. This proves (7.9).

Next, we let k, h be as defined in the statement of the lemma, and we assume without loss of generality that $\int \rho_0 = 1$. This means that $\tilde{k}$ is equal in law to $\tilde{h} - (\tilde{h}, \rho_0)$, where ρ_0 is supported away from the origin. By Lemma 6.34 (in fact, Remark 6.35), we then have

$$d_{TV}(k|_{\delta \mathbb{D}_+}, (h + X)|_{\delta \mathbb{D}_+}) \to 0$$

as $\delta \to 0$, where X is random but independent of h, with the law of $\mathfrak{h}(0) + (\tilde{h}, \rho_0)$. Now let $\varepsilon > 0$ be arbitrary. First, for $\delta > 0$ small enough, we can construct a coupling of k and $h + X$ so that they agree on $\delta \mathbb{D}_+$ with probability at least $1 - \varepsilon$. By taking C large enough, we may also assume that $(\varphi_{C+X}, h^{C+X}|_K)$ and $(\psi_C, k^C|_K)$ depend only on $h|_{\delta \mathbb{D}_+}$ and $k|_{\delta \mathbb{D}_+}$ on this event. Since this gives a coupling under which $(\psi_C, k^C|_K)$ and $(\varphi_{C+X}, h^{C+X}|_K)$ agree with probability at least $1 - \varepsilon$, this implies that

$$d_{TV}\left((\psi_C, k^C|_K), (\varphi_{C+X}, h^{C+X}|_K)\right) \to 0$$

as $C \to \infty$. We conclude by (7.7) and the triangle inequality, since X is almost surely finite. □

As an example of application of this result, we mention that the quantum wedge field $\hat{h}^{\text{wedge}}$ with parameter $\alpha < Q$ (in the unit circle embedding) has a well-defined Liouville bulk measure $\mathcal{M}_{\hat{h}^{\text{wedge}}}$ and boundary measure $\mathcal{V}_{\hat{h}^{\text{wedge}}}$, since it can be coupled with arbitrarily high probability to a Neumann GFF (with a given logarithmic singularity) plus a constant. Note that these measures are locally finite and atomless almost surely, by the results of Chapter 3 (with base measure σ incorporating the log singularity). It is not, however, obvious that $\hat{h}^{\text{wedge}}$ assigns finite mass to any neighbourhood of 0, and thus has a canonical embedding as discussed in Example 7.3. It is not too hard to see (and this will be made explicit below) that this would follow from Theorem 6.42 if $\hat{h}$ were a Neumann GFF in $\mathbb{H}$ with average 0 on the upper unit semicircle. But we must also consider the log-singularity at the origin; the next lemma shows that this does not cause the mass to explode.

Lemma 7.13 *Let $\hat{h}^{\text{wedge}}$ be the representative field of a quantum wedge in the unit circle embedding. Then the bulk measure $\mathcal{M}$ associated with $\hat{h}^{\text{wedge}}$ has $\mathcal{M}(\mathbb{D} \cap \mathbb{H}) < \infty$ almost surely.*

Proof It is more convenient (and equivalent) to prove that the bulk measure $\mathcal{M}$ associated to h^{wedge} defined in S as in Definition 7.7 satisfies $\mathcal{M}([0, \infty) \times (0, \pi)) < \infty$ almost surely. For $n \geq 0$ we let S_n be the strip $[n, n+1] \times (0, \pi)$ and set $0 \leq q < \min(1, 1/\gamma) \in [0, 1)$.

We first claim that for each fixed $n \geq 0$, we have

$$\mathbb{E}[\mathcal{M}(S_n)^q] < \infty. \tag{7.10}$$

Equivalently, let us show that $\mathbb{E}[\hat{\mathcal{M}}(\hat{S}_n)^q] < \infty$, where $\hat{\mathcal{M}}$ is the measure associated to $\hat{h}^{\text{wedge}}$, and $\hat{S}_n$ is the image of S_n by the map $z \mapsto -e^{-z}$, that is, the upper annulus centred at 0 of radii e^{-n} and e^{-n-1}, respectively. Since $n \geq 0$ is fixed, the logarithmic singularity at the origin of $\hat{h}^{\text{wedge}}$ is of no relevance, and it suffices to show that $\mathbb{E}[\mathcal{M}_h(\hat{S}_n)^q] < \infty$, where $\mathcal{M}_h$ is the GMC associated to a Neumann GFF h in $\mathbb{H}$, normalised to have zero unit upper-circle average. We will prove this by making use of Theorem 6.42. To do this, we will map the upper half plane to the unit disc using the map $z \mapsto \psi(z) = (z-i)/(z+i)$. The image of h under this map is a Neumann GFF in $\mathbb{D}$, say $\tilde{h}$, normalised to have a certain average (along a chord extending from i to $-i$) equals zero. The image of $\hat{S}_n$ under this map is a certain subdomain, say $\tilde{S}_n$, of the unit disc $\mathbb{D}$, made of four arcs of circle (two arcs of the unit circle and two hyperbolic geodesics, respectively). Since $|\psi'|$ is bounded away from 0 and $+\infty$ on $\hat{S}_n$, it suffices to show that $\mathbb{E}[\mathcal{M}_{\tilde{h}}(\tilde{S}_n)^q] < \infty$. Although $\tilde{h}$ is normalised differently from the Neumann GFF on the unit disc in Theorem 6.42, but this is of no relevance: the two fields differ by a fixed Gaussian random variable, so that we can conclude by Hölder's inequality and Theorem 6.42. This proves (7.10).

Now, by definition, since we restrict to the strip with real positive part, $h^{\text{wedge}} = h^{\text{wedge}}_{\text{rad}} + h^{\text{wedge}}_{\text{circ}}$ where the two summands are independent, and $h^{\text{wedge}}_{\text{rad}}(z) = B_{2\Re(z)} + (\alpha - Q)\Re(z)$ for B a standard Brownian motion. Setting $m_n = \sup\{h^{\text{wedge}}_{\text{rad}}(z) : z \in S_n\}$ we therefore have, by subadditivity (since

7.1 Convergence of random surfaces   271

$0 \leq \gamma \leq 1$),

$$\mathbb{E}[\mathcal{M}([0,\infty) \times (0,\pi))^q | h_{\text{rad}}^{\text{wedge}}] \leq \sum_{n \geq 0} e^{qm_n} \mathbb{E}[\mathcal{M}_{\text{circ}}(S_n)^q],$$

where $\mathcal{M}_{\text{circ}}$ is the chaos measure associated to $h_{\text{circ}}^{\text{wedge}}$; that is,

$$\lim_{\varepsilon \to 0} \varepsilon^{\gamma^2/2} e^{\gamma (h_{\text{circ}}^{\text{wedge}})_\varepsilon(z)} dz.$$

By (7.10), we have that $\mathbb{E}[\mathcal{M}_{\text{circ}}(S_n)^q] < \infty$ for all $n \geq 0$. On the other hand, this quantity cannot depend on $n \geq 0$ by translation invariance of $h_{\text{circ}}^{\text{wedge}}$). Thus $\mathbb{E}[\mathcal{M}_{\text{circ}}(S_n)^q] = c < \infty$ for all $n \geq 0$. Moreover, since m_n is the maximum of a Brownian motion with drift on $[n, n+1]$ for each n, we have (since $B_t/t \to 0$ almost surely as $t \to \infty$),

$$\sum_{n \geq 0} e^{qm_n} < \infty \quad \text{almost surely.}$$

Putting these together, we obtain that $\mathbb{E}[\mathcal{M}([0,\infty) \times (0,\pi))^q | h_{\text{rad}}^{\text{wedge}}] < \infty$ almost surely, and hence $\mathcal{M}([0,\infty) \times (0,\pi)) < \infty$ almost surely as well.   □

Hence, we obtain the following strengthening of Theorem 7.11. We emphasise that we are making use of the strong convergence (i.e. in total variation distance) here, which allows us to couple things so that they are actually equal (when restricted to compacts) with high probability. We are also using that for a quantum surface parametrised by $\mathbb{H}$ with marked points at 0 and ∞, as in Example 7.3, the scaling map that determines the canonical parametrisation only depends on the field in a neighbourhood of the origin with unit LQG area.

Corollary 7.14 *(i) Suppose that $\tilde{h}, h$ are as in Theorem 7.11. If $(\mathbb{H}, h^C, 0, \infty)$ is the canonical description of $(\mathbb{H}, h + C, 0, \infty)$ and $(\mathbb{H}, \hat{h}^{\text{wedge}}, 0, \infty)$ is the canonical description of an α-quantum wedge, then for any $R > 0$, $h^C|_{R\mathbb{D}_+} \to \hat{h}^{\text{wedge}}|_{R\mathbb{D}_+}$ in total variation distance as $C \to \infty$.*

(ii) Let $\mathcal{M}_{h^C}, \mathcal{M}_{\hat{h}^{\text{wedge}}}$ be the respective Liouville measures of $h^C, \hat{h}^{\text{wedge}}$ as in (i). Then for any $R > 0$ $\mathcal{M}_{h^C}|_{R\mathbb{D}_+} \to \mathcal{M}_{\hat{h}^{\text{wedge}}}|_{R\mathbb{D}_+}$ in total variation distance as $C \to \infty$.

We remark that the convergence in point (ii) of the above Corollary (with weak convergence rather than total variation) was actually used in some of the earlier work of Sheffield, for example in [She16a], as the definition of convergence in law for quantum surfaces.

7.2 Quantum cones

The quantum wedges discussed above are sometimes referred to as *infinite volume quantum surfaces with boundary*, because their associated GMC measures have infinite mass, and because they are parametrised by simply connected domains with boundary (for example, the upper half plane $\mathbb{H}$ or the strip S). In this section, we will discuss surfaces known as **quantum cones**: these are still infinite volume surfaces but now *without boundary*, and are sometimes referred to as having "the topology" of the sphere rather than the disc. There also important examples of quantum surfaces with *finite volume* (with or without boundary); these are **quantum discs** and **quantum spheres** and will be discussed later.

In fact, the theory of quantum cones is entirely parallel to that of quantum wedges, the only difference being that they are defined on the whole plane or the infinite cylinder rather than on a simply connected domain. These quantum cones are obtained in essentially the same way as the quantum wedges, but starting from a whole plane GFF rather than a Neumann GFF.

Let C be the infinite cylinder $C := \{z = x + iy \in \mathbb{C}: y \in [0, 2\pi i]\}/\sim$, where $\sim$ identifies points x with $x + 2\pi i$ for $x \in \mathbb{R}$. Let $\overline{H}^1(C)$ be the Hilbert space completion, with respect to the Dirichlet inner product $(\cdot, \cdot)_\nabla$, of the set of smooth functions modulo constants on C with finite $(\cdot, \cdot)_\nabla$ norm. We first need the analogue of the radial decomposition for $\overline{H}^1(S)$.

Lemma 7.15 *Let $\overline{\mathcal{H}}_{\mathrm{rad}}(C)$ be the subspace of $\overline{H}^1(C)$ obtained as the closure of smooth functions which are constant on each vertical segment $\{x + iy; y \in [0, 2\pi i]\}$, viewed modulo constants. Let $\mathcal{H}_{\mathrm{circ}}(C)$ be the subspace obtained as the closure of smooth functions which have mean zero on all such vertical segments. Then*

$$\overline{H}^1(C) = \overline{\mathcal{H}}_{\mathrm{rad}}(C) \oplus \mathcal{H}_{\mathrm{circ}}(C).$$

Proof This is similar to the proof of the radial decomposition for $\overline{H}^1(S)$: we leave it to the reader as part of Exercise 7.4. □

Recall that the **whole plane GFF** $\overline{\mathbf{h}}^\infty$ is the random distribution modulo constants on $\mathbb{C}$ (i.e. a continuous linear functional on $\tilde{\mathcal{D}}_0(\mathbb{C})$, the set of $f \in C^\infty(\mathbb{C})$ with compact support and $\int_\mathbb{C} f = 0$) with covariance kernel $G^\infty(x, y) = -\frac{1}{2\pi}\log(|x - y|)$. We denote

$$h^\infty := \sqrt{2\pi}\mathbf{h}^\infty$$

as usual. The (whole plane) GFF on the cylinder C, $\overline{h}^C$, is then defined by

$$\overline{h}^C := \overline{h}^\infty \circ \psi^{-1},$$

7.2 Quantum cones

where $\psi : C \to \mathbb{C}$ is the map $z \mapsto -\log(1/z)$, and the meaning of the above is that $(\overline{h}^C, f) = (\overline{h}^\infty, |\psi'|^2 f \circ \psi)$ for every $f \in C^\infty(C)$ with compact support and $\int_C f = 0$. Due to the covariance structure, similarly to in the Neumann GFF case, we can extend the definitions of $\overline{h}^C, \overline{h}^\infty$, respectively, to be stochastic processes indexed by a larger index sets; namely, the sets $\widetilde{\mathfrak{M}}_\infty^C, \widetilde{\mathfrak{M}}_\infty^\mathbb{C}$ of signed Radon measures on $C, \mathbb{C}$, respectively, whose positive and negative parts $\rho^\pm$ have equal mass and satisfy $\int \log|x-y| \rho^\pm(dx)\rho^\pm(dy) < \infty$.

Just as in the case of the Neumann GFF, Lemma 7.15 means that we can decompose

$$\overline{h}^C = \overline{h}^C_{\text{rad}} + h^C_{\text{circ}}, \tag{7.11}$$

where

- $\overline{h}^C_{\text{rad}}$ and h^C_{circ} are independent;
- $\overline{h}^C_{\text{rad}} = \overline{B}_s$ if $\mathfrak{R}(z) = s$, where $\overline{B}$ is a standard Brownian motion modulo constants;
- h^C_{circ} has mean zero on each vertical segment.

Again we leave the details as an exercise for the reader. Notice that the Brownian motion is run at the standard speed in this decomposition (rather than speed two in the case of the Neumann GFF) since, after mapping to the whole plane, this corresponds circle averages around an interior rather (rather than a boundary point).

This leads us to the definition of an α-quantum cone, again for

$$0 \leq \alpha \leq Q = \frac{2}{\gamma} + \frac{\gamma}{2}.$$

Definition 7.16 Let

$$h^{\text{cone}}_{\text{rad}}(z) = \begin{cases} B_s + (\alpha - Q)s & \text{if } \mathfrak{R}(z) = s \text{ and } s \geq 0 \\ \widehat{B}_{-s} + (\alpha - Q)s & \text{if } \mathfrak{R}(z) = s \text{ and } s < 0, \end{cases} \tag{7.12}$$

where $B = (B_t)_{t \geq 0}$ is a standard Brownian motion, and $\widehat{B} = (\widehat{B}_t)_{t \geq 0}$ is an independent Brownian motion conditioned so that $\widehat{B}_t + (Q - \alpha)t > 0$ for all $t > 0$ (see (7.3) for the meaning of this conditioning).

Let $h^{\text{cone}}_{\text{circ}}$ be a stochastic process indexed by $\mathfrak{M}_\infty^C$, that is independent of $h^{\text{cone}}_{\text{rad}}$ and has the same law as h^C_{circ}. Then we set $h^{\text{cone}} = h^{\text{cone}}_{\text{rad}} + h^{\text{cone}}_{\text{circ}}$ (which, since $h^{\text{cone}}_{\text{rad}}$ is just a continuous function, can again be defined as a stochastic process indexed by $\mathfrak{M}_\infty^C$). We call h^{cone} an α-**quantum cone** field in C.

274 Quantum Wedges and Scale-Invariant Random Surfaces

Remark 7.17 Notice that the speed of the Brownian motion in Definition 7.16 above is 1, rather than the 2 in Definition 7.7. This is, roughly speaking, because the Neumann GFF has double the variance of the whole plane GFF near the real line.

We will again want to view the above definition as being a specific equivalence class representative of a quantum surface with two marked points (that we will also refer to as an α-quantum cone with an abuse of notation). That is, if h is as in Definition 7.16, we will associate with it the quantum surface with two marked points $(\mathbb{C}, h, \infty, -\infty)$. Another quadruple (D, h', a, b) represents the same quantum surface if there is a conformal isomorphism $T: \mathbb{C} \to D$ with $T(\infty) = a$, $T(-\infty) = b$ and $h' = h \circ T^{-1} + Q \log |(T^{-1})'|$ as in Definition 7.1. Similarly to the quantum wedge, any such representative will have finite associated Gaussian multiplicative chaos mass in any neighbourhood of a, and infinite mass in any neighbourhood of b.

One particularly nice equivalence class representative of the α-quantum cone is obtained by conformally mapping to $\mathbb{C}$ using the map $z \mapsto -e^{-z}$ which sends ∞ to 0 and $-\infty$ to ∞. Under this mapping, the vertical segment $\{t + iy : y \in [0, 2\pi]\} \subset C$ mapped to the circle of radius e^{-t} around 0 in $\mathbb{C}$, and the shift from the conformal change of coordinates formula is given by $-Q\Re(z)$. As in the wedge case, the obtained representative $(\mathbb{C}, \hat{h}^{\text{cone}}, 0, \infty)$ of the α-quantum cone is said to be in the *unit circle embedding* and the field restricted to the unit disc $\mathbb{D}$ has the same law as $h^\infty + \alpha \log(1/|z|)$ restricted to $\mathbb{D}$, where h^∞ is a whole plane GFF with additive constant fixed so that its average on $\partial \mathbb{D}$ is equal to 0.

Finally, we can state the analogue of Theorem 7.11, which identifies the quantum cone as a local limit of a whole plane GFF with an additional log singularity of strength α at the origin.

Theorem 7.18 *Fix $0 \leq \alpha < Q$. Then the following hold:*

(i) *Let $\tilde{h}^\infty$ be a whole plane GFF (in $\mathbb{C}$) with additive constant fixed so that $(\tilde{h}, \rho_0)$ is equal to 0 for some signed Radon measure ρ_0 with compact support away from the origin in $\mathbb{C}$, $\int_{\mathbb{C} \times \mathbb{C}} \log |x-y| |\rho_0|(dx) |\rho_0|(dy) < \infty$ and $\rho_0(\mathbb{C}) \neq 0$. Set $h(z) = \tilde{h}^\infty(z) + \alpha \log 1/|z|$. Let h^C be such that $(\mathbb{C}, h^C, 0, \infty)$ is the unit circle embedding of $(\mathbb{C}, h + C, 0, \infty)$, and let $(\mathbb{H}, h^{\text{cone}}, 0, \infty)$ be the unit circle embedding of an α-quantum cone. Then for any $R > 0$, $h^C|_{R\mathbb{D}_+}$ converges in total variation distance to $h^{\text{cone}}|_{R\mathbb{D}_+}$ as $C \to \infty$.*

(ii) *If $(\mathbb{C}, h^{\text{cone}}, 0, \infty)$ is an α-quantum cone, then $(\mathbb{C}, h^{\text{cone}}, 0, \infty)$ and $C \in \mathbb{R}$, then $(\mathbb{C}, h^{\text{cone}} + C, 0, \infty)$ have the same law as quantum surfaces.*

Proof Exercise 7.4. □

7.3 Thin quantum wedges

Recall that for $0 \leq \alpha < Q$ we defined an (equivalence class representative) of the α-quantum wedge to be the random distribution on the infinite strip S whose circular (or angular) part $h_{\text{circ}}^{\text{wedge}}$ is equal in law to the circular part $h_{\text{circ}}^{\text{GFF}}$ of a Neumann GFF on S, and whose radial part $h_{\text{rad}}^{\text{wedge}}$ (which is constant on vertical line segments) is independent of $h_{\text{circ}}^{\text{wedge}}$ and evolves as a speed two infinite Brownian motion B_{2s}, plus a negative drift of $(\alpha - Q)s$ from $+\infty$ to $-\infty$, translated to hit 0 for the first time at time 0 (recall (7.3)). **Thin quantum wedges** are the surfaces obtained when the parameter α is instead taken in the range $(Q, Q + \frac{\gamma}{2})$. We will see that in this case, it is not possible to represent the surface by a single random field defined on $\mathbb{H}$ or S, but the correct definition is rather as a Poisson point process of quantum surfaces, or **beads** of the quantum wedge in the terminology of [DMS21].

Let us now make a useful connection between thick quantum wedges and Bessel processes, in order to motivate the definition of thin quantum wedges.

Definition 7.19 (Bessel process) Let $\delta > 0$. We define the Bessel process of dimension δ started from $x \geq 0$ to be $Z_t = Y_t^{1/2}$, where Y solves the square Bessel stochastic differential equation (SDE), namely

$$dY_t = 2\sqrt{Y_t}\, dB_t + \delta\, dt; \quad Y_0 = x^2. \tag{7.13}$$

(See [RY99, Section 3, Chapter IX] for the existence and uniqueness of solutions to this SDE.)

Applying Itô's formula, we can see that on intervals of time in which Z is not equal to 0, Z satisfies its own SDE:

$$dZ_t = dB_t + \frac{\delta - 1}{2Z_t}\, dt\,; \quad Z_0 = x. \tag{7.14}$$

However, defining Z directly from (7.14) is far from straightforward because of the singularity of the drift term when Z gets close to zero. When $\delta \geq 2$, it is easy to check $Z_t > 0$ for all $t > 0$ and if $\delta > 2$, $Z_t \to \infty$ as $t \to \infty$ with probability one (i.e. Z is transient), and thus (7.14) can be used as the definition of a Bessel process of dimension δ. When $\delta < 2$, Z returns to 0 infinitely often with probability one (see [RY99, Chapter 11]), but 0 is instantaneously reflecting if $\delta > 0$: that is, the Lebesgue measure of the set of times where $Z_t = 0$ is a.s. zero ([RY99, Proposition (1.5), Chapter XI]). When $1 < \delta < 2$, it is still possible to think of the Bessel process of dimension δ as solution of

(7.14), because it can be checked that the integral

$$\frac{\delta-1}{2}\int_0^t \frac{du}{Z_u}$$

converges a.s. for all $t \geq 0$, and is equal to $Z_t - B_t$, where B is the Brownian motion from (7.14). When $\delta \leq 1$, the integral no longer converges and the SDE (7.14) does not make sense on intervals of time during which Z hits zero; in fact, it can be checked that Z is then **not** even a semimartingale when $\delta < 1$. Nevertheless, the law of the Bessel process Z of dimension $\delta > 0$ is uniquely specified by the fact that it is a Markov process on $[0, \infty)$ whose infinitesimal generator coincides on $C^2((0, \infty))$ with

$$\frac{1}{2}\frac{d^2}{dx^2} + \frac{\delta-1}{2x}\frac{d}{dx}$$

(i.e. it satisfies (7.14) away from $x = 0$) and with instantaneous reflection at $x = 0$; see [PY82, (1.a)].

In what follows, we define a function on $(-\infty, \infty)$ modulo translation to be an equivalence class of functions $(-\infty, \infty) \to \mathbb{R}$, where two functions $x(t), x'(t)$ are equivalent if $x(t) = x'(t + a)$ for all t and some $a \in \mathbb{R}$. The next lemma shows that the logarithm of a Bessel process with dimension $\delta \geq 2$, started from 0, reparametrised by its quadratic variation and viewed modulo translation, has the same law as a Brownian motion with positive drift from $-\infty$ to $+\infty$, that is, precisely what arises in the definition of a thick quantum wedge. This will give us an alternative point of view on thick quantum wedges (first pointed out by Duplantier, Miller and Sheffield in [DMS21]) which lends itself well to the generalisation required to treat the thin case of quantum wedges.

Lemma 7.20 *Let $0 \leq \alpha < Q$ and let $(Z_t)_{t \geq 0}$ be a Bessel process of dimension*

$$\delta = \delta_{\text{wedge}}(\alpha) := 2 + \frac{2(Q-\alpha)}{\gamma} \quad (7.15)$$

with $Z_0 = 0$. Consider the process

$$X_t := \tfrac{2}{\gamma} \log(Z_{q(t)}), \quad (7.16)$$

where $q \colon (-\infty, \infty) \to (0, \infty)$ is defined by the requirement that $q(0) = q_0$ for some (arbitrary) $q_0 > 0$ and that the quadratic variation of X satisfies $d[X]_t = 2\,dt$ on $(-\infty, \infty)$. Then as functions on $(-\infty, \infty)$ modulo translation,

$$(X_t)_{t \in \mathbb{R}} \stackrel{(\text{law})}{=} (B_{2t} + (Q-\alpha)t)_{t \in \mathbb{R}},$$

where the right-hand side denotes a Brownian motion with speed two and drift $b = Q - \alpha > 0$, from $-\infty$ to $+\infty$, similar to (7.3).

7.3 Thin quantum wedges

Proof Fix $t_0 \in \mathbb{R}$ and let us condition on $(X_s)_{s \le t_0}$ (set $x = X_{t_0}$). Due to the strong Markov property of Brownian motion, it suffices to prove that

$$dX_t = dB_{2t} + (Q - \alpha) dt \quad ; \quad t \ge t_0.$$

Since $\delta \ge 2$, Z is a solution of the SDE (7.14) for all time. Thus by Itô's formula, if $Y_t := \frac{2}{\gamma} \log(Z_t)$, then

$$dY_t = \frac{2}{\gamma Z_t} dB_t + \frac{\delta - 2}{\gamma Z_t^2} dt = dM_t + \tfrac{1}{2}(Q - \alpha) d[M]_t \quad \text{for all } t,$$

where

$$M_t := \frac{2}{\gamma Z_t} B_t$$

is a continuous local martingale. By definition of X, we therefore have

$$dX_t = d\tilde{M}_t + \tfrac{1}{2}(Q - \alpha) d[\tilde{M}]_t \quad \text{for all } t,$$

where $\tilde{M}$ is a reparametrisation of M such that $d[\tilde{M}]_t = 2 dt$ for all t. By Lévy's characterisation of Brownian motion, it must be that $d\tilde{M}_t = dB_{2t}$. Substituting this into the expression for dX_t concludes the proof. $\square$

Remark 7.21 A similar argument applies when we do not assume $\delta \ge 2$ (i.e. if $\alpha > Q$). However in this case, to avoid conditioning on q_0 being less than the hitting time ζ of 0 by Z, which would affect the law of the process, we need to assume that $Z_0 = x > 0$. The conclusion is that the process $X_t = \frac{2}{\gamma} \log(Z_t)$, reparametrised to have quadratic variation $[X]_t = 2t$ for all time $t \ge 0$, is equal in law to $(B_{2t} + (Q - \alpha)t)_{t \ge 0}$ (with B a standard Brownian motion started from $\frac{2}{\gamma} \log(x)$).

As a consequence of Lemma 7.20, for $0 \le \alpha < Q$, we can equivalently define the (thick) α-quantum wedge to be the doubly marked quantum surface $(S, h, -\infty, +\infty)$, where

$$h^{\text{wedge}} = h_{\text{circ}}^{\text{wedge}} + X_{\mathfrak{R}(\cdot)} \quad \text{considered modulo horizontal translation,} \quad (7.17)$$

X is as defined in (7.16), and $h_{\text{circ}}^{\text{wedge}}$ is independent of X with the law of h_{circ}^S. The reason for rewriting the definition in this way is because, defining X in terms of the $\delta_{\text{wedge}}(\alpha)$-dimensional Bessel process Z, there will be a clear extension to the case $\alpha \in (Q, Q + \gamma)$, corresponding to $\delta_{\text{wedge}}(\alpha) \in (0, 2)$.

This extension will rely on the notion of excursion for the Bessel process, which we now introduce. As already mentioned, even when $\delta \in (0, 2)$ the Bessel process of dimension δ is a strong Markov process for which $a = 0$ is a recurrent point. It was already shown in the seminal work of Itô [Itō72] how to attach to such a Markov process a collection of excursions which forms a Poisson point process. To state this properly requires a notion of **local time** for Z at $a = 0$ which, roughly speaking measures the amount of time spent by Z near $a = 0$. Traditionally ([RY99, Kal21]), local time is constructed for semimartingales and we have already mentioned that the semimartingale property for a Bessel process of dimension $\delta > 0$ fails if $0 < \delta < 1$. Nevertheless, Itô's theory does apply to the whole range of dimensions $\delta \in (0, 2)$, and is based on a notion of local time which is called the Blumenthal–Getoor local time of Z at a. (An alternative would be to use the excursion theory for the squared Bessel process $Y_t = Z_t^2$, since that is both a recurrent process and a semimartingale for all $\delta \in (0, 2)$, see (7.13).)

The upshot is the following, which is both a definition and Itô's result [Itô72] in this case:

Definition 7.22 Let $\delta \in (0, 2)$, and Z be a δ-dimensional Bessel process. Then Z has an associated Itô excursion measure $\nu_\delta^{\mathrm{BES}}$ on the space $\mathcal{E}$ of continuous paths from 0 to 0, equipped with the topology of uniform convergence, and a local time l at 0. It satisfies the classical Itô excursion decomposition with excursion measure $\nu_\delta^{\mathrm{BES}}$. That is, if $(e_i)_{i \geq 1}$ is any enumeration of the countable set of excursions that Z makes from 0, and for $i \geq 1$, t_i is the common value of l on the time interval associated to e_i, then

$$\sum_{i \geq 1} \delta_{(t_i, e_i)}$$

has the distribution of a Poisson point process on $\mathbb{R}_+ \times \mathcal{E}$ with intensity measure $du \otimes \nu_\delta^{\mathrm{BES}}$.

Now suppose that $\alpha \in (Q, Q + \gamma)$, so that $\delta_{\mathrm{wedge}}(\alpha) \in (0, 2)$. The above excursion decomposition means that we can extend the definition, (7.17), of a quantum wedge to this range of α. However, rather than a single surface we will actually get a certain **Poisson point process of quantum surfaces**. We are now ready to define the thin quantum wedge.

Definition 7.23 Let $\alpha \in (Q, Q + \gamma)$ and let Z be a δ-dimensional Bessel process with $\delta = \delta_{\mathrm{wedge}}(\alpha) = 2 + 2(Q - \alpha)/\gamma$. Let $\sum_{i \geq 1} \delta_{(t_i, e_i)}$ be the Poisson point process of excursions of Z, as in Definition 7.22, and for each i define X^i to be the function on $(-\infty, \infty)$ modulo translation given by $(2/\gamma) \log e_i$ parametrised to have infinitesimal quadratic variation $2\, dt$ (as in Lemma 7.20). For each i, let $\mathcal{S}_i$ be the doubly marked quantum surface $(S, h^i, +\infty, -\infty)$ where

$$h^i = h^i_{\mathrm{circ}} + X^i_{\mathfrak{R}(\cdot)} \text{ considered modulo horizontal translation,}$$

and $\{h^i_{\mathrm{circ}}; i \geq 1\}$ is a collection of independent copies of h^S_{circ}, independent of $\{X^i; i \geq 1\}$.

7.3 Thin quantum wedges

We define the (thin) α-quantum wedge to be the Poisson point process

$$\mathcal{W} = \sum_{i \geq 1} \delta_{(t_i, \mathcal{S}_i)}.$$

Remark 7.24 In [DMS21], the definition of thin quantum wedges is only given in the case $\alpha \in (Q, Q + \frac{\gamma}{2})$, corresponding to the case $\delta = \delta_{\text{wedge}}(\alpha) \in (1, 2)$. Indeed, for this range of α, one gets an additional property for the law of the total mass (quantum area) of each surface in the Poisson point process above. This additional property is important in the mating of trees (see Proposition 4.4.4 in [DMS21], and see also the end of the proof of Theorem 9.33 at the very end of Chapter 9); fortunately, in that case, we will see that the relevant value of δ will be $\kappa'/4$ with $\kappa' > 4$, so that indeed $\delta > 1$.

To view a thin quantum wedge $\mathcal{W}$ as a random variable in a nice (Polish) space, it is better to view each point in the Poisson point process (or "bead") $\mathcal{S}_i$, as being embedded in the strip S as above but with some fixed choice of translation for the field (for example, so that the radial part of the field has its maximum value at time 0). In this case, $\mathcal{W}$ can be identified with a random variable in the space of (atomic) measures on $\mathbb{R}_+ \times C(\mathbb{R}, \mathbb{R}) \times H^{-1}_{\text{loc}}$, where the last component describes the circular part of the field.

Remark 7.25 In the thick case, the potential relevance of the notion of quantum wedges is made clear by Theorem 7.11, which shows that the (thick) quantum wedge is the scaling limit obtained by zooming near zero in a Neumann GFF with appropriate logarithmic singularity at zero. There is however no immediate analogue of this result in the thin case, so that the reader may wonder at this point what is the relevance of this notion (or, put it another way, how to tell that the generalisation of thick quantum wedges using the Bessel process point of view given in Lemma 7.20 is the "correct" way to do this generalisation). Such a justification will be provided at least in Chapter 9 where thin quantum wedges are an essential aspect of the mating of trees approach to LQG.

We have now described thick quantum wedges in terms of Bessel processes with dimensions $\delta > 2$ and thin quantum wedges in terms of Bessel processes with dimensions $\delta \in (0, 2)$. One nice consequence of this is a duality between thick and thin wedges corresponding to a duality between Bessel process of dimension δ and dimension $4 - \delta$; see Lemma 7.26.

Note that if $\alpha \in (Q - \gamma, Q)$ so that $\delta_{\text{wedge}}(\alpha) \in (2, 4)$, then

$$4 - \delta_{\text{wedge}}(\alpha) = \delta_{\text{wedge}}(2Q - \alpha)$$

In other words, the duality will be between α- and $\hat{\alpha} = (2Q - \alpha)$-quantum wedges. Note that $\hat{\alpha} \in (Q, Q + \gamma)$ so that $\delta_{\text{wedge}}(\hat{\alpha}) \in (0, 2)$. Let us now describe this more precisely.

Lemma 7.26 *For $\delta \in (0, 2)$, decomposing a Bessel excursion according to its maximum value, we can write*

$$v_\delta^{\text{BES}} = c_\delta \int_0^\infty v_\delta^x x^{\delta-3}\, dx, \tag{7.18}$$

where:

- dx *is Lebesgue measure on* $\mathbb{R}_+$;
- $c_\delta \in (0, \infty)$ *depends only on δ; and*
- *for each $x > 0$, v_δ^x is a probability measure on excursions from 0 to 0 in $\mathbb{R}_+$ with maximum value x. A sample from v_δ^x corresponds to a Bessel process of dimension $4 - \delta$ run until it first hits x, then concatenated with x minus the time reversal of an independent copy of the same process.*

Proof See [PY96, Theorem 1]. Note that the description of v_δ^x for given x follows from Lemma 7.20 and Remark 7.21, plus the fact that conditioned on its maximum value, a Brownian motion with negative drift $-a$ can be written as a Brownian with drift a until it hits this maximum value, and then concatenated with an independent Brownian motion with drift $-a$, conditioned to stay negative. See also Lemma B.10 and [Wil74] for closely related statements. □

Remark 7.27 As a consequence, we see that if $\alpha \in (Q, Q+\gamma)$ then, informally speaking, each of the quantum surfaces making up a (thin) α-quantum wedge (when parametrised by the S) looks locally near $\pm\infty$ like a $(2Q - \alpha)$-quantum wedge in S does near $-\infty$ (the marked point with neighbourhoods of finite quantum area).

7.4 Quantum discs

Having defined the thin quantum wedge for $\alpha \in (Q, Q + \gamma)$ as a Poisson point process of quantum surfaces, it is natural to ask about the "law" of each of these surfaces (although of course this actually corresponds to an infinite measure). This will lead us to the notion of quantum discs below.

Recall that given an excursion e_i of a δ-dimensional Bessel process, we defined $X^i =: X^{e_i}$, a function on $(-\infty, \infty)$ modulo translation, to be given by $(2/\gamma) \log(e_i)$ parametrised to have infinitesimal quadratic variation $2\, dt$. Since the excursion e_i of the Bessel process starts at 0, ends at 0 and has finite

7.4 Quantum discs

maximum value, a natural way of fixing the horizontal translation of X^{e_i} is to require that the maximum is reached at time 0. Let us write Y^{e_i} for this function on $(-\infty, \infty)$ (associated with the excursion e_i).

Definition 7.28 Let $\alpha \in (Q, Q + \gamma)$, ν_δ^{BES} be as described in Definition 7.22 with $\delta = \delta_{\text{wedge}}(\alpha) = 2 + 2(Q - \alpha)/\gamma$, and $\mathbb{P}_{\text{circ}}^S$ be the law of h_{circ}^S (obtained from a Neumann GFF in S by subtracting its average value on each vertical line segment). We define the infinite α-quantum disc measure m_α^{disc} to be the measure on $H_{\text{loc}}^{-1}(S)$ obtained by pushing forward the measure $\nu_\delta^{BES} \otimes \mathbb{P}_{\text{circ}}^S$ to $H_{\text{loc}}^{-1}(S)$, via the map taking (e, h_{circ}) to the field $h_{\text{circ}} + Y_{\Re(\cdot)}^e$.

Remark 7.29 In [DMS21], the definition of $\nu_{\delta(\alpha)}^{BES}$ is extended to a larger range of α, and thus so is the α-quantum disc measure. However, we will stick to the case where the measure can be defined using classical Itō excursion theory; see Definition 7.22.

Remark 7.30 With this definition, if $\alpha \in (Q, Q + \gamma)$, the α-quantum disc measure corresponds to the measure "describing" the individual quantum surfaces appearing in a (thin) α-quantum wedge. Indeed, recalling Definitions 7.22 and 7.23, we see that an equivalent definition of the α-quantum wedge is as a Poisson point process

$$\mathcal{W} = \sum_{i \geq 1} \delta_{(t_i, S_i)}$$

with intensity $du \otimes \hat{m}_\alpha^{\text{disc}}$, where $\hat{m}_\alpha^{\text{disc}}$ is the pushforward of m_α^{disc} by the map taking $h \in H_{\text{loc}}^{-1}(S)$ to the doubly marked quantum surface $(S, h, -\infty, \infty)$.

By Remark 7.27, near each of the marked points, a sample from (some suitably conditioned version of) the α-quantum disc measure looks locally like a (thick) $(2Q - \alpha)$-quantum wedge at its apex (i.e. near the marked point which has neighbourhoods of finite quantum mass). Or in other words, it looks locally like a free boundary Gaussian free field plus a $(2Q - \alpha)$-log singularity. (Of course this statement is informal on many levels!)

Notice that, due to Brownian scaling, a Bessel excursion with maximum x (i.e. sampled from ν_δ^x with the notation of Lemma 7.26) is equal in law to x times a Bessel process with maximum 1 (i.e. sampled from ν_δ^1) modulo time change. However, since under the map $e \mapsto X^e$, we reparametrise time anyway (so that the infinitesimal quadratic variation is exactly $2\,dt$) we see that the law of Y^e when e is sampled from ν_δ^x is equal to the law of $((2/\gamma) \log x + Y^e)$ when e is sampled from ν_δ^1. Hence from the decomposition in Lemma 7.26, it follows

282   *Quantum Wedges and Scale-Invariant Random Surfaces*

that for any non-negative measurable function F on $H^{-1}_{\mathrm{loc}}(S)$:

$$m^{\mathrm{disc}}_\alpha(F) = c_\delta \int_0^\infty \mathbb{P}^S_{\mathrm{circ}} \otimes v^1_\delta \big(F\big(h_{\mathrm{circ}} + \tfrac{2}{\gamma}\log(x) + Y^e_{\mathfrak{R}(\cdot)}\big)\big) x^{\delta-3}\, \mathrm{d}x, \quad (7.19)$$

remembering that $\delta = \delta_{\mathrm{wedge}}(\alpha) = 2 + (2/\gamma)(Q - \alpha)$.

From the description of v^1_δ in Lemma 7.26, we see that if e is sampled from v^1_δ and Y^e is as described above, then $Y^e_0 = 0$, and $(Y^e_t)_{t \geq 0}, (Y^e_{-t})_{t \leq 0}$ are independent, each having the law of $(2/\gamma)\log Z$ reparametrised to have quadratic variation $2t$ at time t, where Z is a Bessel process of dimension $4 - \delta_{\mathrm{wedge}}(\alpha) > 2$. By Lemma 7.20 we can rephrase this as follows.

Lemma 7.31 *Let $\alpha \in (Q, Q + \gamma)$ and define $(Y_t)_{t \in \mathbb{R}}$ by setting $Y_0 := 0$ and*

- $Y_t = B_{2t} + (Q - \alpha)t$ for $t > 0$
- $Y_t = \hat{B}_{-2t} + (Q - \alpha)(-t)$ for $t < 0$,

where $B, \hat{B}$ are independent standard linear Brownian motions defined for $t \in [0, \infty)$, started from 0 and conditioned that $B_{2t} + (Q-\alpha)t$ (resp. $\hat{B}_{2t} + (Q-\alpha)t$) is negative for all $t > 0$. Then, if e is sampled from v^1_δ,

$$(Y_t)_{t \in \mathbb{R}} \stackrel{(d)}{=} (Y^e_t)_{t \in \mathbb{R}}.$$

As promised, let us now justify that quantum discs really are *finite volume* quantum surfaces (with boundary).

Lemma 7.32 *For $\alpha \in (Q, Q + \gamma)$, $\mathcal{M}^\gamma_h(S) < \infty$ and $\mathcal{V}^\gamma_h(\partial S) < \infty$ for m^{disc}_α-almost every h.*

Proof We will verify the statement about $\mathcal{M}$; leaving the boundary case as Exercise 7.8. By (7.19), it suffices to show that if h_{circ} is sampled from $\mathbb{P}^S_{\mathrm{circ}}$ (i.e. has the law of h^S_{circ}) and e is sampled independently from v^1_δ, then

$$\mathcal{M}_{h_{\mathrm{circ}} + Y^e_{\mathfrak{R}(\cdot)}}(S) < \infty$$

almost surely.

This follows from the same proof as Lemma 7.13, using that under v^1_δ, Y^e is a two-sided Brownian motion with negative drift and $Y^e_0 = 0$: see Lemma 7.31. □

Conditioned quantum discs. Recall (7.19), which provides the decomposition

$$m^{\mathrm{disc}}_\alpha(F) = c_\delta \int_0^\infty \mathbb{P}^S_{\mathrm{circ}} \otimes v^1_\delta(F(h_{\mathrm{circ}} + \tfrac{2}{\gamma}\log(x) + Y^e_{\mathfrak{R}(\cdot)})) x^{\delta-3}\, \mathrm{d}x$$

(for F non-negative and measurable on $H^{-1}_{\mathrm{loc}}(S)$), of the α-quantum disc measure on H^{-1}_{loc}. Now we know that m^{disc}_α is supported on quantum surfaces with finite

7.4 Quantum discs

quantum mass and boundary length, we can use the above decomposition to describe the pushforward of m_α^{disc} via the map $h \mapsto \mathcal{M}_h^\gamma(S)$ or $h \mapsto \mathcal{V}_h^\gamma(\partial S)$ very precisely. Indeed, we have (for example, working with the boundary length $\mathcal{V}$):

$$m_\alpha^{\text{disc}}(\mathcal{V}_h^\gamma(\partial S) \in A)$$
$$= c_\delta \int_0^\infty \mathbb{P}_{\text{circ}}^S \otimes \nu_\delta^1 (\mathcal{V}_{h_{\text{circ}}+Y_{\mathfrak{R}(\cdot)}^e+(2/\gamma)\log(x)}^\gamma (\partial S) \in A) x^{\frac{2(Q-\alpha)}{\gamma}-1} \, dx.$$

Notice that

$$\mathcal{V}_{h_{\text{circ}}+Y_{\mathfrak{R}(\cdot)}^e+(2/\gamma)\log(x)}^\gamma (\partial S) = x \mathcal{V}_{h_{\text{circ}}+Y_{\mathfrak{R}(\cdot)}^e}^\gamma (\partial S)$$

for each x, so if we make the change of variables $u = x \mathcal{V}_{h_{\text{circ}}+Y_{\mathfrak{R}(\cdot)}^e}^\gamma (\partial S)$ the above becomes

$$m_\alpha^{\text{disc}}(\mathcal{V}_h^\gamma(\partial S) \in A)$$
$$= c_\delta \int_0^\infty \mathbb{P}_{\text{circ}}^S \otimes \nu_\delta^1 \left(\mathbf{1}_{\{u \in A\}} u^{\frac{2(Q-\alpha)}{\gamma}-1} (\mathcal{V}_{h_{\text{circ}}+Y_{\mathfrak{R}(\cdot)}^e}^\gamma (\partial S))^{-\frac{2(Q-\alpha)}{\gamma}} \right) du$$
$$= c_\delta \, \mathbb{P}_{\text{circ}}^S \otimes \nu_\delta^1 \left((\mathcal{V}_{h_{\text{circ}}+Y_{\mathfrak{R}(\cdot)}^e}^\gamma (\partial S))^{-\frac{2(Q-\alpha)}{\gamma}} \right) \int_{u \in A} u^{\frac{2(Q-\alpha)}{\gamma}-1} \, du.$$

This yields the following conclusion.

Lemma 7.33 *Let $\alpha \in (Q, Q + \gamma)$. The pushforward of m_α^{disc} under the map $h \mapsto \mathcal{V}_h^\gamma(\partial S)$ is a constant multiple of the measure $u^{-1+2\gamma^{-1}(Q-\alpha)} \, du$ on $[0, \infty)$. Similarly, the pushforward of m_α^{disc} under the map $h \mapsto \mathcal{M}_h^\gamma(S)$ is a constant multiple of the measure $u^{-1+\gamma^{-1}(Q-\alpha)} \, du$ on $[0, \infty)$.*

To conclude the section, we are going to decompose the α-quantum disc measure according to quantum boundary length and quantum area. It turns out to have a remarkable property: conditioned on the quantum boundary length or quantum area, if we subtract the correct constant from the field so that the area or boundary length becomes one, the law of the resulting field *does not* depend on the mass or area that we conditioned on.

In the case of boundary length, we have the following:

Proposition 7.34 *For $\alpha \in (Q, Q + \gamma)$*

$$m_\alpha^{\text{disc}} = c_\delta \int_0^\infty \mathbb{P}_\alpha^{\text{disc},u} u^{\frac{2(Q-\alpha)}{\gamma}-1} \, du, \tag{7.20}$$

where $\mathbb{P}_\alpha^{\text{disc},u}$ is a probability measure on $H_{\text{loc}}^{-1}(S)$, such that for $\mathbb{P}_\alpha^{\text{disc},u}$ every h, the boundary Gaussian multiplicative chaos measure $\mathcal{V}_h^\gamma(\partial S)$ is well defined and satisfies

$$\mathcal{V}_h^\gamma(\partial S) = u.$$

Moreover, if we write $h_\alpha^{\mathrm{disc},1}$ for the field on S with the law of

$$h_{\mathrm{circ}} + Y_{\mathfrak{R}(\cdot)}^e - \tfrac{2}{\gamma} \log \mathcal{V}_{h_{\mathrm{circ}}+Y_{\mathfrak{R}(\cdot)}^e}^\gamma (\partial S) \text{ weighted by } \left(\mathcal{V}_{h_{\mathrm{circ}}+Y_{\mathfrak{R}(\cdot)}^e}^\gamma (\partial S)\right)^{-\tfrac{2(Q-\alpha)}{\gamma}},$$

where $h_{\mathrm{circ}} \stackrel{(\mathrm{law})}{=} h_{\mathrm{circ}}^S$, and e is an independent Bessel excursion of dimension $\delta_{\mathrm{wedge}}(\alpha)$ conditioned on taking maximum value 1 (i.e. with law v_δ^1, so that Y^e is as described in Lemma 7.31), then

$$h_\alpha^{\mathrm{disc},u} := h_\alpha^{\mathrm{disc},1} + \tfrac{2}{\gamma} \log(u)$$

is a sample from $\mathbb{P}_\alpha^{\mathrm{disc},u}$.

Definition 7.35 We call the quantum surface $(S, h_\alpha^{\mathrm{disc},1}, -\infty, +\infty)$ (when $h_\alpha^{\mathrm{disc},1}$ has law $\mathbb{P}_\alpha^{\mathrm{disc},1}$) a unit boundary length α-quantum disc; and, when $h_\alpha^{\mathrm{disc},u}$ has law $\mathbb{P}_\alpha^{\mathrm{disc},u}$, we call $(S, h_\alpha^{\mathrm{disc},u}, -\infty, +\infty)$ an α-quantum disc with boundary length u.

With an abuse of terminology, we will also sometimes refer to just the field $h_\alpha^{\mathrm{disc},1}$ or $h_\alpha^{\mathrm{disc},u}$ as a unit boundary length α-quantum disc, or an α-quantum disc with boundary length u. Let us emphasise that an α-quantum disc with boundary length u has the same law as a unit boundary length α-quantum disc with a constant $\tfrac{2}{\gamma} \log(u)$ added to the field.

Proof of Proposition 7.34 Let F be a non-negative measurable function on $H_{\mathrm{loc}}^{-1}(S)$. To prove the proposition, it suffices to show that

$$m_\alpha^{\mathrm{disc}}\left(F\left(h - \tfrac{2}{\gamma} \log \mathcal{V}_h^\gamma(\partial S)\right) \mathbf{1}_{\{\mathcal{V}_h^\gamma(\partial S) \in A\}}\right) = m_\alpha^{\mathrm{disc}}(\mathcal{V}_h^\gamma(\partial S) \in A) \mathbb{P}_\alpha^{\mathrm{disc},1}(F),$$

where $\mathbb{P}_\alpha^{\mathrm{disc},1}$ is the law of $h_\alpha^{\mathrm{disc},1}$, as described in the statement of the proposition. Applying the change of variables $u = x \mathcal{V}_{h_{\mathrm{circ}}+Y_{\mathfrak{R}(\cdot)}^e}^\gamma (\partial S)$, we have (very similarly to before):

$$m_\alpha^{\mathrm{disc}}\left(F\left(h - \tfrac{2}{\gamma} \log \mathcal{V}_h^\gamma(\partial S)\right) \mathbf{1}_{\{\mathcal{V}_h^\gamma(\partial S) \in A\}}\right)$$

$$= c_\delta \, \mathbb{P}_{\mathrm{circ}}^{\mathrm{GFF}} \otimes v_\delta^1 \left(\frac{F\left(h_{\mathrm{circ}} + Y_{\mathfrak{R}(\cdot)}^e - \tfrac{2}{\gamma} \log \mathcal{V}_{h_{\mathrm{circ}}+Y_{\mathfrak{R}(\cdot)}^e}^\gamma (\partial S)\right)}{(\mathcal{V}_{h_{\mathrm{circ}}+Y_{\mathfrak{R}(\cdot)}^e}^\gamma (\partial S))^{2\gamma^{-1}(Q-\alpha)}} \right) \int_A u^{\tfrac{2(Q-\alpha)}{\gamma}-1} du$$

$$= \frac{\mathbb{P}^S_{\text{circ}} \otimes \nu^1_\delta \left(F\left(h_{\text{circ}} + Y^e_{\mathfrak{R}(\cdot)} - \frac{2}{\gamma} \log \mathcal{V}^\gamma_{h_{\text{circ}}+Y^e_{\mathfrak{R}(\cdot)}}(\partial S) \right) (\mathcal{V}^\gamma_{h_{\text{circ}}+Y^e_{\mathfrak{R}(\cdot)}}(\partial S))^{-\frac{2(Q-\alpha)}{\gamma}} \right)}{\mathbb{P}^S_{\text{circ}} \otimes \nu^1_\delta \left((\mathcal{V}^\gamma_{h_{\text{circ}}+Y^e_{\mathfrak{R}(\cdot)}}(\partial S))^{-\frac{2(Q-\alpha)}{\gamma}} \right)}$$

$\times m^{\text{disc}}_\alpha (\mathcal{V}^\gamma_h(\partial S) \in A).$

The result then follows from the definition of $\hat{h}^{\text{disc},1}_\alpha$. □

Similarly, we can make sense of a *unit area quantum disc*.

Proposition 7.36 *For $\alpha \in (Q, Q + \gamma)$*

$$m^{\text{disc}}_\alpha = \hat{c}_\delta \int_0^\infty \hat{\mathbb{P}}^{\text{disc},a}_\alpha a^{\frac{(Q-\alpha)}{\gamma}-1} \, da, \quad (7.21)$$

where $\hat{\mathbb{P}}^{\text{disc},a}_\alpha$ is a probability measure on $H^{-1}_{\text{loc}}(S)$, such that for $\hat{\mathbb{P}}^{\text{disc},u}_\alpha$ every h, the bulk Gaussian multiplicative chaos measure $\mathcal{M}^\gamma_h(S)$ is well defined and satisfies

$$\mathcal{M}^\gamma_h(S) = a.$$

Moreover, if we write $\hat{h}^{\text{disc},1}_\alpha$ for the field on S with the law of

$$h_{\text{circ}} + Y^e_{\mathfrak{R}(\cdot)} - \gamma^{-1} \mathcal{M}^\gamma_{h_{\text{circ}}+Y^e_{\mathfrak{R}(\cdot)}}(S) \text{ weighted by } \left(\mathcal{M}^\gamma_{h_{\text{circ}}+Y^e_{\mathfrak{R}(\cdot)}}(S) \right)^{-\frac{(Q-\alpha)}{\gamma}},$$

where $h_{\text{circ}} \overset{(\text{law})}{=} h^S_{\text{circ}}$, and e is an independent Bessel excursion of dimension $\delta(\alpha)$ conditioned on taking maximum value 1 (i.e. with law ν^1_δ, so that Y^e is as described in Lemma 7.31), then

$$\hat{h}^{\text{disc},u}_\alpha := \hat{h}^{\text{disc},1}_\alpha + \frac{1}{\gamma} \log(u)$$

is a sample from $\hat{\mathbb{P}}^{\text{disc},u}_\alpha$.

Proof The proof is very similar to that of Proposition 7.34, and we leave it as an exercise. □

Definition 7.37 We call the quantum surface $(S, \hat{h}^{\text{disc},1}_\alpha, +\infty, -\infty)$ (when $\hat{h}^{\text{disc},1}_\alpha$ has law $\hat{\mathbb{P}}^{\text{disc},1}_\alpha$) a unit area α-quantum disc, and $(S, \hat{h}^{\text{disc},u}_\alpha, +\infty, -\infty)$ (when $\hat{h}^{\text{disc},u}_\alpha$ has law $\hat{\mathbb{P}}^{\text{disc},u}_\alpha$) an α-quantum disc with quantum area u.

7.5 Quantum spheres

The final quantum surface we will introduce in this chapter is the so-called **quantum sphere**, which has finite area like the quantum disc, but does not have a boundary. It can therefore be thought of as the finite volume analogue of the quantum cone introduced in Section 7.2. As usual we consider the parameter $\gamma \in (0, 2)$ to be fixed from now on, and the definition of quantum surfaces (i.e. the change of coordinates formula) is with respect to this value of γ.

Quantum spheres will be defined for a parameter $\alpha \in (Q, Q + \frac{\gamma}{2})$ (note the difference with the case of quantum discs). The α-quantum sphere will be defined as a doubly marked quantum surface (now with **interior** rather than boundary marked points) which looks, locally near the marked points, like a $(2Q - \alpha)$-quantum cone near its apex, at least in a suitable range of α.

Recall that we defined the the α^*-quantum cone, for $0 < \alpha^* < Q$, to be the doubly marked quantum surface represented by $(C, h_{\text{circ}} + h_{\text{rad}}, +\infty, -\infty)$ where $C := \{z = x + iy : y \in \mathbb{R}/(2\pi\mathbb{Z})\}$ is the infinite cylinder, h_{circ} has the law of the whole plane GFF on C minus its average value on each vertical segment $\{x + iy : y \in [0, 2\pi]\}$, and h_{rad} is independent of h_{circ} with

$$h_{\text{rad}}(z) = \begin{cases} B_s + (\alpha^* - Q)s & \text{if } \Re(z) = s \text{ and } s \geq 0 \\ \widehat{B}_{-s} + (\alpha^* - Q)s & \text{if } \Re(z) = s \text{ and } s < 0, \end{cases}$$

for $B = (B_t)_{t \geq 0}$ is a standard Brownian motion, and $\widehat{B} = (\widehat{B}_t)_{t \geq 0}$ is an independent Brownian motion conditioned so that $\widehat{B}_t + (Q - \alpha^*)t > 0$ for all $t > 0$.

As was the case when describing thin quantum wedges and quantum discs, we can switch perspective slightly, and (equivalently) define the α^*-quantum cone to be the doubly marked quantum surface which, when parametrised by C with marked points at $-\infty, +\infty$ (note the switch in order) is represented by a field with the law of $h_{\text{circ}} + h_{\text{rad}}$ viewed modulo horizontal translation, where

$$h_{\text{rad}}(z) = B_s + (Q - \alpha^*)s \quad \text{for} \quad \Re(z) = s, \quad (7.22)$$

where B is a standard two-sided Brownian motion, independent of h_{circ}, and viewed modulo translation (i.e. $(B_t)_{t \in \mathbb{R}}$ is identified with $(B_{t_0+t})_{t \in \mathbb{R}}$ for any $t_0 \in \mathbb{R}$).

For $\alpha \in (Q, Q + \frac{\gamma}{2})$, if we let

$$\delta_{\text{cone}}(\alpha) = 2 + \frac{4}{\gamma}(Q - \alpha)$$

(notice the factor two difference compared to the quantum disc case), then $\delta_{\text{cone}}(\alpha) =: \delta \in (0, 2)$. Moreover, by Lemma 7.20 and Remark 7.21, if v_δ^1 is the δ-dimensional Bessel excursion measure conditioned on reaching maximum value 1, and e is sampled from v_δ^1, then V^e defined by taking $\frac{2}{\gamma}\log(e)$, reparametrised to reach its maximum at time 0 and to have infinitesimal

7.5 Quantum spheres 287

quadratic variation dt (as described in Lemma 7.20 but with 2 dt replaced by dt), then we have that $V_0 = 0$ and

$$V_t = B_t + (Q - \alpha)t \qquad \text{for } t > 0$$
$$V_t = \hat{B}_{-t} + (Q - \alpha)(-t) \qquad \text{for } t < 0, \qquad (7.23)$$

where $B, \hat{B}$ are independent standard linear Brownian motions defined for $t \in [0, \infty)$, started from 0 and conditioned that $B_t + (Q-\alpha)t$ (resp. $\hat{B}_t + (Q-\alpha)t$) is negative for all $t > 0$.
This leads us to the following definition.

Definition 7.38   Let $\alpha \in (Q, Q + \frac{\gamma}{2})$ and ν_δ^{BES} be the Bessel excursion measure with dimension $\delta = \delta_{\text{cone}}(\alpha) = 2 + \frac{4}{\gamma}(Q - \alpha)$. Let $\mathbb{P}_{\text{circ}}^C$ be the law obtained from a whole plane GFF in C by subtracting its average value on each vertical line segment, as described in (7.11).

We define the infinite α sphere measure m_α^{sphere} to be the measure on $H_{\text{loc}}^{-1}(C)$ obtained by pushing forward the measure $\nu_\delta^{\text{BES}} \otimes \mathbb{P}_{\text{circ}}^C$ to $H_{\text{loc}}^{-1}(C)$, via the map taking (e, h_{circ}) to the field $h_{\text{circ}} + V_{\mathfrak{R}(\cdot)}^e$, where V is constructed from e as described above (7.23).

An analogous remark to Remark 7.29 holds in this case. We also have the following analogue of Proposition 7.34

Proposition 7.39   *For $\alpha \in (Q, Q + \frac{\gamma}{2})$*

$$m_\alpha^{\text{sphere}} = c_\delta^* \int_0^\infty \mathbb{P}_\alpha^{\text{sphere},u} u^{\frac{2(Q-\alpha)}{\gamma} - 1} \, du, \qquad (7.24)$$

where c_δ^* is a constant, and $\mathbb{P}_\alpha^{\text{sphere},u}$ is a probability measure on $H_{\text{loc}}^{-1}(C)$, such that for $\mathbb{P}_\alpha^{\text{sphere},u}$-almost every h, the bulk Gaussian multiplicative chaos measure $\mathcal{M}_h^\gamma(C)$ is well defined and satisfies

$$\mathcal{M}_h^\gamma(C) = u.$$

Moreover, if we write $h_\alpha^{\text{sphere},1}$ for the field on C with the law of

$$h_{\text{circ}} + V_{\mathfrak{R}(\cdot)}^e - \frac{1}{\gamma} \log \mathcal{M}_{h_{\text{circ}}+V_{\mathfrak{R}(\cdot)}^e}^\gamma (C) \text{ weighted by } \left(\mathcal{M}_{h_{\text{circ}}+V_{\mathfrak{R}(\cdot)}^e}^\gamma (C)\right)^{-\frac{2(Q-\alpha)}{\gamma}},$$

where h_{circ} is distributed according to $\mathbb{P}_{\text{circ}}^C$ and e is an independent Bessel excursion of dimension $\delta_{\text{cone}}(\alpha)$ conditioned on taking maximum value 1, then

$$h_\alpha^{\text{sphere},u} := h_\alpha^{\text{sphere},1} + \frac{1}{\gamma} \log(u)$$

is a sample from $\mathbb{P}_\alpha^{\text{sphere},u}$.

Proof The proof closely mirrors that of Proposition 7.34, and we leave it as an exercise. □

Definition 7.40 When $h_\alpha^{\text{sphere},1}$ has law $\mathbb{P}_\alpha^{\text{sphere},1}$, we call the doubly marked quantum surface $(C, h_\alpha^{\text{sphere},1}, -\infty, +\infty)$ a unit area α-quantum sphere; and when $h_\alpha^{\text{sphere},u}$ has law $\mathbb{P}_\alpha^{\text{sphere},u}$, we call $(C, h_\alpha^{\text{sphere},u}, -\infty, +\infty)$ an α-quantum sphere with area u.

From (7.23) and (7.22) with $\alpha^* = 2Q - \alpha$, we see that, at least informally speaking, an α-quantum sphere looks locally, near each of its marked points, like an α^*-quantum cone at its marked point with neighbourhoods of finite quantum area.

7.6 Special cases

Theorems 7.11 and 7.18 tell us that the α-quantum wedge and α-quantum cone can be obtained as local limits of Neumann and whole plane GFFs, respectively, at boundary (respectively bulk) points where a deterministic α-log singularity is added to the field. On the other hand, we know by Girsanov's theorem, similarly to Theorem 2.4, that if we take a Neumann (respectively whole plane) GFF and sample a point from the boundary (respectively bulk) γ Liouville measures, this is closely related to sampling a point from Lebesgue measure and then adding a γ-log singularity to the field at this point. As such, the γ-quantum wedge and the γ-quantum cones are particularly important examples of quantum surfaces. Indeed, we will see them appear prominently in the key theorems of Chapters 8 and 9. The corresponding special parameters in the case of discs and spheres are, by the duality discussed in Sections 7.4 and 7.5, when $\alpha = 4/\gamma$.

Remark 7.41 (Weights) In [DMS21], an alternative parametrisation of wedges, cones, discs and spheres is used, in terms of their so-called **weight** W. The reason for this is that parameterising by weight behaves well (in fact additively) under operations of cutting and welding surfaces; we will see such operations in Theorems 8.31, 9.24, 9.26 and 9.29. The conversion from α to W goes as follows:

- **Wedge**: $W = \gamma(\frac{2}{\gamma} + Q - \alpha)$;
- **Cone**: $W = 2\gamma(Q - \alpha)$;
- **Disc**: $W = \gamma(\frac{2}{\gamma} + \alpha - Q)$;
- **Sphere**: $W = 2\gamma(\alpha - Q)$.

For the special cases mentioned above we thus have:

- **Wedge**: $\alpha = \gamma \Rightarrow W = 2$;
- **Cone**: $\alpha = \gamma \Rightarrow W = 4 - \gamma^2$;
- **Disc**: $\alpha = 4/\gamma \Rightarrow W = 2$;
- **Sphere**: $\alpha = 4/\gamma \Rightarrow W = 4 - \gamma^2$.

7.7 Equivalence of quantum and Liouville spheres

In this section, we show that the notion of quantum sphere introduced in this chapter actually coincides with the unit volume Liouville sphere (coming from Liouville CFT) defined in Chapter 5, in the following sense.

Theorem 7.42 (Equivalence of spheres) *Fix $\alpha \in (Q, Q + \frac{\gamma}{2})$ and suppose that h is sampled from $\mathbb{P}_\alpha^{\text{sphere},1}$ (defined above Definition 7.40). Given h, let $w \in C$ be sampled from $\mathcal{M}_h$, normalised to be a probability measure. Let*

$$h' = h \circ \psi_w^{-1} + Q \log |(\psi_w^{-1})'(\cdot)|,$$

where $\psi: C \to \hat{C}$ sends $-\infty \mapsto 0$, $\infty \mapsto \infty$ and $w \mapsto 1$. Then h' has the law of the unit volume Liouville sphere $h_{\beta,\mathbf{z}}^{L,1}$ from Definition 5.40 with $\beta = (2Q - \alpha, 2Q - \alpha, \gamma)$ and $\mathbf{z} = (0, \infty, 1)$.

Remark 7.43 The restriction of α to the range $\alpha \in (Q, Q + \frac{\gamma}{2})$ is not only to guarantee the existence of the unit area quantum sphere, but also to guarantee that the insertion parameter $\beta = (2Q - \alpha, 2Q - \alpha, \gamma)$ satisfies the Seiberg bounds, which is necessary in order to define $h_{\beta,\mathbf{z}}^{L,1}$. In the special case mentioned above, when $\alpha = 4/\gamma$ (this is possible if $\gamma \in (\sqrt{2}, 2)$), this produces the unit volume Liouville sphere with all weights equal to γ.

Theorem 7.42 was first shown in the case $\alpha = 4/\gamma$ by Aru, Huang and Sun [AHS17] (this is stated for $\gamma \in (0, 2)$, but as already noted the restriction to $\gamma \in (\sqrt{2}, 2)$ is important in our framework; an extension to smaller values γ is possible using Remarks 5.41 and 7.29 but would require more work). The more general statement above is implicit in the work of Ang, Holden and Sun in [AHS24]. The proof below is based on [AHS24]; compared to [AHS17], a key idea is to not work directly with the law of the unit volume objects (which are hard to manipulate directly owing to the singularity of the conditioning) and instead work in the setting of infinite measures from which these laws arise. Since we already know the corresponding disintegration statement with respect to the "law" of the volume in both cases, this will greatly simplify the analysis (although it lends itself less to the probabilistic intuition).

290   *Quantum Wedges and Scale-Invariant Random Surfaces*

Proof of Theorem 7.42 As mentioned above, it will be more convenient to work in the setting of infinite measures, where both spheres are more canonically defined. To this end, our first step will be to define two natural infinite measures M^S and M^L (corresponding to the quantum surface and Liouville CFT perspectives, respectively) and reduce the proof to showing that these measures are identical (up to a deterministic multiplicative constant).

We start with the measure M^S on $C \times H^{-1}(\hat{\mathbb{C}})$, defined by

$$M^S(F(h)f(w)) :=$$

$$\int_{\mathbb{R}} m_\alpha^{\text{sphere}} \left(\int_C f(w) F(h^t \circ \psi_w^{-1} + Q \log |(\psi_w^{-1})'(\cdot)|) \mathcal{M}_{h^t}(\mathrm{d}w) \right) \mathrm{d}t$$

for non-negative Borel functions F on $H^{-1}(\hat{\mathbb{C}})$ and f on $\mathbb{R}$, where for $t \in \mathbb{R}$, h^t denotes the field $h(\cdot + t)$. That is, we "sample" the field h on C according to the infinite quantum sphere measure m_α^{sphere}, and "independently sample" a real number t from Lebesgue measure. Then we horizontally shift the field by t, choose $w \in C$ according to the quantum area measure associated with the shifted field h^t, and finally take the image of h^t after conformally mapping C to $\hat{\mathbb{C}}$, sending $-\infty \mapsto 0$, $\infty \mapsto \infty$ and $w \mapsto 1$. Note that this is equivalent to simply choosing w according to the quantum area measure associated with h and mapping h to $\hat{\mathbb{C}}$ (as we will justify and use below), but we want the measure M^S to be represented in the form above. The reason for this is because, under m_α^{sphere}, the horizontal translation is fixed so that the maximum of the field is obtained at $\Re(\cdot) = 0$, while this is not the case for the Liouville field.

Next we define M^L on $H^{-1}(\hat{\mathbb{C}}) \times C$ by

$$M^L(F(h)f(w)) = m_{\beta,\mathbf{z}}^L(F) \int_C f(w) \, \mathrm{d}w,$$

where $\beta = (2Q - \alpha, 2Q - \alpha, \gamma)$, $\mathbf{z} = (0, \infty, 1)$ and $m_{\beta,\mathbf{z}}^L(F)$ is as defined in Remark 5.43. The measure $m_{\beta,\mathbf{z}}^L$ is an infinite measure on $H^{-1}(\hat{\mathbb{C}})$ which defines the random area Liouville CFT sphere with insertions of strength $2Q - \alpha$ at 0 and ∞, and an insertion of strength γ at 1.

We will show that M^S and M^L are proportional to one another; let us first explain why this yields the proof of the theorem. Observe that by changing variables $x = w + t$ (with $u \in C, t \in \mathbb{R}$), we have that (for arbitrary non-negative measurable functions F and f)

$$M^S(F(h)f(w))$$

$$= \int_{\mathbb{R}} m_\alpha^{\text{sphere}} \left(\int_C f(w) F(h^t \circ \psi_w^{-1} + Q \log |(\psi_w^{-1})'(\cdot)|) \mathcal{M}_{h^t}(\mathrm{d}w) \right) \mathrm{d}t$$

7.7 Equivalence of quantum and Liouville spheres

$$= m_\alpha^{\text{sphere}} \left(\int_C \left(\int_\mathbb{R} f(x-t) \, dt \right) F(h \circ \psi_x^{-1} + Q \log |(\psi_x^{-1})'(\cdot)|) \, \mathcal{M}_h(dx) \right),$$

since $h^t \circ \psi_w^{-1} = h \circ \psi_x^{-1}$ and $|(\psi_w^{-1})'(\cdot)| = |(\psi_x^{-1})'(\cdot)|$. So, for example, choosing $f_0(z) = p(\mathfrak{R}(z))$ with $\int_\mathbb{R} p(y) \, dy = 1$ with p non-negative and measurable, we get that

$$M^S(F(h)f_0(w))$$

$$= m_\alpha^{\text{sphere}} \left(\int_C F(h \circ \psi_w^{-1} + Q \log |(\psi_w^{-1})'(\cdot)|) \mathcal{M}_h(dw) \right)$$

$$= c \int_{u>0} u^{\frac{2(Q-\alpha)}{\gamma}} \mathbb{P}_\alpha^{\text{sphere},u} \left(\int_C F(h \circ \psi_w^{-1} + Q \log |(\psi_w^{-1})'|) \frac{\mathcal{M}_h(dw)}{u} \right) du$$

by (7.39), where $c = c_\delta^*$ is a deterministic constant.

Now consider M^L. By Remark 5.43, we have that

$$M^L(F(h)f_0(w)) = c' \int_{u>0} u^{\frac{2(Q-\alpha)}{\gamma}} \mathbb{E}(F(h_{\beta,\mathbf{z}}^{L,u})) \, du, \quad (7.25)$$

where c' is another constant (depending only on γ and α) and $h_{\beta,\mathbf{z}}^{L,u}$ is the volume u-Liouville sphere from Chapter 5 (equivalently $h_{\beta,\mathbf{z}}^{L,1} + \gamma^{-1} \log(u)$, where $h_{\beta,\mathbf{z}}^{L,1}$ is the unit volume Liouville sphere).

Therefore, if M^S and M^L are proportional to one another, it follows that the law of $h_{\beta,\mathbf{z}}^{L,u}$ and that of $h \circ \psi_w^{-1} + Q \log |(\psi_w^{-1})'(\cdot)|)$ when (h, w) is sampled from $\mathbb{P}_\alpha^{\text{sphere},u} \mathcal{M}_h(dw)/u$, are proportional to each other for Lebesgue almost all $u > 0$. Noting that both are probability measures (indeed, the total $\mathcal{M}_h$-mass of C under $\mathbb{P}_\alpha^{\text{sphere},u}$ is a.s. equal to u), and noting that the dependence on u is continuous, it follows that these two laws are equal to one another for all $u > 0$. We thus obtain the desired statement by taking $u = 1$.

It therefore remains to prove that M^S and M^L are equal as (infinite) measures on $H^{-1}(\hat{\mathbb{C}}) \times C$, up to a multiplicative constant. We now state three key claims, whose proofs we postpone to the end.

Claim 7.44 (An identity for shifted Brownian motions with drift)

$$\int_\mathbb{R} m_\alpha^{\text{sphere}}(F(h^t)) \, dt = b_1 \int_\mathbb{R} e^{(2Q-2\alpha)c} P^C(F(h+c)) \, dc$$

for some deterministic constant b_1, where under the probability measure P^C, $h = h_{\text{circ}} + B_{\mathfrak{R}(\cdot)} + (Q-\alpha)|\mathfrak{R}(\cdot)|$, B is a two-sided Brownian motion equal to 0 at 0, and h_{circ} has the law $\mathbb{P}_{\text{circ}}^C$, as in the Definition, 7.38, of the unit volume quantum sphere.

292   *Quantum Wedges and Scale-Invariant Random Surfaces*

Claim 7.45 (A version of Girsanov for infinite measures)

$$\int_{\mathbb{R}} e^{(2Q-2\alpha)c} P^C \left(\int_C f(w) F(h+c) \mathcal{M}_{h+c}(\mathrm{d}w) \right) \mathrm{d}c$$
$$= \int_C f(w) e^{\gamma(Q+\gamma/2-\alpha)|\Re(w)|} \int_{\mathbb{R}} e^{(2Q-2\alpha+\gamma)c} P^C(F(h+\gamma G(\cdot,w)+c)) \, \mathrm{d}c \, \mathrm{d}w,$$
(7.26)

where G is the covariance kernel of $h_{\text{circ}} + B_{\Re(\cdot)}$.

Claim 7.46 (An application of the Weyl anomaly) *There exists a constant $b_2 \neq 0$ depending only on γ and α, such that for each fixed $w \in C$, and for all non-negative Borel functions F on $H^{-1}_{\text{loc}}(C)$*

$$m^L_{\beta,\mathbf{z}}(F(h \circ \psi_w + Q \log |\psi'_w|)) =$$
$$b_2 e^{\gamma(Q+\gamma/2-\alpha)|\Re(w)|} \int_{\mathbb{R}} e^{(2Q-2\alpha+\gamma)c} P^C(F(h+\gamma G(\cdot,w)+c)) \, \mathrm{d}c$$

(Note that since h is viewed as an element of $H^{-1}(\hat{\mathbb{C}})$ under $m^L_{\beta,\mathbf{z}}$, which is a subset of $H^{-1}_{\text{loc}}(\mathbb{C})$, it is indeed the case that $h \circ \psi_w + Q \log |\psi'_w| \in H^{-1}_{\text{loc}}(C)$.)

Let us check how this claims imply the desired proportionality result between M^L and M^S, which will be obtained in (7.27). For non-negative, measurable functions F and f, set $\tilde{F}(h) = F(h) \int_C f(w) \mathcal{M}_h(\mathrm{d}w)$. Then

$$\int_{\mathbb{R}} m^{\text{sphere}}_\alpha \left(\int_C f(w) F(h^t) \mathcal{M}_{h^t}(\mathrm{d}w) \right) \mathrm{d}t$$
$$= \int_{\mathbb{R}} m^{\text{sphere}}_\alpha (\tilde{F}(h^t)) \, \mathrm{d}t$$
$$= b_1 \int_{\mathbb{R}} e^{(2Q-2\alpha)c} P^C(\tilde{F}(h+c)) \, \mathrm{d}c \quad \text{(by Claim 7.44)}$$
$$= b_1 \int_{\mathbb{R}} e^{(2Q-2\alpha)c} P^C \left(F(h+c) \int_C f(w) \mathcal{M}_{h+c}(\mathrm{d}w) \right) \mathrm{d}c$$
$$= \frac{b_1}{b_2} \int_C f(w) m^L_{\beta,\mathbf{z}}(F(h \circ \psi_w + Q \log |\psi'_w|)) \, \mathrm{d}w,$$

by Claims 7.45 and 7.46. Since this holds for measurable non-negative functions f and F, a monotone class argument shows that for all jointly measurable non-negative functions G of h and w,

$$\int_{\mathbb{R}} m^{\text{sphere}}_\alpha \left(\int_C G(h^t, w) \mathcal{M}_{h^t}(\mathrm{d}w) \, \mathrm{d}t \right)$$
$$= \frac{b_1}{b_2} \int_C f(w) m^L_{\beta,\mathbf{z}}(G(h \circ \psi_w + Q \log |\psi'_w|, w)) \, \mathrm{d}w.$$

7.7 Equivalence of quantum and Liouville spheres

Applying this identity with $G(h, w) = F(h \circ \psi_w^{-1} + Q \log |(\psi_w^{-1})'|) f(w)$ shows that

$$\int_{\mathbb{R}} m_\alpha^{\text{sphere}} \left(\int_C F(h^t \circ \psi_w^{-1} + Q \log |(\psi_w^{-1})'|) f(w) \right) \mathcal{M}_{h^t}(\mathrm{d}w) \, \mathrm{d}t$$
$$= \frac{b_1}{b_2} \int_C m_{\beta, \mathbf{z}}^L (F(h) f(w)) \, \mathrm{d}w.$$

Referring to the definition of these measures, this means

$$M^S(F(h)f(w)) = \frac{b_1}{b_2} M^L(F(h)f(w)) \quad (7.27)$$

for arbitrary non-negative and measurable functions F, f. As discussed, this completes the proof. □

Proof of Claim 7.44 Recall that

$$m_\alpha^{\text{sphere}}(F) = \mathbb{P}_{\text{circ}}^C \otimes \nu_\delta^{\text{BES}}(F(h_{\text{circ}} + V_{\mathfrak{R}(\cdot)}^e)),$$

where in the above, h_{circ} has law $\mathbb{P}_{\text{circ}}^C$ and V^e is defined from the excursion e (sampled from the Bessel excursion measure ν_δ^{BES} with $\delta = \delta_{\text{cone}}(\alpha) = 2 + (4/\gamma)(Q - \alpha) \in (0, 2)$ for our range of values of $\alpha \in (Q, Q + \frac{\gamma}{2})$) as $(2/\gamma) \log(e)$ with time parametrised to have infinitesimal quadratic variation $\mathrm{d}t$ and horizontal translation fixed so its maximum value occurs at time 0.

By invariance of the law of h_{circ} under horizontal translations, the proof of this claim therefore amounts to showing that the measure η on the space of continuous functions from $\mathbb{R}$ to $\mathbb{R}$ defined by

$$\eta(A) = \int_{\mathbb{R}} e^{2(Q-\alpha)c} \mathbb{P}(B_\cdot + (Q - \alpha)| \cdot | + c \in A) \, \mathrm{d}c$$

and the measure $\tilde{\eta}$ defined by

$$\tilde{\eta}(A) = \int_{\mathbb{R}} \nu_\delta^{\text{BES}} (V_{\cdot + t}^e \in A) \, \mathrm{d}t$$

are equal up to a multiplicative constant. Above and for the rest of this proof, we use the notation $\mathbb{P}$ and $\mathbb{E}$ for the law of two-sided Brownian motion B started from 0 at time 0 (i.e. a standard Brownian motion run forward from time 0 joined with an independent standard Brownian motion run backwards from time 0).

To show the equivalence of these measures, we first show that the "marginal law" of the function at time 0 under η and $\tilde{\eta}$ are the same, up to deterministic multiplicative constant. For this, observe that for any $C \in \mathbb{R}$, if we let $\tau_C = \{\inf s \in \mathbb{R} : V_s^e = C\}$ then the ν_δ^{BES} law of $(V_{\tau_C + s}^e)_{s \geq 0}$ conditioned on $\sup e \geq \exp(\gamma C/2)$ is that of $(2/\gamma) \log(e^{(C)})$ parametrised to have quadratic variation s

at time s, where $e^{(C)}$ is a δ-dimensional Bessel process, started from $\exp(\gamma C/2)$ and killed upon hitting zero. This is straightforward to see in the case of Itô's excursion theory (see also [RV19, Lemma 3.4]). By Lemma 7.20, the ν_δ^{BES} law of $(V^e_{\tau_C+s})_{s\geq 0}$ conditioned on $\sup e \geq \exp(\gamma C/2)$ is simply the law of $(C + B_s + (Q - \alpha)s)_{s\geq 0}$ where B is a standard Brownian motion started from 0 at time 0. In particular, if A_C is the set of functions which are $\geq C$ at time 0, then by definition of $\tilde\eta$,

$$\tilde\eta(A_C) = \int_{\mathbb{R}} \nu_\delta^{\text{BES}}(V^e_t \geq C)\, dt$$

$$= \int_{\mathbb{R}} \nu_\delta^{\text{BES}}(V^e_t \geq C, \sup e \geq \exp(\gamma C/2), \tau_C \leq t)\, dt$$

$$= \int_0^\infty \nu_\delta^{\text{BES}}(V^e_{t+\tau_C} \geq C, \sup e \geq \exp(\gamma C/2))\, dt \qquad (7.28)$$

by Fubini and changing variables $t \to t + \tau_C$. Hence

$$\tilde\eta(A_C) = \int_0^\infty \mathbb{P}(B_t + (Q - \alpha)t \geq 0)\nu_\delta^{\text{BES}}(\sup e \geq \exp(\gamma C/2))\, dt$$

$$= \nu_\delta^{\text{BES}}(\sup e \geq \exp(\gamma C/2)) \int_0^\infty \mathbb{P}(B_t \geq (\alpha - Q)t)\, dt$$

$$\propto \int_{e^{\gamma C/2}}^\infty x^{\delta-3}\, dx \times \int_0^\infty \int_{(\alpha-Q)\sqrt{t}}^\infty \frac{1}{\sqrt{2\pi}} e^{-x^2/2}\, dx\, dt \quad \text{(by Lemma 7.26)}$$

$$= \frac{e^{2(Q-\alpha)C}}{\frac{4}{\gamma}(\alpha - Q)} \frac{1}{2(\alpha - Q)^2} \quad \text{(by Fubini and using the value of } \delta\text{)}$$

$$\propto \eta(A_C), \qquad (7.29)$$

where the implied constants of proportionality above do not depend on C. In other words, if we push forward the measures η and $\tilde\eta$ via the map $X \mapsto X_0$, then the resulting infinite measures on $\mathbb{R}$ are multiples of one another.

Furthermore, for any $C \in \mathbb{R}$, and any non-negative measurable functions F and G, by the same argument as above,

$$\tilde\eta(F((X_{-s})_{s\geq 0})G((X_r - X_0)_{r\geq 0}))$$

$$= \int_{\mathbb{R}} \nu_\delta^{\text{BES}}(F((V^e_{t-s})_{s\geq 0})G((V^e_{t+r} - V^e_t)_{r\geq 0})\mathbf{1}_{V^e_t \geq C})\, dt$$

$$= \int_0^\infty \nu_\delta^{\text{BES}}(F((V^e_{\tau_C+t-s})_{s\geq 0})G((V^e_{\tau_C+t+r} - V^e_{\tau_C+t})_{r\geq 0})\mathbf{1}_{V^e_{\tau_C+t} \geq C})\, dt.$$

Now, as noted previously, the ν_δ^{BES} law of $(V^e_{\tau_C+r})_{r\geq 0}$ conditioned on $\sup e \geq \exp(\gamma C/2)$ is that of $(C + B_r + (Q - \alpha)r)_{r\geq 0}$ where B is a standard Brownian motion started from 0 at time 0, independent of $(V^e_{\tau_C-s})_{s\geq 0}$. Using the Markov

7.7 Equivalence of quantum and Liouville spheres

property at time t of the above Brownian motion with drift, we can therefore rewrite the final expression above as

$$\mathbb{E}((G(B_r + (Q-\alpha)r)_{r\geq 0}) \int_0^\infty \nu_\delta^{\text{BES}}(F((V^e_{\tau_C+t-s})_{s\geq 0})\mathbf{1}_{V^e_{\tau+t}\geq C})\,dt.$$

Considering the special case where $G = 1$, we deduce that

$$\tilde{\eta}(F((X_{-s})_{s\geq 0})G((X_r - X_0)_{r\geq 0}))$$
$$= \tilde{\eta}(F((X_{-s})_{s\geq 0}))\mathbb{E}((G(B_r + (Q-\alpha)r)_{r\geq 0})$$
$$= \tilde{\eta}(F((X_{-s})_{s\geq 0}))\eta(G(X_r - X_0)_{r\geq 0}).$$

In other words (and somewhat informally), conditionally on the value of the function at time 0 under $\tilde{\eta}$, the future evolution is the same as under η, and it is independent of the value at time 0 and the evolution before time 0. Since both $\tilde{\eta}$ and η are manifestly invariant under reversal of time, this shows that the conditional laws of the evolution under η and under $\tilde{\eta}$, given the value at time 0, are equal. Putting this together with (7.29) completes the proof of the claim. □

Proof of Claim 7.45 For each $c \in \mathbb{R}$, applying Theorem 3.16 to the field $h - (Q-\alpha)|\mathfrak{R}(\cdot)|$ under P^C (whose distribution is that of $h_{\text{circ}} + B_{\mathfrak{R}(\cdot)}$), we see that

$$P^C\left(\int_C f(w)F(h+c)M_{h+c}(dw)\right)$$
$$= e^{\gamma c}P^C\left(\int_C f(w)e^{\gamma(Q-\alpha)|\mathfrak{R}(\cdot)|}F(h+c)M_{h-(Q-\alpha)|\mathfrak{R}(\cdot)|}(dw)\right)$$
$$= e^{\gamma c}\int_C f(w)F(h+c+\gamma G(\cdot,w))\sigma_\gamma(dw)),$$

where G is (as in the statement of Claim 7.45) the covariance kernel of the field $h_{\text{circ}}(\cdot) + B_{\mathfrak{R}(\cdot)}$, and $\sigma_\gamma(dw)$ is the measure $A \mapsto \mathbb{E}[M_{h_{\text{circ}}+B_{\mathfrak{R}(\cdot)}}(A)]$ for A a Borel subset of C. We can compute the expected mass of this GMC as follows:

$$\sigma_\gamma(dw) = \lim_{\varepsilon \to 0} \varepsilon^{\frac{\gamma^2}{2}} e^{\frac{\gamma^2}{2}\text{Var}(h_\varepsilon(w)+B_{\mathfrak{R}(w)})}\,dw.$$

The covariance of the field $h^*(\cdot) := h_{\text{circ}}(\cdot) + B_{\mathfrak{R}(\cdot)}$ is easy to compute from the one of h^c on $\hat{\mathbb{C}}$ via the exponential map $w \in C \mapsto e^w \in \hat{\mathbb{C}}$; by Lemma 5.30, we get

$$\text{Var}(h^*_\varepsilon(w)) = \log(1/\varepsilon) - \log|e^w| + 2\log(|e^w| \vee 1) + o(1)$$
$$= \log(1/\varepsilon) + |\mathfrak{R}(w)| + o(1).$$

Thus,
$$\sigma_\gamma(dw) = e^{\frac{\gamma^2}{2}|\Re(w)|}.$$

Substituting this into the above and integrating over c completes the proof of the claim. □

Proof of Claim 7.46 Recall that $\mathbf{z} = (0, \infty, 1)$. We first use Möbius invariance, Theorem 5.27 and the specialisation to one marked point at ∞ (5.70), to observe that if we set $\phi_w : \hat{\mathbb{C}} \to \hat{\mathbb{C}}$ to be the Möbius map $z \mapsto z/(\exp(w))$

$$m^L_{\beta,\mathbf{z}}(F(h \circ \phi_w) - Q \log |\phi'_w|))$$
$$= |\phi'_w(0)|^{-2\Delta_2 Q - \alpha}|\phi'_w(1)|^{-2\Delta_\gamma}|\phi'_w(\infty)|^{2\Delta_2 Q - \alpha} m^L_{\beta,\exp(w)\mathbf{z}}(F)$$
$$= |\exp(w)|^{2\Delta_\gamma} m^L_{\beta,\exp(w)\mathbf{z}}(F).$$

Now by Remark 5.44 with $z = \exp(w)$ we have that $m^L_{\beta,\exp(w)\mathbf{z}}(F)$ is equal to a multiple C (not depending on w) of

$$(|\exp(w)| \vee 1)^{-4\Delta_\gamma + 2\gamma(2Q - \alpha)} |\exp(w)|^{-\gamma(2Q-\alpha)} \times$$
$$\int_{u>0} \mathbb{E}(F(\tilde{h}^c + \gamma^{-1}(\log u - \log \mathcal{M}_{\tilde{h}^c}(\mathbb{C}))) \mathcal{M}_{\tilde{h}^c}(\mathbb{C})^{-s}) u^{s-1} du, \quad (7.30)$$

where $s = \frac{\sum_i \alpha_i - 2Q}{\gamma} = \frac{2Q - 2\alpha + \gamma}{\gamma}$,

$$\tilde{h}^c = h^c + (2Q - 2\alpha) \log(|\cdot| \vee 1) - (2Q - \alpha) \log |\cdot| + 2\pi \gamma G^c(\exp(w), \cdot), \quad (7.31)$$

and h^c is the whole plane GFF with zero average on the unit circle. Combining the powers of $|\exp(w)|$ and $|\exp(w)| \vee 1$, noting that $|\exp(w)| = e^{\Re(w)}$, and applying the change of variables $u = \mathcal{M}_{\tilde{h}^c}(\mathbb{C}) e^{\gamma c}$ in the integral, after multiple cancellations, we reach the expression

$$m^L_{\beta,\mathbf{z}}(F(h \circ \phi_w - Q \log |\phi'_w|))$$
$$= e^{(-2\Delta_\gamma + \gamma(2Q-\alpha))|\Re(w)|} \int_\mathbb{R} e^{(2Q - 2\alpha + \gamma)c} \mathbb{E}(F(\tilde{h}^c + c)) dc$$

with $\tilde{h}^c$ as above.

From here, notice that $\psi_w = \phi_w \circ \exp$, so that $\psi'_w(\cdot) = \exp(\cdot) \phi'_w \circ \exp(\cdot)$. Thus if we let

$$\tilde{F}(h) = F(h \circ \exp + Q \log |\exp(\cdot)|) = F(h \circ \exp + Q \Re(\cdot)|),$$

we can write the left-hand side of the identity in Claim 7.46 as

$$m^L_{\beta,\mathbf{z}}(F(h \circ \psi_w - Q \log |\psi'_w|)) = m^L_{\beta,\mathbf{z}}(\tilde{F}(h \circ \phi_w - Q \log |\phi'_w|))$$

$$= e^{(-2\Delta_\gamma+\gamma(2Q-\alpha))|\Re(w)|} \int_\mathbb{R} e^{(2Q-2\alpha+\gamma)c} \mathbb{E}(\tilde{F}(\tilde{h}^c + c)) \, dc$$

$$= e^{(-2\Delta_\gamma+\gamma(2Q-\alpha))|\Re(w)|} \int_\mathbb{R} e^{(2Q-2\alpha+\gamma)c} \mathbb{E}(F(\tilde{h}^c \circ \exp(\cdot) + Q\Re(\cdot) + c)) \, dc.$$

Note furthermore using (7.31) that

$$\tilde{h}^c \circ \exp(\cdot) + Q\Re(\cdot) = h_{\text{circ}} + B_{\Re(\cdot)} + (Q-\alpha)|\Re(\cdot)| + 2\pi\gamma G^c(\exp(w), \exp(\cdot)),$$

where h_{circ} and B are as in the statement of the claim, and $2\pi G^c(\exp(\cdot), \exp(\cdot))$ is the covariance of $h^c \circ \exp = h_{\text{circ}} + B_{\Re(\cdot)}$, and is therefore equal to G by definition. Combining this with the fact that $-2\Delta_\gamma + \gamma(2Q-\alpha) = -\gamma Q + \gamma^2/2 + 2\gamma Q - \gamma\alpha = \gamma(Q + \gamma/2 - \alpha)$, we obtain the statement of the claim. □

7.8 Exercises

7.1 Let $D = \{z : \arg(z) \in [0, \theta]\}$ be the (Euclidean) wedge of angle θ, and suppose that $\theta \in (0, 2\pi)$. Let h be a Neumann GFF in D. Show that by zooming in (D, h) near the tip of the wedge, we obtain a thick quantum wedge with $\alpha = Q(\theta/\pi - 1)$ (which satisfies $\alpha < Q$ if $\theta < 2\pi$).

7.2 Show that Theorem 7.11(i) remains true in the sense of convergence in distribution with respect to the topology of uniform convergence on compacts (as opposed to total variation) if we replace h by $h = \tilde{h} + \alpha \log(1/|\cdot|) + \varphi$, where $\tilde{h}$ is a Neumann GFF on $\mathbb{H}$ with some fixed additive constant and φ is a function which is independent of $\tilde{h}$ and continuous at 0. That is, show that if h is as above then as $C \to \infty$, the surfaces $(\mathbb{H}, h + C, 0, \infty)$ converge to an α-thick wedge in distribution.

7.3 ([DMS21, Proposition 4.2.5].) Show the following characterisation of quantum wedges. Fix $\alpha < Q$ and suppose that h is a fixed representative of a quantum surface that is parametrised by $\mathbb{H}$. Suppose that the following hold:

(i) The law of $(\mathbb{H}, h, 0, \infty)$ (as a quantum surface with two marked points 0 and ∞) is invariant under the operation of multiplying its area by a constant. That is, if we fix $C \in \mathbb{R}$, then $(\mathbb{H}, h + C/\gamma, 0, \infty)$ has the same law as $(\mathbb{H}, h, 0, \infty)$.

(ii) The total variation distance between the law of h restricted to $B(0, r)$ and the law of an α-quantum wedge field h_{wedge} (in its unit circle embedding in $\mathbb{H}$) restricted to $B(0, r)$ tends to 0 as $r \to 0$.

Then $(\mathbb{H}, h, 0, \infty)$ has the law of an α-quantum wedge; more precisely h has the law of h_{wedge}.

298   Quantum Wedges and Scale-Invariant Random Surfaces

7.4 (**Quantum cones.**) Verify Lemma 7.15 and give a proof of Theorem 7.18. State and prove the analogue of Exercise 7.1.

7.5 Show that a δ-dimensional Bessel process starting from zero cannot satisfy the SDE (7.14) on $[0, t]$ when $\delta = 1$.

7.6 Prove that a Bessel process enjoys the Brownian scaling property: if Z is a δ-dimensional Bessel process with $Z_0 = x > 0$, then for all $\lambda > 0$, $(Z_{\lambda t}/\sqrt{\lambda})_{t \geq 0}$ is a δ-dimensional Bessel process started from $x/\sqrt{\lambda}$.

Show that the converse is also true; if Z solves the SDE $dZ_t = \sigma(Z_t) dB_t + \beta(Z_t) dt$ until the first hitting time of zero, with σ, β locally Lipschitz on $(0, \infty)$, and if Z satisfies the above Brownian scaling property, then $\sigma(x) \equiv \sigma$ is constant and $\beta(x) \propto 1/x$ for all $x > 0$.

7.7 For an excursion e (i.e. a continuous path from an interval $(0, \zeta)$ to $(0, \infty)$, with $\lim_{t \to 0} e(t) = \lim_{t \to \zeta} e(t) = 0$), set

$$I(e) = \int_0^\zeta \frac{dt}{e(t)}.$$

Verify (using the Brownian scaling property of a Bessel process, see previous exercise) that

$$\nu_\delta^{\text{BES}}(I(e) \in dx) \propto x^{\delta-3} dx.$$

Deduce that $\sum_{i:\, t_i \leq 1} I(e_i) < \infty$ if and only if $\delta > 1$ where (t_i, e_i) is a Poisson point process of intensity $dt \otimes \nu_\delta^{\text{BES}}$. Explain how this is related to the fact that the Bessel SDE (7.14) can only be solved for $\delta > 1$.

7.8 Prove Lemma 7.32 for the total quantum length of ∂S, and Proposition 7.36 for the unit area quantum disc.

7.9 (**Quantum spheres.**) Give a proof of Proposition 7.39.

8
SLE and the Quantum Zipper

In this chapter, we discuss some fundamental results due to Sheffield [She16a] that have the following flavour.

1. Theorem 8.1: An SLE_κ curve has a 'nice' coupling with $e^{\gamma h}$, when h is a certain variant of the Neumann GFF. This coupling can be formulated as a Markov property analogous to the domain Markov properties inherent to random maps. It makes the conjectures about convergence of random maps toward Liouville quantum gravity plausible, and in particular justifies that the "correct" relationship between κ and γ is $\gamma = \min(\sqrt{\kappa}, \sqrt{16/\kappa})$.
2. Theorem 8.9: An SLE_κ curve can be endowed with a random measure which can roughly be interpreted as $e^{\gamma h} \, d\lambda$ for $d\lambda$ a natural length measure on the curve. In fact, the measure $d\lambda$ is in itself hard to define, and the exponent γ needs to be changed slightly from $\sqrt{\kappa}$ to take into account the quantum scaling exponent of the SLE curve – see [BSS23] for a discussion – so we will not actually take this route to define the measure. We will instead use the notion of quantum boundary length. This has the advantage that measures on *either side* of the SLE_κ curve can be defined without difficulty, but we will have to do a fair bit of work to show that they are the same.
3. Theorem 8.31: An SLE_κ curve divides the upper half plane into two independent random surfaces, glued according to boundary length. Thus, SLE curves are solutions of natural random *conformal welding problems*. In fact, the existence of such solutions from a complex analytic view point is a highly non-trivial problem.

Note: We collect some relevant background material on SLE in Appendix A. Readers unfamiliar with the theory may wish to refer to this now.

8.1 SLE and GFF coupling: domain Markov property

Here we describe one of the two couplings between the GFF and SLE. This was first stated in the context of Liouville quantum gravity (although presented slightly differently from here) in [She16a]. However, ideas for a related coupling go back to two seminal papers by Schramm and Sheffield [SS13] on the one hand, and Dubédat [Dub09b] on the other.

Notational remarks:

- In what follows, we will use the multiplicative normalisation for our Neumann GFFs as on the left-hand side of (6.31). That is, such that its covariance in the bulk grows like log (rather than $(2\pi)^{-1}$ log)) near the diagonal.
- Unless stated otherwise, in what follows the use of bars (for example, $\bar{h}$) indicates a distribution that is considered modulo constants.

Let $\bar{h}$ be a Neumann GFF on $\mathbb{H}$ (viewed modulo constants). Let $\kappa > 0$ and let

$$\gamma = \min\left(\sqrt{\kappa}, \sqrt{\tfrac{16}{\kappa}}\right) = \begin{cases} \sqrt{\kappa}; & \text{if } \kappa \leq 4 \\ \sqrt{\tfrac{16}{\kappa}}; & \text{if } \kappa \geq 4. \end{cases} \quad (8.1)$$

Set

$$\bar{h}_0 = \bar{h} + \varphi \quad \text{where } \varphi(z) = \frac{2}{\sqrt{\kappa}} \log |z|; \quad z \in \mathbb{H}. \quad (8.2)$$

Hence $\bar{h}_0$ is a Neumann GFF from which we have *subtracted* (rather than added) a logarithmic singularity at zero. (The reason for the choice of multiple $2/\sqrt{\kappa}$ will become clear only gradually.)

Let $\eta = (\eta_t)_{t \geq 0}$ be an independent chordal SLE_κ curve in $\mathbb{H}$, going from 0 to ∞ and parametrised by half plane capacity, where $\kappa = \gamma^2$. Let g_t be the unique conformal isomorphism $g_t : \mathbb{H}\setminus\{\eta_s\}_{s \leq t} \to \mathbb{H}$ such that $g_t(z) = z+2t/z+o(1/z)$ as $z \to \infty$ (we will call g_t the Loewner map). Then

$$\frac{\mathrm{d}g_t(z)}{\mathrm{d}t} = \frac{2}{g_t(z) - \xi_t}; \quad z \notin \{\eta_s\}_{s \leq t},$$

where $(\xi_t)_{t \geq 0}$ is the Loewner driving function of η, and has the law of $\sqrt{\kappa}$ times a standard one-dimensional Brownian motion. Let $\tilde{g}_t(z) = g_t(z) - \xi_t$ be the *centred* Loewner map.

Theorem 8.1 *Let $T > 0$ be deterministic, and set*

$$\bar{h}_T = \bar{h}_0 \circ \tilde{g}_T^{-1} + Q \log |(\tilde{g}_T^{-1})'|, \text{ where } Q = \frac{2}{\gamma} + \frac{\gamma}{2}.$$

8.1 SLE and GFF coupling: domain Markov property

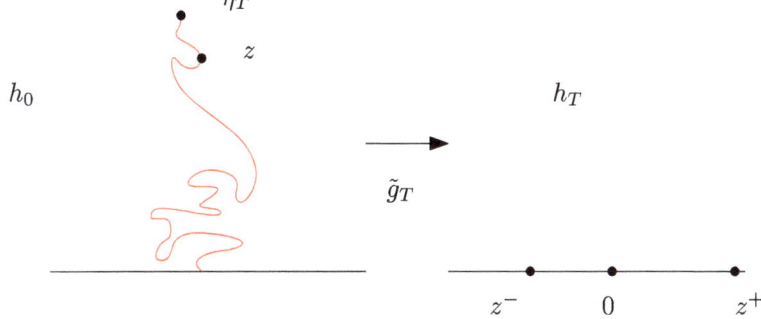

Figure 8.1 Start with the field $\bar{h}_0$ and an independent SLE_κ curve run up to some time T. After mapping $\bar{h}_0$, restricted to the complement of the curve H_T, by the Loewner map $\tilde{g}_T$ and applying the change of coordinate formula, we obtain a distribution modulo constants $\bar{h}_T$ in $\mathbb{H}$ which by the theorem has the same law as $\bar{h}_0$. This is a form of Markov property for random surfaces.

Then $\bar{h}_T$ defines a distribution in $\mathbb{H}$ modulo constants which has the same law as $\bar{h}_0$.

(Recall the meaning of ∘ when dealing with generalised functions:

$$(\bar{h}_0 \circ \tilde{g}_T^{-1}, \rho) = (\bar{h}_0, |(\tilde{g}_T^{-1})|^2 (\rho \circ \tilde{g}_T^{-1}))$$

for any test function ρ.)

Remark 8.2 Here we have started with a field $\bar{h}_0$ with a certain law (described in (8.2)) and a curve η which is *independent* of $\bar{h}_0$. However, η is *not* independent of $\bar{h}_T$. In fact, we will see later on that $\bar{h}_T$ entirely *determines* the curve $(\eta_s)_{0 \le s \le T}$. More precisely, we will see in Theorem 8.9 that when we apply the map $\tilde{g}_T$ to the curve $(\eta_s)_{0 \le s \le T}$, the boundary lengths (measured with $\bar{h}_T$) of the two intervals to which η is mapped by $\tilde{g}_T$ must agree: that is, on Figure 8.1, the γ quantum lengths (with respect to $\bar{h}_T$) of $[z^-, 0]$ and $[0, z^+]$ are the same. (Note that these quantum lengths are only defined up to a multiplicative constant, but their ratio is well defined, so this statement makes sense.) Then, in Theorem 8.29, we will show that given $\bar{h}_T$, the curve $(\eta_s)_{0 \le s \le T}$ is determined by the requirement that $\tilde{g}_T^{-1}$ maps intervals of equal quantum length to identical pieces of the curve η. This is the idea of **conformal welding** (we are welding $\mathbb{H}$ to itself by welding together pieces of the positive and negative real line that have the same quantum length).

Remark 8.3 Suppose that instead of starting with $\bar{h}_0$, viewed modulo constants, we took h_0 to be an equivalence class representative of $\bar{h}_0$ with an additive constant fixed in some arbitrary way (so that $(h_0, \rho_0) = 0$ for some deterministic $\rho_0 \in \mathcal{M}_N$ with $\int \rho_0 = 1$). Then $h_T := h_0 \circ \tilde{g}_T^{-1} + Q \log |(\tilde{g}_T^{-1})'|$ would be such that

$$h_T - (h_T, \rho_0) \stackrel{(\text{law})}{=} h_0 \quad \text{as distributions}.$$

In other words, the laws of h_T and h_0 would differ by a random constant.

Remark 8.4 The proof of the theorem (and the statement which can be found in Sheffield's paper [She16a, Theorem 1.2]) involves the (centred) reverse Loewner flow f_t rather than, for a fixed t, the map $\tilde{g}_t^{-1}$. In this context, the theorem is equivalent to saying that

$$\bar{h}_T = \bar{h}_0 \circ f_T + Q \log |f_T'|, \quad \text{where } Q = \frac{2}{\gamma} + \frac{\gamma}{2}.$$

Moreover, in this case, the theorem is also true if T is a bounded stopping time (for the underlying reverse Loewner flow). The current formulation of Theorem 8.1 has been chosen because the usual forward Loewner flow is a simpler object and more natural in the context of the Markovian interpretation discussed below. On the other hand, the formulation in terms of the reverse flow will be the most useful when we actually come to prove things in this section.

Discussion and interpretation. Let $H_T = \mathbb{H} \setminus \{\eta_s\}_{0 \leq s \leq T}$ and let h_0 be an equivalence class representative of $\bar{h}_0$ (defined by (8.2)), with the additive constant fixed in some arbitrary way. In the language of random surfaces, Theorem 8.1 (more precisely, Remark 8.3) states that the random surface $(H_T, h_0|_{H_T}, \eta(T), \infty)$ has the same distribution, *up to multiplying areas by a random constant*, as $(\mathbb{H}, h_0, 0, \infty)$. This is because h_T is precisely obtained from h_0 by mapping its restriction to H_T through the centred Loewner map $\tilde{g}_T$ and applying the change of coordinates formula. The meaning of "up to multiplying areas by a random constant" corresponds to the fact that the laws of h_T and h_0 differ by a random constant: see Remark 8.3.

In other words, suppose we start with a surface described by $(\mathbb{H}, h_0, 0, \infty)$. Then we explore a small portion of it using an independent SLE_κ, started where the logarithmic singularity of the field is located (here it is important to assume that γ and κ are related by (8.1)). In this exploration, what is the law of the surface that remains to be discovered after some time T? The theorem states that, after zooming in or out by a random amount,[1] this law is the same as

---
[1] Recall from Section 7.1 that we can view the addition of a constant to the field describing a

8.1 SLE and GFF coupling: domain Markov property

the original one. Hence the theorem can be seen as a **Markov property for Liouville quantum gravity**.

The fact that this invariance only holds up to additive constants for the field, or multiplicative constants for the area measure, is because the Neumann GFF is only really uniquely defined modulo constants. *A more natural result comes if one replaces the Neumann GFF by a quantum wedge*, which is scale invariant by definition (meaning that if one adds a constant to the field, its law as a quantum surface does not change). In this context, we have a similar Markov property, but only if the exploration is stopped when the quantum boundary length of the curve reaches a given value: see Theorems 8.9 and 8.16. Of course at the moment, however, we do not even know that the quantum boundary length of SLE is well defined – this will be addressed in Section 8.2.

Connection with the discrete picture. This Markov property is to be expected from the discrete side of the story. To see this, consider for instance the uniform infinite half plane triangulation (UIHPT) constructed by Angel and Curien [AS03, Ang03, AC15]. This is obtained as the local limit of a uniform planar map with a large number of faces and a large boundary, rooted at a uniform edge along the boundary. One can further add a critical site percolation process on this map by colouring vertices black or white independently with probability 1/2 (as shown by Angel, this is indeed the critical value for percolation on such a map). We make an exception for vertices along the boundary, where those to the left of the root edge are coloured in black, and those to the right in white. This generates an interface and it is possible to use that interface to discover the map. Such a procedure is called *peeling* and was used with great efficacy by Angel and Curien [AC15] to study critical percolation on the UIHPT. The important point for us is that conditionally on the map being discovered up to a certain point using this peeling procedure, it is straightforward to see that the rest of the surface that remains to be discovered also has the law of the UIHPT. An analogue also exists for FK models with $q \in (0, 4)$ in place of critical percolation.

This suggests that a nice coupling between the GFF and SLE should exist, recalling the discussion of Section 4.2. However, identifying the exact analogue in the continuum requires a little thought. First, observe that if one embeds the UIHPT into the upper half plane with the distinguished root edge sent to 0, there is a freedom in how the upper half plane is scaled. Roughly, it can be specified how many triangles should be mapped into the upper unit semidisc. The natural scaling limit to consider is then the one that arises by letting this

---

random surface, equivalently multiplying the area measure for the random surface by a constant, as "zooming" in or out of the surface.

number of triangles go to infinity, and rescaling the counting measure on faces appropriately. Note that such a scaling limit will be a "scale invariant" random surface by definition. Indeed, it is expected to be the ($\gamma = \sqrt{8/3}$) LQG measure associated with a certain quantum wedge.

In fact, it is known that in the abstract "Gromov–Hausdorff–Prokhorov topology", the UIHPQ[2] equipped with its natural area measure converges under the rescaling described above to a metric measure space known as the Brownian half plane [BMR19, GM17]. Furthermore, the aforementioned quantum wedge can be equipped with metric in such a way that it agrees in law with the Brownian half plane as a metric measure space. Conjectures also hold for other models of maps, and correspondingly, for wedges associated with different values of γ. This explains (arguably) why the most natural Markov property is actually the one that holds for quantum wedges.

Proof of Theorem 8.1 First, the idea is to use the *reverse Loewner flow* rather than the ordinary Loewner flow $g_t(z)$ and its centred version $\tilde{g}_t(z) = g_t(z) - \xi_t$. Recall that while $\tilde{g}_t(z) \colon H_t \to \mathbb{H}$ satisfies the SDE:

$$d\tilde{g}_t(z) = \frac{2}{\tilde{g}_t(z)}\, dt - d\xi_t,$$

in contrast, the reverse Loewner flow is the map $f_t \colon \mathbb{H} \to H_t := f_t(\mathbb{H})$ defined by the SDE:

$$df_t(z) = -\frac{2}{f_t(z)}\, dt - d\xi_t.$$

Note the change of signs in the dt term, which corresponds to a change in the direction of time. This Loewner flow is building the curve from the ground up rather than from the tip. More precisely, in the ordinary (forward) Loewner flow, an unusual increment for $d\xi_t$ will be reflected in an unusual behaviour of the curve near its tip at time t. But in the reverse Loewner flow, this increment is reflected in an unusual behaviour near the origin. Furthermore, by using the fact that for any fixed time $T > 0$, the process $(\xi_T - \xi_{T-t})_{0 \le t \le T}$ is a Brownian motion with variance κ run for time $T > 0$, the reader can check that $f_T = \tilde{g}_T^{-1}$ in distribution. Note that this is not necessarily true if T is a stopping time: we will see an example of this later on.

Lemma 8.5 *Suppose that $\gamma > 0$ and $\kappa > 0$ are arbitrary. For $z \in \mathbb{H}$, let*

$$M_t = M_t(z) := \frac{2}{\sqrt{\kappa}} \log |f_t(z)| + Q \log |f_t'(z)|; \qquad Q = \frac{2}{\gamma} + \frac{\gamma}{2}.$$

Then for any fixed z, $(M_t(z); t \ge 0)$ is a continuous local martingale (with

---

[2] Quadrangulation rather than triangulation here.

8.1 SLE and GFF coupling: domain Markov property

respect to the filtration generated by ξ) if and only if $\gamma^2 = \kappa$ or $\gamma^2 = 16/\kappa$. (Thus if also $\gamma < 2$, this holds if and only if γ and κ are related via (8.1)). Furthermore, if $z, w \in \mathbb{H}$, then the quadratic cross variation between $M(z)$ and $M(w)$ satisfies

$$d[M(z), M(w)]_t = 4\Re\left(\frac{1}{f_t(z)}\right)\Re\left(\frac{1}{f_t(w)}\right) dt.$$

In particular,

$$[M(z)]_t \le \frac{4}{(\Im(z))^2} t \tag{8.3}$$

for all t.

Proof Set $Z_t = f_t(z)$. Then $dZ_t = -2/Z_t\, dt - d\xi_t$. Set $M_t^* = \frac{2}{\sqrt{\kappa}} \log f_t(z) + Q \log f_t'(z)$, so that $M_t = \Re(M_t^*)$. Applying Itô's formula, we see that

$$d\log Z_t = \frac{dZ_t}{Z_t} - \frac{1}{2}\frac{d[\xi]_t}{Z_t^2}$$
$$= -\frac{d\xi_t}{Z_t} + \frac{1}{Z_t^2}(-2 - \kappa/2)\, dt.$$

To obtain $df_t'(z)$ we differentiate $df_t(z)$ with respect to z; the term $d\xi_t$ does not contribute to the derivative in z (since it is the same driving function ξ for different values of z). We find that

$$df_t'(z) = 2\frac{f_t'(z)}{Z_t^2}\, dt,$$

and therefore

$$d\log f_t'(z) = \frac{df_t'(z)}{f_t'(z)} = \frac{2}{Z_t^2}\, dt.$$

Putting the two pieces together, we find that

$$dM_t^* = -\frac{2\, d\xi_t}{\sqrt{\kappa} Z_t} + \frac{2}{Z_t^2}\left(\frac{1}{\sqrt{\kappa}}(-2 - \kappa/2) + Q\right) dt. \tag{8.4}$$

The dt term vanishes if and only if $2/\sqrt{\kappa} + \sqrt{\kappa}/2 = Q$. Clearly this happens if and only if $\gamma = \sqrt{\kappa}$ or $\gamma = \sqrt{16/\kappa}$.

Furthermore, taking the real part in (8.4), if z, w are two points in the upper half plane $\mathbb{H}$, then the quadratic cross variation between $M(z)$ and $M(w)$ is a process which can be identified as

$$d[M(z), M(w)]_t = 4\Re\left(\frac{1}{f_t(z)}\right)\Re\left(\frac{1}{f_t(w)}\right) dt,$$

and so Lemma 8.5 follows.

To prove (8.3), note that

$$d[M(z)]_t = 4\left(\Re\left(\frac{1}{f_t(z)}\right)\right)^2 = 4\frac{(\Re(f_t(z)))^2}{|f_t(z)|^4}$$
$$\leq \frac{4}{|f_t(z)|^2} \leq \frac{4}{(\Im(f_t(z)))^2} \leq \frac{4}{(\Im(z))^2},$$

where we use the fact that for reverse SLE, $t \mapsto \Im(f_t(z))$ is increasing with time (as can be seen readily using the reverse Loewner equation for instance). □

One elementary but tedious calculation shows that if

$$G_t(z,w) = G_N^{\mathbb{H}}(f_t(z), f_t(w)) = -\log(|f_t(z) - \overline{f_t(w)}|) - \log(|f_t(z) - f_t(w)|)$$

then $G_t(z,w)$ is a finite variation process and furthermore:

Lemma 8.6 (Hadamard's formula) *We have that*

$$dG_t(z,w) = -4\Re(\frac{1}{f_t(z)})\Re(\frac{1}{f_t(w)})\,dt.$$

In particular, $d[M(z), M(w)]_t = -dG_t(z,w)$.

Proof This is proved in [She16a, Section 4]. We encourage the reader to skip the proof here, which is included only for completeness. (However, the result itself will be quite important in what follows.)

Set $X_t = f_t(z)$ and $Y_t = f_t(w)$. From the definition of the Neumann Green function,

$$dG_t(x,y) = -d\log(|X_t - \overline{Y}_t|) - d\log(|X_t - Y_t|)$$
$$= -\Re(d\log(X_t - \overline{Y}_t)) - \Re(d\log(X_t - Y_t)).$$

Now, $dX_t = (2/X_t)\,dt - d\xi_t$ and $dY_t = (2/Y_t) - d\xi_t$ so taking the difference

$$d(X_t - Y_t) = \frac{2}{X_t}\,dt - \frac{2}{Y_t}\,dt = 2\frac{Y_t - X_t}{X_t Y_t}\,dt$$

and so

$$d\log(X_t - Y_t) = -\frac{2}{X_t Y_t}\,dt; \quad d\log(X_t - \overline{Y}_t) = -\frac{2}{X_t \overline{Y}_t}\,dt.$$

Thus we get

$$dG_t(x,y) = -2\Re\left(\frac{1}{X_t Y_t} + \frac{1}{X_t \overline{Y}_t}\right)dt. \tag{8.5}$$

8.1 SLE and GFF coupling: domain Markov property

Now, observe that for all $x, y \in \mathbb{C}$,

$$\frac{1}{xy} + \frac{1}{\overline{xy}} = \frac{\overline{xy} + \overline{x}y}{|xy|^2} = \frac{\overline{x}(\overline{y}+y)}{|xy|^2} = \frac{2\Re(y)}{|xy|^2}\overline{x}.$$

Therefore, plugging into (8.5):

$$dG_t(x, y) = -4\frac{\Re(X_t)\Re(Y_t)}{|X_tY_t|^2} = -4\Re\left(\frac{1}{X_t}\right)\Re\left(\frac{1}{Y_t}\right)$$

as desired. □

Equipped with the above two lemmas, we prove Theorem 8.1. Set $\overline{h}_0 = \overline{h} + \varphi = \overline{h} + \frac{2}{\sqrt{\kappa}}\log|z|$, and let $(f_t; t \geq 0)$ be an independent reverse Loewner flow as above. Define

$$\overline{h}_t = \overline{h}_0 \circ f_t + Q\log|f'_t|.$$

Then, viewed as a distribution modulo constants, we claim that:

$$\overline{h}_t \text{ has the same distribution as } \overline{h}_0. \tag{8.6}$$

Let ρ be a test function with zero average, so $\rho \in \overline{\mathcal{D}}(\mathbb{H})$. To prove (8.6), it suffices to check that $(\overline{h}_t, \rho)$ is a Gaussian with mean (φ, ρ) and variance as in (6.8), that is, $\sigma^2 = \int \rho(dz)\rho(dw)G(z, w)$ where $G(z, w)$ is a valid choice of covariance for the Neumann GFF in $\mathbb{H}$.

To do this, we take conditional expectations given $\mathcal{F}_t = \sigma(\xi_s, s \leq t)$ (note that f_t is measurable with respect to $\mathcal{F}_t$), and obtain

$$\mathbb{E}[e^{i(\overline{h}_t, \rho)} \mid \mathcal{F}_t] = \mathbb{E}[e^{i(\overline{h}_0 \circ f_t + Q\log|f'_t|, \rho)} \mid \mathcal{F}_t]$$
$$= e^{i(\frac{2}{\sqrt{\kappa}}\log|f_t| + Q\log|f'_t|, \rho)} \times \mathbb{E}(e^{i(\overline{h} \circ f_t, \rho)} \mid \mathcal{F}_t)$$
$$= e^{iM_t(\rho)}\mathbb{E}[e^{i(\overline{h} \circ f_t, \rho)} \mid \mathcal{F}_t], \tag{8.7}$$

where

$$M_t(\rho) = \int_{\mathbb{H}} M_t(z)\rho(z)\,dz. \tag{8.8}$$

Note that by Fubini's theorem for conditional expectations, and (8.3) (which implies that $\mathbb{E}(M_t(z)^2) \leq Ct$ for some constat depending only on the support of ρ in $\mathbb{H}$), $M_t(\rho)$ is in fact a martingale.

Now we evaluate the term in the conditional expectation in (8.7). By definition of $\overline{h} \circ f_t$, the term $(\overline{h} \circ f_t, \rho)$ can be computed almost surely by changing variables, that is, is equal to $(\overline{h}, \rho_t)$, where the corresponding "integration" takes place on $f_t(\mathbb{H}) = H_t$ and where

$$\rho_t(z) = |(f_t^{-1})'(z)|^2 \rho \circ f_t^{-1}(z).$$

We may view H_t as a subset of $\mathbb{H}$, and note that the test function ρ_t, which is defined a priori on H_t, can be extended to $\mathbb{H}$ by setting it to zero on $\mathbb{H} \setminus H_t = K_t$. Then this test function also has mean zero on $\mathbb{H}$ (by change of variable), and we deduce that $(\bar{h} \circ f_t, \rho) = (\bar{h}, \rho_t)$ is Gaussian with mean zero and variance

$$\mathrm{Var}(\bar{h}, \rho_t)$$
$$= \iint_{\mathbb{H}^2} \rho_t(z)\rho_t(w) G_N^{\mathbb{H}}(z,w) \, dz \, dw$$
$$= \int_{H_t}\int_{H_t} G_N^{\mathbb{H}}(z,w) |(f_t^{-1})'(z)|^2 |(f_t^{-1})'(w)|^2 (\rho \circ f_t^{-1})(z)(\rho \circ f_t^{-1})(w) \, dz \, dw$$
$$= \int_{\mathbb{H}}\int_{\mathbb{H}} G_t(z,w) \rho(z) \rho(w) \, dz \, dw,$$

by change of variables, where we recall that $G_t(z,w) := G_N^{\mathbb{H}}(f_t(z), f_t(w))$. Hence, bearing in mind our notation for $M_t(\rho)$ in (8.8), we deduce that

$$\mathbb{E}(e^{i(\bar{h}_t, \rho)} \mid \mathcal{F}_t) = e^{iM_t(\rho)} \times e^{-\frac{1}{2} \iint \rho(z)\rho(w) G_t(z,w) \, dz \, dw}. \tag{8.9}$$

Next, we claim that

$$d[M(\rho)]_t = \int \rho(z)\rho(w) d[M(z), M(w)]_t \, dz \, dw. \tag{8.10}$$

To see this, we need to verify that

$$Q_t := M_t(\rho)^2 - \int \rho(z)\rho(w) [M(z), M(w)]_t \, dz \, dw$$
$$= \int \rho(z)\rho(z) \left(M_t(z) M_t(w) - [M(z), M(w)]_t \right) dz \, dw$$

is a local (and in fact a true) martingale. Indeed, from (8.3) and the Kunita–Watanabe inequality, we know that $M_t(z)M_t(w) - [M(z), M(w)]_t$ is a true martingale, so

$$\mathbb{E}[Q_t | \mathcal{F}_s] = \int \rho(z)\rho(w) \mathbb{E}\left(M_t(z)M_t(w) - [M(z), M(w)]_t \mid \mathcal{F}_s \right) dz \, dw$$
$$= \int \rho(z)\rho(w) \left(M_s(z)M_s(w) - [M(z), M(w)]_s \right) dz \, dw.$$

Here in the first line we have used Fubini's theorem for conditional expectations, whose use is once again justified by (8.3) and the Kunita–Watanabe inequality. This proves (8.10).

Therefore, by Lemma 8.6,

$$\int \rho(x)\rho(y) G_t(x,y) \, dx \, dy = \int \rho(z)\rho(w) G_N^{\mathbb{H}}(z,w) \, dz \, dw - [M(\rho)]_t.$$

Combining with (8.9) finally implies that

$$\mathbb{E}(e^{i(\overline{h}_t,\rho)}) = e^{-\frac{1}{2}\int \rho(z)\rho(w) G_N^{\mathbb{H}}(z,w)\,\mathrm{d}z\,\mathrm{d}w} \mathbb{E}(e^{iM_t(\rho)+\frac{1}{2}[M(\rho)]_t}).$$

To conclude we observe that by Itô's formula, $e^{iM_t(\rho)+\frac{1}{2}[M(\rho)]_t}$ is an exponential local martingale, and it is not hard to see that it is a true martingale ($[M(\rho)]_t \lesssim \int |\rho(z)||\rho(w)| G_N^{\mathbb{H}}(z,w)$, which is finite for all t). We deduce that the expectation in the right-hand side above is equal to $\mathbb{E}(e^{i(M_0,\rho)}) = e^{i(\varphi,\rho)}$, and therefore

$$\mathbb{E}(e^{i(h_t,\rho)}) = e^{-\frac{1}{2}\int \rho(z)\rho(w) G_N^{\mathbb{H}}(z,w)\,\mathrm{d}z\,\mathrm{d}w} e^{i(\varphi,\rho)}.$$

This proves (8.6). Arguing that f_t and $\tilde{g}_t^{-1}$ have the same distribution finishes the proof of the theorem. □

Remark 8.7 As mentioned earlier, since the proof relies on martingale computation and the optional stopping theorem, the theorem remains true if T is a (bounded) stopping time for the *reverse* Loewner flow.

Remark 8.8 This martingale is obtained by taking the real part of a certain complex martingale. Taking its imaginary part (in the case of the forward flow) gives rise to the imaginary geometry developed by Miller and Sheffield in a striking series of papers [MS16a, MS16b, MS16c, MS17].

8.2 Quantum length of SLE

We start with one of the main theorems of this section, which allows us, given a chordal SLE$_\kappa$ curve and an independent Neumann GFF, to define a notion of quantum length of the curve unambiguously. The way this is done is by mapping the curve down to the real line with the centred Loewner map $\tilde{g}_t$, and using the quantum boundary measure $\mathcal{V}$ (associated with the image of the GFF via the change of coordinates formula) to define the length. However, when we map away the curve using the map $\tilde{g}_t$, each point of the curve corresponds to two points on the real line (except for the tip of the curve which is sent to the origin since we consider the centred map). Hence, to define the length of the curve unambiguously, we first need to know that measuring the length on one side of 0 almost surely gives the same answer as measuring the length on the other side of 0.

This is basically the content of the next theorem. For ease of proof, the theorem is stated in the case where h is not a Neumann GFF but rather the field of a certain wedge. However, we will see (Corollary 8.11) that this is no loss of generality.

Theorem 8.9 *Let $0 < \gamma < 2$ and let $(\mathbb{H}, h, 0, \infty)$ be an α-quantum wedge in the unit circle embedding (see Remark 7.10), with $\alpha = \gamma - 2/\gamma$. Let ζ be an independent SLE_κ with $\kappa = \gamma^2$. Let $\tilde{g}_t$ be the (half plane capacity parametrised) centred Loewner flow for ζ, fix $t > 0$, and consider the distribution $h_t = h \circ \tilde{g}_t^{-1} + Q \log |(\tilde{g}_t^{-1})'|$ as before. Let $\mathcal{V}_{h_t}$ be the boundary Liouville measure on $\mathbb{R}$ associated with the distribution h_t. Finally, given a point $z \in \zeta([0,t])$, let $z^- < z^+$ be the two images of z under $\tilde{g}_t$. Then*

$$\mathcal{V}_{h_t}([z^-, 0]) = \mathcal{V}_{h_t}([0, z^+]),$$

almost surely for all $z \in \zeta([0,t])$.

Remark 8.10 By Remark 8.3 and the fact that the slit domain formed by an SLE_κ with $\kappa < 4$ is almost surely Hölder continuous, we see that a Neumann GFF (with arbitrary normalisation) plus a $(\gamma - 2/\gamma)$ log-singularity in such a slit domain does satisfy the conditions of Definition 6.41. That is, the quantum boundary length on either side of the curve is well defined by mapping down to the real line. Since a $(\gamma - 2/\gamma)$-quantum wedge in the unit circle embedding has the same law when restricted to $B(0, 1) \cap \mathbb{H}$ as such a Neumann GFF (with normalisation fixed so that it has mean value 0 on the upper unit semicircle), this implies that the field h of the above theorem also satisfies the conditions of Definition 6.41, at least when restricted to $B(0, 1)$. Scale invariance implies that this holds when the field is restricted to any large disc. In other words, the boundary Liouville measure $\mathcal{V}_{h_t}$ for h_t is well defined.

Corollary 8.11 *Theorem 8.9 is still true when h is replaced by a Neumann GFF on $\mathbb{H}$, with arbitrary normalisation. Indeed, by the discussion in the previous remark, it is true until the curve exits the upper unit semidisc, when the normalisation for the GFF is such that it has average 0 on the upper unit semicircle. This extends to arbitrary normalisations, since two Neumann GFFs with different normalisations (can be coupled so that they) differ by a random additive constant. Finally, scaling removes the need to restrict to the unit semidisc.*

Definition 8.12 The quantity

$$\mathcal{V}_{h_t}([\zeta(s)^-, 0]) = \mathcal{V}_{h_t}([0, \zeta(s)^+])$$

is called the **quantum length** of $\zeta([s,t])$ in the wedge $(\mathbb{H}, h, 0, \infty)$.

False proof of Theorem 8.9 The following argument does not work but helps explain the idea and why wedges are a useful notion. Let ζ be the infinite SLE_κ curve parametrised by half plane capacity. Let $L(t) = \mathcal{V}_{h_t}([\zeta(t)^-, 0])$ be the quantum length of left-hand side of the curve ζ up to time t (measured by computing the boundary quantum length on the left of zero after applying the map $\tilde{g}_t$) and likewise, let $R(t)$ be the quantum length of the right-hand side of ζ. Then it is *tempting* (but wrong) to think that, because SLE is stationary via the domain Markov property, and the Neumann GFF is invariant by Theorem 8.1, $L(t)$ and $R(t)$ form processes with stationary increments. If that were the case, we would conclude from Birkhoff's ergodic theorem for stationary increment processes that $L(t)/t$ converges almost surely to a possibly random constant, and $R(t)/t$ converges also to a random constant. We would deduce that $L(t)/R(t)$ converges to a possibly random constant. Finally, we would argue that this constant cannot be random because of tail triviality of SLE (i.e. of driving Brownian motion) and in fact must be one by left-right symmetry. On the other hand by scale invariance, the distribution of $L(t)/R(t)$ is constant. Hence we would deduce that $L(t) = R(t)$.

This proof is wrong on at least two counts: first of all, it is not true that $L(t)$ and $R(t)$ have stationary increments. This does not hold, for instance, because h loses its stationarity (i.e. the relation $h_T = h_0$ in distribution does not hold) as soon as a normalisation is fixed for the Neumann GFF. Likewise, the scale invariance does not hold in this case. This explains the importance of the concept of wedges, for which scale invariance holds by definition, as well as a certain form of stationarity (see Theorem 8.16). These properties allow us to make the above proof rigorous.

8.3 Proof of Theorem 8.9

Essential to the proof of Theorem 8.9 is the definition of two stationary processes: the *capacity zipper* and the *quantum zipper*. As in the original paper of Sheffield [She16a], once the existence and stationarity of these processes is proven, Theorem 8.9 follows relatively easily (in fact, using a similar argument to the "false proof" above).

In order to simplify notation in what follows, whenever f is a conformal isomorphism and h is a distribution or distribution modulo constants, we write

$$f(h) := h \circ f^{-1} + Q \log |(f^{-1})'|. \quad (8.11)$$

From now on, we assume that $\gamma \in (0, 2)$ is fixed, $Q = Q_\gamma$, and $\kappa = \gamma^2$. Recall that $\overline{\mathcal{D}}_0'(\mathbb{H})$ denotes the space of distributions modulo constants on $\mathbb{H}$, and we write $C([0, \infty), \mathbb{H})$ for the space of continuous functions from $[0, \infty)$ to $\mathbb{H}$.

Theorem 8.13 (Capacity zipper) *There exists a two-sided **stationary** process $(\overline{h}^t, \eta^t)_{t \in \mathbb{R}}$, taking values in $\overline{\mathcal{D}}_0'(\mathbb{H}) \times C([0, \infty); \mathbb{H})$, such that:*

312   *SLE and the Quantum Zipper*

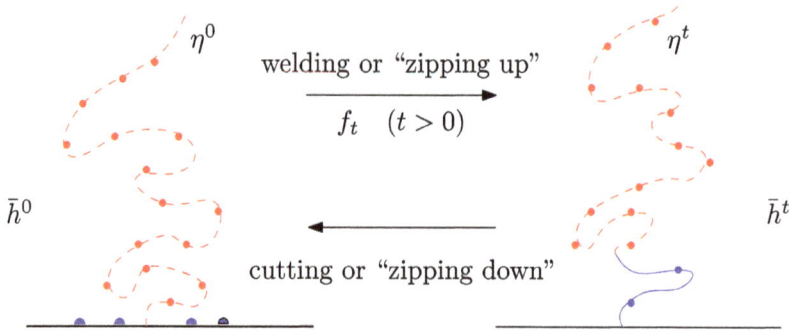

Figure 8.2 The capacity zipper

- **(Marginal law)** $(\overline{h}^0, \eta^0)$ *has the law of a Neumann GFF (modulo constants) plus the function* $\varphi(z) = \frac{2}{\gamma} \log |z|$, *together with an independent* SLE_κ;
- **(Positive time)** *there exists a family of conformal isomorphisms* $(f_t)_{t \geq 0}$: $\mathbb{H} \to \mathbb{H} \setminus \eta^t([0, t])$, *whose (marginal) law is that of a reverse* SLE_κ *Loewner flow parametrised by capacity, and such that* $\overline{h}^t|_{\mathbb{H} \setminus \eta^t([0,t])} = f_t(\overline{h}^0)$ *and* $\eta^t([t, \infty)) = f_t(\eta^0)$ *for all* $t \geq 0$;
- **(Negative time)** *for* $t < 0$, *if* $\tilde{g}_{-t}$ *is the centred Loewner map corresponding to* $\eta^0([0, -t])$ *then* $\eta^t = \tilde{g}_{-t}(\eta^0)$ *and* $\overline{h}^t = \tilde{g}_{-t}(\overline{h}^0)$.

Thus given a field $\overline{h}^0$ and an independent SLE_κ infinite curve η^0, we can either "zip it up" (weld it to itself) to obtain the configuration $(\overline{h}^t, \eta^t)$ for some $t > 0$, or "zip it down" (cut it open along η^0) to obtain the configuration $(\overline{h}^t, \eta^t)$ for some $t < 0$ (see Figure 8.2). Beware that the relation between time t and time 0 is opposite to that of Theorem 8.1 – hence the change in notation from subscripts to superscripts for the time index.

Also note that for $t > 0$, $\overline{h}^t|_{\mathbb{H} \setminus \eta^t([0,t])}$ uniquely defines $\overline{h}^t$ as a distribution modulo constants on $\mathbb{H}$ (since $\eta^t([0, t])$ is independent of $\overline{h}^t$ and has Lebesgue measure zero). The term "capacity" in the definition refers to the fact that in any positive time t, we are zipping up a curve with $2t$ units of half plane capacity.

Remark 8.14 Note that the capacity zipper of Theorem 8.13 is defined to be a process taking values in $\overline{\mathcal{D}}_0'(\mathbb{H}) \times C([0, \infty), \mathbb{H})$. However, we can also define from $(\overline{h}^0, \eta^0, (f_t)_{t \geq 0})$ a version $(\tilde{h}^t, \eta^t)_{t \geq 0}$ of the capacity zipper indexed by positive times and taking values in $\mathcal{D}_0'(\mathbb{H}) \times C([0, \infty), \mathbb{H})$. That is, so that the field at any time is a distribution, not just a distribution modulo constants.

8.3 Proof of Theorem 8.9

To do this, we can just fix a normalisation of $\tilde{h}^0$ to obtain $\tilde{h}^0 \in \mathcal{D}'_0(\mathbb{H})$, and then for $t > 0$ set $(\tilde{h}^t, \eta^t) := (f_t(\tilde{h}^0), \mathbb{H} \setminus f_t(\mathbb{H} \setminus \eta^0))$. Note that this process will be no longer stationary: for given t, $\tilde{h}^t$ will have the law of $\tilde{h}^0$ plus a random constant.

Now we move on to the definition of the **quantum zipper**. For this, we need the notion of doubly marked surface curve pair. This is just an extension of the definition of doubly marked surface, when the surface comes together with a chordal curve. More precisely, suppose that for $i = 1, 2$, D_i is a simply connected domain with marked boundary points (a_i, b_i), h_i is a distribution in D_i, and η_i is a simple curve (considered up to time reparametrisation) from a_i to b_i in D_i.

Definition 8.15 We say that $(D_1, h_1, a_1, b_1, \eta_1)$ and $(D_2, h_2, a_2, b_2, \eta_2)$ are equivalent if there exists a conformal isomorphism $f: D_1 \to D_2$ such that $h_2 = f(h_1)$, $a_2 = f(a_1)$, $b_2 = f(b_1)$ and $\eta_2 = f(\eta_1)$. A doubly marked surface curve pair (from here on in just *surface curve pair*) is an equivalence class of (D, h, a, b, η) under this equivalence relation.

Theorem 8.16 (Quantum zipper) *There exists a two-sided process*

$$((\mathbb{H}, h^t, 0, \infty, \zeta^t))_{t \in \mathbb{R}},$$

(which we will sometimes write as $(h^t, \zeta^t)_{t \in \mathbb{R}}$ with a slight abuse of notation), which is **stationary** *as a process of surface-curve pairs, and such that:*

- $(\mathbb{H}, h^0, 0, \infty)$ *is a quantum wedge in the unit circle embedding;*
- (h^0, ζ^0) *has the law, as a surface curve pair, of a $(\gamma - 2/\gamma)$-quantum wedge together with an independent SLE_κ;*
- *for any $t > 0$, if ζ^0 is parametrised by half plane capacity,*

$$\sigma(t) := \inf\{s \geq 0 \colon \mathcal{V}_{h^0}\left(RHS \text{ of } \zeta^0([0, s])\right) \geq t\},$$

if $\tilde{g}_{\sigma(t)}$ is the centred Loewner map sending $\mathbb{H} \setminus \zeta^0[0, \sigma(t)]$ to $\mathbb{H}$, then we have that $h^{-t} = \tilde{g}_{\sigma(t)}(h^0)$ and $\zeta^{-t} = \tilde{g}_{\sigma(t)}(\zeta^0)$.

Note that by stationarity, this defines the law of the process for all time (positive and negative).

So this is a similar picture to that of the capacity zipper (moving backwards in time corresponds to "cutting down" and hence moving forward in time corresponds to "zipping up") but now a segment of ζ^0 with *right h^0* LQG boundary length t is cut out between times 0 and $-t$. Hence the name "quantum zipper": the dynamic is parametrised by (right) quantum boundary length. Note that it makes sense to talk about the right boundary length of a segment of η, by conformally mapping to the upper half plane and applying the change of the coordinate formula (see Remark 8.10). Also note the difference with the capacity zipper: here h^0 is a distribution (not a distribution modulo constants) while the stationarity is in the sense of quantum surface curve pairs.

Assuming for now that Theorem 8.16 holds, we make the following claim.

Claim 8.17 *For any fixed t, the $\mathcal{V}_{h^0}$ boundary length of the* left-hand side *of $\zeta^0([0, \sigma(t)])$ is also equal to t.*

This means that the parametrisation is really, unambiguously, by quantum boundary length. It also immediately implies Theorem 8.9.

Proof of Claim 8.17, and hence Theorem 8.9, given Theorem 8.16 Denote by $L(t)$ the $\mathcal{V}_{h^0}$ boundary length of the left-hand side of $\zeta^0[0, \sigma(t)]$, so our aim is to show that $L(t) \equiv t$. We begin by making the following observations.

- By stationarity of the quantum zipper, we have that $(L(s + t) - L(s))_{t \geq 0}$ is equal in distribution to $(L(t))_{t \geq 0}$ for any fixed $s \geq 0$.
- By scale invariance of SLE_κ and the invariance property of quantum wedges (Theorem 7.11),

$$\frac{L(t)}{t} \stackrel{(d)}{=} L(1)$$

for any $t > 0$, and for any $A > 0$ and $s < t$,

$$\frac{L(At)}{At} - \frac{L(As)}{As} \stackrel{(d)}{=} \frac{L(t)}{t} - \frac{L(s)}{s}.$$

The first point means that we can apply Birkhoff's ergodic theorem [Kal21, Theorem 25.6]

$$\frac{L(n)}{n} \to X = \mathbb{E}(L(1) \mid \mathcal{I}) \text{ almost surely as } n \to \infty \text{ in } \mathbb{N}, \tag{8.12}$$

where $\mathcal{I}$ is the σ-field generated by invariant sets under the shift map

$$(L(1), L(2), \ldots) \mapsto (L(2), L(3), \ldots).$$

Note that the theorem is often stated under the assumption that $\mathbb{E}(|L(1)|)$ is finite, but the conclusion is also true if we only know $L(1) \geq 0$ almost surely: in this case, conditional expectation can always be defined using monotone convergence, and the left-hand side in (8.12) converges to infinity on the event that the conditional expectation is infinite; see [Kal21, Theorem 25.6] for the proof.

Note also that since $L(n)/n$ converges to X in distribution, and since $L(n)/n$

8.3 Proof of Theorem 8.9

is equal in distribution to $L(1) < \infty$ almost surely, we see that in fact $X < \infty$ almost surely. We may then deduce that

$$\frac{L(t)}{t} - \frac{L(s)}{s} = 0 \text{ almost surely}$$

for any fixed $s, t \in \mathbb{Q}$ with $s \leq t$. Indeed, the law m of this difference is equal to that of $L(At)/At - L(As)/As$ for any A, and by taking a sequence $A_k \uparrow \infty$ such that $A_k t \in \mathbb{N}$, $A_k s \in \mathbb{N}$ for all k, we obtain a sequence of random variables all having law m, which by (8.12) tend to 0 as $k \to \infty$. Hence, with probability one we have that

$$\frac{L(t)}{t} = X \text{ for all } t \in \mathbb{Q} \qquad (8.13)$$

(where X is as in (8.12)). In particular, we have that

$$X = \lim_{t \downarrow 0, t \in \mathbb{Q}} \frac{L(t)}{t}.$$

Now by definition, the above limit (and therefore the random variable X) is measurable with respect to the σ-algebra

$$\mathcal{T} = \bigcap_{\varepsilon > 0} \sigma((h^0 - h^0_\varepsilon)|_{B(0,\varepsilon) \cap \mathbb{H}}, \zeta^0|_{B(0,\varepsilon) \cap \mathbb{H}})$$

(here h^0_ε is the ε-semicircle average of h^0 about the origin and can be subtracted since $L(t)/t$ is not affected by adding a constant to the field). On the other hand, since the h^0 right/left quantum boundary lengths along ζ^0 almost surely do not have atoms at $0^\pm$, X is also measurable with respect to

$$\sigma(\mathcal{A}) \; ; \; \mathcal{A} = \bigcup_{\varepsilon > 0} \sigma((h^0 - h^0_\varepsilon)|_{B(0,1) \setminus B(0,\varepsilon)}, \zeta^0|_{B(0,1) \setminus B(0,\varepsilon)}).$$

Hence the proof will be complete if we can show $\mathcal{T} \cap \sigma(\mathcal{A})$ is trivial, because then X must be almost surely constant, and by symmetry, this constant must be equal to 1.

For this final step, since $\mathcal{A}$ is a π-system, it suffices to show that for any $\varepsilon_0 > 0$, $A_0 \in \mathcal{T}$ and

$$A \in \sigma(h^0 - h^0_{\varepsilon_0}|_{B(0,1) \setminus B(0,\varepsilon_0)}, \zeta^0|_{B(0,1) \setminus B(0,\varepsilon_0)}),$$

we have $\mathbb{P}(A \cap A_0) = \mathbb{P}(A)\mathbb{P}(A_0)$. However, this follows by independence of h^0 and ζ^0, since the driving function of ζ^0 is a Brownian motion, and by Lemma 6.34. □

The rest of this section will be dedicated to proving Theorems 8.13 and 8.16. In fact, the latter is straightforward to obtain from Theorem 8.1. The idea to then deduce Theorem 8.16 is to reparametrise time according to right quantum boundary length and appropriately "zoom in" at the whole capacity zipper picture at the origin. This step, however, is somewhat technical.

8.3.1 The capacity zipper

In this section, we prove Theorem 8.13. That is, we construct the stationary two-sided capacity zipper, using the coupling theorem, Theorem 8.1.

Let $\overline{h}_0$ be as in the original Theorem 8.1 (i.e. $\overline{h}_0$ has the distribution (8.2)), and let $\eta = \eta_0$ be an independent infinite SLE_κ curve from 0 to ∞. As in the coupling theorem, set $\overline{h}_t = \overline{h}_0 \circ \tilde{g}_t^{-1} + Q \log |(\tilde{g}_t^{-1})'|$, where $\tilde{g}_t$ is the centred Loewner map corresponding to $\eta_0([0,t])$ for each t, and let η_t be the image by $\tilde{g}_t$ of the initial infinite curve $\eta = \eta_0$. Then Theorem 8.1 says that $\overline{h}_t = \overline{h}_0$ in distribution, and in fact we can also see that the joint distribution $(\overline{h}_t, \eta_t)$ is identical to that of $(\overline{h}_0, \eta_0)$.

For $0 \le t \le T$, let $\overline{h}^t = \overline{h}_{T-t}$, and let $\eta^t = \eta_{T-t}$. Then it is an easy consequence of Theorem 8.1 that the following lemma holds:

Lemma 8.18 *The laws of the process $(\overline{h}^t, \eta^t)_{0 \le t \le T}$ (with values in $\overline{\mathcal{D}}'(\mathbb{H}) \times C([0,\infty))$ are consistent as T increases.*

By Lemma 8.18, and applying Kolmogorov's extension theorem, it is obvious that there is a well-defined process $(\overline{h}^t, \eta^t)_{0 \le t < \infty}$ whose restriction to $[0,T]$ agrees with the process described above. Hence for $t > 0$, starting from $\overline{h}^0$ and an infinite curve η^0, there is a well-defined, possibly random, procedure giving rise to $(\overline{h}^t, \eta^t)$, that we want to view as "welding" together parts of the positive and negative real lines, or "zipping up". The dynamic on the field is obtained by applying the change of coordinates formula to $\overline{h}^0$, with respect to a flow $(f_s)_{s \le t}$ that has the *marginal* law of a reverse Loewner flow, but we stress that here the reverse Loewner flow is not independent of $\overline{h}^0$ (rather, it will end up being uniquely determined by $\overline{h}^0$, while $(f_s)_{s \le t}$ will be independent of $\overline{h}^t$).

But we could also go in the other direction, cutting $\mathbb{H}$ along η^0, as in Theorem 8.1. Indeed we could define, for $t < 0$ this time, a field $\overline{h}^t$ by considering the centred Loewner flow $(\tilde{g}_{|t|})_{t<0}$ associated to the infinite curve η^0, and setting

$$\overline{h}^t = \overline{h}^0 \circ \tilde{g}_{|t|}^{-1} + Q \log |(\tilde{g}_{|t|}^{-1})'| \quad (t < 0).$$

We can also, of course, get a new curve η^t for $t < 0$ by pushing η^0 through the map $\tilde{g}_{|t|}$. This gives rise to the two-sided stationary process $(\overline{h}^t, \eta^t)_{t \in \mathbb{R}}$ of Theorem 8.13.

Remark 8.19 An equivalent way to define this process would be as follows.

8.3 Proof of Theorem 8.9

Start from the set-up of Theorem 8.1: thus $\bar{h}_0$ is a field distributed as in (8.2), and η_0 an independent infinite SLE_κ curve. Set $\bar{h}_t = \bar{h}_0 \circ \tilde{g}_t^{-1} + Q \log |(\tilde{g}_t^{-1})'|$ as before, and $\eta_t = g_t(\eta^0 \setminus \eta_0[0,t])$. Then Theorem 8.1 tells us that $(\bar{h}_t, \eta_t)_{t \geq 0}$ is a stationary process, so we can consider the limit as $t_0 \to \infty$ of $(\bar{h}_{t_0+t}, \eta_{t_0+t})_{t \geq -t_0}$, which defines a two-sided process. The capacity zipper process $(\bar{h}^t, \eta^t)_{t \in \mathbb{R}}$ can then be defined as the image of this process under the time change $t \mapsto -t$.

8.3.2 The quantum zipper

Note: We recall the notation

$$f(h) := h \circ f^{-1} + Q \log |(f^{-1})'|. \tag{8.14}$$

It will be used repeatedly in what follows.

In this subsection, we show the existence and stationarity of the quantum zipper: Theorem 8.16. In what follows, we will usually take our quantum wedges to be in the **unit circle embedding** $(\mathbb{H}, h, 0, \infty)$ (recall that the law of $h - \alpha \log(1/|z|)$ restricted to the upper unit semidisc is then just that of a Neumann GFF with the additive constant fixed so that its average on the upper unit semicircle is equal to zero).

The key to the proof of Theorem 8.16 is the following:

Proposition 8.20 *Let $(h, \zeta) = ((h, \mathbb{H}, 0, \infty), \zeta)$ be a $(\gamma - 2/\gamma)$-quantum wedge in the unit circle embedding, together with an independent SLE_κ. If ζ is parametrised by half plane capacity, let σ be the smallest time such that the $\mathcal{V}_h$ boundary length of the right-hand side of $\zeta([0, \sigma])$ exceeds 1.[3] Let g_σ be the centred Loewner map from $\mathbb{H} \setminus \zeta([0, \sigma]) \to \mathbb{H}$. Then $(g_\sigma(h), g_\sigma(\zeta))$ is equal in law to (h, ζ) as a surface curve pair. That is, if ψ is the unique conformal isomorphism such that $(\psi \circ g_\sigma)(h)$ is in the unit circle embedding, then*

$$(\psi \circ g_\sigma(h), \psi \circ g_\sigma(\zeta)) \stackrel{(d)}{=} (h, \eta).$$

In words: if we start with a $(\gamma - 2/\gamma)$-quantum wedge and an independent SLE_κ, and "zip" down by one unit of right quantum boundary length, the law of the resulting quantum surface curve pair does not change.

---

[3] Recall that to measure this boundary length, we map the right-hand side of the curve down to an interval $[0, x]$ of the positive real line using the centred Loewner map. Then we take the quantum boundary length of $[0, x]$ with respect to the field defined by applying the change of coordinates formula to h with respect to this map.

Proof of Theorem 8.16 given Proposition 8.20 Note that there is nothing special about the choice to zip down by quantum boundary length 1 in Proposition 8.20. Indeed, we could replace one by any other $t > 0$ and would obtain the result. Then the existence and stationarity of the quantum zipper follows in the same way that Theorem 8.13 followed from Theorem 8.1 (see Section 8.3.1). □

The proof of Proposition 8.20 is quite tricky and consists of several steps.

Step 1: Reweighting. We write $\mathbb{P}$ for the law of $(\tilde{h}^t, \eta^t)_{t \geq 0}$, the capacity zipper as in Remark 8.14, where the constant for $\tilde{h}^0$ has been fixed so that its unit semicircle average around the point 10 is equal to 0 (this is fairly arbitrary, apart from the fact that the measure is supported a good distance away from the origin). We can extend this to define a law $\mathbf{P}$ on $\mathcal{D}'(\mathbb{H}) \times C([0, \infty), \mathbb{H}) \times [1, 2]$, by setting $\mathbf{P} := \mathbb{P} \times \mathrm{Leb}_{[1,2]}$ (so a sample from $\mathbf{P}$ consists of a capacity zipper $(\tilde{h}^t, \eta^t)_{t \geq 0}$ as just described, plus a point Z chosen independently from Lebesgue measure on $[1, 2]$). Define

$$c(z) := \mathbb{E}_{\mathbf{P}}(e^{\frac{\chi}{2}\tilde{h}^0_\delta(z)} \delta^{\gamma^2/4}) \text{ for } z \in [1, 2],$$

which by Theorem 6.36 does not depend on $\delta > 0$ and is a smooth function on $z \in [1, 2]$.

We want to study the joint law of the capacity zipper plus a quantum boundary length typical point (in $[1, 2]$). In fact, this is much easier to do if we reweight the law of the field $\tilde{h}^0$. To this end, we define a family of laws $(\mathbf{Q}_\varepsilon)_{\varepsilon > 0}$ by setting

$$\frac{d\mathbf{Q}_\varepsilon}{d\mathbf{P}} = \frac{e^{\frac{\chi}{2}\tilde{h}^0_\varepsilon(Z)} \varepsilon^{\frac{\gamma^2}{4}}}{\int_{[1,2]} c(z)\, dz} =: \frac{e^{\frac{\chi}{2}\tilde{h}^0_\varepsilon(Z)} \varepsilon^{\frac{\gamma^2}{4}}}{c([1,2])} \qquad (8.15)$$

for each ε.

Under $\mathbf{Q}_\varepsilon$, the marginal law of $(\tilde{h}^t, \eta^t)_{t \geq 0}$ is its $\mathbb{P}$ law weighted by

$$\frac{\mathcal{V}_{\tilde{h}^0_\varepsilon}([1,2])}{c([1,2])}.$$

Moreover, given $(\tilde{h}^t, \eta^t)_{t \geq 0}$ the point Z is sampled from the ε-approximate measure $\mathcal{V}_{\tilde{h}^0_\varepsilon}$ (restricted to $[1, 2]$ and normalised to be a probability measure). Therefore, since $\mathcal{V}_{\tilde{h}^0_\varepsilon}([1,2]) \to \mathcal{V}_{\tilde{h}^0}([1,2])$ in $\mathcal{L}^1$ as $\varepsilon \to 0$ and the measure $\mathcal{V}_{\tilde{h}^0_\varepsilon}$ converges weakly in probability to $\mathcal{V}_{\tilde{h}^0}$, we can deduce that

$$\mathbf{Q}_\varepsilon \Rightarrow \mathbf{Q}$$

as $\varepsilon \to 0$ where $\mathbf{Q}$ is the measure described by (a) and (b) of Lemma 8.21.

8.3 Proof of Theorem 8.9

This reweighting is analogous to the argument used to describe the GFF viewed from a Liouville typical point – see Theorem 2.4. As in this proof, we can reverse the order in which $(\tilde{h}^0, \eta^0)$ and Z are sampled, and this leads to the alternative description given by points (c) to (e) in the following lemma.

Lemma 8.21 *Under* $\mathbf{Q}$*, the following are true:*

(a) *the marginal law of* $(\tilde{h}^t, \eta^t)_{t \in \mathbb{R}}$ *is given by* $\mathcal{V}_{\tilde{h}^0}([1,2])/c([1,2])\, d\mathbb{P}$ *(and is therefore absolutely continuous with respect to* $\mathbb{P}$*);*
(b) *conditionally on* $(\tilde{h}^t, \eta^t)_{t \in \mathbb{R}}$*,* Z *is chosen uniformly from* $\mathcal{V}_{\tilde{h}^0}$ *on* $[1,2]$*;*
(c) *the marginal law of* Z *on* $[1,2]$ *has density* $c(z)/c([1,2])$ *with respect to Lebesgue measure;*
(d) *conditionally on* Z*, for every* $0 \le t \le \tau_Z$ *(where* τ_Z *is the first time that* $f_t(Z) = 0$*) the law of* $\eta^t([0,t])$ *is that of a reverse* $SLE_\kappa(\kappa, -\kappa)$ *curve with force points* $(Z, 10)$*, run up to time* t*;*
(e) *for any* $t \ge 0$*, conditionally on* $\{Z, (f_s)_{0 \le s \le t}\}$*, we have that the conditional law of* $\tilde{h}^t$ *as a distribution modulo constants, on the event* $t \le \tau_Z$*, is that of*

$$\overline{h} + \frac{2}{\gamma}\log(|\cdot|) + \frac{\gamma}{2}G_N^{\mathbb{H}}(\cdot, f_t(Z)) - \frac{\gamma}{2}\int G^{\mathbb{H}}(\cdot, f_t(y))\rho_{10,1}(dy)$$

where $\overline{h}$ *has the law of a Neumann GFF (modulo constants) that is independent of* $(f_s)_{0 \le s \le t}$*. Here for* $x \in \mathbb{R}, \delta > 0$*,* $\rho_{x,\delta}$ *denotes uniform measure on the upper semicircle of radius* δ *around* x*.*

Remark 8.22 The force point at 10 in (d) and the final term in the expression for $\tilde{h}^t$ in (e) make these descriptions look rather complicated. However, we will really be interested in taking $t = \tau_Z$ and looking at $(\tilde{h}^t, \eta^t)$ in small neighbourhoods of the origin. In such a setting, as we will soon see, these terms will have asymptotically negligible contribution to the behaviour. The only features in the descriptions (d) and (e) that are genuinely important, are the force point of weight κ at Z, and the function $(2/\gamma)\log(|\cdot|) + (\gamma/2)G(\cdot, f_t(Z))$.

Proof (a) and (b) define the measure $\mathbf{Q}$ (see discussion above the lemma) and (c) follows since this is true under $\mathbf{Q}_\varepsilon$ for every $\varepsilon > 0$.

For (d), we first claim that for any $t \ge 0$ and for any measurable function F of $(f_s; s \le t)$ we have

$$\mathbf{Q}_\varepsilon(F(f_s; s \le t)\mathbf{1}_{\{t \le \tau_{Z-\varepsilon}\}})$$
$$= \frac{\mathbf{P}(F(f_s; s \le t)\mathbf{1}_{\{t \le \tau_{Z-\varepsilon}\}}e^{\frac{\gamma}{2}(M_t(Z)-M_t(10))-\frac{\gamma^2}{8}[M(Z)-M(10)]_t})}{\mathbf{P}(e^{\frac{\gamma}{2}(M_t(Z)-M_t(10))-\frac{\gamma^2}{8}[M(Z)-M(10)]_t})}. \quad (8.16)$$

To see this, we note that by definition $\tilde{h}^0 = \tilde{h}^t \circ f_t + Q \log |(f_t)'|$ and, due to the normalisation we chose for $\tilde{h}^0$, $\tilde{h}^0_\varepsilon = (\tilde{h}^0, \overline{\rho}^\varepsilon_Z) := (\tilde{h}^0, \rho_{Z,\varepsilon} - \rho_{10,1})$. Therefore

$$\mathbf{P}(e^{\frac{\chi}{2}\tilde{h}^\varepsilon_0(Z)} \mid Z, (f_s; s \le t))$$
$$= e^{\frac{\chi}{2}(M_t, \rho_{Z,\varepsilon} - \rho_{10,1})} \mathbf{P}(e^{\frac{\chi}{2}(\tilde{h}^t \circ f_t - (2/\gamma) \log(|\cdot|) \circ f_t, \rho_{Z,\varepsilon} - \rho_{10,1})} \mid Z, (f_s; s \le t))$$

where, because the average value of $\rho_{Z,\varepsilon} - \rho_{10,1}$ is equal to 0,

$$(\tilde{h}^t \circ f_t - (2/\gamma) \log(|\cdot|)) \circ f_t, \rho_{Z,\varepsilon} - \rho_{10,1})$$

depends only on the equivalence class modulo constants of $\tilde{h}^t \circ f_t - (\frac{2}{\gamma}) \log(|\cdot|)$. Moreover, by stationarity of the capacity zipper, this law is that of a Neumann GFF $\overline{h}$ (modulo constants) that is independent of $(f_s; s \le t)$. Thus we are reduced to doing a simple Gaussian computation. This is very similar to what was carried out in the proof of Theorem 8.1 and yields that

$$\mathbf{P}(e^{\frac{\chi}{2}(\tilde{h}^t \circ f_t - (2/\gamma) \log(|\cdot|) \circ f_t, \rho_{Z,\varepsilon} - \rho_{10,1})} \mid Z, (f_s; s \le t)) = e^{-\frac{\chi^2}{8}[(M, \rho_{Z,\varepsilon} - \rho_{10,1})]_t}.$$

We may also note that when $t \le \tau_{Z-\varepsilon}$, M_t can be extended by Schwarz reflection to a harmonic function on a domain containing $B(Z, \varepsilon)$ and $B(10, 1)$, and so by the mean value theorem $(M_t, \overline{\rho}_{Z,\varepsilon} - \rho_{10,1}) = M_t(Z) - M_t(10)$. Similarly, on the event that $t \le \tau_{Z-\varepsilon}$, $[(M, \rho_{Z,\varepsilon} - \rho_{10,1})]_t = [M(Z) - M(10)]_t$. Equation (8.16) then follows by definition of $\mathbf{Q}_\varepsilon$ and conditioning.

Next, recall from the proof of Lemma 8.5 that $dM^*_r = -(2/(\gamma f_r(z))) \, dW_r$, where W is the driving function of $(f_r)_r$ (and is a Brownian motion run at speed γ^2). Hence, by (8.16) and the Cameron–Martin–Girsanov theorem, we have that (under $\mathbf{Q}_\varepsilon$, conditionally on Z and up to time $\tau_{Z-\varepsilon}$), $W_t - \frac{\chi}{2}[W, M(Z) - M(10)]_t$ is a (speed γ^2) Brownian motion, or equivalently

$$dW_t = \gamma \, dB_t - \gamma^2 \Re\left(\frac{1}{f_t(Z)}\right) dt + \gamma^2 \Re\left(\frac{1}{f_t(10)}\right) dt.$$

Since this does not depend on ε, the same must hold under $\mathbf{Q}^Z = \mathbf{Q}(\cdot \mid Z)$, at least up to time $\tau_{Z-\varepsilon}$. However, as $\varepsilon > 0$ was arbitrary, it in fact holds until time τ_Z. Since this is exactly the equation satisfied by the driving function of an $\mathrm{SLE}_\kappa(\kappa, -\kappa)$ process with force points at $(Z, 10)$, we conclude the proof of (d).

Finally, we deal with (e). For this, we use the same rewriting of $\tilde{h}^\varepsilon_0$ as above to see that

$$\mathbf{Q}^\varepsilon(F(\tilde{h}^t) \mid (f_s)_{0 \le s \le t}, Z) = \frac{\mathbf{P}(F(\tilde{h}^t)e^{\frac{\chi}{2}(\tilde{h}^t, (\rho_{Z,\varepsilon} - \rho_{10,1}) \circ f_t^{-1})} \mid (f_s)_{0 \le s \le t}, Z)}{\mathbf{P}(e^{\frac{\chi}{2}(\tilde{h}^t, (\rho_{Z,\varepsilon} - \rho_{10,1}) \circ f_t^{-1})} \mid (f_s)_{0 \le s \le t}, Z)} \tag{8.17}$$

for any bounded measurable function F of $\tilde{h}^t$ modulo constants. On the

8.3 Proof of Theorem 8.9

other hand, recall that, under $\mathbf{P}$, $\tilde{h}^t$ viewed modulo constants is independent of $(f_s)_{s \leq t}$ and is distributed like a Neumann GFF plus the function $(2/\gamma) \log | \cdot |$ (modulo constants). Thus, by the Cameron–Martin–Girsanov theorem applied conditionally on $(Z, (f_s)_{s \leq t})$, the law of $\tilde{h}^t$ under $\mathbf{Q}_\varepsilon$ and conditionally on $(Z, (f_s)_{s \leq t})$, considered modulo constants, is that of a Neumann GFF (modulo constants) *plus* the function $(2/\gamma) \log | \cdot |$, *plus* the function $w \mapsto \int G_N^{\mathbb{H}}(w, y)(\overline{\rho}_Z^\varepsilon \circ f_t^{-1})(dy)$. Now, for any $t \leq \tau_{Z-\varepsilon}$ and any $w \in \mathbb{H} \setminus f_t(B(Z, \varepsilon) \cap \mathbb{H})$, we have $\int G_N^{\mathbb{H}}(w, y)(\overline{\rho}_Z^\varepsilon \circ f_t^{-1})(dy) = G_N^{\mathbb{H}}(w, f_t(Z))$, and so on the set $\mathbb{H} \setminus f_t(B(Z, \varepsilon) \cap \mathbb{H})$ we can write (as distributions modulo constants)

$$\tilde{h}^t \stackrel{(d)}{=} \overline{h} + \frac{2}{\gamma} \log | \cdot | + \frac{\gamma}{2} G_N^{\mathbb{H}}(\cdot, f_t(Z)) - \frac{\gamma}{2} \int G_N^{\mathbb{H}}(\cdot, f_t(y))\rho_{10}^1(dy), \quad (8.18)$$

where the equality in distribution holds under $\mathbf{Q}_\varepsilon$ conditionally on Z and $(f_s; s \leq t)$, and where $\overline{h}$ is as described in the statement of (e). Taking a limit as $\varepsilon \to 0$, we obtain the result. □

Corollary 8.23 *Taking* $t \nearrow \tau_Z$ *in the previous lemma, we see that under* $\mathbf{Q}^Z = \mathbf{Q}(\cdot \mid Z)$, $(\tilde{h}^{\tau_Z}, \eta^{\tau_Z})$ *can be described as follows:*

- $\eta^{\tau_Z}([0, \tau_Z])$ *has the law of a reverse* $SLE_\kappa(\kappa, -\kappa)$, *with force points at* $(Z, 10)$, *and run until the point Z reaches* 0;
- *given* $\eta^{\tau_Z}([0, \tau_Z])$, *we have*

$$\tilde{h}^{\tau_Z} \stackrel{(d)}{=} \overline{h} + (\gamma - 2/\gamma) \log(1/|\cdot|) - \frac{\gamma}{2} \int G_N^{\mathbb{H}}(\cdot, f_{\tau_Z}(y))\rho_{10,1}(dy),$$

where the equality is an equality of distributions modulo constants, and $\overline{h}$ has the law of a Neumann GFF that is independent of $(f_s)_{0 \leq s \leq \tau_Z}$.

Remark 8.24 Taking $t \nearrow \tau_Z$ rigorously in Lemma 8.21 requires some justification, since the statement of (e) is actually for deterministic $t \geq 0$.

To do this, we first consider $\tau := \tau_{\{Z-\delta\}}$ for arbitrary $\delta > 0$. Then from Lemma 8.21(e) we have that, for any deterministic $k, n \geq 0$, the conditional law of $\tilde{h}^{k/n}$ given Z and $(f_s)_{0 \leq s \leq k/n}$, on the event that $\tau \in (\frac{k-1}{n}, \frac{k}{n}]$, is that of

$$\overline{h} + F_{f_{k/n}, Z}(\cdot);$$

$$F_{f_{k/n}, Z}(\cdot) := \frac{2}{\gamma} \log(|\cdot|) + \frac{\gamma}{2} G_N^{\mathbb{H}}(\cdot, f_t(Z)) - \frac{\gamma}{2} \int G_N^{\mathbb{H}}(\cdot, f_t(y))\rho_{10,1}(dy).$$
(8.19)

Now write $\tau^{(n)} := k/n$ for the unique $k \in \mathbb{N}$ such that $\tau \in (\frac{k-1}{n}, \frac{k}{n}]$. Then for arbitrary continuous functionals H_1, H_2, H_3 (defined on appropriate spaces) taking values in $[0, 1]$, we have

$$\mathbf{Q}[H_1(\tilde{h}^\tau)H_2((f_s)_{0\le s\le \tau})H_3(Z)]$$
$$= \lim_{n\to\infty} \mathbf{Q}[H_1(\tilde{h}^{\tau^{(n)}})H_2((f_s)_{0\le s\le \tau^{(n)}})H_3(Z)]$$
$$= \lim_{n\to\infty} \sum_k \mathbf{Q}[H_1(\tilde{h}^{k/n})H_2((f_s)_{0\le s\le k/n})H_3(Z)\mathbf{1}_{\tau^{(n)}=k/n}]$$
$$= \lim_{n\to\infty} \sum_k \mathbf{Q}[H_1(\overline{h} + F_{f_{k/n},Z}(\cdot))H_2((f_s)_{0\le s\le k/n})H_3(Z)\mathbf{1}_{\tau^{(n)}=k/n}]$$
$$= \lim_{n\to\infty} \mathbf{Q}[H_1(\overline{h} + F_{f_{\tau^{(n)}},Z}(\cdot))H_2((f_s)_{0\le s\le \tau^{(n)}})H_3(Z)]$$
$$= \mathbf{Q}[H_1(\overline{h} + F_{f_\tau,Z}(\cdot))H_2((f_s)_{0\le s\le \tau})H_3(Z)],$$

where the third equality follows from (8.19), the first and final by continuity, and the second and fourth by definition of $\tau^{(n)}$. In other words, Lemma 8.21(e) holds for the random time $t = \tau = \tau_{\{Z-\delta\}}$ for any $\delta > 0$. We can similarly take $\delta \to 0$ to obtain the statement for $t = \tau_Z$.

We will use this to show that when we zoom in at this weighted capacity zipper at time τ_Z, we obtain a field and curve whose joint law is that in the statement of Proposition 8.20.

Step 2: Zooming in to get a wedge and an independent SLE. Suppose that η is a simple curve from 0 to ∞ in $\mathbb{H}$, considered up to time reparametrisation, and that $K \subset \mathbb{H}$ is compact with non-empty interior. In what follows, by η restricted to K, we mean the trace of η run up to the first time that it exits the set K (which does not depend on the choice of time parametrisation). If $h \in \mathcal{D}'_0(\mathbb{H})$, by h restricted to K, we mean the restriction to the interior of K, in the standard sense of restriction of distributions in the standard sense of restriction of distributions.

Lemma 8.25 *Let* $((\tilde{h}^t, \eta^t)_{0\le t\le \tau_Z}, Z)$ *be sampled from* $\mathbf{Q}$. *Let* φ_C *be the unique conformal isomorphism* $\mathbb{H} \to \mathbb{H}$ *such that* $(\mathbb{H}, \varphi_C(\tilde{h}^{\tau_Z} + C), 0, \infty)$ *is the unit circle embedding of* $(\mathbb{H}, \tilde{h}^{\tau_Z} + C, 0, \infty)$. *Then for any* $K \subset \mathbb{H}$ *compact, the law of* $(\varphi_C(\tilde{h}^{\tau_Z}+C), \varphi_C(\eta^{\tau_Z}))$ *restricted to* K *converges in total variation distance to the law of* (h, ζ) *restricted to* K, *where* (h, ζ) *is as in Proposition 8.20.*

Note that $\{(\varphi_C(\tilde{h}^{\tau_Z} + C), \varphi_C(\eta^{\tau_Z})) : C > 0\}$ is completely determined by $(\tilde{h}^{\tau_Z}, \eta^{\tau_Z}, Z)$.

For the proof of Lemma 8.25, let us first explain why zooming in (i.e. applying φ_C to η^{τ_Z}) produces an SLE_κ curve ζ as $C \to \infty$.

Lemma 8.26 *Let* $K \subset \mathbb{H}$ *be compact, let* Z *be sampled from its* $\mathbf{Q}$ *law, and let* $(\eta^t)_{0\le t\le \tau_Z}$ *be a reverse* $SLE_\kappa(\kappa, -\kappa)$ *curve with force points at* $(Z, 10)$, *run*

8.3 Proof of Theorem 8.9

until $f_t(Z)$ reaches 0. Let $\phi_R(z) = Rz$ for $R > 0$. Then $\phi_R(\eta^{\tau_Z})|_K$ converges in total variation distance to $\zeta|_K$ as $R \to \infty$, where ζ has the law of an SLE_κ curve from 0 to ∞ in $\mathbb{H}$.

Proof It is equivalent to prove that $\eta^{\tau_Z}|_{\delta\mathbb{D}_+}$ converges in total variation distance to $\zeta|_{\delta\mathbb{D}_+}$ as $\delta \downarrow 0$, where we recall that $\mathbb{D}_+ := B(0,\delta) \cap \mathbb{H}$ for $\delta > 0$.

To show this, since reverse Loewner evolutions grow "from the base" rather than the tip, we only need to concentrate on the evolution of the curve on a time interval just preceeding τ_Z. So, if $(f_t)_t$ is the reverse flow associated with $(\eta^t)_t$, we first pick $a > 0$ small and run f_t until the first time τ_Z^a that $f_{\tau_Z^a}(Z) = a$. Note that by Lemma B.6, $f_T(10) > 1$. Moreover by the same lemma, the total variation distance between the flow $(f_{\tau_Z^a+t})_{0<t\leq\tau_Z-\tau_Z^a}$ and the flow $(\tilde{f}_t)_{0\leq t\leq\tau_a}$ of a reverse $SLE_\kappa(\kappa)$ $\tilde{\eta}$ with a single force point at a run until $\tilde{f}_t(a) = 0$, converges to 0 as $a \to 0$. (To apply Lemma B.6 directly, we need to rescale by $1/a$, sending $a \mapsto 1$ and $f_{\tau_Z^a}(10) \mapsto f_{\tau_Z^a}(10)/a > 1/a$.)

But for $(\tilde{f}_t)$ we can use the time reversal symmetry of $SLE_\kappa(\rho)$ – Corollary B.11 – which says that the curve generated the reverse Loewner flow $(\tilde{f}_t)_{0\leq t\leq\tau_a}$ is that of an ordinary forward SLE_κ run until an almost surely positive time Λ_a. Roughly speaking, this concludes the proof since zooming in at the origin for this segment of forward SLE_κ produces an infinite SLE_κ as desired.

Let us now put all these pieces together rigorously. Given $\varepsilon > 0$, we first choose $a > 0$ small enough that $(f_t)_{t\leq\tau_Z}$ can be coupled with the hybrid flow - that is, $(f_t)_{t\leq\tau_Z^a}$ concatenated with $(\tilde{f}_t)_{t\leq\tau_a}$ - so that they agree with probability $\geq 1 - \varepsilon$. Then for $\delta > 0$ small enough, we also have a sub-event of probability $\geq 1 - 2\varepsilon$ that $\tilde{f}_{\tau_a}^{-1}(B(0,\delta))$ does not intersect η^{τ_Z}. In other words, on this subevent, we have that $\eta^{\tau_Z}|_{\delta\mathbb{D}_+} = \tilde{\eta}^{\tau_a}|_{\delta\mathbb{D}_+}$, where $\tilde{\eta}^{\tau_a}$ has the law of a reverse $SLE_\kappa(\kappa)$ with a force point at a, run until the reverse Loewner image of a reaches the origin. Finally, we use that for fixed a, the total variation distance between $\tilde{\eta}^{\tau_a}|_{\delta\mathbb{D}_+}$ and $\zeta|_{\delta\mathbb{D}_+}$ converges to 0 as $\delta \to 0$. So choosing δ small enough we can couple $\eta^{\tau_Z}|_{\delta\mathbb{D}_+}$ and $\zeta|_{\delta\mathbb{D}_+}$ so that they agree with probability $\geq 1 - 3\varepsilon$. Since $\varepsilon > 0$ was arbitrary, this proves the desired convergence in total variation distance. □

Since we know that under **Q**, $\tilde{h}^{\tau_Z}$ is essentially a Neumann GFF plus an appropriate log-singularity, independent of η^{τ_Z} we want to conclude the proof of Lemma 8.25 using Theorem 7.11, which says that zooming in at the origin for such a field produces a quantum wedge. However, the law of $\tilde{h}^{\tau_Z}$ also includes a random additive constant and a harmonic function (see Corollary 8.23), both of which in fact depend on the curve η^{τ_Z} itself. This dependence could be problematic when trying to describe what happens to the curve η^{τ_Z} when we zoom in. We thus appeal to a more refined version of Theorem 7.11, namely Lemma 7.12.

Proof of Lemma 8.25 Corollary 8.23 and Lemma 7.12 imply that almost surely, conditionally on $(f_t)_{t \le \tau_Z}$, the conditional law of $(\varphi_C, \varphi_C(\tilde{h}^{\tau_Z} + C)|_K)$ becomes arbitrarily close, in total variation distance as $C \to \infty$, to the law of $(\psi_C, \psi_C(h' + C)|_K)$ where h' is a Neumann GFF normalised to have average 0 on $\partial \mathbb{D}_+$ plus a $(\gamma - 2/\gamma)$ log singularity at the origin, and ψ_C is the scaling map such that $\psi_C(h' + C)$ is the unit circle embedding of $h' + C$. It is not hard to see that this implies

$$d_{\mathrm{TV}}((\eta^{\tau_Z}, \varphi_C, \varphi_C(\tilde{h}^{\tau_Z} + C)|_K); (\eta^{\tau_Z}, \psi_C, \psi_C(h' + C)|_K)) \to 0$$

as $C \to \infty$, where in the second triple above, $(\psi_C, \psi_C(h' + C))$ is independent of η^{τ_Z}. In turn, this implies that

$$d_{\mathrm{TV}}((\varphi_C(\eta^{\tau_Z})|_K, \varphi_C(\tilde{h}^{\tau_Z} + C)|_K); (\psi_C(\eta^{\tau_Z})|_K, \psi_C(h' + C)|_K)) \to 0$$

as $C \to \infty$. Now by Theorem 7.11, as $C \to \infty$, the law of $\psi_C(h' + C)|_K$ converges (in total variation) to the law of the quantum wedge h appearing in the statement of Lemma 8.25. Furthermore, since ψ_C and h' are independent of η^{τ_Z}, Lemma 8.26 implies that

$$d_{\mathrm{TV}}((\psi_C(\eta^{\tau_Z})|_K, \psi_C(h' + C)|_K); (\zeta|_K, h|_K)) \to 0$$

as $C \to \infty$, where ζ is an SLE_κ independent of h. Applying the triangle inequality concludes the proof. □

Step 3: Stationarity. In Step 2, we have shown that if one zooms in at the capacity zipper with reweighted law $\mathbf{Q}$ at time τ_Z, then one obtains a field/curve pair having the distribution of (h, ζ) as in Proposition 8.20. In this step, we will prove that the operation of "zipping down right quantum boundary length one" does not change this law, and hence prove Proposition 8.20 (see Figure 8.3).

Given a sample $((\tilde{h}^t, \eta^t)_{0 \le t \le \tau_Z}, Z)$ from $\mathbf{Q}$, and $C > 0$, let $Z_C \in [0, Z]$ be such that $\mathcal{V}_{\tilde{h}^0}([Z_C, Z]) = e^{-C\gamma/2}$. If this is not possible (i.e., if $\mathcal{V}_{\tilde{h}^0}([0, Z]) < e^{-C\gamma/2}$), set $Z_C = 0$. Set $\tau_C = \tau_{Z_C}$ and let ϕ_C be the unique conformal isomorphism such that $(\mathbb{H}, \phi_C(\tilde{h}^{\tau_C} + C), 0, \infty)$ is the unit circle embedding of $(\mathbb{H}, \tilde{h}^{\tau_C} + C, 0, \infty)$.

Recall the notation g_σ, ψ from Proposition 8.20.

Lemma 8.27 *For any $K \subset \mathbb{H}$ compact with non-empty interior, $(\phi_C(\tilde{h}^{\tau_C} + C), \phi_C(\eta^{\tau_C}))$ restricted to K converges in total variation distance to $(\psi \circ g_\sigma(h), \psi \circ g_\sigma(\zeta))$ restricted to K, as $C \to \infty$.*

8.3 Proof of Theorem 8.9

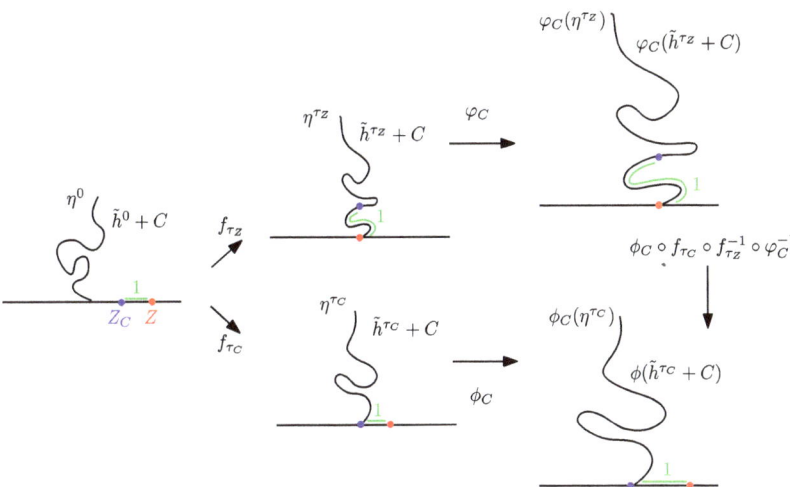

Figure 8.3 All the marked quantum boundary lengths (with respect to the field indicated on the relevant diagram) are equal to one. This is by definition of the conformal isomorphisms $f_{\tau Z}, f_{\tau C}, \phi_C$ and φ_C. Recall that $f(h)$ is obtained from h by applying the conformal change of coordinates formula which preserves quantum boundary length.

Lemma 8.28 *For any $K \subset \mathbb{H}$ compact, $(\phi_C(\tilde{h}^{\tau_C} + C), \phi_C(\eta^{\tau_C}))$ restricted to K converges in total variation distance to (h, ζ) restricted to K, as $C \to \infty$.*

Proof of Proposition 8.20 Lemmas 8.27 and 8.28 tell us that for any K, we can couple (h, ζ) and $(\psi \circ g_\sigma(h), \psi \circ g_\sigma(\zeta))$ together so that they agree when restricted to K with as high probability as we like. Thus their laws, when restricted to K, must agree. Since K was arbitrary, we can conclude. $\square$

Proof of Lemma 8.27 Let $\varepsilon > 0$ be arbitrary. First observe that we can choose $K_\varepsilon \subset \mathbb{H}$ compact so that $K \subset g_\sigma(K_\varepsilon)$ with probability greater than $1 - \varepsilon$. By Lemma 8.25, for large enough C we can also couple $(\varphi_C(\tilde{h}^{\tau z} + C), \varphi_C(\eta^{\tau z}))$ and (h, ζ) such that with probability $> 1 - \varepsilon$ they are equal in K_ε. We may also (by taking C large enough) require that $Z_C \neq 0$ on this event. Then, since on this event we have that

$$(\phi_C(\tilde{h}^{\tau_C} + C), \phi_C(\eta^{\tau_C})) = (\psi \circ g_\sigma(h), \psi \circ g_\sigma(\zeta))$$

(this is clear since these pairs are obtained from $(\varphi_C(\tilde{h}^{\tau z} + C), \varphi_C(\eta^{\tau z}))$ and (h, ζ), respectively, by zipping down 1 unit of right quantum boundary length and applying a conformal isomorphism so as to be in the unit circle parametrisation) the result follows. $\square$

Proof of Lemma 8.28 For this, observe that if μ is the law of a uniform point in $[0, A]$ for $A > 0$, and ν is the law of $U - \varepsilon$ for $U \sim \mu$, then the total variation distance between ν and μ tends to 0 as $\varepsilon \to 0$. This means that we can couple the **Q** laws of $(Z, (\tilde{h}^t, \eta^t)_{t \geq 0})$ and $(Z_C, (\tilde{h}^t, \eta^t)_{t \geq 0})$ such that they are equal with probability tending to 1 as $C \to \infty$ (by Lemma 8.21(b), definition of Z_C and the fact that the $\tilde{h}^0$ boundary length of $[1, 2]$ is finite almost surely). Hence we can couple the **Q** laws of $(\varphi_C(\tilde{h}^{\tau_Z} + C), \varphi_C(\eta^{\tau_Z}))$ and $(\phi_C(\tilde{h}^{\tau_C}), \phi_C(\eta^{\tau_C}))$ so they are equal with probability tending to 1 as $C \to \infty$. Since the former law converges to that of (h, ζ) as $C \to \infty$ (Lemma 8.25), the same therefore holds for the latter. □

8.3.3 Uniqueness of the welding

Consider the **capacity zipper** $(\tilde{h}^t, \eta^t)_{t \in \mathbb{R}}$ of Remark 8.14 (where the additive constant for $\tilde{h}^0$ is fixed). The (reverse) Loewner flow associated to $(\eta^t)_{t \geq 0}$ has the property that it zips together intervals of $\mathbb{R}_+$ and $\mathbb{R}_-$ with the same $\mathcal{V}_{\tilde{h}^0}$ quantum length by Theorem 8.9. It is natural to wonder if this actually determines the reverse flow. That is to ask: could there be any other Loewner flow with the property that intervals of identical quantum length on either side of zero are being zipped together?

We will now show that the answer to this question is no, and hence the Loewner flow for $t \geq 0$ is entirely determined by $\tilde{h}^0$.

Theorem 8.29 *Let $(\tilde{h}^t, \eta^t)_{t \in \mathbb{R}}$ be a capacity zipper as in Remark 8.14, with reverse Loewner flow $(f_t)_{t \geq 0}$. Then for $t > 0$, the following holds almost surely. If $\hat{f}_t : \mathbb{H} \to \hat{H}_t := \hat{f}_t(\mathbb{H})$ is a conformal isomorphism such that:*

- *$\hat{H}_t$ is the complement of a simple curve $\hat{\eta}^t$,*
- *$\hat{f}_t$ has the hydrodynamic normalisation $\lim_{z \to \infty} \hat{f}_t(z) - z = 0$;*
- *$\hat{f}_t$ has the property that $\hat{f}_t(z^-) = \hat{f}_t(z^+)$ as soon as $\mathcal{V}_{\tilde{h}^0}([z^-, 0]) = \mathcal{V}_{\tilde{h}^0}([0, z^+])$ and $f_t(z^-) \in \mathbb{H} \cup \{0\}$;*

then $\hat{f}_t = f_t$ and $\hat{\eta}^t = \eta^t$. In particular, the reverse Loewner flow $(f_t)_{t \geq 0}$ is determined by $\tilde{h}^0$ only (and hence $((\tilde{h}^t, \eta^t))_{t \geq 0}$ is entirely determined by $(\tilde{h}^0, \eta^0)$).

Proof Before we go any further, we recall from the definition of the capacity zipper in Theorem 8.13, that we only have defined the reverse Loewner flow as being coupled to $\tilde{h}^0$ in a certain way specified by the application of Kolmogorov's theorem. Usually, proving that objects coupled to a GFF are determined by it can be quite complicated (for example, this is the case in the set-up of imaginary geometry, or when making sense of level lines of the GFF).

Here the proof will turn out to be quite simple, given some classical results from the literature. Indeed consider

$$\phi = \hat{f}_t \circ f_t^{-1}.$$

A priori, ϕ is a conformal isomorphism on $f_t(\mathbb{H}) = H_t$, and its image is $\phi(H_t) = \hat{H}_t$. However, because of our assumptions on $\hat{f}_t$ (and the properties of f_t), the definition of ϕ can be extended unambiguously to all of $\mathbb{H}$. Moreover when we do so, the extended map is a homeomorphism of $\mathbb{H}$ onto $\mathbb{H}$, which is conformal off the curve $\eta'([0,t])$. Thus the theorem will be proved if we can show that any such map must be the identity. In the terminology of complex analysis, this is equivalent to asking that the curve $\eta'([0,t])$ is a *removable* set. Now, by a result of Rohde and Schramm [RS05], the complement H_t of the curve is almost surely a Hölder domain for $\kappa < 4$ (or $\gamma < 2$), and by a result of Jones and Smirnov [JS00] it follows that $\eta'([0,t])$ is a removable set. Hence the theorem follows. □

Remark 8.30 By the same argument, it also holds that for the quantum zipper $(h^t, \zeta^t)_{t \in \mathbb{R}}$ of Theorem 8.16 (h^t, ζ^t) is almost surely determined by (h^0, ζ^0) for any $t > 0$. In the language of conformal welding, (h^t, ζ^t) is obtained from (h^0, ζ^0) by welding the interval on the left of 0 with h^0 quantum length t to the interval on the right of 0 with h^0 quantum length t (and pushing through ζ^0 by the resulting conformal isomorphism).

8.4 Slicing a wedge with an SLE

In this section, we complement our previous discussion with the following remarkable theorem due to Sheffield [She16a]. This result is fundamental to the theory developed in [DMS21], where the main technical tool is a generalisation of the result below.

Suppose we are given a $(\gamma - 2/\gamma)$-quantum wedge $(\mathbb{H}, h, 0, \infty)$ in some embedding, and an independent SLE_κ curve η with $\kappa = \gamma^2 < 4$. Then the curve η slices the wedge into two surfaces (see Figure 8.4). The result below says that *as quantum surfaces* these are independent and that they are both γ-thick wedges.

Theorem 8.31 *Suppose we are given an $(\gamma-2/\gamma)$-quantum wedge $(\mathbb{H}, h, 0, \infty)$ in the unit circle embedding, and an independent SLE_κ curve η with $\kappa = \gamma^2 < 4$. Let D_1, D_2 be the two connected components of $\mathbb{H}\setminus\eta$, whose boundaries contain the negative and positive real lines, respectively. Let $h_1 = h|_{D_1}$ and $h_2 = h|_{D_2}$. Then the two surfaces $(D_1, h_1, 0-, \infty)$ and $(D_2, h_2, 0+, \infty)$ are independent γ-quantum wedges.*

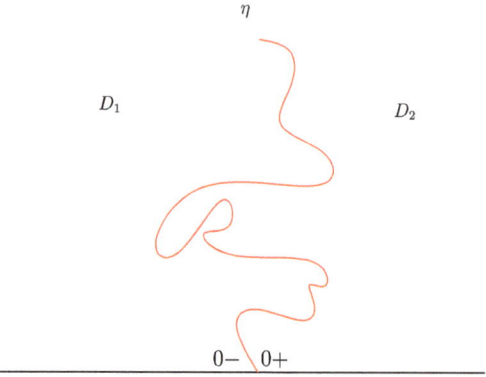

Figure 8.4 An independent SLE slices an $(\gamma - 2/\gamma)$-thick wedge into two independent γ-thick wedges.

Remark 8.32 This does *not* imply that the fields, or generalised functions, h_1 and h_2 are independent. It is a statement about the two doubly marked surfaces $(D_1, h_1, 0-, \infty)$ and $(D_2, h_2, 0+, \infty)$. So what it does say, for example, is that if $\tilde{h}_1$ and $\tilde{h}_2$ are the fields corresponding to the unit circle embeddings of these surfaces, then $\tilde{h}_1$ and $\tilde{h}_2$ are independent.

Remark 8.33 By the same argument as in Section 8.3.3, the surfaces

$$(D_1, h_1, 0-, \infty) \text{ and } (D_2, h_2, 0+, \infty)$$

determine h and η in the following sense. Suppose that $(\mathbb{H}, \tilde{h}_1, 0, \infty)$ and $(\mathbb{H}, \tilde{h}_2, 0, \infty)$ are the two unit circle embeddings of these surfaces, and that $(\hat{f}_1, \hat{f}_2, \hat{\eta})$ are such that:

- $\hat{\eta}$ is a simple curve from 0 to ∞;
- $\hat{f}_1$ (resp. $\hat{f}_2$) is a conformal isomorphism from $\mathbb{H}$ to the left-hand side (resp. right-hand side) of $\hat{\eta}$;
- $\hat{f}_1, \hat{f}_2$ extend to $\mathbb{R}$ in such a way that for any $x^\pm \in \mathbb{R}_\pm$ with $\mathcal{V}_{\tilde{h}^1}([0, x^+]) = \mathcal{V}_{\tilde{h}^2}([x^-, 0])$ we have $\hat{f}_1(x^+) = \hat{f}_2(x^-)$.

Then if $\hat{h}$ is defined by setting it equal to $\hat{f}_1(\tilde{h}_1)$ (resp. $\hat{f}_2(\tilde{h}_2)$) on the left-hand side (resp. right-hand side) of $\hat{\eta}$, we have that with probability one, $(\hat{h}, \hat{\eta}) = (\phi(h), \phi(\eta))$ for some simple scaling map $\phi \colon z \mapsto az$.

We also remark that the choice of embedding for the $(\gamma - 2/\gamma)$-wedge in Theorem 8.31 does not matter, which can be argued as follows. Suppose that $(\mathbb{H}, h, 0, \infty)$ is some parametrisation of a $(\gamma - 2/\gamma)$-quantum wedge and that η

8.4 Slicing a wedge with an SLE

is an SLE_κ that is independent of h. Then there exists a scaling map $\varphi \colon \mathbb{H} \to \mathbb{H}$ such that $(\mathbb{H}, \varphi(h), 0, \infty)$ is the unit circle embedding of the quantum wedge. Since φ is independent of η and SLE is scale invariant, $\varphi(\eta)$ is an SLE_κ that is independent of $\varphi(h)$. Thus, applying Theorem 8.31, we see that the two quantum surfaces obtained by slicing $\varphi(h)$ along $\varphi(\eta)$ are two independent γ quantum wedges. On the other hand, these surfaces are by definition equivalent to the two surfaces obtained by slicing h along η. This means that the latter pair also have the law (as doubly marked quantum surfaces) of two independent γ quantum wedges.

Proof of Theorem 8.31 It is clear from the definition that $(D_1, h_1, 0-, \infty)$, $(D_2, h_2, 0+, \infty)$ almost surely have finite LQG areas in neighbourhoods of 0− and 0+, respectively, and infinite LQG areas in neighbourhoods of ∞. Therefore, we can define unique conformal isomorphisms $\phi_1 \colon D_1 \to \mathbb{H}$ sending $0- \to 0$ and $\infty \to \infty$ and $\phi_2 \colon D_2 \to \mathbb{H}$ sending $0+ \to 0$ and $\infty \to \infty$, so that $(\mathbb{H}, \phi_i(h_i), 0, \infty)$ gives LQG area one to the upper unit semidisc $B(0, 1) \cap \mathbb{H}$ for $i = 1, 2$. Recall that we refer to $\phi_i(h_i)$ as the canonical description of the surface $(D_i, h_i, 0\pm, \infty)$, and we continue to use the "change of coordinate" notation (8.11) for conformal isomorphisms applied to fields. It clearly suffices to show that for any large semidisc $K \subset \mathbb{H}$, $(\phi_1(h_1)|_K, \phi_2(h_2)|_K)$ agrees in law with $(h_1^{\text{wedge}}|_K, h_2^{\text{wedge}}|_K)$ where h_1^{wedge} and h_2^{wedge} are independent, and each has the law of the canonical description of a γ-quantum wedge. (The reason we choose to work with the canonical description rather than the unit circle embedding here is simply to avoid any ambiguity concerning the a priori existence of the maps ϕ_1 and ϕ_2.)

To show this equality in law, we need to appeal to the results of the Section 8.3: in particular Lemma 8.25 and Theorem 8.9. Consider the process $((\tilde{h}^t, \eta^t)_{t\geq 0}, Z)$ under the law $\mathbf{Q}$ from Lemma 8.21, and in this set-up, let Y denote the point to the left of zero such that the $\tilde{h}^0$ boundary length of $[Y, 0]$ is equal to that of $[0, Z]$. Write h_Z^C for the canonical description of $(H_Z, \tilde{h}^0 + C, Z, \infty)$ and h_Y^C for the canonical description of $(H_Y, \tilde{h}^0 + C, Z, \infty)$ where H_Z and H_Y are the connected components of $\mathbb{H} \setminus \tilde{\eta}^0$ containing Z and Y, respectively. Combining Lemma 8.25 and Theorem 8.9 gives that:

Claim 8.34 *We can couple pairs of fields with*

- *the joint law of (h_Y^C, h_Z^C) under $\mathbf{Q}$, and*
- *the joint law of $(\phi_1(h_1), \phi_2(h_2))$ described in the first paragraph,*

so that they agree when restricted to K, with probability arbitrarily close to one as $C \to \infty$.

Proof of claim (See Figure 8.5). First we observe that one (slightly convoluted!) way to sample a pair with the law of (h_Y^C, h_Z^C) under $\mathbf{Q}$ is to:

(1) consider the "zipper" $((\tilde{h}^t, \eta^t)_{t \geq 0}, Z)$ under $\mathbf{Q}$ and apply the conformal isomorphism $f_{\tau_Z}^C$ that zips up Z to 0 and then scales $\mathbb{H}$ so that $f_{\tau_Z}^C(\tilde{h}^0)$ is in the unit circle embedding;
(2) then restrict the field $f_{\tau_Z}^C(\tilde{h}^0 + C)$ to the left and right of $f_{\tau_Z}^C(\eta^0)$ and apply conformal isomorphisms from these left- and right-hand sides to $\mathbb{H}$, such that the resulting fields (under the change of coordinates formula) are the canonical descriptions of these two surfaces.

Here we are using the fact, due to Theorem 8.9, that Y is zipped up to 0 at exactly the same time as Z.

On the other hand, Lemma 8.25 says that we can couple $(f_{\tau_Z}^C(\tilde{h}^0+C), f_{\tau_Z}^C(\eta^0))$ with (h, η) as in the statement of the present theorem, so that they agree in any large semidisc K', with probability arbitrarily close to one as $C \to \infty$. Consequently, if we restrict the field $f_{\tau_Z}^C(\tilde{h}^0 + C)$ to the left and right of $f_{\tau_Z}^C(\eta^0)$, and apply conformal isomorphisms as in the second step of the previous bullet point, then the resulting pair of fields can be coupled with $(\phi_1(h_1), \phi_2(h_2))$ so that they agree when restricted to K with arbitrarily high probability.

Combining these two paragraphs yields the claim. □

So, with the claim in hand, it actually suffices to show that we can couple (h_Y^C, h_Z^C) with $(h_1^{\text{wedge}}, h_2^{\text{wedge}})$ (recall that the latter are a pair of independent γ-wedge fields in their canonical descriptions) so that their restrictions to K agree with probability arbitrarily close to 1 as $C \to \infty$. The idea is that when C is very large, the restrictions of h_Y^C and h_Z^C to K will correspond to images – under the conformal change of coordinates (8.11) – of $\tilde{h}^0 + C$ restricted to very tiny neighbourhoods of Z and Y. Roughly speaking, these restrictions become independent in the limit as the size of the neighbourhoods goes to 0, and furthermore, the field near Z (and by symmetry near Y) converges to a γ-quantum wedge field.

To be more precise, let us consider a sample $(\tilde{h}^0, Z)$ from $\mathbf{Q}$, together with a field $\tilde{h} = \hat{h} + (\gamma - 2/\gamma) \log(|\cdot|^{-1})$, where $\hat{h}$ is a Neumann GFF normalised to have average 0 on the upper unit semicircle that is *independent* of $\tilde{h}^0$. Then we have the following:

Lemma 8.35 *As above, let $(\tilde{h}^0, Z)$ have their* $\mathbf{Q}$ *joint law, and let $\tilde{h} = \hat{h} + (\gamma - 2/\gamma) \log(|\cdot|^{-1})$, where $\hat{h}$ is a Neumann GFF normalised to have average 0 on the upper unit semicircle that is independent of $\tilde{h}^0$. Then the total variation*

8.4 Slicing a wedge with an SLE   331

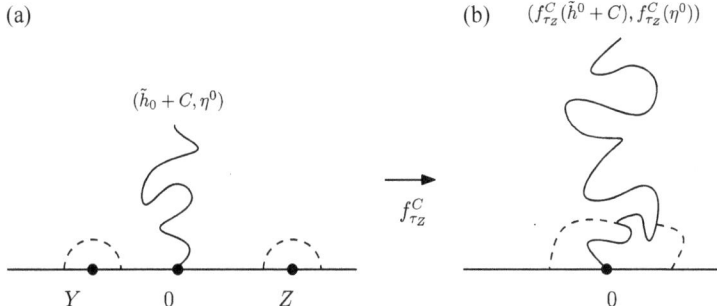

Figure 8.5 The surfaces to the left and right of η^0 on (a) (defined using the field $\tilde{h}^0 + C$ and marked at (Y, ∞) and (Z, ∞)) have canonical descriptions given by $(\mathbb{H}, h_Y^C, 0, \infty)$ and $(\mathbb{H}, h_Z^C, 0, \infty)$. So the same is true, by definition, for the surfaces to the left and right of the curve on (b) (defined using the field $f_{\tau_Z}^C(\tilde{h}^0 + C)$). But Lemma 8.25 says that for C large, the joint law of the field and curve on the (b) is very close to that of (h, η) from the statement of Theorem 8.31. So, the law of the canonical descriptions of the surfaces to the left and right of the curve is very close to that of $(\mathbb{H}, \phi_1(h_1), 0, \infty)$, $(\mathbb{H}, \phi_2(h_2), 0, \infty)$. Hence we can approximate the joint law of $(\phi_1(h_1), \phi_2(h_2))$ by that of (h_Y^C, h_Z^C) for C large.

distance between

$$\left(\tilde{h}^0|_{\overline{B(Z,\varepsilon)\cap\mathbb{H}}}, \tilde{h}^0|_{\mathbb{H}\setminus B(Z,1)}, \mathcal{V}_{\tilde{h}^0}[1,Z], \mathcal{V}_{\tilde{h}^0}[Z,2]\right)$$

and

$$\left(\tilde{h}(\cdot - Z)|_{\overline{B(Z,\varepsilon)\cap\mathbb{H}}}, \tilde{h}^0|_{\mathbb{H}\setminus B(Z,1)}, \mathcal{V}_{\tilde{h}^0}[1,Z], \mathcal{V}_{\tilde{h}^0}[Z,2]\right)$$

converges to 0 as $\varepsilon \to 0$.

In words, this says that *conditionally* on $\tilde{h}^0$ outside of $B(Z, 1)$ *and* on the boundary lengths $\mathcal{V}_{\tilde{h}^0}[1, Z], \mathcal{V}_{\tilde{h}^0}[Z, 2]$, the law of $\tilde{h}^0$ restricted to $\overline{B(Z, \varepsilon) \cap \mathbb{H}}$ is very close in total variation distance to the field $\tilde{h}$ recentred at Z and restricted to $\overline{B(Z, \varepsilon) \cap \mathbb{H}}$.

Before proving the lemma, let us first see how it allows us to conclude the proof of the theorem. From now on, we assume that $K \subset \mathbb{H}$ is large, fixed semidisc. Consider a pair $(h_Y^C, \tilde{h}^C)$ where h_Y^C has its **Q** law, and $\tilde{h}^C$ is independent of h_Y^C having the law of the canonical description of $(\mathbb{H}, \tilde{h} + C, 0, \infty)$. The consequence of Lemma 8.35 is that by taking ε very small and then C sufficiently large, we can couple the joint law of (h_Y^C, h_Z^C) with that of the pair $(h_Y^C, \tilde{h}^C)$, so that the fields agree when restricted to K with probability arbitrarily close to one. Since the law of $\tilde{h}^C|_K$ converges in total variation distance to $h_2^{\text{wedge}}|_K$ as $C \to \infty$, see Corollary 7.14, this means that we can couple (h_Y^C, h_Z^C) with $(h_Y^C, h_2^{\text{wedge}})$ (where the latter pair are independent) so that they agree when restricted to K with arbitrarily high probability as $C \to \infty$.

To finish the proof, we observe that by symmetry, h_Y^C has the same law as h_Z^C for each C. Since the argument above clearly gives that $h_Z^C|_K \to h_2^{\text{wedge}}|_K$ in total variation distance as $C \to \infty$, it must therefore also be the case that $h_Y^C|_K$ converges in total variation distance to $h_1^{\text{wedge}}|_K$ as $C \to \infty$. Thus $(h_Y^C, h_2^{\text{wedge}})$ can be coupled with $(h_1^{\text{wedge}}, h_2^{\text{wedge}})$ so that the fields agree when restricted to K with arbitrarily high probability as $C \to \infty$. Putting this together with the previous paragraph, we obtain the desired result. □

Proof of Lemma 8.35 We first claim that for any $\delta > 0$,

$$d_{\text{TV}}\left((\tilde{h}^0|_{\overline{B(Z,\varepsilon)\cap\mathbb{H}}}, \tilde{h}^0|_{\mathbb{H}\setminus B(Z,\delta)}), (\tilde{h}(\cdot - Z)|_{\overline{B(Z,\varepsilon)\cap\mathbb{H}}}, \tilde{h}^0|_{\mathbb{H}\setminus B(Z,\delta)})\right) \to 0 \quad (8.20)$$

as $\varepsilon \to 0$. Indeed, by Lemma 8.21, the $\mathbf{Q}^Z$ (i.e. $\mathbf{Q}(\cdot|Z)$) law of $\tilde{h}^0$ recentred at Z is that of $h' + (\gamma - 2/\gamma)\log(|\cdot|^{-1}) + \mathfrak{h}$, where h' is a Neumann GFF normalised to have average 0 on the upper unit semicircle centred at 10 and $\mathfrak{h}$ is a harmonic function that independent of h' and is deterministically bounded in $B(Z, 1)$. Hence (8.20) follows from Lemma 6.34 and Remark 6.35.

We will now extend this in the following way. We are going to show that the law of $\mathcal{V}_{\tilde{h}^0}([1, Z])$ is basically the same (when ε is small enough) whether we condition on $\tilde{h}^0$ restricted to $\mathbb{H} \setminus B(Z, 1)$ *and* $\overline{\mathbb{H} \cap B(Z, \varepsilon)}$, or just restricted to $\mathbb{H} \setminus B(Z, 1)$: see (8.21). The basic idea for the proof is that, given the restriction of $\tilde{h}^0$ to $\mathbb{H} \setminus B(Z, 1)$, the restriction of $\tilde{h}^0$ to $\overline{\mathbb{H} \cap B(Z, \varepsilon)}$ has a very tiny influence on the boundary length of $[1, Z]$ when ε is small. On the other hand, there is quite a bit of variation in the boundary length coming from sources completely independent of $\tilde{h}^0|_{\overline{B(Z,\varepsilon)\cap\mathbb{H}}}$. To argue this rigorously, we will use the Fourier decomposition of the free field, similarly to the argument [She16a].

We take $\delta > 0$ small, and fix a function ϕ that is smooth, positive and supported in the upper unit semidisc of radius $\delta/4$ centred at $Z - 3\delta/2$, with $(\phi, \phi)_\nabla = 1$ (see Figure 8.6). Let us write $U := [Z - 7\delta/4, Z - 5\delta/4]$, $U^C = [1, Z] \setminus U$ (see Figure 8.6). This will be non-empty with arbitrarily high probability if δ is small enough, so let us assume from now on that Z is such that this is the case. Then by Definitions 6.3 and 6.21, we can decompose $\tilde{h}^0 = X\phi + h$ where X is Gaussian and h is independent of X.

Next, we observe that due to the decomposition of $\tilde{h}^0$, the conditional law of $\mathcal{V}_{\tilde{h}^0}(U)$ given $h|_{\mathbb{H}\setminus B(Z,\delta)}$ almost surely has smooth density $F^{h|_{\mathbb{H}\setminus B(Z,\delta)}}$ with respect to Lebesgue measure: indeed, given h restricted to $\mathbb{H} \setminus B(Z, \delta)$, $\mathcal{V}_{\tilde{h}^0}(U)$

8.4 Slicing a wedge with an SLE

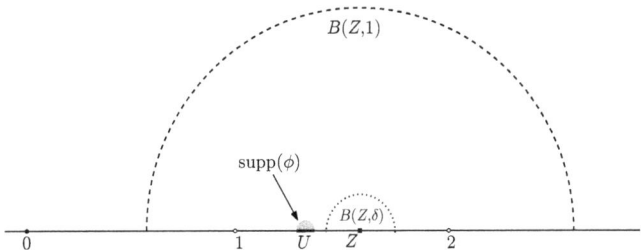

Figure 8.6 Illustration of the random point Z and $U = [Z - 7\delta/4, Z - 5\delta/4]$

is almost surely smooth and increasing in X. In particular,

the conditional law of $\mathcal{V}_{\tilde{h}^0}([1, Z])$ given h has density
$$\propto F^{h|_{\mathbb{H}\setminus B(Z,\delta)}}(\cdot - \mathcal{V}_h(U^c))$$

with respect to Lebesgue measure. Using the fact that F is smooth, that $\mathcal{V}_h$ almost surely does not have an atom at Z, and (8.20) applied with $\delta' \ll \delta$, we may deduce from this that for any $x \in \mathbb{R}$, the quantity

$$\mathbb{E}(F^{h|_{\mathbb{H}\setminus B(Z,\delta)}}(x - \mathcal{V}_h(U^c)) \mid h|_{\mathbb{H}\setminus B(Z,\delta)})$$
$$- \mathbb{E}(F^{h|_{\mathbb{H}\setminus B(Z,\delta)}}(x - \mathcal{V}_h(U^c)) \mid h|_{\mathbb{H}\setminus B(Z,\delta)}, h|_{\overline{B(Z,\varepsilon)\cap\mathbb{H}}})$$

tends to 0 almost surely as $\varepsilon \to 0$. This is important because it means that

$$d_{\mathrm{TV}}\big(\mathcal{L}(\mathcal{V}_{\tilde{h}^0}([1, Z]) \mid h|_{\mathbb{H}\setminus B(Z,\delta)}),$$
$$\mathcal{L}(\mathcal{V}_{\tilde{h}^0}([1, Z]) \mid h|_{\mathbb{H}\setminus B(Z,\delta)}, h|_{\overline{B(Z,\varepsilon)\cap\mathbb{H}}})\big) \to 0 \qquad (8.21)$$

in probability $\varepsilon \to 0$ (where $\mathcal{L}(Y_1|Y_2)$ denotes the law of Y_1 conditioned on Y_2). In fact, since

$$h|_{\overline{B(Z,\varepsilon)\cap\mathbb{H}}} = \tilde{h}^0|_{\overline{B(Z,\varepsilon)\cap\mathbb{H}}},$$

and by combining with (8.20) this actually means that

$$d_{\mathrm{TV}}\Big(\big(\mathcal{V}_{\tilde{h}^0}([1, Z]), \tilde{h}^0|_{\overline{B(Z,\varepsilon)\cap\mathbb{H}}}, h|_{\mathbb{H}\setminus B(Z,\delta)}\big),$$
$$\big(\mathcal{V}_{\tilde{h}^0}([1, Z]), \tilde{h}(\cdot - Z)|_{\overline{B(Z,\varepsilon)\cap\mathbb{H}}}, h|_{\mathbb{H}\setminus B(Z,\delta)}\big)\Big) \to 0$$

in probability $\varepsilon \to 0$. This is extends with exactly the same argument (but a little more notation) to the same statement with $\mathcal{V}_{\tilde{h}^0}([1, Z]), \mathcal{V}_{\tilde{h}^0}([Z, 2])$ in place of just $\mathcal{V}_{\tilde{h}^0}([1, Z])$. Putting this together with the fact that $h = \tilde{h}^0$ outside of $B(Z, 1)$ completes the proof. □

9
Liouville Quantum Gravity as a Mating of Trees

9.1 Orientation

In this chapter, we take forward the ideas developed in Chapter 8 and obtain a beautiful and important description of the way that a certain quantum cone can be explored by an independent variant of SLE called *space-filling SLE*. This description has many important implications, and we already emphasise the following points.

- On the one hand, this shows that a quantum cone, considered as a random surface decorated with a designated space-filling path, can rigorously be described as the "mating" (i.e. gluing or welding) of two correlated (infinite) continuum random trees. This is the so-called **peanosphere** description of a quantum cone. It also has analogues for other quantum surfaces; see Section 9.8.
- On the other hand, this construction is the direct continuum analogue of the discrete bijection due to Sheffield in [She16b] for random planar maps weighted by the self dual Fortuin–Kasteleyn percolation model, as was presented in Chapter 4. Hence a particular consequence of this work (as developed in [GMS19, GS17] and [GS15]) is that, at least in the so-called peanosphere sense (which is a relatively weak notion of convergence), these random planar maps can be proven to converge to quantum cones.

This very fruitful approach was developed in the seminal paper of Duplantier, Miller and Sheffield [DMS21]. Since this paper is long and difficult, we will not aim to present complete proofs of their main theorems; rather, we will state precise results and hope to convey some of the key ideas that are used in their proofs.

To state the main theorems, we must first explain the construction of the aforementioned space-filling SLE. The details of the construction are not straightforward, and in fact rely on a whole other body of work (the so-called *imaginary geometry* of Miller and Sheffield, [MS16a, MS16b, MS16c] and especially [MS17]), which falls outside of the scope of this book.

Note: Although we will give a complete and self contained introduction to whole plane SLE (Section 9.2), and space-filling SLE_κ for $\kappa \geq 8$ (Section 9.3), the construction of space-filling SLE_κ in the case $\kappa \in (4, 8)$ (Section 9.4) will be explained rather than fully justified.

In Section 9.5, we will state a cutting/welding theorem for (thick and thin) quantum wedges, analogous to but more complicated than the welding statement of Theorem 8.31, which is crucial to the "mating of trees" theorem of [DMS21], which we will state in Section 9.6. In Section 9.7, we will discuss the implications of this theorem, in relation to the two bullet points above. In Section 9.8, we will give a proof of the main theorem in the case $\kappa' \in (4, 8)$, admitting the welding theorems from Section 9.5 and a stationarity statement (analogous to but more complicated than the stationarity of the quantum zipper in Proposition 8.20). This proof also partially covers the case $\kappa' \geq 8$, up to a certain step that we will explain properly in that section.

9.2 Whole plane SLEκ and $\text{SLE}_\kappa(\rho)$

9.2.1 Whole plane SLE_κ

Definition 9.1 (Whole plane SLE_κ) For $\kappa > 0$, whole plane SLE_κ in $\mathbb{C}$ from 0 to ∞ is defined to be the collection of maps $(g_t)_{t \in \mathbb{R}}$ that solve the *whole plane Loewner equation* for each $z \in \mathbb{C} \setminus \{0\}$:

$$\partial_t g_t(z) = g_t(z) \frac{U_t + g_t(z)}{U_t - g_t(z)}; \quad \lim_{t \to -\infty} e^t g_t(z) = z; \quad t \in (-\infty, \zeta(z)), \quad (9.1)$$

where $U_t = e^{i\sqrt{\kappa} B_t}$, B is a standard *two-sided* Brownian motion, and for each $z, \zeta(z) := \inf\{t \in \mathbb{R} \colon U_t = g_t(z)\}$.

We emphasise that the map g_t in the definition above is defined for all $t \in \mathbb{R}$, not just for $t \geq 0$ as is the case for chordal or radial Loewner chains (see Appendices A and C).

For a given realisation of B, existence and uniqueness of $g_t(z)$ for each $z \in \mathbb{C} \setminus \{0\}$ and $t \in (-\infty, \zeta(z))$ follows from standard ODE theory. If $K_t := \overline{\{z \colon \zeta(z) \leq t\}}$ for $t \in \mathbb{R}$ are the *whole plane Loewner hulls* generated by B, then it can be shown that $g_t(z)$ is indeed a conformal isomorphism from

336   *Liouville Quantum Gravity as a Mating of Trees*

$\hat{\mathbb{C}} \setminus K_t \to \hat{\mathbb{C}} \setminus \overline{\mathbb{D}}$ and that $\mathrm{cap}(K_t) := \lim_{z \to \infty} z/g_t(z) = e^t$ for each t; see for example [Law05].

In fact, many more properties can be deduced immediately from the following connection to *radial* SLE_κ. In some sense, the following lemma suggests that whole plane SLE_κ should be viewed as a bi-infinite time version of radial SLE_κ.

Lemma 9.2 *Let $(g_t, K_t)_{t \in \mathbb{R}}$ be the conformal isomorphisms and Loewner hulls associated to a driving function $(U_t)_{t \in \mathbb{R}}$ as in (9.1). Then for any $t_0 \in \mathbb{R}$, $\tilde{K}_t := g_{t+t_0}(K_{t+t_0} \setminus K_t)$ has the law of a radial Loewner evolution in $\hat{\mathbb{C}} \setminus K_{t_0}$ from the point $g_{t_0}^{-1}(U_{t_0})$ to ∞. More precisely, the hulls $1/g_{t+t_0}(K_{t+t_0} \setminus K_t)$ for $t \geq 0$ are described by a radial Loewner evolution in $\mathbb{D}$ whose driving function is given by $(\overline{U}_{t_0+t})_{t \geq 0}$ (the complex conjugate function of $(U_{t_0+t})_{t \geq 0}$).*

In particular, if $(K_t)_{t \in \mathbb{R}}$ are the whole plane hulls associated to an SLE_κ and $t_0 \in \mathbb{R}$, then $(K_{t_0+t})_{t \geq 0}$ are the hulls of a radial SLE_κ in $\mathbb{C} \setminus K_{t_0}$ from $g_{t_0}^{-1}(U_{t_0})$ to ∞.

Proof It suffices to check that if $\tilde{g}_s := 1/g_s(1/\cdot)$ for $s \in \mathbb{R}$ and $\hat{g}_t := \tilde{g}_{t+t_0} \circ \tilde{g}_{t_0}^{-1}$ for $t > 0$ (so that $\hat{g}_t$ is the unique conformal isomorphism from $\mathbb{D} \setminus \{1/g_{t+t_0}(K_{t+t_0} \setminus K_t)\}$ to $\mathbb{D}$ with $\hat{g}_t'(0) = e^t$ for each $t \geq 0$) then $(\hat{g}_t)_{t \geq 0}$ satisfies the radial Loewner equation (C.1) with driving function given by $\overline{U}_{t+t_0}$. This follows from a simple calculation using (9.1), which we leave to the reader (note that (C.1) and (9.1) are identical, apart from the time domain and the "initial" conditions). □

It therefore follows from the corresponding results for radial SLE_κ (see Section C.2) that for each $\kappa > 0$, and given $(B_t)_{t \in \mathbb{R}}$, there almost surely exists a continuous non-self-crossing curve $\gamma : (-\infty, \infty) \to \mathbb{C}$ such that the unique conformal isomorphism g_t from the unbounded connected component of $\mathbb{C} \setminus \gamma((-\infty, t])$ to $\mathbb{C} \setminus \mathbb{D}$ with $g_t(\infty) = \infty$ and $g_t'(\infty) > 0$, solves (9.1) (and in fact has $g_t'(\infty) = e^{-t}$) as in Definition 6.25. The curve starts at 0 in the sense that $\lim_{t \to -\infty} \gamma(t) = 0$ and is transient, that is, $\lim_{t \to \infty} \gamma(t) = \infty$. It also follows that whole plane SLE_κ has the same distinct phases as radial (and chordal) SLE_κ: it is a simple curve for $\kappa \leq 4$, is self intersecting but non-self-crossing and non-space-filling for $\kappa \in (4, 8)$, and is space filling for $\kappa \geq 8$.

The scaling property of Brownian motion also implies that if γ is a whole plane SLE_κ from 0 to ∞ and $a \in \mathbb{C} \setminus \{0\}$, then $a\gamma$ (with time reparametrised appropriately) also has the law of a whole plane SLE_κ from 0 to ∞. This means that the following definition makes sense.

Definition 9.3 Let $z_1, z_2 \in \hat{\mathbb{C}}$. Whole plane SLE_κ from z_1 to z_2 is defined to be the image of whole plane SLE_κ from 0 to ∞, under a Möbius transformation

9.2 Whole plane SLEκ and SLE$_\kappa(\rho)$

sending from 0 to z_1 and from ∞ to z_2. The law of this process does not depend on the choice of Möbius transformation.

With this definition, it is immediate that as a family indexed by $z_1, z_2 \in \hat{\mathbb{C}}$, whole plane SLE$_\kappa$ from z_1 to z_2 is Möbius invariant (in law). For instance, the whole plane SLE$_\kappa$ from ∞ to 0 is obtained by applying the Möbius inversion $\psi(z) = 1/z$ to the hulls $(K_t)_{t \in \mathbb{R}}$ of Definition 9.1. In doing so, we obtain hulls $\tilde{K}_t = \psi(K_t), t \in \mathbb{R}$; note that the parametrisation of $\tilde{K}_t$ is then such that the capacity seen from 0 of $\tilde{K}_t$ is always equal to e^t. In other words,

$$\mathrm{CR}(0; \mathbb{C} \setminus \tilde{K}_t) = e^{-t}, \tag{9.2}$$

where we recall that $\mathrm{CR}(x, D)$ stands for the conformal radius of x in D. In this sense, $(\tilde{K}_t)_{t \in \mathbb{R}}$ is just simply parametrised by **log conformal radius**.

We caution that invariance of whole plane SLE$_\kappa$ from 0 to ∞ under the inversion map $z \mapsto 1/z$ does not mean that it is *reversible*. That is, it is *a priori* not obvious that the image of whole plane SLE$_\kappa$ from 0 to ∞ under the inversion map coincides in law with its time-reversal. In fact, while chordal SLE$_\kappa$ is reversible for $\kappa \in (0, 8]$ (this was proved by Zhan in [Zha08b] for $\kappa \in (0, 4]$ and extended to $\kappa \in (4, 8]$ in [MS16c] using the tools of imaginary geometry), chordal SLE$_\kappa$ is known to be non-reversible for $\kappa > 8$. (This is credited in [MS17] to unpublished work of Rohde and Schramm.) Nevertheless, Viklund and Wang [VW24] conjectured reversibility of whole-plane SLE$_\kappa$ for $\kappa > 8$ on the basis of an invariance of a certain large deviation functional in the limit $\kappa \to \infty$; this remarkable conjecture was proved in the no less remarkable work of Ang and Yu [AY23].

9.2.2 Whole plane SLE$_\kappa(\rho)$

In this section, we will discuss the definition of SLE$_\kappa(\rho)$ for $\rho > -2$. We will only consider the case of one "weight" ρ, and the initial force point will be (in some sense) be the same as the starting point.

To do this, we need the following lemma.

Lemma 9.4 ([MS17], Proposition 2.1) *Suppose that $\kappa > 0, \rho > -2$ and that $(\tilde{U}_s, \tilde{V}_s)_{s \geq 0}$ solves (C.2) (with $m = 1$)[1] and some choice of $\tilde{U}_0, \tilde{V}_0 \in \partial \mathbb{D}$. There exists a unique time stationary law on continuous processes $(U_t, V_t)_{t \in \mathbb{R}}$, taking values on $\partial \mathbb{D} \times \partial \mathbb{D}$, for which $(U_t, V_t)_{t \geq t_0}$ is equal to the limit in law and in total variation distance of $(\tilde{U}_{t+T}, \tilde{V}_{t+T})_{t \geq t_0}$ as $T \to -\infty$ for any $t_0 \in \mathbb{R}$. This law does not depend on the choice of $\tilde{U}_0, \tilde{V}_0 \in \partial \mathbb{D}$.*

---
[1] That is, $(\tilde{U}_s)_s$ is the driving function for a radial SLE$_\kappa(\rho)$ from $\tilde{U}_0$ to 0, with force point initially at $\tilde{V}_0$, and $(\tilde{V}_s)_s$ is the evolution of $\tilde{V}_0$ under the Loewner flow.

Sketch of proof The idea behind this lemma is simple. Let us write $U_t = e^{i\xi_t}$ and $V_t = e^{i\psi_t}$, where $(\xi_t)_{t\geq 0}$ and $(\psi_t)_{t\geq 0}$ are uniquely defined by continuity. Then it can be checked that, analogous to the Bessel equation (A.3), the angle difference $\theta_t = \psi_t - \xi_t$ satisfies

$$d\theta_t = \frac{\rho+2}{2} \cot(\theta_t)\, dt + \sqrt{\kappa}\, dB_t. \tag{9.3}$$

Recall that $\cot(\theta) \sim 1/\theta$ as $\theta \to 0^+$ so this diffusion looks like a Bessel diffusion of dimension $1 + 2(\rho+2)/\kappa > 1$ near zero, and the same is true as $\theta \to 2\pi^-$, with the drift now repelling θ_t away from 2π. Even when the dimension of the Bessel process is such that these two boundary values are touched by the diffusion, the process $(\theta_t)_{t\geq 0}$ always takes values in $[0, 2\pi]$ and thus has a unique invariant distribution. This is the desired law. □

Definition 9.5 Let $(\overline{U}_t, \overline{V}_t)_{t\in\mathbb{R}}$ have the stationary law in Lemma 9.4 for some $\kappa > 0, \rho > -2$. Whole plane $\mathrm{SLE}_\kappa(\rho)$ from 0 to ∞ is defined to be the family of whole plane Loewner hulls generated by $(U_t)_{t\in\mathbb{R}}$ via the whole plane Loewner equation, as described in Definition 9.1.

We use $\overline{U}_t$ and $\overline{V}_t$ in the definition instead of U_t and V_t even though this does not change the resulting law, but we do so in order to be consistent with Lemma 9.2. Indeed, is immediate from the definition and from Lemma 9.2 that given $(U_t, V_t)_{t\in(-\infty,t_0]}$ for fixed $t_0 \in \mathbb{R}$, the associated whole plane $\mathrm{SLE}_\kappa(\rho)$ from 0 to ∞ from time t_0 onwards has the law of a radial $\mathrm{SLE}_\kappa(\rho)$ in $\hat{\mathbb{C}} \setminus K_t$, from $g_{t_0}^{-1}(U_{t_0})$ to ∞ and with marked point at $g_{t_0}^{-1}(V_{t_0})$. (Its driving function is exactly equal to $(\overline{U}_{t+t_0})_{t\geq 0}$). In particular, for every $\kappa > 0$, there almost surely exists a continuous curve $(\gamma(t))_{t\in\mathbb{R}}$ such that $g_t^{-1}(\hat{\mathbb{C}} \setminus \overline{\mathbb{D}}) = \hat{\mathbb{C}} \setminus K_t$ is the unbounded connected component of $\hat{\mathbb{C}} \setminus \gamma((-\infty, t])$ for each t, and as with ordinary whole plane SLE_κ, it satisfies $\lim_{t\to-\infty} \gamma(t) = 0$ and $\lim_{t\to\infty} \gamma(t) = \infty$ (see for example [Law13]).

Whole plane $\mathrm{SLE}_\kappa(\rho)$ is also scale invariant: if $a \in \mathbb{C} \setminus \{0\}$ and γ is a whole plane $\mathrm{SLE}_\kappa(\rho)$ from 0 to ∞, then $a\gamma$ has the same law (modulo time parametrisation) as a whole plane $\mathrm{SLE}_\kappa(\rho)$. This again allows us to define whole plane $\mathrm{SLE}_\kappa(\rho)$ from z_1 to z_2 with $z_1 \neq z_2 \in \hat{\mathbb{C}}$ in a consistent way.

Definition 9.6 Let $z_1, z_2 \in \hat{\mathbb{C}}$ and $\rho > -2$. Whole plane $\mathrm{SLE}_\kappa(\rho)$ from z_1 to z_2 is defined to be the image of whole plane SLE_κ from 0 to ∞, under a Möbius transformation sending 0 to z_1 and ∞ to z_2. The law of this process does not depend on the choice of Möbius transformation.

9.2.3 Whole plane $\text{SLE}_\kappa(\kappa - 6)$

We end this section on whole plane $\text{SLE}_\kappa(\rho)$ with a short discussion about some properties of the curve in the special case $\rho = \kappa - 6$. These will be needed in the construction of space-filling SLE_κ.

Lemma 9.7 (Target invariance of $\text{SLE}_\kappa(\kappa - 6)$) *Suppose that $\kappa > 4$ and $b_1, b_2 \in \mathbb{C}$. Then it is possible to couple a whole plane $\text{SLE}_\kappa(\kappa - 6)$ curve from 0 to b_1 in $\mathbb{C}$, and from 0 to b_2 in $\mathbb{C}$ so that they coincide until b_1, b_2 are contained in separate components of the complement of the curve, and afterwards evolve independently.*

Proof This follows from the target invariance of radial $\text{SLE}_\kappa(\kappa - 6)$ (Lemma C.7) and the relationship between whole plane and radial SLE (Lemma 9.2). This requires discovering a small (in terms of diameter, say) part of either whole plane curves and taking a limit as the diameter shrinks to zero; the details are left to the reader. □

Let $\kappa \geq 8$. The next statement shows that the whole plane $\text{SLE}_\kappa(\kappa - 6)$ from 0 to ∞ does not fill the whole plane (even though, for example, chordal SLE_κ in $\mathbb{H}$ does fill the whole of $\mathbb{H}$). Let K denote the hull generated by η: this is the set of points for which solving the Loewner equation *is not* possible for all times. Equivalently $K = \bigcup_{t \in \mathbb{R}} K_t$, with $K_t = \mathbb{C} \setminus D_t$, and D_t the unique unbounded component of $\mathbb{C} \setminus \eta(-\infty, t]$.

Lemma 9.8 *Suppose $\kappa \geq 8$. The hull K of a whole plane $\text{SLE}_\kappa(\kappa - 6)$ curve η from 0 to ∞ is not all of $\mathbb{C}$. Moreover, η is transient: $\eta(t) \to \infty$ as $t \to \infty$.*

Note that this is in contrast with say, chordal SLE_κ for $\kappa \geq 8$, which eventually swallows every point of the upper half plane.

Proof To see that D is not empty, suppose that we discover a small chunk $\eta(-\infty, t_0)$ of the whole plane $\text{SLE}_\kappa(\kappa - 6)$ from 0 to ∞. The future of this curve is, by Lemma 9.2 a radial $\text{SLE}_\kappa(\kappa - 6)$ in the complement of the hull generated by $\eta(-\infty, t_0)$, started at $\eta(t_0)$ and targeted at ∞. Its force point is determined by V_{t_0}; more precisely it is given by $z_0 = g_{t_0}^{-1}(V_{t_0})$, where V_{t_0} is as in Definition 9.5. By changing coordinates (i.e. Lemma C.5), we can also view it as a *chordal* $\text{SLE}_{\kappa'}$ (with no force point) but targeted at z_0 and run until it hits ∞ (which is just some interior point of the domain in which this chordal $\text{SLE}_{\kappa'}$ lives). In particular, the hull generated by the curve η does not contain all of $\mathbb{C}$.

Transience is shown in from [Law13]; alternatively, it follows from the above argument and elementary properties of chordal $\text{SLE}_{\kappa'}$. □

The following result is in some sense elementary but also very useful conceptually (and also technically, as we will see below). It states that whole plane $\text{SLE}_\kappa(\kappa - 6)$ from ∞ to 0 can be viewed as the infinite volume limit of standard *chordal* SLE_κ in a large domain between two arbitrary boundary points, and stopped when it reaches zero. (We will see later in the chapter that this has a useful implication for *space-filling* $\text{SLE}_{\kappa'}$: namely, space-filling $\text{SLE}_{\kappa'}$ is the infinite volume limit of the same curve, *without* stopping it when it reaches zero. See Theorem 9.16).

In order to state this result, we need to discuss the topology for which this convergence holds. This will be the topology of uniform convergence on intervals of the form $[t_0, \infty)$ for every $t_0 \in \mathbb{R}$ (we leave it to the reader to check this defines a metric space, in fact a complete separable metric space, although this is not needed here). In other words, η_n converges to η_t in this topology if for all $\varepsilon > 0$, for all $t_0 \in \mathbb{R}$, there exists n_0 such that $|\eta_n(t) - \eta(t)| \leq \varepsilon$ for all $n \geq n_0$ and $t \geq t_0$. Since this is a metrizable topology (and in fact a Polish metrizable one, as mentioned above), it makes sense to talk about convergence in distribution with respect to this topology.

Let $(\eta(t))_{t \in \mathbb{R}}$ denote a whole plane SLE_κ from ∞ to 0, and recall from (9.2) that η is parametrised so that $\text{CR}(0, \mathbb{C} \setminus \eta((-\infty, t])) = e^{-t}$ for all t.

Lemma 9.9 *Suppose that $\kappa > 4$ and let D_n be a sequence of simply connected domains such that $D_n \subset D_{n+1}$ and $\bigcup_{n \geq 0} D_n = \mathbb{C}$. For each n, let a_n, b_n be two prime ends of D_n, and let η_n denote a chordal SLE_κ in D_n from a_n to b_n. Then as $n \to \infty$, the law of η_n converges to the law of η, a whole plane SLE_κ from ∞ to 0, in the sense described above.*

In fact, let $\varepsilon > 0$ and $t_0 \in \mathbb{R}$ be given. Then for all $n \geq n_0(\varepsilon, t_0)$ large enough, there exists a coupling between η_n and η, and an event A_n of probability at least $1 - \varepsilon$ for this coupling, on which $\eta_n(t) = F_n(\eta(t))$ for every $t \geq t_0$, where F_n is a conformal isomorphism defined in a neighbourhood U of $\eta(t_0, \infty)$ satisfying $|F_n(w) - w| \leq \varepsilon$ for every $w \in U$.

Proof It suffices to prove the second claim, since this clearly implies the first. Let $\varepsilon > 0$ and let $t_0 \in \mathbb{R}$. Let $R = Ce^{-2t_0}/\varepsilon$, where C will be made precise later (it will in fact be allowed to depend on ε) and define t_1 via $e^{-t_1} = R$. Observe that by the change of coordinate formula (Lemma C.5), until hitting zero η_n has the same law as a radial $\text{SLE}_\kappa(\kappa - 6)$ from a_n to 0 with force point at b_n (although since $\rho = \kappa - 6$, by the target invariance property of Lemma C.7, the precise location of this force point will is not relevant except to know that η_n does not separate 0 from b_n until reaching 0). Let g_n denote the conformal isomorphism from $D_n \setminus \eta_n((-\infty, t_1])$ to $\mathbb{D}$ with $g_n(0) = 0$ and $g'_n(0) > 0$ and also let g denote the conformal isomorphism from $\mathbb{C} \setminus \eta((-\infty, t_1])$ to $\mathbb{D}$ with $g(0) = 0$

9.2 Whole plane SLE$_\kappa$ and SLE$_\kappa(\rho)$

and $g'(0) > 0$. Let $\tilde{\eta}_n = g_n(\eta_n([t_1, \infty)])$ and let $\tilde{\eta} = g(\eta([t_1, \infty)))$. Note that both $\tilde{\eta}_n$ and $\tilde{\eta}$ are radial Loewner evolutions in $\mathbb{D}$, whose driving functions we denote, respectively, by $(U^n_{t_1+t})_{t \geq 0}$ and $(U_{t_1+t})_{t \geq 0}$. Note that $(U_t)_{t \geq t_1}$ has the equilibrium law of Lemma 9.4. Note also that, using the convergence to equilibrium in Lemma 9.4, we can choose n_0 large enough so that not only does D_n contain the ball of radius t_1 for all $n \geq n_0$, but in fact, for all $n \geq n_0$, we can couple η_n and η so that $U^n_t = U_t$ for all $t \geq t_1$ on an event of probability at least $1 - \varepsilon/2$. Let A'_n denote this event.

Also choose a constant $k = k(\varepsilon) > 0$ large enough so that with probability at least $1 - \varepsilon/2$, $\eta(t_0, \infty)$ stays in a ball of radius ke^{-t_0}. Let A_n denote the intersection of this event with A'_n (and note that A_n has probability at least $1 - \varepsilon$). It remains to show that on A_n, $\eta_n([t_0, \infty))$ and $\eta([t_0, \infty))$ are uniformly close. This will follow from well known distortion estimates, for example from Proposition 3.26 in [Law05]. Indeed, from this proposition, we know that there exists a constant $C = C_{1/2}$ such that for any function f defined on the unit disc which is analytic and one to one with $f(0) = 0$ and $f'(0) = 1$,

$$|f(z) - z| \leq C_{1/2}|z^2| \tag{9.4}$$

for $|z| \leq 1/2$. Now consider the map

$$F_n = g_n^{-1} \circ g \colon \mathbb{C} \setminus \eta((-\infty, t_1]) \to D_n \setminus \eta_n((-\infty, t_1]),$$

and observe that $\eta_n(t_1, \infty)$ is obtained from $\eta(t_1, \infty)$ by mapping it through F_n. So it suffices to prove that F_n is close to the identity on the relevant region. Let $R' = e^{-t_1}/4 = R/4$. By Koebe's quarter theorem, the domain where F_n is defined contains at least $B(0, R')$. The map $z \mapsto F_n(zR')/R'$ is therefore analytic and one to one on the unit disc, fixing zero and having unit derivative at zero. Hence by (9.4), we deduce that for $r > 0$ and $w \in B(0, r)$

$$|F_n(w) - w| \leq C_{1/2}r^2/R'.$$

Choosing $r = ke^{-t_0}$ and keeping in mind that $R' = e^{-t_1}/4$ and $e^{-t_1} = R = Ce^{-2t_0}/\varepsilon$, with C to be determined, this means that

$$|F_n(w) - w| \leq C_{1/2}k^2\varepsilon/C. \tag{9.5}$$

for $w \in B(0, ke^{-t_0})$. We obtain the desired result by taking $C = C_{1/2}k^2$: indeed, on the event A_n, for $t \geq t_0$, $\eta_n(t) = F_n(\eta(t))$ and $\eta(t) \in B(0, ke^{-t_0})$ so the use of (9.5) is justified. Consequently, $|\eta_n(t) - \eta(t)| \leq \varepsilon$ for all $t \geq t_0$ on the event A_n. □

Remark 9.10 Note that Lemma 9.9 also holds if the curves are parametrised by Lebesgue area rather than log conformal radius (with respect to time zero).

342   *Liouville Quantum Gravity as a Mating of Trees*

In this case, the convergence holds uniformly on compact time intervals (rather than on sets of the form $[t_0, \infty)$). Indeed, the second claim of the lemma shows that the two curves are equal with high probability, up to a small uniform distortion.

Likewise, the complement of $\eta_n(-\infty, \infty)$ in D_n converges to the complement of $\eta(-\infty, \infty)$ in $\mathbb{C}$ in a very strong sense. For instance, the proof shows that given any neighbourhood U of the origin, we can couple η_n and η so that with probability arbitrarily close to 1 as $n \to \infty$, $D_n \setminus \eta_n(-\infty, \infty)$ is the image of $\mathbb{C} \setminus \eta(-\infty, \infty)$ under a conformal isomorphism F_n (defined on a larger domain, including U) and is arbitrarily close to the identity on U.

It will also be useful to describe the boundary of $K = \bigcup_{t \in \mathbb{R}} K_t$, which is non-empty (by Lemma 9.8 in the case $\kappa \geq 8$ and by the corresponding property of radial SLE_κ in the case $\kappa \in (4, 8)$). Since the description of the boundary requires talking about both the value κ and the dual parameter $16/\kappa$, we switch to the notation where $\kappa' > 4$ and $\kappa = 16/\kappa' < 4$. In fact, we will only give a description of the boundary in the case $\kappa' \geq 8$.

Lemma 9.11   *Let $\kappa' \geq 8$ and let $\kappa = 16/\kappa' < 2$. Let η denote a whole plane $\text{SLE}_{\kappa'}(\kappa' - 6)$ from ∞ to 0. Then the boundary of K has the same law as $\eta_L(-\infty, \infty) \cup \eta_R(-\infty, \infty)$, where η_L, η_R are defined as follows:*

- *η_L is a whole plane $\text{SLE}_\kappa(2 - \kappa)$ from 0 to ∞ (note that η_L is a simple curve by our assumption that $\kappa' \geq 8 > 6$ and Lemma A.11)*
- *Given η_L, η_R is a chordal $\text{SLE}_\kappa(-\kappa/2, -\kappa/2)$ from 0 to ∞ in $\mathbb{C} \setminus \eta_L(-\infty, \infty)$ with force points on either side of the starting point 0 (note that since $\kappa' \geq 8$, $-\kappa/2 \geq \kappa/2 - 2$ and so by Lemma A.11, η_R does not hit any part of η_L).*

Proof   We expect that such a statement might follow from known duality arguments for chordal $\text{SLE}_{\kappa'}$ via Lemma 9.9 (see, for example, [Zha08a, Dub09a]). However, we could not find such a result in the literature. Nonetheless, this description can be deduced from Theorems 1.4 and 1.6 in [MS17].   □

Remark 9.12   Furthermore, given a whole plane $\text{SLE}_{\kappa'}(\kappa' - 6)$ curve η from ∞ to 0, with $\kappa' \geq 8$, it is possible to unambiguously associate to it two curves η_L and η_R, whose union is the boundary of η, and such that η_L lies to the *left* of the curve as we traverse it from ∞ to zero, while η_R lies to its right as we traverse it from ∞ to zero (this is a topological property of curves – which are oriented by definition – in two dimensions). The distribution of (η_L, η_R) is as specified above. Interestingly however, the joint distribution of (η_L, η_R) is the same as that of (η_R, η_L).

9.3 Space-filling SLE in the case $\kappa' \geq 8$

In order to state the mating of trees theorem, we first explain the definition and construction of a space-filling version of SLE in the whole plane (from ∞ to ∞). We first fix $\kappa' \geq 8$ and stick with the convention that κ' denotes a parameter greater than 4, that will take the value $16/\gamma^2$ when our curves are coupled with γ Liouville quantum gravity. The notation κ is reserved for the dual parameter $\kappa = 16/\kappa' \in (0, 4)$. In fact, when $\kappa' \geq 8$, the whole plane $\text{SLE}_{\kappa'}$, whose definition and properties we have studied in the sections above, already fills the entire hull that it generates. As a result, the construction is much simpler in this case than when $\kappa' \in (4, 8)$. We note that on the LQG side (when we eventually couple our space-filling curve with γ LQG), choosing $\kappa' \geq 8$ amounts to restricting γ to the interval $(0, \sqrt{2}]$ (which essentially corresponds to the L^2 phase of GMC). Unfortunately, it is the interval $\gamma \in [\sqrt{2}, 2)$ which is believed to correspond to scaling limits of random planar maps weighted by the self dual FK percolation model described in Chapter 4.

9.3.1 Definition of space-filling $\text{SLE}_{\kappa'}$, ($\kappa' \geq 8$)

Let $\kappa' \geq 8$. The whole plane $\text{SLE}_{\kappa'}$ defined in Section 9.2.2 is a curve $(\tilde{\eta}_t)_{t \in (-\infty, \infty)}$ which "starts" at zero (meaning $\lim_{t \to -\infty} \tilde{\eta}_t = 0$) and is targeted at infinity. For the mating of trees theorem, however, we will need to define a curve from ∞ to ∞, which visits zero at time 0; this will make it possible for the "past" and "future" of the curve with respect to 0 to play symmetric roles, which turns out to be an important feature of the theory.

We therefore cannot directly use $\tilde{\eta}$ as our space-filling SLE. Instead we proceed in two steps. Let η^- denote a whole plane $\text{SLE}_{\kappa'}(\kappa' - 6)$ from ∞ to 0, as defined in Definition 9.6. Let $K^- = \eta^-((-\infty, \infty))$, and let $D^- = \mathbb{C} \setminus K^-$. We will use K^- as the "past" of time zero, and the closure of D^- will be the future. The following property of η^- will motivate the definition of the space-filling curve coming below.

Lemma 9.13 *Let $\kappa' \geq 8$ and let τ be any almost surely finite stopping time for η^- (with respect to the filtration generated by the curve η^- itself, parametrised so that the conformal radius of 0 in the complement of the curve is e^{-t}).[2] Then the complement, D_τ^-, of $\eta^-(-\infty, \tau]$ in $\mathbb{C}$, is an unbounded simply connected set with probability 1. Moreover, given $\eta^-(s), s \leq \tau$], the law of $\eta^-|_{[\tau, \infty)}$ is that of a radial $\text{SLE}_{\kappa'}$ in D_τ^- from $\eta^-(\tau)$ to 0, parametrised by minus log conformal radius seen from 0.*

[2] Or equivalently, the filtration generated by the pair $(U_t, V_t)_{t \in \mathbb{R}}$ of Definition 9.5 after applying a Möbius inversion.

Proof Let $\eta(t)$ denote a whole plane $\text{SLE}_{\kappa'}(\kappa'-6)$ from 0 to ∞ (thus the laws of η^- and η are related to each other by Möbius inversion). We may assume without loss of generality that τ is a stopping time for η. Let D_τ denote the complement of $\eta(-\infty, \tau]$. To prove the lemma, it suffices to show that: (1) D_τ is simply connected; (2) contains points arbitrarily close to zero; and (3) given $\eta(-\infty, \tau]$, the rest of the curve η is distributed as a radial $\text{SLE}_{\kappa'}$ in D_τ from $\eta(\tau)$ to 0, parametrised by logarithmic capacity (seen from infinity). By changing coordinates (i.e. Lemma C.5), (3) is equivalent to saying that the conditional law of $\eta([\tau, \infty))$ is that of a chordal $\text{SLE}_{\kappa'}(\kappa'-6)$ in D_τ from $\eta(\tau)$ to ∞, with force point at 0.

Let $(\overline{U}_t, \overline{V}_t)_{t \in \mathbb{R}}$ be the stationary (radial) process of Lemma 9.4 defining the whole plane curve η. For $t \in \mathbb{R}$, let K_t be the hull generated by $\eta((-\infty, t])$, and let g_t be the unique conformal isomorphism from $\hat{\mathbb{C}} \setminus K_t \to \hat{\mathbb{C}} \setminus \overline{\mathbb{D}}$ with $g_t(\infty) = \infty$ and $g_t'(\infty) = e^{-t}$ (i.e. $g_t(z) = ze^{-t} + O(1)$ as $z \to \infty$). Then from the strong Markov property of (U, V) and the relationship between whole plane and radial $\text{SLE}_{\kappa'}(\rho)$ (specifically the discussion just below Definition 9.6 in the appendix) we learn given $\eta(-\infty, \sigma]$ for any η-stopping time $\sigma \in \mathbb{R}$, the remainder of η is a radial $\text{SLE}_{\kappa'}(\kappa'-6)$ in $\hat{\mathbb{C}} \setminus K_\sigma$, targeted at ∞, and with force point located at $g_\sigma^{-1}(V_\sigma)$. (Equivalently by Lemma C.5, it is a chordal $\text{SLE}_{\kappa'}$ targeted at $g_\sigma^{-1}(V_\sigma)$). Since $\kappa' \geq 8$, this means that the curve η is in the "space-filling phase". In particular, $D_\tau = \mathbb{C} \setminus K_\tau$ almost surely, and by properties of the Loewner evolution, D_τ is almost surely simply connected.

To see point (2) – that D_τ contains points arbitrarily close to zero – we simply observe that $D_\tau \supset D_\infty$ which itself satisfies this property by Lemma 9.11. Indeed, the boundary of η is given by an explicit pair of SLE curves and therefore the complement of η, i.e. D_∞, contains points arbitrarily close to zero as desired. In fact this argument shows that D_∞ is a Jordan domain for which all boundary points (including 0) correspond to a unique prime end in the language of [Pom92]. As a consequence, note that 0 corresponds to a unique prime 5 in D_τ as well, and not just in D_∞, which is a consequence of transience and a zero-one argument left to the reader. This will be required below.

Finally, we are left to show (3), which by the discussion above (with $\sigma = \tau$)), boils down to proving that $z_\tau = g_\tau^{-1}(V_\tau) = 0$.

To see this, fix a sequence $t_n \to -\infty$. As discussed above, given (U_t, V_t) for $t \leq t_n$, the conditional law of η after time t_n is that of a chordal $\text{SLE}_{\kappa'}$ in D_{t_n}, from $\eta(t_n)$ to z_{t_n}, reparametrised according to log capacity seen from ∞. In particular, it will not hit z_{t_n} again after time t_n, which means that z_t stays constant after time t_n. Consequently, we have that $z_\tau = z_{t_n}$ almost surely. Since $n \geq 1$ was arbitrary, and z_{t_n} lies on the boundary of $\eta((-\infty, t_n])$ (a set

9.3 Space-filling SLE in the case $\kappa' \geq 8$

of diameter tending deterministically to 0), it follows that $z_{t_n} \to 0$ and thus $z_\tau = 0$ as desired.

Note also that this argument implies that U_t and V_t are determined by $\eta(-\infty, t)$, since U_t is the driving of the Loewner evolution (explicitly, $U_t = g_t(\eta(t))$) and $V_t = g_t(0)$. Therefore the filtrations generated by (U, V) and by the curve are indeed equal, as claimed in the Lemma. □

In particular, this makes the following definition possible (and natural).

Definition 9.14 Let $\kappa' \geq 8$. Given a whole plane $\text{SLE}_{\kappa'}(\kappa' - 6)$ from ∞ to 0 which we denote by η^-, let η^+ denote a (conditionally independent) chordal $\text{SLE}_{\kappa'}$ in $\mathbb{C} \setminus \eta^-(\mathbb{R})$ from 0 to ∞. By definition, the whole plane **space-filling** $\text{SLE}_{\kappa'}$ from ∞ to ∞, is the curve η obtained by concatenating η^- and η^+, and then reparametrising time so that $\eta(0) = 0$ and $\text{Leb}(\eta([0, t])) = |t|$; that is, so that η is parametrised by its (Lebesgue) area.

Indeed, it can be checked that both η^- and η^+ cover an area of positive Lebesgue measure in any finite-time interval, and that this area is in fact a continuous function of the length of the interval (this is well known for η^+ by properties of chordal $\text{SLE}_{\kappa'}$, and for η^- it can be deduced from Lemma 9.13.) This means that such a continuous reparametrisation by Lebesgue area is indeed possible.

Given Lemma 9.13, it is not surprising that the space-filling SLE curve we have just defined is stationary in a strong sense; however, there are some subtleties in justifying this because of the way the curve is parametrised (since we know at time 0 it must visit 0, and visit exactly an area of size t in any interval of length t). A precise statement of this sort will be given (but not proved) in Lemma 9.31 a bit later on.

For now, we formulate a useful Markov property.

Lemma 9.15 Let $\kappa' \geq 8$, let η be a space-filling $\text{SLE}_{\kappa'}$ from ∞ to ∞, let U be a non-empty bounded subset of $\mathbb{C}$, and let τ be the first time that η enters U. Then conditionally on $(\eta(t), t \leq \tau)$, the rest of the curve $(\eta(\tau + t))_{t \geq 0}$ is, up to a change of time parametrisation, a chordal $\text{SLE}_{\kappa'}$ in $\hat{\mathbb{C}} \setminus \eta((-\infty, \tau])$ from $\eta(\tau)$ to ∞.

Proof Let $g \colon \hat{\mathbb{C}} \setminus \eta((-\infty, \tau]) \to \hat{\mathbb{H}}$ be the unique conformal isomorphism with $g(\eta(\tau)) = 0$ and $g(z)/z \to 1$ as $z \to \infty$. Let $\tilde{\eta}$ be the image of $(\eta(\tau + t))_{t \geq 0}$ under g, reparametrised by half plane capacity (i.e. so that the infinite connected component of $\mathbb{H} \setminus \tilde{\eta}([0, t])$ has half plane capacity $2t$ for $t \geq 0$). The lemma is equivalent to the fact that, conditionally on $(\eta(t), t \leq \tau)$ $\tilde{\eta}$ has the law of a chordal $\text{SLE}_{\kappa'}$ in $\mathbb{H}$ from 0 to ∞.

There are two events to consider. On the event that $0 \in \eta((-\infty, \tau))$, let $\tau_0 := \inf\{t: 0 \in \eta((-\infty, t])\}$, so that $\tau_0 \leq \tau$ and $U \subset \hat{\mathbb{C}} \setminus \eta((-\infty, \tau_0])$. Then $(\eta(t))_{t \geq \tau_0}$ is by definition a chordal SLE$_{\kappa'}$ in $\hat{\mathbb{C}} \setminus \eta((-\infty, \tau_0])$ from $\eta(\tau_0)$ to ∞, reparametrised by Lebesgue area. τ is simply the first time that this curve enters U, and the Markov property of chordal SLE$_{\kappa'}$ implies the desired statement in this case.

On the event that $0 \notin \eta((-\infty, \tau])$, write η' for η but in its usual whole plane Loewner evolution parametrisation, and τ' for the firs time it enters U. Notice that the sigma-fields generated by $(\eta'(t), t \leq \tau')$ and $(\eta(t), t \leq \tau)$ are the same. Therefore, by Lemma 9.13, and after reparameterising by half plane capacity, $\tilde{\eta}$ has the law of a chordal SLE$_{\kappa'}$ targeted at ∞ up until the first time it hits $g(0)$. After this time, by definition, it has the law of a chordal SLE$'_\kappa$ in the remaining domain, targeted at ∞. But this two-step description gives exactly the law of a chordal SLE$_\kappa$ from 0 to ∞ in $\mathbb{H}$. Thus $\tilde{\eta}$ has this law, as required. □

One consequence is that any fixed point $z \in \mathbb{C}$ is almost surely not a double point of the space-filling SLE$_{\kappa'}$ curve η, since this is true of chordal SLE$_{\kappa'}$ (by the Markov property of chordal SLE$_{\kappa'}$ and properties of Bessel processes).

9.3.2 Space-filling SLE as an infinite volume limit of chordal SLE ($\kappa' \geq 8$)

The following description of space-filling SLE is extremely useful for the intuition: it says that we can view space-filling SLE$_{\kappa'}$ as the infinite volume limit of standard, chordal SLE$_{\kappa'}$ in a domain D_n tending to infinity between two arbitrary prime ends of D_n. This point of view is sometimes taken as a definition of space-filling SLE$_{\kappa'}$, although we were not able to find a reference for the existence of such a limit in the literature.

Theorem 9.16 *Let D_n be a sequence of simply connected domains such that $0 \in D_n \subset D_{n+1}$ and $\bigcup_{n \geq 0} D_n = \mathbb{C}$. For each n, let a_n, b_n be two prime ends of D, and let η_n denote a chordal SLE$_{\kappa'}$ in D_n from a_n to b_n, parametrised by Lebesgue area with $\eta_n(0) = 0$. Then as $n \to \infty$, the law of η_n converges to the law of η, a space-filling SLE$_{\kappa'}$ from ∞ to ∞, for the topology of uniform convergence on compact intervals of time.*

Proof The proof of the theorem follows almost directly from Lemma 9.9 (see also Remark 9.10). Indeed let η_n, η be as in the theorem, and let $K_n^- = \eta_n((-\infty, 0])$ (resp $K^- = \eta((-\infty, 0])$). Lemma 9.9 shows that $\eta_n|_{(-\infty, 0]}$ converges weakly to $\eta|_{(-\infty, 0]}$, uniformly on compact time intervals. Furthermore, given K_n^-, $\eta_n(0, \infty)$ is a chordal SLE$_{\kappa'}$ in $D_n \setminus K_n^-$, while given K^-, $\eta(0, \infty)$ is

9.3 Space-filling SLE in the case $\kappa' \geq 8$ 347

a chordal $\mathrm{SLE}_{\kappa'}$ in $\mathbb{C} \setminus K^-$. Moreover, by Remark 9.10, we can couple $D_n \setminus K_n^-$ and $\mathbb{C} \setminus K^-$ so that for any fixed neighbourhood U of the origin, with probability arbitrarily close to 1 as $n \to \infty$, $D_n \setminus K_n^-$ is the image of $\mathbb{C} \setminus K^-$ under a conformal isomorphism F_n (defined on a larger domain, including U), that is arbitrarily close to the identity on U. This immediately implies the desired convergence. □

Remark 9.17 (Reversibility of (whole plane) space-filling $\mathrm{SLE}_{\kappa'}$) Although we will not need, it can be checked that this theorem implies the reversibility of whole plane, space-filling $\mathrm{SLE}_{\kappa'}$ (this is the only kind of space-filling SLE discussed in this book). This is not entirely straightforward because chordal $\mathrm{SLE}_{\kappa'}$ is *not* exactly reversible when $\kappa' \geq 8$. Instead, let us sketch the argument here (we emphasise the rest of the arguments in this chapter do not depend on this reversibility). The time reversal of an $\mathrm{SLE}_{\kappa'}$ from a_n to b_n is a chordal $\mathrm{SLE}_{\kappa'}(\rho, \rho)$ with $\rho = \kappa'/2 - 4$ and the two force points located on either side of b_n ([MS17]). However, as $n \to \infty$, the effect of these force points vanishes when we concentrate on a bounded window around zero. Indeed, even though the location of the force points changes whenever the chord swallows a force point, these remain constantly on the boundary of D_n and thus uniformly far away from the bounded window.

9.3.3 Alternative construction from a branching SLE ($\kappa' \geq 8$))

Let $Q = \{z_i\}_{i \geq 1}$ denote a countable dense set in $\mathbb{C}$. It is not hard to see that space-filling path η that we have just defined almost surely induces an order on Q: indeed let us say that

$$z_i \preceq_\eta z_j \tag{9.6}$$

if and only if η visits z_i before z_j. This is almost surely an order, since if $z_i \preceq_\eta z_j$ and $z_j \preceq_\eta z_i$ then either $z_i = z_j$ or z_i is a double point of η, where the latter event has probability zero (simultaneously for all i) by Lemma 9.15.

Let us suppose that $z_0 = 0$, and call the **past of 0** the set $K_Q^-(0) = \{z_i : z_i \preceq 0\}$. Likewise let us call the **future of 0** the set $K_Q^+(0) = \{z_i : 0 \preceq_\eta z_i\}$. Both these sets can be described directly using the whole plane $\mathrm{SLE}_{\kappa'}(\kappa' - 6)$ curve η^- from ∞ to 0: namely, $K_Q^-(0) = K^- \cap Q$ with K^- the hull of η^-, and $K_Q^+(0) = Q \setminus K_Q^-(0)$.

We will now give an equivalent description of the (law of the) ordering $\preceq_\eta$ on Q defined in (9.6), in terms of what is known as **branching $\mathrm{SLE}_{\kappa'}(\kappa' - 6)$**. Conversely, this gives us an alternative (implicit) description of the law of the space-filling curve η in terms of such branching $\mathrm{SLE}_{\kappa'}(\kappa' - 6)$, which provides a useful alternative point of view.

Branching SLE$_{\kappa'}(\kappa'-6)$. We first give a definition, valid for every $\kappa' > 4$, of the branching SLE$_{\kappa'}(\kappa'-6)$ (branching SLE$_{\kappa'}(\rho)$ only makes sense in the case when the weight ρ of the force point is equal to $\kappa' - 6$, since target invariance is a key part of the definition). Recall that by Lemma 9.7, given two points z and w in $\mathbb{C}$, it is possible to couple a whole plane SLE$_{\kappa'}(\kappa'-6)$ from ∞ to z, and w, respectively, in such a way that the two curves coincide (up to reparametrisation) up until z and w are separated from one another by the curve, after which the evolution of the two curves is independent. This coupling can immediately be extended to the dense countable set Q: that is, for each point $z_i \in Q$, we have a whole plane SLE$_{\kappa'}(\kappa'-6)$ curve η_{z_i} from ∞ to z_i, and the joint law of η_{z_i} and η_{z_j} is as described above for all pairs i, j.

A concrete inductive construction when $\kappa' \geq 8$ goes as follows. Start with a whole plane SLE$_{\kappa'}(\kappa'-6)$ from ∞ to z_1, and call it η_{z_1}. Now consider z_2. If η_{z_1} visits z_2 (at time τ_{z_2}, say) then we define η_{z_2} to be $\eta_{z_1}((-\infty, \tau_{z_2}])$, up to reparametrisation. Otherwise, we run an independent radial SLE$_{\kappa'}(\kappa'-6)$ in $\mathbb{C} \setminus \eta_{z_1}(\mathbb{R})$ from z_1 to z_2, with force point at ∞, and call the concatenation of η_{z_1} and this additional curve. Now we proceed inductively as follows. Suppose that $\eta_{z_1}, \ldots, \eta_{z_n}$ have been constructed and that for each $1 \leq i \neq j \leq n$, either η_{z_i} is a subcurve of η_{z_j} or the other way around; let η_{z_m} denote the maximal curve. We construct $\eta_{z_{n+1}}$ as follows. If z_{n+1} is visited by η_{z_m}, at time $\tau_{z_{n+1}}$ say, then $\eta_{z_{n+1}} = \eta_{z_m}((-\infty, \tau_{z_{n+1}}])$ (up to reparametrisation). If not, then we append to η_{z_m} an independent radial SLE$_{\kappa'}(\kappa'-6)$ in $\mathbb{C} \setminus \eta_{z_m}(\mathbb{R})$ from z_m to z_{n+1} with force point at ∞. The validity of this construction is justified simply by the strong Markov property of whole plane SLE$_{\kappa'}(\rho)$ (see the discussion Definition 9.6) and the target invariance of Lemma 9.7.

Ordering from a branching SLE$_{\kappa'}(\kappa'-6)$ when $\kappa' \geq 8$. We now return to the ordering of Q associated with a branching SLE$_{\kappa'}(\kappa'-6)$, and assume that $\kappa' \geq 8$. Let $\{\eta_{z_i}\}_{i \geq 1}$ be a branching SLE$_{\kappa'}(\kappa'-6)$ and for each $i \geq 1$, let $K^-(z_i)$ denote the hull of η_{z_i}, and let $K_Q^-(z_i) = K^-(z_i) \cap Q$. We can use this to define an order on Q almost surely: we say that $z_i \preceq_b z_j$ (the b stands for branching) if $z_i \in K^-(z_j)$. It is not hard to see that this is indeed almost surely an order on Q: for instance, to check transitivity, one simply notes that since $\kappa' \geq 8$, z_i becomes separated from z_j if and only if η_{z_j} hits z_i on its way to z_j, almost surely. Transitivity follows immediately, as does antisymmetry.

We can now verify that the two orders $\preceq_\eta$ and $\preceq_b$ on Q coincide in law.

Lemma 9.18 *Let $\kappa' \geq 8$. There is a coupling of a space-filling SLE$_{\kappa'}$, η, and a branching SLE$_{\kappa'}(\kappa'-6)$, $(\eta_{z_i})_{z_i \in Q}$ such that $\eta_{z_i} = \eta((-\infty, \tau_{z_i}])$ for each*

9.3 Space-filling SLE in the case $\kappa' \geq 8$

$z_i \in Q$, where τ_{z_i} is the first time that η visits z_i. In particular, in this coupling, $z_i \preceq_\eta z_j$ if and only if $z_i \preceq_b z_j$.

Proof Indeed if η is a space-filling $\text{SLE}_{\kappa'}$, and τ_{z_i} is the first time that η visits z_i, then the collection $\eta_{z_i} := \eta((-\infty, \tau_{z_i}])$ ($z_i \in Q$) has the law of a branching $\text{SLE}_{\kappa'}(\kappa' - 6)$, up to reparametrisation of the curves. This follows from the inductive construction defining the branching $\text{SLE}_{\kappa'}(\kappa' - 6)$ on the one hand, and the Markov property of space-filling $\text{SLE}_{\kappa'}$ proved in Lemma 9.15. □

9.3.4 Imaginary geometry ordering: continuum trees ($\kappa' \geq 8$)

There is another, perhaps slightly more geometric, description of the ordering defined by the branching $\text{SLE}_{\kappa'}(\kappa' - 6)$ which can be phrased simply in terms of the left and right boundaries of each branch η_{z_i}, as defined in Lemma 9.11. Recall from this lemma that if $z \in \mathbb{C}$ is fixed and η_z is a whole plane $\text{SLE}_{\kappa'}(\kappa' - 6)$ from ∞ to z, then the boundary of η has the law of the union of two curves η_z^L and η_z^R, where η_z^L has the law of an $\text{SLE}_\kappa(2 - \kappa)$ and, given η_z^L, η_z^R has the law of a chordal $\text{SLE}_\kappa(-\kappa/2, -\kappa/2)$ in the complement of η_z^L from 0 to ∞ and with force points on either side of zero. (Recall that $\kappa = 16/\kappa'$ and that in the case of the whole plane curve η_z^L, we don't need to specify the location of the force point of weight $2 - \kappa'$, which is in some sense the same as the starting point, i.e. ∞ in this case). In fact, these two curves η_z^L and η_z^R are determined unambiguously by η_z, see Remark 9.12.

Remark 9.19 It can be checked that the joint laws of the curves $\{\eta_z^L\}_{z \in Q}$ and $\{\eta_z^R\}_{z \in Q}$ are identical to what Miller and Sheffield, [MS17], call the family of "flow lines" of a Gaussian free field h in the whole plane with respective angles $-\pi/2$ and $\pi/2$. This will not be needed in the following but, together with the discussion below, it explains why our definition of space-filling $\text{SLE}_{\kappa'}$ coincides with that given in [MS17] and [DMS21].

Continuing with this geometric definition, take two points $z_i, z_j \in Q$, and consider their associated left boundary paths (say). That is, the two curves $\eta_{z_i}^L$ and $\eta_{z_j}^L$, where we now view them as starting from z_i and z_j, respectively, and targeted at ∞. One can see from the inductive construction of the branching $\text{SLE}_{\kappa'}(\kappa' - 6)$ that these two paths necessarily merge eventually. We will see that the way these paths merge actually determines the ordering between z_i and z_j. Indeed, let us say that

$$z_i \preceq_{\text{IG}} z_j \text{ if and only if } \eta_{z_i}^L \text{ merges with } \eta_{z_j}^L \text{ from the } \textit{left}. \quad (9.7)$$

See Figure 9.1. Equivalently, we can use the right boundaries; in this case

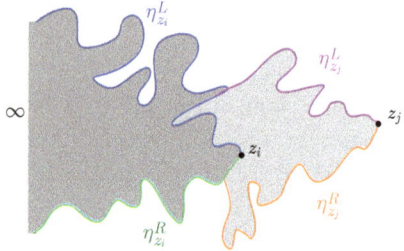

Figure 9.1 Illustration of a branching whole plane $SLE_{\kappa'}(\kappa'-6)$ $\{\eta_{z_i}^-\}_{z_i \in Q}$. The range of $\eta_{z_i}^-$ (the whole plane $SLE_{\kappa'}(\kappa'-6)$ branch from ∞ to z_i) is shaded dark grey, and the range of $\eta_{z_j}^-$ (the whole plane $SLE_{\kappa'}(\kappa'-6)$ branch from ∞ to z_j) is shaded light grey. Their left and right outer boundaries are coloured in purple/blue and orange/green, respectively. In this situation $z_i \preceq z_j$, since $\eta_{z_i}^L$ merges with $\eta_{z_j}^L$ from the left (equivalently, $\eta_{z_i}^R$ merges with $\eta_{z_j}^R$ from the right).

$z_i \preceq_{IG} z_j$ if and only if $\eta_{z_i}^R$ merges with $\eta_{z_j}^R$ from the right. We refer to this ordering as the **imaginary geometry ordering**. It is not hard to check that (for topological reasons)

$$z_i \preceq_{IG} z_j \text{ if and only if } z_i \preceq_b z_j. \tag{9.8}$$

By Lemma 9.18, the space-filling curve η can be uniquely recovered from this ordering: it is the unique curve (up to reparametrisation) which traverses the points z_i in an order compatible with (9.7). This is the definition used in [DMS21], but we stress that it is not obvious at all why such a (continuous) curve should exist at all (in [DMS21] the existence of the curve is imported from the theory of imaginary geometry and in particular [MS17]).

Let us conclude this subsection by describing another, more heuristic, way to think of how the boundary curves $\{\eta_z^L\}_{z \in Q}$ (or $\{\eta_z^R\}_{z \in Q}$) determine the space-filling curve η.

Continuum trees. Observe that the merging property of the curves $\{\eta_z^L\}_{z \in Q}$ described above means that they define a topological tree $\mathcal{T}^L$ embedded in the plane. This tree is simply the union of all the paths $\{\eta_z^L\}_{z \in Q}$ (by topological tree $\mathcal{T}$, we simply mean that for every pair of points $z, w \in \mathcal{T}$ there is a unique simple continuous path going from z to w up to reparametrisation). Likewise, the right boundary curves $\{\eta_z^R\}_{z \in Q}$ define a topological tree $\mathcal{T}^R$ embedded in the plane. These two trees are dual to one another in the sense that, for instance, a curve in $\mathcal{T}^L$ cannot cross another curve in $\mathcal{T}^R$.

9.4 Space-filling SLE for $\kappa' \in (4, 8)$

These two trees $\mathcal{T}^L$ and $\mathcal{T}^R$ can be thought of as the continuum analogues, and indeed should be the scaling limits, of the two canonical trees arising from Sheffield's bijection described in Chapter 4 for random planar maps weighted by the self-dual Fortuin–Kasteleyn percolation model. The space-filling $SLE_{\kappa'}$ defined above can then be thought of as the Peano curve "snaking" in between these two trees.

9.3.5 Summary of the constructions for $\kappa' \geq 8$

We have now seen several equivalent viewpoints of (whole plane) space-filling $SLE_{\kappa'}$, which can be used as alternative equivalent definitions depending on the properties one cares about. As these points of views are quite different from one another, we summarise what we have just done in Table 9.1.

Table 9.1 *Summary of the different constructions*

Direct construction	Branching ordering	Imaginary geometry ordering
Whole plane $SLE_{\kappa'}$, followed by chordal $SLE_{\kappa'}$	$z_i \preceq_b z_j$ if and only if $z_i \in \eta_{z_j}^-$	$z_i \preceq_m z_j$ if and only if $\eta_{z_i}^L$ merges with $\eta_{z_j}^L$ from *left*.
Definition 9.14	Lemma 9.18	(9.8)

We also recall that we proved that space-filling $SLE_{\kappa'}$ can be obtained as the infinite volume limit of chordal $SLE_{\kappa'}$ in large domains from one arbitrary boundary point (prime end) to another, see Theorem 9.16. This too could be used as a definition.

It will be useful to contrast these definitions with the definitions we will give in the case where $\kappa' \in (4, 8)$, which is much more delicate (and about which we will consequently prove less in this book).

9.4 Space-filling SLE for $\kappa' \in (4, 8)$

As hinted at above, the definition of space-filling $SLE_{\kappa'}$ is considerably easier when $\kappa' \geq 8$ than when $\kappa' \in (4, 8)$: the reason for this is that chordal $SLE_{\kappa'}$ is already space-filling, so that the definition requires little modifications. By contrast when $\kappa' \in (4, 8)$ a chordal $SLE_{\kappa'}$ is self-touching but not space-filling, and the definition of the space-filling versions (chordal or whole plane) of $SLE_{\kappa'}$ requires sophisticated tools (note that the space-filling version of SLE does *not* exist for $\kappa < 4$, and the case $\kappa = 4$ is very delicate and will be partly discussed later on). In the case when $\kappa \in (4, 8)$ let us say right from the start,

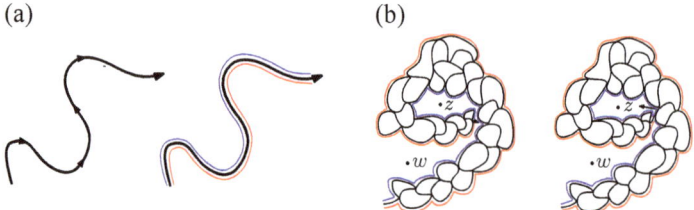

Figure 9.2 (a) The colouring of the two sides of a planar curve. (b) When z and w are first separated, z is in a monochromatic component, while w is in a bichromatic component (note that only part of the boundary of the component containing w has been drawn, in fact this will be a finite, bounded component with probability 1).

to help the intuition, that the space-filling version of $\text{SLE}_{\kappa'}$ is believed to be the scaling limit of the discrete space-filling paths associated to decorated planar maps defined in Chapter 4.

To this day the only tools which have been developed to define space-filling $\text{SLE}_{\kappa'}$ in the case where $\kappa' \in (4, 8)$ are those coming from the theory of *imaginary geometry* developed in [MS16a, MS16b, MS16c] and especially [MS17]. In order to avoid going into such technical details, we have opted for a presentation of those aspects of the definition which do not rely on imaginary geometry and can be understood without familiarity or knowledge of this theory. The downside of this approach, however, is that the proof of the theorem defining space-filling $\text{SLE}_{\kappa'}$ as a continuous curve will not be included in this book.

Disclaimer: we warn the reader that throughout Section 9.4 we intend to provide some explanations which we believe to be useful, but **these should *not* be considered fully rigorous proofs**.

Instead of providing a direct construction of space-filling $\text{SLE}_{\kappa'}$ for $\kappa' \in (4, 8)$, we will define an ordering of a dense set of points in $\mathbb{C}$, in the manner of columns 2 and 3 in Table 9.1, and we will check that these two orderings are consistent with one another. However, we will not verify that this ordering is associated with a (unique, up to translation of time) continuous curve in the sense that points are traversed by the curve in the order specified above. We will, however, make a precise statement and give references for the proof.

9.4.1 Colouring

For both constructions, it will be essential to have a notion of **colouring** of the two sides of an $\text{SLE}_{\kappa'}$ curve. Suppose η is such a curve (or in fact, more generally, suppose η is *any* planar curve): we can colour the points immediately to the left of the curve with one colour (say, blue) and the points immediately to its right with another colour (say, red). This choice of colours is made to match the conventions we adopted in Chapter 4 when discussing discrete space-filling paths on planar maps (recall, for example, Figure 4.8).

Formally, the colouring is a function defined on the *prime ends* of $D_t = \mathbb{C} \setminus \eta((-\infty, t])$ to $\{0, 1\}$, for every $t \in \mathbb{R}$. We leave it to the reader to check that, when the curve is not space-filling, this assignment of colours is *consistent* as t varies: that is, a prime end for D_s also corresponds to a prime end for D_t when $s < t$, and its colour at time s also matches its colour at time t. (The assignment of colours can be defined precisely using Loewner theory, but we will leave the description at this informal level.) Because of the consistency of the colours, we can refer unambiguously to points on the left-hand side of the curve (blue) and points on its right-hand side (red). When the curve is space-filling, the set of prime ends of $\mathbb{C} \setminus \eta((-\infty, t])$ depends on the time t, but their colours, provided they exists, do not.

Now, suppose that η takes values in D (where D could be a simply connected domain or could also be $\mathbb{C}$). Then, for $t \in \mathbb{R}$, each connected component C of $D \setminus \eta(-\infty, t)$ is either:

- **monochromatic** if all the boundary points of C (i.e. all its prime ends) have the same colour;
- or **bichromatic** (sometimes polychromatic), otherwise.

See Figure 9.2 for an illustration. If $t \in \mathbb{R}$ is such that there exist two points z, w in the same connected component of $D \setminus \eta((-\infty, s])$ for all $s < t$, but in separate connected components of $D \setminus \eta((-\infty, t])$, we call t a **disconnection time**. A useful observation is that at every such disconnection time, either the new components of $D \setminus \eta((-\infty, t])$ containing z will be monochromatic and the one containing w will be bichromatic, or vice versa. The behaviour of the space-filling $\text{SLE}_{\kappa'}$ path at this disconnection time will depend on which of these two situations occurs. Such a distinction is to be anticipated, bearing in mind the connection with the construction of the discrete path coming from Sheffield's bijection on decorated planar maps.

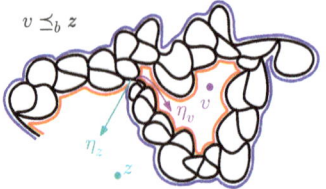

Figure 9.3 The order defined by a branching $\text{SLE}_{\kappa'}$, $\kappa' \in (4, 8)$.

9.4.2 Branching ordering, $\kappa' \in (4, 8)$

We start with the branching ordering. We content ourselves with giving the definition in the whole plane (the chordal definition is completely analogous, we will outline how to adapt it to this case at the end). Let Q denote a fixed dense countable set, and let $\{\eta_z\}_{z \in Q}$ denote a branching $\text{SLE}_{\kappa'}(\kappa' - 6)$, which we defined in Section 9.3.3.

Definition 9.20 (Branching ordering) Fix $z \in Q$ and $w \in Q$ with $z \neq w$. Let us say that $w \preceq_b z$ if, at the time that w is disconnected from z by η_z, then w belongs to a *monochromatic* component.

Equivalently, we could consider the branch η_w targeted at w. Then $w \preceq_b z$ if and only if z belongs to a *polychromatic* component at the time when η_w disconnects z from w. By convention, we take $z \preceq_b z$ for any $z \in Q$. See Figure 9.3 for an illustration. We leave it to the reader to check this does almost surely define a total order on Q.

For $z \in Q$, let $K_Q^-(z) = \{w \in Q; w \preceq_b z\}$ be the past of z (restricted to Q), and let $\mathcal{K}_z^-$ denote its closure; this is the "past" of $z \in Q$. One can check that for a given $w \in Q$ and $z \in Q$ with $z \neq w$, the event $\{w \preceq z\}$ is measurable. It is not hard to deduce that $\mathcal{K}_z^-$ is also a random variable (on closed sets equipped with Hausdorff topology, say).

9.4.3 Imaginary geometry ordering, $\kappa' \in (4, 8)$

We now wish to define the analogue of the curves η_z^L and η_z^R in the case $\kappa' \in (4, 8)$, which was introduced in the case $\kappa' \geq 8$ in Section 9.3.4 in order to introduce the alternative ("imaginary geometry") description of the space-filling curve. Recall that when $\kappa' \geq 8$, the past $\mathcal{K}_z^-$ coincides with the trace of the branch towards z, η_z, of the branching $\text{SLE}_{\kappa'}(\kappa' - 6)$. There was therefore no difficulty in talking about the boundary of the past, which is simply the boundary of η_z and whose law is thus described by Lemma 9.11.

9.4 Space-filling SLE for $\kappa' \in (4, 8)$

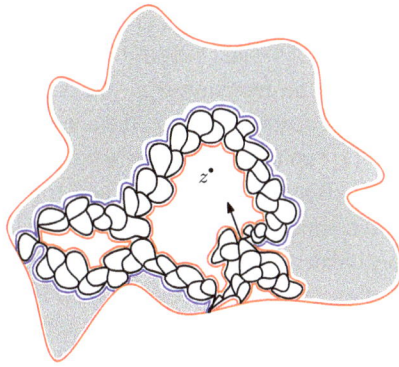

Figure 9.4 The three bichromatic components cut off by η_z between times (s, t) are shaded in grey.

As everything else, the situation is more complicated when $\kappa' \in (4, 8)$. Indeed, the past does not coincide with the trace of η_z or even the filling in of the trace of η_z (which one could define by filling in the components that are disconnected by η_z and do not contain z). The issue is that some of these components, namely the bichromatic ones, will in fact be part of the future of the space-filling curve.

We give the informal definition now. Consider the bichromatic components disconnected from z by η_z, with the order they inherit from $\preceq_b$. That is, for $w \in Q$, let $C_z(w)$ be the component containing w when w is disconnected from z by η_z, and consider the set $\mathcal{S} = \{C_z(w) \colon w \in Q, z \preceq_b w\}$ which can be ordered as follows: $C_z(w_1) \preceq C_z(w_2)$ if $w_1 \preceq_b w_2$. Note that even though w is not uniquely associated to its component $C_z(w)$, the above order is consistently defined.

It can be checked that, almost surely, the boundary of each component $C = C_z(w)$ for some $w \in Q$ with $z \preceq w$, consists of exactly two monochromatic arcs of opposite colours, which can be parametrised by curves. We call these, respectively, $\eta^L(C)$ and $\eta^R(C)$. The curves can be given a direction, which corresponds to the reverse chronological order with which these points were visited by the original curve η_z. Equivalently, C lies to the right of $\eta^L(C)$ and to the left of $\eta^R(C)$. See Figure 9.5. Then, by definition η_z^L and η_z^R is the result of the concatenation of these arcs $\eta^L(c)$ and $\eta^R(C)$ as C varies across the set $\mathcal{S}$, in the order defined above. Note that it is not obvious (but true) that these give continuous simple curves, although this can be understood at a heuristic level by drawing enough pictures, and by considering the following situation.

Let $s \in \mathbb{R}$ be a time at which some $w \in Q$ is disconnected by η_z, and suppose that $z \preceq_b w$ so w belongs to a bichromatic component, whereas z belongs to the monochromatic one, call it D, and suppose without loss of generality that D is coloured red. Consider the evolution of η_z after time s, until the first time $t > s$ where the component containing z (i.e. the connected component of the complement of $\eta_z((-\infty, t])$ containing z) does not share a positive proportion of the boundary with D. Between the times s and t, the curve η_z creates a number of bichromatic components which will be added to the set S. These components are simply created by the hits of η_z to the boundary of D, but only those where the boundary is on the left curve when it hits it. These form a connected sequence of components, whose right boundary arcs will come from the boundary of D, and left boundary arc will come from the curve itself. See Figure 9.4.

A theorem from Miller and Sheffield [MS17] shows that η_z^L and η_z^R are indeed curves and describes their joint law:

Lemma 9.21 *Let $\kappa' \in (4, 8)$ and let $\kappa = 16/\kappa' \in (2, 4)$. Let $\eta^L = \eta_z^L, \eta^R = \eta_z^R$ be as above. Then, almost surely, η_z^L, η_z^R are continuous curves. Furthermore,*

- η_L *is a whole plane* $SLE_\kappa(2 - \kappa)$ *from z to ∞ (note that η_L is a simple curve when $\kappa' \geq 6$ but not when $\kappa' \in (4, 6)$, cf. Lemma A.11)*
- *Given η_L, η_R is a chordal $SLE_\kappa(-\kappa/2, -\kappa/2)$ from z to ∞ in $\mathbb{C} \setminus \eta_L(-\infty, \infty)$ with force points on either side of the starting point z. (Note that since $\kappa' < 8$, $-\kappa/2 < \kappa/2 - 2$ and so by Lemma A.11, η_R hits both sides of η_L).*

Although the definitions of η_z^L and η_z^R are much more complicated in the case $\kappa' \in (4, 8)$ than in the case $\kappa' \geq 8$, the above description is formally exactly the same in both cases, see Lemma 9.11. As in that result, the joint law of η^L and η^R is actually symmetric: (η^R, η^L) has the same law of (η^L, η^R). Hence η^R is also a whole plane $SLE_\kappa(2 - \kappa)$ from z to ∞ and in particular is simple precisely when $\kappa' \in [6, 8)$. Note that η^L and η^R hit themselves when $\kappa' \in (4, 6)$ but not when $\kappa' \in [6, 8)$, but they *always* hit each other.

Remark 9.22 It can also be checked that the curves η^L and η^R have the same law as a pair of so-called flow lines ([MS17]) of a whole plane Gaussian free field with respective angles $-\pi/2, \pi/2$. As in the case $\kappa' \geq 8$, this will not be needed in any proof in the following.

Having defined the curves η_z^L and η_z^R for $z \in Q$, we can finally give the description of the "imaginary geometry" ordering induced by the branching $SLE_{\kappa'}(\kappa' - 6)$, $\kappa' \in (4, 8)$. Take two points $z, w \in Q$, and consider their associated left-boundary paths (say), η_z^L and η_w^L. Since the two branches going

9.4 Space-filling SLE for $\kappa' \in (4, 8)$

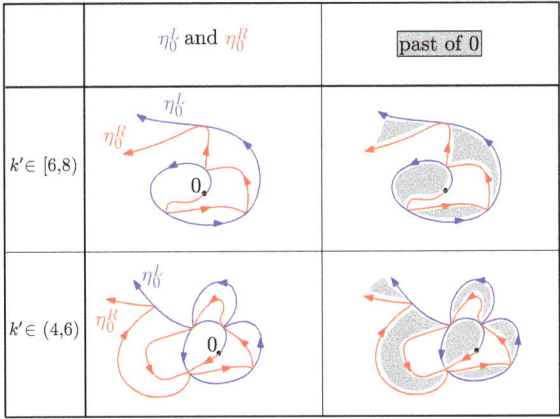

Figure 9.5 The left and right boundary of the space-filling curve when $\kappa' \in (4, 8)$.

to z and w coincide for sufficiently negative times (up to parametrisation), it is straightforward to check that η_z^L and η_w^L must also merge eventually.

Let us say

$$w \preceq_{\mathrm{IG}} z \text{ if and only if } \eta_w^L \text{ merges with } \eta_z^L \text{ from the } \textit{left}. \tag{9.9}$$

Then we claim this order is identical to the branching order: that is,

$$w \preceq_{\mathrm{IG}} z \text{ if and only if } w \preceq_b z. \tag{9.10}$$

This is in fact easier to check in the case $\kappa' \in (4, 8)$ than in the case $\kappa' \geq 8$, as here one can use the fact that the component of w disconnected by η_z is bounded, forcing the paths η_z^L and η_w^L to coincide once they both leave this component.

Thus, the branching and imaginary geometry orders that are associated to the branching $\mathrm{SLE}_{\kappa'}(\kappa' - 6)$ coincide. What is left to say is that both of these orders define a unique continuous, space-filling curve $(\eta(t))_{t \in \mathbb{R}}$. This is the content of the following theorem, which can be derived from results in [MS17] (which, roughly speaking, applies because of Remark 9.22).

Theorem 9.23 *Let $\kappa' \in (4, 8)$. There almost surely exists a unique curve $(\eta(t))_{t \in \mathbb{R}}$ which is space-filling and is continuous, such that $\eta(0) = 0$ and $\mathrm{Leb}\,\eta(s, t) = |t - s|$ for any $s, t \in \mathbb{R}$, and for every $z, w \in \mathbf{Q}$, if t_z (resp. t_w) is the first time that η visits z (resp. w), then*

$$z \preceq w \text{ if and only if } t_z \leq t_w.$$

Furthermore, the curve η does not depend on the choice of Q. η is called the (whole plane) space-filling $SLE_{\kappa'}$ from ∞ to ∞.

The space-filling $SLE_{\kappa'}$ is invariant under translation, rotation and scaling (up to time-reparametrisation). It is also, in fact, invariant under Möbius inversion (again, up to reparametrisation) and thus under all Möbius transformations of the Riemann sphere.

9.5 Cutting and welding theorems

We are now ready to describe the framework of [DMS21] and to state some of the main theorems. The ultimate goal is to describe the exploration of a γ-quantum cone with an independent space-filling $SLE_{\kappa'}$ path, where γ and κ' are related by

$$\kappa' = 16/\gamma^2,$$

so $\kappa = 16/\kappa' = \gamma^2$ as was already the case in Chapter 8. We will present the results covering both cases $\kappa' \in (4, 8)$ and $\kappa' \geq 8$ here; we thus only assume in what follows that $\kappa' > 4$, and that we are given the existence and continuity of the space-filling $SLE_{\kappa'}$ from ∞ to ∞ in $\mathbb{C}$.

Note: The proofs of the results in this section fall outside of the scope of this book, although we provide references to the proofs in the literature. They can be proved using similar tools to the proof of Theorem 8.31 in Chapter 8, but the proofs are more involved. In Section 9.8, we will explain how they lead to the main result of [DMS21].

Let h denote the field of a γ-quantum cone, as defined in Definition 7.16, but embedded in $(\mathbb{C}, 0, \infty)$ via the map $w \in C \mapsto z = -e^{-w} \in \mathbb{C}$ (so a neighbourhood of zero has finite γ LQG mass, but a neighbourhood of ∞ has infinite γ LQG mass). Recall that we say that h is a *unit circle embedding* of a γ-quantum cone.

Let η be an independent space-filling $SLE_{\kappa'}$ curve from ∞ to ∞. A priori, η comes parametrised so that $\eta(0) = 0$ and $\text{Leb}(\eta(s,t)) = t - s$ for any $s < t$. However, it is crucial in the theorem below to reparametrise η by its quantum area: that is, we define a reparametrisation η' of η such that

$$\mu_h(\eta'(s,t)) = t - s \tag{9.11}$$

for any $s < t$, where μ_h is the γ Liouville measure (or area measure) associated

9.5 Cutting and welding theorems

to h, that is, $\mu_h(\mathrm{d}x) = \lim_{\varepsilon \to 0} \varepsilon^{\gamma^2/2} e^{\gamma h_\varepsilon(x)} \, \mathrm{d}x$, where the limit is in probability (or almost surely along an appropriate subsequence) and $h_\varepsilon(x)$ is some regularisation of h at scale ε, as described in Chapter 3.

It is not obvious, but nonetheless true, that one can reparametrise η so that (9.11) holds. This follows from the fact that almost surely, for any $s < t$, $\eta(s, t)$ contains an open ball, and any open ball has positive mass (since it contains a ball centred at a point with rational coordinates and with rational radius). Likewise, the reparametrisation η' is such that there are no intervals of constancy, which follows from the fact that μ_h has no atoms almost surely (itself a consequence of, for example, Exercise 3.4).

Let $\eta(-\infty, 0] = K_0^-$ denote the past of zero, and let η_0^L and η_0^R denote its left and right boundaries (which, we recall are given by a pair of SLE curves whose joint law is specified by Lemma 9.11 for $\kappa' \geq 8$ and Lemma 9.21 for $\kappa' \in (4, 8)$). The first result below states that the boundaries η_0^L and η_0^R divide the quantum cone into two regions (namely, the past and the future of zero), on which the restrictions of h define two independent quantum wedges with parameter $\alpha = 3\gamma/2$ (in the terminology favoured by [DMS21], the wedges have "weight" $W = 2 - \gamma^2/2$).

Note that when $\kappa' \geq 8$, the wedge parameter satisfies $\alpha \leq Q$ and is therefore "thick" in the terminology of Chapter 7, whereas $\alpha > Q$ is "thin" for $\kappa' \in (4, 8)$. Recall that such a thin wedge corresponds to an ordered collection of beads; these beads correspond precisely to the bichromatic components created by the branch η_0 (targeted at 0) of the branching $\mathrm{SLE}_{\kappa'}(\kappa' - 6)$.

The first theorem we present describes the surface that one obtains by "cutting" the quantum cone $(\mathbb{C}, h, 0, \infty)$ with just one half of the boundary of η_0, say η_0^L (see Figure 9.6 for an illustration).

Theorem 9.24 *Suppose $\kappa' > 4$, and let h and η be as described just above* (9.11).

Let D denote $\mathbb{C} \setminus \eta_0^L(\mathbb{R})$. Then $\mathcal{W} = (D, h|_D, 0, \infty)$ has the law of a quantum wedge of parameter $\alpha = 2\gamma - 2/\gamma$, which is thick when $\gamma^2 \leq 8/3$, equivalently $\kappa' \geq 6$; and is thin when $\kappa' \in (4, 6)$.

Remark 9.25 This is [DMS21, Theorem 1.2.4].

When $\kappa' \in (4, 6)$, η_0^L touches itself almost surely, as already noted in Lemma 9.21. Thus $D = \mathbb{C} \setminus \eta_0^L(\mathbb{R})$ is not simply connected but consists instead of a countable (ordered) collection of simply connected domains. In this case, the theorem has to be understood as saying that the restriction of h to these domains form a wedge of parameter $\alpha = 2\gamma - 2/\gamma$. This corresponds to the fact that the wedge is thin in this case. When we cut along the second half of the boundary, the result is the following (see Figure 9.6 for an illustration).

Theorem 9.26 *Suppose $\kappa' > 4$, and let h and η be as described just above* (9.11).
Let D^- denote the interior of K_0^- and let $D^+ = \mathbb{C} \setminus K_0^-$. Let $\mathcal{W}^- = (D^-, h|_{D^-}, 0, \infty)$ and let $\mathcal{W}^+ = (D^+, h|_{D^+}, 0, \infty)$. Then $\mathcal{W}^-$ and $\mathcal{W}^+$ are independent quantum wedges with parameter $\alpha = 3\gamma/2$.

Remark 9.27 This is [DMS21, Theorem 1.2.1].

Once again, the way to read this theorem properly depends on the value of κ'. Indeed when $\kappa' \in (4, 8)$, neither D^- nor D^+ are simply connected, instead they consist of ordered collections of simply connected domains. In that case the theorem states that the restriction of h to these domains form independent quantum wedges, each with parameter $\alpha = 3\gamma/2$ (this corresponds to the thin case precisely when $\kappa' \in (4, 8)$).

Remark 9.28 A consequence of this statement and the description of the conditional law of η_0^R given η_0^L (provided in Lemma 9.11) is that if one takes a quantum wedge $\mathcal{W} = (\mathbb{H}, \tilde{h}, 0, \infty)$ with parameter $\alpha = 2\gamma - 2/\gamma$ and cuts it with an independent chordal $\text{SLE}_\kappa(-\kappa/2, -\kappa/2)$ curve from 0 to ∞ with force points on either side of zero, then the restriction of $\tilde{h}$ to the complement of this curve defines two independent quantum wedges $\mathcal{W}^-$ and $\mathcal{W}^+$ with parameter $\alpha = 3\gamma/2$, as in the theorem. In reality, the proof of the theorem goes in the converse direction: that is, one first establishes this fact and Theorem 9.24 in order to deduce Theorem 9.26. However, it is latter which is the key input for the mating of trees theorem.

The identification of $\mathcal{W}$ and of $\mathcal{W}^-, \mathcal{W}^+$ as quantum wedges means we can talk about their boundary length measures. Theorems 9.24 and 9.26 can be complemented by a result showing that the boundary lengths naturally match with one another along η_0^L and η_0^R. In other words, the quantum cone $(\mathbb{C}, h, 0, \infty)$ can be viewed as a conformal welding of $\mathcal{W}$ with itself (where points on either side of 0 of equal boundary length are identified with one another) and can also be viewed as a conformal welding of $\mathcal{W}^-$ with $\mathcal{W}^+$, where points at equal (signed) distance from 0 along the boundary in $\mathcal{W}^-$ and $\mathcal{W}^+$ are identified with one another.

Put it another way, using Remark 9.28, we can conformally weld $\mathcal{W}^-$ and $\mathcal{W}^+$ along just one half of their boundary (identifying points at equal distance from 0) to get $\mathcal{W}$. Subsequently, we can conformally weld the two halves of

9.5 Cutting and welding theorems   361

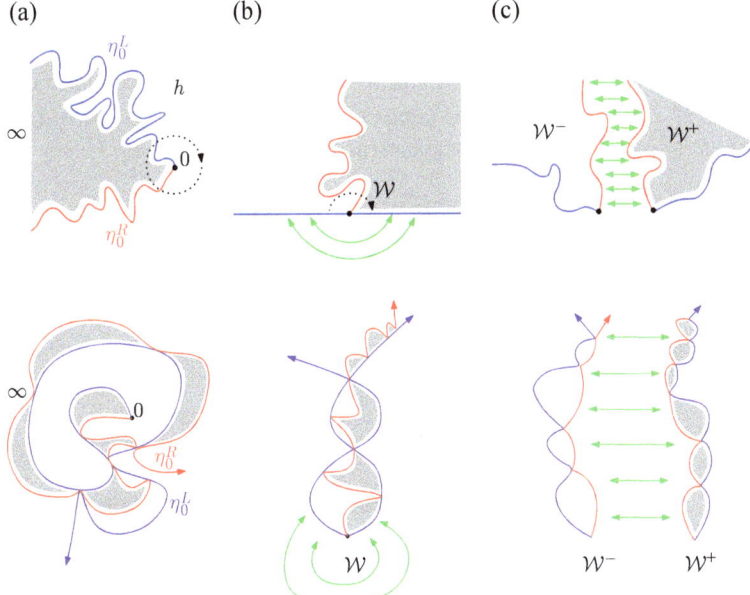

Figure 9.6 Illustrations of the cutting and welding theorems when $\kappa' \geq 8$ (top) and when $\kappa' \in (4, 8)$ (bottom). To get from (a) to (b), one "cuts" along the curve η_0^L – more precisely, considers the quantum cone field h in $D = \mathbb{C} \setminus \eta_0^L$, and views $\mathcal{W} := (D, h, 0, \infty)$ as a quantum surface. Theorem 9.24 says that $\mathcal{W}$ has the law of a quantum wedge with parameter $\alpha = 2\gamma - 2/\gamma$ (we have conformally mapped D to the upper half plane in (b), which illustrates a different embedding, or equivalence class representative in the sense of doubly marked quantum surfaces, of $\mathcal{W}$). To get from (b) to (c), one cuts further along the image of η_0^R, and considers the restriction of the field to either side, to define a pair of quantum surfaces $\mathcal{W}^-, \mathcal{W}^+$. Theorem 9.26 says that these are independent quantum wedges with parameter $\alpha = 3\gamma/2$. The welding theorem, Theorem 9.29 (which is stated only in the case $\kappa' \geq 8$) describes what happens when goes from (c) to (a) in the above pictures.The operation at each step is illustrated by the identification, or "welding", of points at equal quantum boundary length away from the black dot (this identification is depicted with green arrows).

the boundary of $\mathcal{W}$ to obtain the quantum cone $(\mathbb{C}, h, 0, \infty)$. This can all be encapsulated in the following welding theorem, which for simplicity we only state in the case $\kappa' \geq 8$.

Theorem 9.29 *In the same settings as Theorems 9.24 and 9.26, let $\kappa' \geq 8$ and let g^+ (resp. g^-) be a conformal isomorphism from D^+ (resp. D^-) to $\mathbb{H}$ sending 0 to 0 and ∞ to ∞, with η_0^L being mapped to $(-\infty, 0]$ and η_0^R being mapped to $[0, \infty)$. Let h^+ (resp. h^- denote the image of $h|_{D^+}$ under g^+ (resp.*

of $h|_{D^-}$ under g^-) under the change of coordinate formula (2.11), and let $\mathcal{V}_{h^+}$ (resp. $\mathcal{V}_{h^-}$) denote the boundary length measure of h^+ (resp. h^-) in $\mathbb{H}$.

Almost surely the following statement holds for all points z on either η_0^L or η_0^R. Let z^+ (resp. z^-) denote the image of z under g^+ (resp. g^-) in $\mathbb{R}$. Let I^+ (resp. I^-) denote the interval between 0 and z^+ (resp. z^-). Then

$$\mathcal{V}_{h^+}(I^+) = \mathcal{V}_{h^-}(I^-).$$

Remark 9.30 This follows from [DMS21, Theorems 1.2.1 and 1.2.4].

Note that in particular, Theorem 9.29 allows us to unambiguously define the quantum length of any measurable portion of η_0^L or η_0^R with respect to the quantum cone $(\mathbb{C}, h, 0, \infty)$.

9.6 Statement of the mating of trees theorem

As in the Section 9.5, we let $(\mathbb{C}, h, 0, \infty)$ be a γ-quantum cone and η be an independent space-filling $\text{SLE}_{\kappa'}$ ($\kappa' = 16/\gamma^2$), parametrised by Lebesgue area in such a way that $\eta(0) = 0$. We let η' be the reparametrisation of its quantum area μ_h relative to time 0 (which induces a dependence between h and η').

We have explained above how Theorem 9.29 allows us to unambiguously define the quantum length of any measurable portion of η_0^L or η_0^R with respect to the quantum cone $(\mathbb{C}, h, 0, \infty)$. In order to state the mating of trees theorem we will also need to measure the quantum lengths (with respect to $(\mathbb{C}, h, 0, \infty)$) of the curves η_z^L and η_z^R, when z is of the form $z = \eta'(t)$ for $t \in \mathbb{R}$ fixed. (Recall that when $\kappa' \geq 8$, $z = \eta'(t)$, η_z^L, η_z^R are the left and right boundaries of $\eta'[0, t]$, and when $\kappa' \in (4, 8)$ they are slightly more complicated to define, see Section 9.4.3, but still correspond to the left and right boundaries of $\eta'[0, t]$ in an appropriate sense). The following key lemma shows that the quantum cone decorated with the space-filling path η', viewed from $\eta'(t)$, is in fact stationary, and this (in particular) implies that the quantum lengths described above are well defined.

To state the lemma, recall the notion of a curve decorated random surface $[(D, h, a, b); \eta]$ from Definition 8.15. The stationarity will be in the sense of such objects.

Lemma 9.31 Let $t \in \mathbb{R}$, and let $z = \eta'(t)$. Then the law of the curve decorated surface $[(\mathbb{C}, h, z, \infty); \eta'(t + \cdot)]$ is the same as that of $[(\mathbb{C}, h, 0, \infty); \eta'(\cdot)]$.

Remark 9.32 For this statement it is crucial that η' is parametrised by its quantum area.

9.6 Statement of the mating of trees theorem

To spell out what the statement really says, we warn the reader that it would be incorrect to say that the joint law of $(h(z + \cdot), \eta'(t + \cdot))$ is the same as that of (h, η'). Indeed, the laws of h and $h(z + \cdot)$ are *not* the same *as fields*; we would have to applying a random rescaling to $h(z + \cdot)$ for this to be the case. Nonetheless, the objects in the lemma have the same law *as curve decorated random surfaces*.

Note: One can prove Lemma 9.31 in a similar manner to the proof of Proposition 8.20 in Chapter 8, but we will not provide the details in this book (we direct the interested reader to [DMS21, Proof of Lemma 8.1.3]). We will instead focus, in Section 9.8, on how this leads to the proof of the main theorem of [DMS21] (Theorem 9.33).

We now turn to the statement of one of the main theorems of this chapter. Let h and η' be as above. Let us define a process $(L_t, R_t)_{t \in \mathbb{R}}$ as follows. Informally, L_t tracks the change in the length of the left (outer) boundary of $\eta'(t)$, relative to time zero, whereas R_t tracks the same change but for the right (outer) boundary. To define it formally, fix $s < t$, and let $w = \eta'(s), z = \eta'(t)$. Then we define the increment

$$L_t - L_s := \mathcal{V}_h(\eta_z^L \setminus \eta_w^L) - \mathcal{V}_h(\eta_w^L \setminus \eta_z^L), \quad (9.12)$$

and make the same definition for R_t except that η^L is replaced with η^R in all occurrences. If we also set $L_0 = R_0 = 0$, then (9.12) specifies a unique two-sided process $(L_t, R_t)_{t \in \mathbb{R}}$. Note that the meaning of the random variables in (9.12) measuring the lengths of various boundary curves is provided by Theorem 9.29 (see the discussion immediately below that theorem) and the stationarity of Lemma 9.31. With these definitions, we can finally state the main theorem below.

Theorem 9.33 *Let h, η' be as above and let $(L_t, R_t)_{t \in \mathbb{R}}$ denote the boundary length process (9.12). There exists $a > 0$ depending solely on $\gamma \in (0, 2)$ such that $(L_{at}, R_{at})_{t \in \mathbb{R}}$ is a two-sided correlated Brownian motion in $\mathbb{R}^2$, with*

$$\mathrm{Var}(L_{at}) = \mathrm{Var}(R_{at}) = |t|; \qquad \mathrm{Cov}(L_{at}, R_{at}) = -\cos\left(\frac{4\pi}{\kappa'}\right)|t|.$$

Observe that the Brownian motions are negatively correlated for $\kappa' \geq 8$, positively correlated when $\kappa' \in (4, 8)$, and independent when $\kappa' = 8$ (which corresponds to the case of the uniform spanning tree).

Remark 9.34 The value of the constant a appearing in the statement of that theorem was unknown for some time, until a recent work of Ang, Rémy and Sun [ARSar] who computed it using tools coming from Liouville conformal field theory.

9.7 Discussion and uniqueness

Theorem 9.33 should be compared with Theorem 4.13. In that theorem, we showed that the scaling of the left and right relative boundary lengths of a space-filling path (these are precisely the hamburger and cheeseburger counts) exploring the infinite volume random planar map weighted by the self-dual critical Fortuin–Kasteleyn percolation model, is also given by a pair of correlated Brownian motions. Identifying the limiting covariance of Theorem 4.13 with that in Theorem 9.33 gives a relation between q and γ (or equivalently κ') which is the same as the one announced in Table 4.1.

$$q = 2 + 2\cos(\gamma^2 \pi/2) = 2\cos^2(4\pi/\kappa').$$

This is consistent with the physics prediction discussed in Chapter 4.

9.7.1 A mating of trees?

Before we explain some of the key steps going into the proof of Theorem 9.33, we spend some time explaining why this theorem is related to a "mating of trees". The word "mating" (i.e. glueing) originates from the field of complex dynamics. It was coined by Douady and Hubbard [Dou83] who spoke of matings of polynomials to describe a way to glue together two (connected and locally connected) filled Julia sets along their boundaries. In a sense, Theorem 9.33 describes a similar construction where the role of the Julia sets is played by two infinite continuum random trees (CRT).

Let us first briefly explain the notion of Continuous Random Tree, originally due to Aldous [Ald93]; we refer to the lecture notes by Le Gall [LG05] for a much more complete discussion and additional references, including in particular the history and applications of this important subject. Traditionally, the theory is defined from a Brownian excursion $(e_t)_{0 \le t \le 1}$; the resulting continuum tree would then be a compact metric space. However, for our purposes, it will be more natural to consider infinite volume analogues of this CRT, in which case the Brownian path defining the tree is simply a (real valued) two-sided Brownian motion $(B_t)_{t \in \mathbb{R}}$. The definition is simply the following and can be made path by path, that is almost surely given a fixed continuous function

9.7 Discussion and uniqueness

$f \colon \mathbb{R} \to \mathbb{R}$ (which will later be taken to have the law of the two-sided Brownian motion B). Given $s, t \in \mathbb{R}$, let us define an equivalence relation $\sim_f$

$$s \sim_f t \text{ if } f(s) = f(t) = \inf_{u \in [s,t]} f(u). \tag{9.13}$$

(Here one can have $s \leq t$ or $t \leq s$). It is easy to see that this defines an equivalence relation. By definition, the (infinite) continuous random tree associated to B is simply equal to the quotient space $\mathcal{T}_f = \mathbb{R}/\sim_f$. We can turn $\mathcal{T}_f$ into a metric space by considering, for $s, t \in \mathbb{R}$ (say with $s \leq t$ without loss of generality),

$$d_f(s,t) = f(s) + f(t) - 2 \inf_{u \in [s,t]} f(u).$$

(Note that $d_f(s,t) = 0$ if and only if $s \sim_f t$, as required for a metric). This metric turns $\mathcal{T}_f$ into what is known as a real or $\mathbb{R}$-tree: that is, any two simple curves σ_1 and σ_2 in $\mathcal{T}$ (i.e. injective continuous maps from $[0, 1]$ to $\mathcal{T}_f$) with same starting and endpoints must be reparametrisations of one another.

A convenient way of visualising the tree $\mathcal{T}_f$ associated to f is to imagine painting the underside of the graph of f with glue, and then squishing this graph horizontally (see below for a more precise description). Indeed the points that are glued with one another in this process correspond exactly to those that are identified via (9.13). This suggests another way of describing $\mathcal{T}_f$ which will here be more natural. Consider the portion of the (t, x) plane lying below the graph of f: that is,

$$\Gamma_f = \{(t,x) \in \mathbb{R}^2 : x \leq f(t)\}. \tag{9.14}$$

Define an equivalence relation $\approx_f$ on Γ_f as follows: for every $s, t \in \mathbb{R}$ such that $s \sim_f t$ put a horizontal segment between $(s, f(s))$ and $(t, f(t))$ (note this segment lies entirely in Γ_f by definition of $\sim_f$) and identify all the points of Γ_f lying on this segment; these identifications describe the equivalence classes of $\approx_f$. Now, Γ_f inherits a topological structure from $\mathbb{R}^2$, thus turning the quotient space $\Gamma_f / \approx_f$ into a topological space. Furthermore, the equivalence classes of $\approx_f$ are clearly in bijection with those of $\sim_f$, hence

$$(\Gamma_f / \approx_f) \quad = \quad \mathcal{T}_f,$$

in the sense, for example, that these two topological spaces are homeomorphic.

Coming back to Theorem 9.33, let $(L_t)_{t \in \mathbb{R}}$ and $(R_t)_{t \in \mathbb{R}}$ denote the correlated two-sided Brownian motions describing the relative left and right boundary lengths associated with the quantum cone h and the space-filling path η' as in Theorem 9.33. As mentioned above, each of L and R separately encode an infinite CRT, which we denote by $\mathcal{T}_L$ and $\mathcal{T}_R$. In addition, the space-filling path

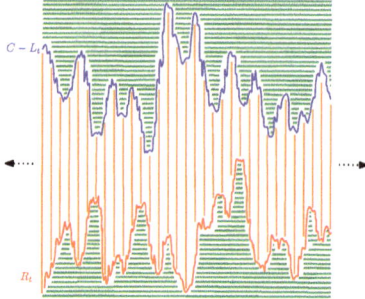

Figure 9.7 The glueing of two CRT produces a topological surface (plane or sphere) equipped with a space-filling path; that is, a **peanosurface**.

η' gives a natural way to identify (and hence glue) points on $\mathcal{T}_L$ and $\mathcal{T}_R$. More precisely, for $t \in \mathbb{R}$, let $\ell(t) \in \mathcal{T}_L$ denote the point of $\mathcal{T}_L$ corresponding to time t (i.e. the equivalence class of t for $\sim_L$). Similarly, let $r(t) \in \mathcal{T}_R$ denote the point of $\mathcal{T}_R$ corresponding to time t (the equivalence class of t for $\sim_R$). Since $\eta'(t)$ visits both $\ell(t)$ and $r(t)$, it is natural to identify $\ell(t)$ and $r(t)$. Somewhat miraculously this identification can be seen to give rise to a topological surface M (in fact a topological plane) in the sense that M is a topological space, almost surely homeomorphic to the plane.

Let us explain this construction in more detail. In our infinite volume setting, we will in fact first describe a finite volume approximation. To this end, we fix $T > 0$ and consider the restriction of L and R to $[-T, T]$. Pick a constant $C > 0$ (depending on T as well as on L and R) such that $C > \sup_{|t| \le T} L_t + \sup_{|t| \le T} R_t$. Consider the graphs of $(R_t)_{-T \le t \le T}$ and $(C - L_t)_{-T \le t \le T}$, drawn simultaneously as in Figure 9.7. By our choice of C these two graphs do not intersect, and in fact the graph $C - L$ sits entirely above the graph of R; beyond this the value of C will not matter. Consider the closed rectangle $\mathcal{R}$ of the (t, x)-plane containing the graphs of R and $C - L$; that is,

$$\mathcal{R} = \left\{ (t, x) \in \mathbb{R}^2 \colon -T \le t \le T, \inf_{|u| \le T} R_u \le x \le \sup_{|u| \le T} C - L_u \right\}.$$

This rectangle inherits a topological structure from $\mathbb{R}^2$. We will now consider an equivalence relation $\cong$ on $\mathcal{R}$, defined as follows. On the underside of the graph R (restricted to $[-T, T]$), we draw the horizontal segments $\approx_R$ as explained in (9.14). On the upperside of the graph of $C - L$ we can draw the analogous horizontal segments (see Figure 9.7). To these horizontal segments, we add a

9.7 Discussion and uniqueness

vertical segment joining (t, R_t) to $(t, C - L_t)$ for each $-T \le t \le T$. Having drawn these horizontal segments, the equivalence relation $\cong$ is defined by identifying any two points lying on the same horizontal segment and any two points lying on the same vertical segments.

While most equivalence classes in $\cong$ consist of just one segment (vertical), it is possible for an equivalence class to contain more than one segment. For instance, if $s \sim_R t$ then at least three segments are identified with one another: the horizontal segment containing the (s, R_s) and (t, R_t) but also the vertical segments containing these two points. In principle, doing this construction with arbitrary pairs of continuous functions, we could have equivalence classes with arbitrary many segments. However, it is possible to see that, when L and R are correlated Brownian motions, an equivalence class has at most five segments almost surely (of which two are then horizontal, corresponding to a branch point in the CRT, and three are vertical). Importantly, no equivalence class consists of four segments forming a rectangle (two vertical and two horizontal segments).

The equivalence relation $\cong$ on $\mathcal{R}$ is furthermore *topologically closed*: that is, if $x_n \cong y_n$ and $x_n \to x$, $y_n \to y$ then necessarily we have $x \cong y$. For such closed equivalence relations, there is a very nice criterion due to Moore [Moo25] (see Milnor [Mil04] for a more modern formulation), which can be used to check whether the quotient space retains the topology of $\mathcal{R}$ (i.e. a closed disc here). Namely, no equivalence class should disconnect $\mathcal{R}$ into more than one connected component; indeed such an equivalence class would correspond to a pinch point in $\mathcal{R}/\cong$, and would prevent the quotient from being homeomorphic to a closed disc. Here it can be checked this is almost surely the case, precisely because no equivalence class may consist of a rectangle. See [DMS21] for details of these arguments.

The identification of $\mathcal{T}_R$ with $\mathcal{T}_L$ over $[-T, T]$ thus gives us a topological space $\mathcal{R}/\cong$, which is homeomorphic to a closed disc. This closed disc also comes equipped with a natural space-filling path (call it $\tilde{\eta}(t), t \in [-T, T]$), which at time $t \in [-T, T]$ visits the equivalence class corresponding to the vertical segment joining (t, R_t) with $(t, C - L_t)$. The pair $(\mathcal{R}/\cong, \eta)$ is what we call a **peanosurface** (here a closed "peanodisc"). Sending $T \to \infty$ in the natural way gives us an "infinite volume" version of this construction.

The upshot is that Theorem 9.33 gives us access to a peanosurface, constructed as above from the glueing of the two trees $\mathcal{T}_L$ and $\mathcal{T}_R$ associated to the relative left and right boundary length of the decorated quantum cone. At this point, the parallel with Theorem 4.10 coming from Sheffield's bijection for FK-weighted random planar maps should be clear. Indeed these discrete planar maps could also be described as a glueing of two discrete trees whose scaling limit is given by two correlated CRTs (Theorem 4.13).

9.7.2 Uniqueness

Theorem 9.33 and the above discussion make it clear that we can associate to a quantum cone h, decorated by a space-filling SLE path η', a peanosurface which is obtained from the glueing of two infinite correlated CRTs. The parallel with the discrete theory described in Chapter 4 raises the following question: do the processes $(L_t, R_t)_{t \in \mathbb{R}}$ characterise the pair (h, η') uniquely? This question is natural because, in the discrete, there is a bijection between the trees and the decorated map. Remarkably, this turns out to be the case, as stated in Theorem 1.4.3 of [DMS21].

Theorem 9.35 *In the setting of Theorem 9.33, the pair $(L_t, R_t)_{t \in \mathbb{R}}$ almost surely determines (h, η') uniquely up to a rotation of the plane. That is, suppose that (h_1, η_1') and (h_2, η_2') are two quantum cones (with a unit circle embedding) defined on the same probability space, and η_i' is a space-filling $SLE_{\kappa'}$ independent of h_i parametrised by the respective quantum area ($i = 1, 2$). Let $(L_t^i, R_t^i)_{t \in \mathbb{R}}$ ($i = 1, 2$) be their associated left and right relative boundary lengths, and suppose also that $L_t^1 = L_t^2$ and $R_t^1 = R_t^2$ for all $t \in \mathbb{R}$. Then (h_2, η_2') is obtained from (h_1, η_1') by applying a fixed rotation around the origin.*

When (h_1, η_1') and (h_2, η_2') are defined on the same probability space, the four-dimensional process (L^1, R^1, L^2, R^2), is a priori (as will follow from the proof described below) a generic four-dimensional two-sided Brownian motion with some correlation matrix. The assumption that $L_t^1 = L_t^2$ and $R_t^1 = R_t^2$ for all $t \in \mathbb{R}$ corresponds to the assumption that this correlation matrix is block diagonal.

The proof of this result is highly technical and we will therefore not cover it here. Instead we refer the reader to Section 9 of [DMS21].

9.8 Some elements of the proof of Theorem 9.33

We now have all the tools in hand to begin the proof of Theorem 9.33 per se, given the stationarity of Lemma 9.31 and the cutting theorem of Theorem 9.26. The proof consists of two fairly distinct steps.

Step 1. Show that $(L_t, R_t)_{t \in \mathbb{R}}$ has stationary and independent increments as well as the Brownian scaling property and is therefore a two-sided

9.8 Some elements of the proof of Theorem 9.33

Brownian motion with some covariance matrix. This actually works for all $\kappa' > 4$ and not just $\kappa' \in (4, 8)$.

Step 2. Identify the covariance matrix using the notion of *cone times*. This argument we will present comes from [DMS21] and only works for $\kappa' \in (4, 8)$; in the case $\kappa' \geq 8$ a related but more complicated argument was given separately by Holden, Gwynne, Miller and Sun in [GHMS17].

Proof of Step 1 for all $\kappa' > 4$ Define a filtration $\mathcal{F}_t$ by considering, for any $t \in \mathbb{R}$, the sigma-algebra generated by $\eta(s)$, $s \leq t$ and $h|_{\eta(-\infty,t)}$. Let D_t^- denote the interior of $\eta(-\infty, t)$ and let D_t^+ denote the interior of $\eta(t, \infty)$. Observe that by the cutting theorem (Theorem 9.26), the doubly marked quantum surfaces

$$\mathcal{W}_t^\pm = (D_t^\pm, h|_{D_t^\pm}, \eta(t), \infty)$$

are quantum wedges of parameter $\alpha = 3\gamma/2$ with $\mathcal{W}_t^+$ independent of $\mathcal{W}_t^-$. Indeed, for $t = 0$, this follows from the second bullet point in that theorem, and for other values of $t \in \mathbb{R}$, the same can be deduced from the stationarity of the quantum cone viewed from $\eta(t)$ (Lemma 9.31). Recall that this means that if $g_t^\pm$ is a map from $D_t^\pm$ to $\mathbb{H}$ sending $\eta(t)$ to 0 and fixing ∞ with some scaling chosen so that the resulting fields are in the unit circle embedding, then the fields $g_t^\pm(h)$ obtained from h by applying the change of coordinates formula are independent fields in $\mathbb{H}$ (recall also that this does not require considering these fields as being defined modulo additive constant), with laws of a (thick) quantum wedge as specified in Chapter 7.

Recall also that by Lemma 9.13, given $\mathcal{F}_t$, the curve $(\eta(t+s))_{s \geq 0}$ is just a chordal $\mathrm{SLE}_{\kappa'}$ in its domain D_t^+, from $\eta(t)$ to ∞. The key observation is that the increments of (L, R) over $[t, \infty)$ can be described intrinsically in terms of the surface $\mathcal{W}_t^+$. More precisely, since the boundary length can be computed by conformally changing the coordinates, we can compute the conditional law given $\mathcal{F}_t$ of the increment $(L_{t+u} - L_t, R_{t+u} - R_t)$ for $u \geq 0$ as follows:

- Take a quantum wedge of the appropriate parameter $\alpha = 3\gamma/2$ embedded in $\mathbb{H}$, and consider a chordal $\mathrm{SLE}_{\kappa'}$ curve η in $\mathbb{H}$ from 0 to ∞, reparametrised by Liouville area, and run it for u units of time.
- Compute the relative boundary lengths of $\mathbb{H} \setminus \eta(0, u)$ to the left and right of $\eta(u)$, compared to those of $\mathbb{H}$, left and right of zero (note that these could be both positive or negative!)

As the reader can see, this description is independent of $\mathcal{F}_t$ and depends only on u. This immediately gives the desired independence and stationarity of the increments.

To conclude Step 1, it remains to check that $(L_t, R_t)_{t \in \mathbb{R}}$ obeys the Brownian scaling property. Namely, if $\lambda > 0$, we want to show that

$$\frac{1}{\sqrt{\lambda}}(L_{\lambda t}, R_{\lambda t})_{t \in \mathbb{R}}$$

has the same law as (L, R). Informally, this will follow from the fact that the volume (which parametrises L and R) scales like $e^{\gamma h}$, while the length, which gives the values of L and R, scales like $e^{(\gamma/2)h}$. More precisely, recall that if $C \in \mathbb{R}$, the quantum cone $(\mathbb{C}, h, 0, \infty)$ and $(\mathbb{C}, h+C, 0, \infty)$ have the same laws as quantum surfaces. Let (L^C, R^C) denote the process of left and right boundary lengths associated to the field $h + C$ along the curve η parametrised by μ_{h+C}. Since $h + C$ and h define the same quantum surfaces in law, and because η has the scale-invariance property and is independent of h,

(L^C, R^C) has the same law as the original process (L, R).

On the other hand, it is clear that L^C can be obtained from L simply by time changing and scaling: more precisely,

$$L_t^C = \frac{1}{\sqrt{\lambda}} L_{\lambda t}$$

with $\lambda = e^{\gamma C}$, since quantum areas for $h + C$ are multiplied by λ, and quantum lengths of $h + C$ are multiplied by $\sqrt{\lambda}$. The same holds for R as well, which concludes the proof of Brownian scaling and thus of Step 1. □

By symmetry of L and R, Step 1 implies that (for all $\kappa' > 4$) we can write

$$\begin{pmatrix} L_t \\ R_t \end{pmatrix} = a \begin{pmatrix} \sin(\theta) & -\cos(\theta) \\ 0 & 1 \end{pmatrix} \begin{pmatrix} X_t \\ Y_t \end{pmatrix}, \quad (9.15)$$

where $(X_t, Y_t)_{t \in \mathbb{R}}$ is a standard two-sided planar Brownian motion (started at the origin and with independent coordinates), $a > 0$ is such that $\text{Var}(L_1) = \text{Var}(R_1) = a^2$ and $\theta \in [0, \pi]$ is such that $\text{Cov}(L_1, R_1) = -a^2 \cos(\theta)$.

Proof of Step 2 for $\kappa' \in (4, 8)$. This step consists of identifying θ in (9.15), and the method we present will only work for the case $\kappa' \in (4, 8)$, as will become clear shortly. This range of κ' corresponds to $\theta \in (\pi/2, \pi)$, equivalently, $-\cos(\theta) = a^{-2} \text{Cov}(L_1, R_1) > 0$. That is, the case where L and R are positively correlated.

The argument we present will use the notion of *cone times*. We say that t is a local θ-cone time for a process $(X_s, Y_s)_{s \in \mathbb{R}}$ if there exists $\varepsilon > 0$ such that (X_s, Y_s) remains in the set $(X_t + Y_t) + C_\theta$ for all $s \in [t, t + \varepsilon]$, where $C_\theta = \{z \in \mathbb{C} : \arg(z) \in [0, \theta]\}$. It is straightforward to see that if (L, R) and

9.8 Some elements of the proof of Theorem 9.33

(X, Y) are related by (9.15), then the set of local $\pi/2$-cone times for (L, R) correspond precisely to the set of local θ-cone times for (X, Y).

The key idea is to identify θ using a result of Evans, [Eva85], which states that the almost sure Hausdorff dimension of the set of local θ-cone times of (X, Y) is equal to 0 for $\theta \in [0, \pi/2]$, and equal to $1 - \pi/2\theta$ for $\theta \in (\pi/2, \pi)$. On the other hand, as we will explain below, it is possible to compute the almost sure Hausdorff dimension of the set of local $\pi/2$-cone times for (L, R), using the definition of (L, R) in terms of space-filling SLE on a quantum cone. This will be equal to 0 when $\kappa' \geq 8$, and $1 - \kappa'/8$ when $\kappa' \in (4, 8)$. Hence, we learn nothing if $\kappa' \geq 8$, but for $\kappa' \in (4, 8)$, we see that necessarily:

$$1 - \frac{\kappa'}{8} = 1 - \frac{\pi}{2\theta}, \text{ equivalently } \theta = \frac{4\pi}{\kappa'},$$

as required.

So, it remains to argue that when $\kappa' \in (4, 8)$ the Hausdorff dimension of the set of local $\pi/2$-cone times of (L, R) is almost surely equal to $1 - \kappa'/8$. In fact, since $(L_s, R_s)_{s \in \mathbb{R}}$ and $(\hat{L}_s, \hat{R}_s)_{s \in \mathbb{R}} := (L_{-s}, R_{-s})_{s \in \mathbb{R}}$ are identical in law, it suffices to consider the set $\mathcal{A}$ of local $\pi/2$-cone times for $(\hat{L}, \hat{R})$ and show that the Hausdorff dimension of $\mathcal{A}$ is almost surely equal to $1 - \kappa'/8$.

We are going to identify the Hausdorff dimension of the set $\mathcal{A}$ with the Hausdorff dimension of a set of times determined by the geometry of η'. To set up for this, first notice that by the definition of $\pi/2$-cone times, if s is a local $\pi/2$-cone time for $(\hat{L}, \hat{R})$, then there exists some $t \in \mathbb{Q}$ with $(\hat{L}_r, \hat{R}_r) \in (\hat{L}_s, \hat{R}_s) + C_{\pi/2}$ for all $r \in (s, t)$. This implies that $(L_{-t+u} - L_{-t}, R_{-t+u} - R_{-t})_{u \geq 0}$ has a simultaneous running infimum at $u = t - s$. Conversely, for any $t \in \mathbb{Q}$, each simultaneous running infima of $(L_{-t+u} - L_{-t}, R_{-t+u} - R_{-t})_{u \geq 0}$ corresponds to a local $\pi/2$ cone time for $(\hat{L}, \hat{R})$. Thus, we can write $\mathcal{A} = \bigcup_{q \in \mathbb{Q}} \{-q - \mathcal{A}_q\}$, where $\mathcal{A}_q$ is the set of simultaneous running infima of $(L_{q+u} - L_q, R_{q+u} - R_q)_{u \geq 0}$. Notice also that by the stationarity of Lemma 9.31, the almost sure Hausdorff dimension of $\mathcal{A}_q$ does not depend on $q \in \mathbb{Q}$. In particular, this implies that the Hausdorff dimensions of $\mathcal{A}$ and $\mathcal{A}_0$, say, are equal almost surely. We claim that

$$\mathcal{A}_0 = \{s > 0 \colon \eta'(s) \in \eta_0^L \cap \eta_0^R\} \tag{9.16}$$

with probability 1. To see this, recall from Section 9.4 that η_0^L and η_0^R are the concatenation left and right boundaries of an ordered collection of simply connected domains, ordered consistently with the ordering of $(\eta'(r))_{r<0}$. But we can also reverse this ordering, and view η_0^L and η_0^R as the boundaries of an ordered collection of simply connected domains forming $\mathbb{C} \setminus \eta'((-\infty, 0]) = \eta'((0, \infty))$, ordered according to when they are visited by $(\eta'(s))_{s>0}$. It is not hard to convince oneself that L has a running infimum at time $r > 0$ if and only if $\eta'(r) = z \in \eta_0^L$, and $r = \sup\{u: \eta'(u) = z\}$. The analogous statement holds when L is replaced by R. Thus, (L, R) have a simultaneous running infima at $s > 0$ if and only if $\eta'(s) = z \in \eta_0^L \cap \eta_0^R$ and $s = \sup\{u: \eta'(u) = z\}$. It also follows from the definitions that points of $\eta_0^L \cap \eta_0^R$ are visited exactly once by η', and so in this case (L, R) have a simultaneous running infima at $s > 0$ if and only if $\eta'(s) = z \in \eta_0^L \cap \eta_0^R$.

Next, we recall the statement of Theorem 9.26. This says that if $(\mathbb{C}, h, 0, \infty)$ is the γ-quantum cone used to define (L, R), then viewed as a quantum surface, $(\eta'[0, \infty]), h, 0, \infty)$ has the law of a quantum wedge with parameter $\alpha = 3\gamma/2$. This is a thin wedge for $\kappa' \in (4, 8)$, meaning that it is an ordered collection of quantum surfaces, and this ordered collection of surfaces correspond precisely to the ordered collection of simply connected domains described in the above paragraph. The intervals of time on which $\eta'(r) \notin \eta_0^L \cap \eta_0^R$ correspond precisely to the intervals during which $\eta'|_{[0,\infty)}$ visits one of these domains. By definition of the parametrisation of η', the lengths of these intervals are exactly the quantum areas of the quantum surfaces making up the thin wedge. It therefore follows from Lemma 7.33 (and an additional scaling property together with the finiteness of a certain moment, see Proposition 4.4.4 in [DMS21]) that the ordered collection of lengths of these intervals are equal in law to the durations of excursions away from 0, for a Bessel process of dimension $\delta = \kappa'/4$. It turns out that the finite moment assumption boils down to requiring that $\delta >$, which fortunately is the case when $\delta = \kappa'/4$ and $\kappa' > 4$. By classical excursion theory, see for instance [Ber96, Chapter III], it follows that $\mathcal{A}_0$ is the range of a $1 - \delta/2$ stable subordinator, and the Hausdorff dimension of such a set is almost surely equal to $1 - \delta/2 = 1 - \kappa'/8$. □

The concludes the proof of Theorem 9.33.

Appendix A
Chordal Loewner Chains and Chordal SLE

The aim of this appendix is to collect some relevant background material on Schramm–Loewner evolutions (SLE), primarily to accompany Chapters 8 and 9. For a much more detailed and pedagogical exposition, the reader is referred to [BN11, Kem17, Law05]. The presentation here most closely follows [BN11].

A.1 Chordal Loewner chains

Complex analysis basics. First, we fix some basic notation and terminology.

- $K \subset \mathbb{H}$ is said to be a *compact $\mathbb{H}$-hull* if it is bounded and $H := \mathbb{H} \setminus K$ is a simply connected domain.
- For any such hull, by the Riemann Mapping Theorem, one can choose a conformal isomorphism $g_K : H \to \mathbb{H}$ such that $g_K(z) - z \to 0$ as $z \to \infty$. In fact, one can prove that for this g_K, the expansion $g_K(z) = z + \frac{a_K}{z} + O(|z|^{-2})$ holds as $z \to \infty$ for some $a_K \geq 0$. We call g_K the *Loewner map* of K.
- a_K is known as the *half plane capacity* of K and denoted by $\mathrm{hcap}(K)$.
- In some sense, the half plane capacity measures the size of the hull K, when "viewed from infinity". In particular, the half plane capacity increases as a hull increases: if $K \subset K'$ are two complex $\mathbb{H}$ hulls, then $\mathrm{hcap}(K) \leq \mathrm{hcap}(K')$.

Loewner Chains. A Loewner chain is a family $(K_t)_{t \geq 0}$ of *increasing* ($K_s \subsetneq K_t$ for $s \leq t$) compact $\mathbb{H}$-hulls which satisfy a *local growth property*: for any $T \geq 0$,

$$\sup_{s,t \in [0,T], |s-t| \leq h} \mathrm{rad}\left(g_{K_s}(K_t \setminus K_s)\right) \to 0 \text{ as } h \to 0.$$

Here the radius of a hull means the radius of the smallest semicircle in which it can be inscribed. For such a chain, one can show that the half plane capacity is

Chordal Loewner Chains and Chordal SLE

Figure A.1 A Loewner chain drawn up to two times. (a) a time before $\zeta(z)$; (b) just after $\zeta(z)$.

a strictly increasing bijection from $[0, \infty) \to [0, \infty)$, so we can always assume (by convention) that time is parametrised so that $\mathrm{hcap}(K_t) = 2t$ for all t.

Theorem A.1 (Loewner's theorem) *Loewner discovered that such chains (parametrised by half plane capacity) are in bijection with continuous real valued functions via the following correspondence.*

- *Given $(K_t)_{t \geq 0}$ a Loewner chain, there is a unique point*

$$\xi_t \in \overline{\bigcap_{h > 0} g_{K_t}(K_{t+h} \setminus K_t)}$$

for each $t \geq 0$. The function $(\xi_t)_{t \geq 0}$ is continuous and real valued and is called the driving function *of $(K_t)_{t \geq 0}$.*

- *Given $(\xi_t)_{t \geq 0}$ a continuous real-valued function, define, for each $z \in \mathbb{H}$, $g_t(z)$ to be the maximal solution to the* Loewner equation

$$\frac{\partial g_t(z)}{\partial t} = \frac{2}{g_t(z) - \xi_t}, \quad g_0(z) = z \quad (A.1)$$

which exists on some time interval $[0, \zeta(z)]$ by classical ODE theory. Let $K_t = \{z \in \mathbb{H} : \zeta(z) \leq t\}$. Then $(K_t)_{t \geq 0}$ is a Loewner chain with driving function ξ_t. Moreover, $g_t = g_{K_t}$ for all t.

We call $(g_t)_{t \geq 0}$ the (forward) *Loewner flow*. $\zeta(z)$ is the time that the growing hull K_t "swallows" the point z. See Figure A.1.

Remark A.2 Continuous curves $(\gamma(t))_{t\geq 0} =: (\gamma_t)_{t\geq 0}$ in $\mathbb{H}$ which do not cross themselves and have $|\gamma_t| \to \infty$ as $t \to \infty$ provide examples of Loewner chains. More precisely, one defines $H_t = \mathbb{H} \setminus K_t$ for each t to be the connected component of $\mathbb{H} \setminus \gamma([0, t])$ containing ∞. Then $(K_t)_{t\geq 0}$ is an increasing family of compact $\mathbb{H}$-hulls satisfying the local growth property. The map g_t sends the tip of the curve, γ_t, to the point ξ_t (where g_t is extended by "continuity", or more precisely, this holds in the sense of prime ends).

A.2 Chordal SLE_κ

Chordal SLE_κ processes, for $\kappa > 0$, were introduced by Oded Schramm [Sch00] as a family of potential scaling limits for interfaces in critical statistical physics models. As we will soon see, they satisfy two very natural properties that make them appropriate candidates for such limits: conformal invariance and a certain domain Markov property.

It turns out ([Sch00]) that these two properties actually *characterise* SLE_κ as a one-parameter family, which means that there really can be no other candidates. On the other hand, proving convergence of discrete interface models to SLE is typically very challenging. To date, it has been verified for just a few special values of κ; for example, critical percolation interfaces [Smi01] and the loop-erased random walk [LSW04].

Definition A.3 (Chordal SLE in $\mathbb{H}$ from $0 \to \infty$) For $\kappa > 0$, SLE_κ in $\mathbb{H}$ from 0 to ∞ is defined to be the Loewner chain driven by $\xi_t = \sqrt{\kappa} B_t$ where B_t is a standard Brownian motion.

One of the first things to note about SLE is that, due to the scaling property of Brownian motion (B_t has the same law as $\sqrt{t} B_1$ for any t), SLE is itself scale invariant. That is, for any $r \geq 0$ if $(K_t)_{t\geq 0}$ is an SLE_κ process, then the rescaled process $(r^{-1/2} K_{rt})_{t\geq 0}$ also has the law of an SLE_κ. This says that SLE is invariant under conformal isomorphisms of $\mathbb{H}$ that fix 0 and ∞. This allows us to define SLE, by conformal invariance, in any simply connected domain and between any two marked boundary points.

Definition A.4 (Chordal SLE) SLE_κ is a collection $(\mu_{D,a,b})_{D,a,b}$ of laws on Loewner chains, indexed by triples (D, a, b) where D is a simply connected domain and a and b are two marked boundary points. The law $\mu_{\mathbb{H},0,\infty}$ is that given by Definition A.3. For any other triple (D, a, b), $\mu_{D,a,b}$ is defined to be the image of $\mu_{\mathbb{H},0,\infty}$ under (any) conformal isomorphism sending $\mathbb{H}$ to D, 0 to a and ∞ to b.

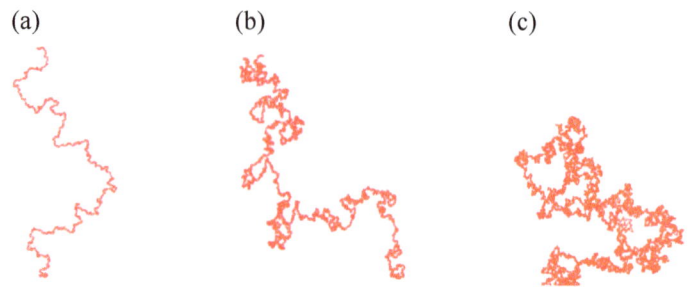

Figure A.2 From (a) to (c): SLE$_2$, SLE$_4$, SLE$_6$. Simulations by Tom Kennedy.

The choice of conformal isomorphism above does not impact the law of the curve up to time-change, since any two such conformal isomorphisms differ by scaling and SLE$_\kappa$ in the upper-half plane from 0 to ∞ is invariant under scaling. See [BN11] for a somewhat more rigorous measure-theoretic framework.

Chordal SLE: properties.

- Chordal SLE$_\kappa$ is generated by a curve γ (in the sense of Remark A.2) for every $\kappa > 0$: due to [RS05] for $\kappa \neq 8$, and [LSW04] for $\kappa = 8$.
- *Conformal invariance:* if γ is an SLE$_\kappa$ in D from a to b and $\psi : D \to D'$ is a conformal isomorphism with $\psi(a) = a'$ and $\psi(b) = b'$, then $\psi(\gamma)$ (up to time reparametrisation) has the law of an SLE$_\kappa$ in D' from a' to b'.
- *Domain Markov property:* if γ is an SLE$_\kappa$ from a to b in D and T is a bounded stopping time that is measurable with respect to γ, then conditionally on $\gamma([0,T])$, writing D_T for the connected component containing b of $D \setminus \gamma([0,T])$, $\gamma([T,\infty))$ has the law of an SLE$_\kappa$ from $\gamma(T)$ to b in D_T.
- It has three distinct *phases*: for $\kappa \in [0,4]$ SLE$_\kappa$ is almost surely generated by a simple (non-self-touching and non-boundary touching) curve; for $\kappa \in (4,8)$, it almost surely hits (but doesn't cross) itself and the boundary of the domain; and for $\kappa \geq 8$, it is almost surely space filling. See Figure A.2.

A.3 Chordal SLE$_\kappa(\underline{\rho})$

It is best to view chordal SLE$_\kappa$ as a family of laws $\mu_{(D,a,b)}$ on random *chords* in the domain D connecting one boundary point a to another boundary point b (where boundary is understood in a conformal sense). Both the conformal invariance and domain Markov properties of chordal SLE are then easily

A.3 Chordal SLE$_\kappa(\underline{\rho})$

formulated through this notation: for instance, the requirement of conformal invariance is that $\mu_{(\phi(D);\phi(a),\phi(b))}$ is the push-forward of the measure $\mu_{D,a,b}$ by the conformal isomorphism ϕ.

It is also very natural to consider random curves in which the domain Markov property and conformal invariance are satisfied only provided we specify additional information such as the location of a specified number of points in the domain or on its boundary. For concrete examples, consider the scaling limits of discrete interface models in which there is a change of boundary conditions at a specified number of points along the boundary: the law of the scaling limit will depend on the location of these special points.

As it turns out (see Remark A.6 for a proof in the case of one marked point on the boundary), such curves are described by variants of SLE$_\kappa$, which have an additional attraction or repulsion from certain *marked points* (also sometimes known as force points) in the domain or on its boundary. These are known as SLE$_\kappa(\underline{\rho})$ and first appeared in [LSW03]; see also [SW05, MS16a]. The vector $\underline{\rho}$ encodes how strong this attraction or repulsion is, and in which direction.

Let $\kappa > 0$. We will take again the upper half plane as a reference domain and $a = 0, b = \infty$ for the start and target points on the boundary. Let $v^1, \ldots, v^m$ be m marked points on the boundary (we will discuss interior points below) and corresponding weights $\rho^1, \ldots, \rho^m \in \mathbb{R}$ are such that

$$\sum_{i \in S} \rho^i \geq -2 \text{ for every } S \subset \{1, \ldots, M\}. \tag{A.2}$$

To define the law $\mu_{(\mathbb{H},0,\infty);(v^1,\ldots,v^m)}$ of SLE$_\kappa(\underline{\rho})$ = SLE$_\kappa(\rho^1, \ldots, \rho^m)$, we proceed as follows. It will be a Loewner chain and hence by Loewner's theorem, can be defined by specifying its driving function. As for ordinary SLE$_\kappa$, this driving function will be a random function closely related to Brownian motion. However, the Brownian motion now comes with a *drift*. This drift will depend on the position of the marked points $(V^1_t, \ldots, V^m_t)_{t \geq 0}$ after applying the Loewner flow $(g_t)_{t \geq 0}$ as follows.

Definition A.5 (SLE$_\kappa(\rho^1, \ldots, \rho^m)$ in $\mathbb{H}$ from 0 to ∞) Suppose that $v^1, \ldots, v^m \in (\mathbb{R} \cup \{\infty\}) \setminus \{0\}$ are distinct and $\rho^1, \ldots, \rho^m$ satisfy the condition (A.2). SLE$_\kappa(\rho^1, \ldots, \rho^m)$ with marked points at $v^1, \ldots, v^m$ is the Loewner chain with driving function $(\xi_t)_{t \geq 0}$ satisfying the following system of SDEs:

$$\begin{aligned} \xi_t &= \sqrt{\kappa} B_t + \sum_i \int_0^t \frac{\rho^i}{\xi_s - V^i_s} ds \\ V^i_t &= v^i + \int_0^t \frac{2}{V^i_s - \xi_s} ds \text{ for } 1 \leq i \leq M. \end{aligned} \tag{A.3}$$

The second equation in (A.3) is simply Loewner's equation describing the evolution V_t^i of the marked point v^i under the Loewner flow, at least until v^i is swallowed by the chain (in fact, the evolution can be extended beyond this point). The first equation in (A.3) describes the driving function of the Loewner flow as usual.

For any value of $\underline{\rho}$, the strong existence and uniqueness of solutions to (A.3) is clear until the first time that one of the V^i collides with ξ, that is, $\sup\{t \geq 0 : \xi_t \neq V_t^i \ 1 \leq i \leq m\}$. This is also the first time that one of the marked points v^i is swallowed by the hull generated by the Loewner chain. In fact, it can be shown (see [MS16a]) that when $\underline{\rho}$ satisfies (A.2), there exists an almost surely continuous Markovian process $(\xi, V^1, \ldots, V^m)$ that satisfies the integrated equation (A.3) for *all* time, and for which the set of times t with $\xi_t = V_t^i$ for some i almost surely has Lebesgue measure 0. It is also shown in [MS16a] that the law of this process is unique. Consequently, the corresponding chordal $\text{SLE}_\kappa(\underline{\rho})$ Loewner chain in Definition A.5 is well defined (for all time).

Remark A.6 In the case of one marked point on the boundary ($m = 1$), the process $(V_t^1 - \xi_t)_{t \geq 0}$ describing the distance between the driving function and the evolution of the marked point, is $\sqrt{\kappa}$ times a Bessel process. When $\underline{\rho} = \rho^1 = \rho$, the dimension of the Bessel process is

$$\delta = 1 + \frac{2(\rho + 2)}{\kappa}. \tag{A.4}$$

This formalises the notion that $\text{SLE}_\kappa(\underline{\rho})$ processes have an additional attraction/repulsion from the marked points.

In fact, for a Loewner chain to satisfy a conformal Markov property with an extra marked point (that is, the property that for any stopping time σ, the future evolution after applying the Loewner map at time σ has the same law as the original process, with the marked point now located at the image of the original marked point) one finds that the difference between the driving function and the evolution of the marked point must be a continuous Markov process satisfying Brownian scaling. This implies that it actually has to be a Bessel process of some dimension. One can take this as an explanation for the form of the SDEs (A.3).

Remark A.7 The definition can also be extended to the case where there are marked points located infinitesimally to the left and/or right of 0 (denoted 0^- and 0^+). This is done by taking a limit in law (with respect to the Carathéodory topology on Loewner chains[1] as one of the marked points approaches 0 from

---

[1] This is the topology for which a sequence of chordal Loewner chains (from 0 to ∞ in $\mathbb{H}$) with

A.3 Chordal SLE$_\kappa$($\underline{\rho}$)

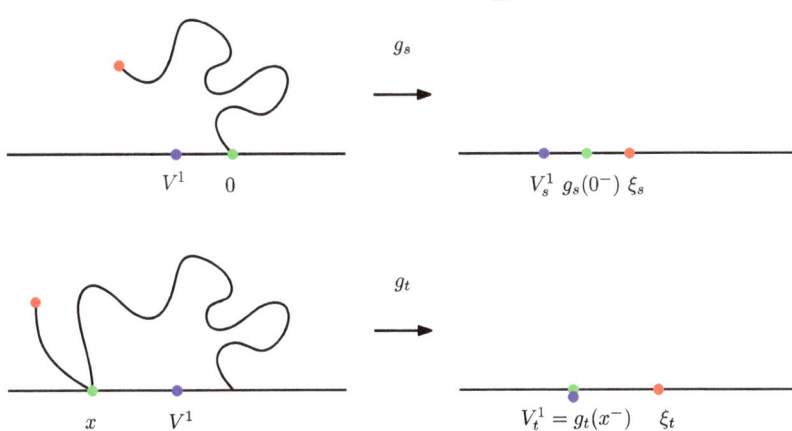

Figure A.3 A schematic picture of an SLE$_\kappa$($\underline{\rho}$) with one marked point, drawn up to two times s, t with $s < t$. At time s the marked point has not been swallowed, but at time t it has. After time t the evolution of V_1^t coincides (by definition) with the evolution under the Loewner flow of the point infinitesimally to the left of x.

the left and/or one of the marked points approaches 0 from the right. Again this gives rise to unique laws on Loewner chains that are defined for all time.

When there is just one marked point, this boils down to starting a Bessel process of positive dimension from zero; in fact by (A.4), the dimension of this Bessel process is greater than 1 when $\rho > -2$. The reason why we assume the dimension to be greater than 1 (and so ρ to be > -2) is to ensure that the integral in (A.3) is convergent. When $\rho \leq -2$, assigning a meaning to this integral is less straightforward, though there are known procedures, including for example a principal value correction, see [She09].

Remark A.8 Definition A.5 can also be extended to include *interior force points*. That is, with some of the $v^i = V_0^i$ located in $\mathbb{H}$ rather than in $\mathbb{R}$. The definition is exactly the same, but in this case, existence and uniqueness of solutions to (A.3) is only guaranteed until the first time that $\xi_t = V_t^i$ for some i such that $v^i \in \mathbb{H}$. As such, the chordal SLE$_\kappa$($\underline{\rho}$) with interior force points is a well-defined random Loewner chain, but only up to the first time that one of the interior force points is "swallowed".

Due to the scaling property of Brownian motion, it follows easily that SLE$_\kappa$($\underline{\rho}$) from 0 to ∞ in $\mathbb{H}$ also satisfies a form of scale invariance. More

Loewner flow $(g_t^n)_{t \geq 0}$ converges to a Loewner chain with Loewner flow $(g_t)_{t \geq 0}$ as $n \to \infty$ if and only if $(g_t^n)^{-1}(z) \to g_t^{-1}(z)$ uniformly on compact subsets of time and subsets of space that are compactly contained in $\mathbb{H}$.

precisely, if $(K_t)_{t\geq 0}$ is an $\text{SLE}_\kappa(\underline{\rho})$ process with force points at $v^1,\ldots,v^m$, then the rescaled process $(r^{-1/2}K_{rt})_{t\geq 0}$ has the law of an $\text{SLE}_\kappa(\underline{\rho})$ process with force points at $r^{-1/2}v^1,\ldots,r^{-1/2}v^m$ for any $r > 0$. This allows us to extend the definition of $\text{SLE}_\kappa(\underline{\rho})$ to arbitrary domains with finitely many marked boundary points.

Definition A.9 ($\text{SLE}_\kappa(\rho^1,\ldots,\rho^m)$ in D from a to b) Suppose that $\rho^1,\ldots,\rho^m$ are as in Definition A.5 and $(D,a,b,v^1,\ldots,v^m)$ is a given domain with $(m+2)$ marked points. Let $\psi : \mathbb{H} \to D$ be a conformal isomorphism sending a to 0 and b to ∞.

$\text{SLE}_\kappa(\rho^1,\ldots,\rho^m)$ from a to b in D with marked points at $v^1,\ldots,v^m$ is defined to be the image under ψ of $\text{SLE}_\kappa(\rho^1,\ldots,\rho^m)$ from 0 to ∞ in $\mathbb{H}$, with marked points at $\psi^{-1}(v^1),\ldots,\psi^{-1}(v^m)$.

This definition also extends to the case of interior force points, with both of the above $\text{SLE}_\kappa(\underline{\rho})$ curves being defined up to the first time that an interior force point is swallowed.

Remark A.10 (Properties) $\text{SLE}_\kappa(\underline{\rho})$ possesses many properties similar to those of SLE_κ, along with some additional features.

- For any $\kappa > 0$ and $\underline{\rho}$ satisfying (A.2), $\text{SLE}_\kappa(\underline{\rho})$ is almost surely generated by a continuous curve γ, with $\gamma(0) = a$ and $\gamma(t) \to b$ as $t \to \infty$: see [MS16a].
- By definition, if $\psi : D \to D'$ is a conformal isomorphism sending

$$(a,b,v^1,\ldots,v^m) \text{ to } (a',b',(v^1)',\ldots,(v^m)'),$$

then the image of $\text{SLE}_\kappa(\underline{\rho})$ from a to b in D with force points at $v^1,\cdots,v^m$ has the law of $\text{SLE}_\kappa(\underline{\rho})$ from a' to b' in D' with force points at $(v^1)',\ldots,(v^m)'$.
- Going back to the set-up in the upper half plane, the processes $(V_t^i)_{t\geq 0}$ from (A.3) describe the evolution of the force points v^i under the Loewner flow. More precisely, for each i and until the first time τ^i that v^i is "swallowed" by the curve, V_t^i is equal to $g_t(v^i)$ (where g_t is continuously extended to the boundary if necessary). After this time, if $v^i \in \mathbb{R}_+$ (respectively $\mathbb{R}_-$), V_t^i will be equal to the image under g_t of the furthest right (resp. furthest left) point on the real line that has been swallowed at time τ^i (see Figure A.3).
- As we have mentioned already, $\text{SLE}_\kappa(\underline{\rho})$ satisfies a domain Markov property, that now involves the marked points. To state this precisely, suppose that γ is an $\text{SLE}_\kappa(\underline{\rho})$ from a to b in D with force points at $v^1,\ldots,v^m$ and that T is a bounded stopping time for γ. Write D_T for the connected component of $D \setminus \gamma([0,T])$ containing b. Then conditionally on $\gamma([0,T])$, $\gamma([T,\infty))$ has the law of an $\text{SLE}_\kappa(\underline{\rho})$ from $\gamma(T)$ to b in D_T, with force points at $V_T^1,\ldots,V_T^m$.

A.3 Chordal SLE$_\kappa(\rho)$

- By inspecting (A.3), it follows that for SLE$_\kappa(\rho)$ in $\mathbb{H}$, putting any weight ρ at the boundary point ∞ does not affect the law of the curve. This observation will be useful when studying the relationship between chordal and *radial* SLE.

Recall that chordal SLE$_\kappa$, $(\gamma(t))_{t \geq 0}$ has three distinct phases. In terms of its interaction with the boundary $\partial \mathbb{H} = \mathbb{R}$, this can be described as follows: if $\kappa \in [0, 4]$, $\gamma([0, \infty)) \cap \mathbb{R} = \emptyset$ almost surely; if $\kappa \in (4, 8)$, $\gamma([0, \infty)) \cap \mathbb{R}$ is almost surely non-empty, unbounded but has Lebesgue measure 0; and if $\kappa \geq 8$, $\gamma([0, \infty)) \cap \mathbb{R} = \mathbb{R}$ almost surely. In the case of SLE$_\kappa(\rho)$, where there is additional attraction or repulsion from force points on $\mathbb{R}$, this behaviour may be modified.

Indeed, consider the case of SLE$_\kappa(\rho)$ with one force point at $v^1 \in \mathbb{R}$ of weight ρ. Then we have already seen that the distance between the driving function and the evolution of v^1 under the Loewner flow, $(V_t^1 - \xi_t)_{t \geq 0}$, is a Bessel process of dimension $1 + 2\kappa^{-1}(\rho + 2)$. This means that if $\rho \geq \frac{\kappa}{2} - 2$ then $(V_t^1 - \xi_t)$ will almost surely be positive for all $t > 0$; that is, the SLE$_\kappa(\rho)$ will almost surely not hit the half-closed interval between v_1 and ∞ at any time $t > 0$. If $\rho < \frac{\kappa}{2} - 2$, then the Bessel process will hit 0, which means that the SLE$_\kappa(\rho)$ *will* hit this half-closed interval. We have proved the following lemma:

Lemma A.11 *Let η be a chordal SLE$_\kappa(\rho)$ with $\kappa > 0$ and $\rho > -2$ for some boundary marked point v. Then η hits v (or more precisely v is swallowed by the hull generated by η) if and only if $\rho < \kappa/2 - 2$.*

For multiple force points, a description of the interaction of SLE$_\kappa(\rho)$ with the real line (depending on $\underline{\rho}$)) can be found in [Dub09a, Lemma 15].

Appendix B
Reverse Loewner Flow and Reverse SLE

B.1 Definitions

Appendix A focused on standard Loewner evolutions, describing increasing families of compact hulls: in nice cases, growing curves. However, these should really be referred to as *forward* Loewner evolutions, because they also have a counterpart: *reverse Loewner evolutions*. A reverse Loewner evolution is no longer a family of hulls that increases in time, but rather a family of hulls where in each infinitesimal increment of time, an infinitesimal new piece of hull is added "at the root". The whole of the previous hull is then conformally mapped to something slightly different (one might envisage the new piece of hull as "pushing" the existing one further into the domain) (see Figure B.1). Note that one cannot therefore speak of a "single curve" associated to a reverse Loewner evolution.

In the following, we will only ever discuss *centred* reverse Loewner evolutions. Informally, this means that new pieces of curve are always added at the origin.

Definition B.1 (Reverse Loewner evolution in $\mathbb{H}$)   Let $(\xi_t)_{t \geq 0}$ be a continuous real-valued function with $\xi_0 = 0$. The solution $(f_t(z))_{t \geq 0, z \in \mathbb{H}}$ to the family of equations

$$\frac{\partial (f_t(z) + \xi_t)}{\partial t} = \frac{-2}{f_t(z)}, \qquad f_0(z) = z; \quad z \in \mathbb{H} \tag{B.1}$$

is called the reverse Loewner flow driven by $(\xi_t)_{t \geq 0}$. In contrast to the forward case, $f_t(z)$ is defined for all $t \geq 0$ and $z \in \mathbb{H}$. This means that f_t defines a conformal isomorphism from $\mathbb{H}$ to some domain H_t for all t (and one can check that $f_t(z) \sim z$ as $z \to \infty$ for each t). $(\mathbb{H} \setminus H_t)_{t \geq 0}$ is called the reverse Loewner evolution driven by $(\xi_t)_{t \geq 0}$.

We will now discuss the (deterministic) relation between forward and reverse

B.1 Definitions

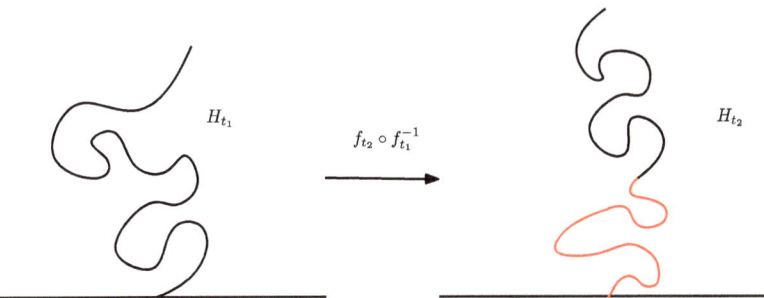

Figure B.1 A reverse Loewner evolution at two times t_1, t_2 with $t_1 < t_2$. In both cases, H_{t_i} is the complement of the curve. One can see that between the two times, a new piece of curve (drawn in red) is added "at the root", and the existing curve (black) is conformally mapped into the domain formed by the complement of the red curve. In contrast, under the forward Loewner flow, new pieces of curve are always added "at the tip" of the existing curve.

Loewner evolutions. For this, it is helpful to consider the centred forward Loewner maps $\tilde{g}_t := g_t - \xi_t$ and associated with a given driving function $(\xi_t)_{t \geq 0}$.

Lemma B.2 (Forward/Reverse flow) *Suppose that $(\tilde{g}_t)_{t \geq 0}$ is the centred forward Loewner flow with driving function $(\xi_t)_{t \geq 0}$. Fix $T > 0$ and write $\hat{\xi}_t = \xi_{T-t} - \xi_T$ for $0 \leq t \leq T$. Let $(\hat{f}_t)_{0 \leq t \leq T}$ be the centred reverse Loewner flow with driving function $(\hat{\xi}_t)_{0 \leq t \leq T}$. Then*

$$\hat{f}_t(z) := \tilde{g}_{T-t} \circ \tilde{g}_T^{-1}(z) \ ; \ t \in [0, T] \, , \, z \in \mathbb{H}.$$

In particular, $\hat{f}_T \equiv \tilde{g}_T^{-1}$.

Proof Since $\tilde{g}_t = g_t - \xi_t$ by definition, the forward Loewner equation (A.1) and then the substitution $t \mapsto T - t$ yields

$$\mathrm{d}(\tilde{g}_t(z)) = \frac{2}{\tilde{g}_t(z)}\,\mathrm{d}t + \mathrm{d}\xi_t \ ; \ \mathrm{d}(\tilde{g}_{T-t}(z)) = -\frac{2}{\tilde{g}_{T-t}(z)} - \mathrm{d}\hat{\xi}_t$$

for every z. Replacing z with $\tilde{g}_T^{-1}(z)$, we may deduce that $\hat{f}_t(z)$ satisfies the reverse Loewner equation (B.1) with driving function $\hat{\xi}$. □

Reverse SLE. Now we have defined reverse Loewner evolutions, reverse SLE_κ is simply defined in the analogous way to forward SLE_κ.

Definition B.3 (Reverse SLE_κ) Reverse SLE_κ for $\kappa > 0$ is the centred reverse Loewner evolution driven by a Brownian motion with diffusivity κ. That is,

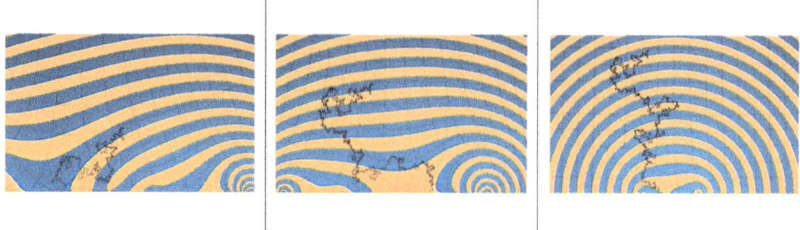

Figure B.2 A reverse SLE$_4$ at three increasing times. The background shows the deformation of the upper half plane under the reverse Loewner flow. Simulation by Henry Jackson.

with driving function $(\xi_t)_{t \geq 0} = (\sqrt{\kappa} B_t)_{t \geq 0}$ where B is a standard Brownian motion. See Figure B.2 for a simulation of SLE_4 at three increasing times.

Definition B.4 (Reverse SLE$_\kappa(\underline{\rho})$ [She16a]) Suppose that $v^1, \ldots, v^m \in \overline{\mathbb{H}}$ and $\rho^1, \ldots, \rho^m$ are real numbers. Reverse SLE$_\kappa(\rho^1, \ldots, \rho^m)$ with force points at $v^1, \ldots, v^m$ is the reverse (centred) Loewner evolution with driving function $(\xi_t)_{t \geq 0}$ satisfying:

$$\xi_t = \sqrt{\kappa} B_t - \sum_i \int_0^t \Re\left(\frac{\rho^i}{f_s(v^i)}\right) ds. \qquad (B.2)$$

It is immediate that this has a unique solution in law, at least until the first time that $f_t(v^i) = 0$ for some i. We will only consider the reverse SLE$_\kappa(\underline{\rho})$ up until this time.

Remark B.5 In the case $m = 1$ and $\rho^1 = \rho$, a straightforward calculation shows that $f_t(v^1)$ is $\sqrt{\kappa}$ times a Bessel process of dimension

$$\delta = 1 + \frac{2(\rho - 2)}{\kappa}.$$

Note the difference with Remark A.6. Roughly speaking, this is because the reverse SLE$_\kappa(\rho)$ generally pulls points towards the origin, while the forward version will be pushes them away (for intuition, consider the case $\rho = 0$ and the way that the flow is defined).

The following properties of reverse SLE$_\kappa(\rho)$ will be needed for a technical discussion in Chapter 8. It says, roughly speaking, that specific SLE$_\kappa(\rho)$ curves are well behaved, in the sense that they do not create massive distortions, and that putting a force point very far away does not affect the law of the evolution

B.1 Definitions

at small times. A reader simply wishing to learn about SLE would be safe to skip this.

Lemma B.6 (1) *Let $(f_t)_{t \leq \tau_1}$ be a reverse $SLE_\kappa(\kappa)$ flow, with a force point at $1 \in \mathbb{R}$ and τ_1 the first time that $f_t(1) = 0$. Then as $R \to \infty$, the probability that $f_{\tau_1}(B(0, R)) \supset B(0, 1)$ tends to 1.*
(2) *Let $(\tilde{f}_t)_{t \leq \tilde{\tau}_1}$ be a reverse $SLE_\kappa(\kappa, -\kappa)$ flow with force points at $(1, R)$ and $\tilde{\tau}_1$ the first time that $\tilde{f}_t(1) = 0$. Then the total variation distance between $(f_t)_{t \leq \tau_1}$ and $(\tilde{f}_t)_{t \leq \tilde{\tau}_1}$ tends to 0 as $R \to \infty$.*
(3) *Let $(\tilde{f}_t)_t$ be a reverse $SLE_\kappa(\kappa, -\kappa)$ flow with force points at $(z, 10)$, and $z \in [1, 2]$. For $a \in (0, 1]$ let $\tilde{\tau}_a$ be the first time that $\tilde{f}_t(z) = a$. Then with probability 1, $f_{\tilde{\tau}_a}(\{w \in \overline{\mathbb{H}}: |w - 10| = 1\}) \subset \overline{\mathbb{H}} \setminus B(0, 1)$.*

Proof (1) Note that $f_t(1)$ is $\sqrt{\kappa}$ times a Bessel process of dimension $3 - (4/\kappa) < 2$ started from 1, and so the time τ_1 is almost surely finite. Moreover, the driving function is continuous up to and including time τ_1, because the integral $\int_0^{\tau_1} f_t(1)^{-1} \, dt$ converges almost surely. This implies that $f_{\tau_1}(z) \to \infty$ as $z \to \infty$ (see Definition B.1), which is the same thing as (1).

(2) For this, we compute the Radon–Nikodym derivative between $(f_t)_{t \leq \tau_1}$ and $(\tilde{f}_t)_{t \leq \tilde{\tau}_1}$ using Girsanov's theorem. Let us write $\mathbb{P}$ for the law of $(f_t)_{t \leq \tau_1}$, and ξ_t for the associated driving function, so for $z \in \mathbb{H}$ we have $df_t(z) = -(2/f_t(z)) \, dt - d\xi_t$ and

$$d\xi_t = \sqrt{\kappa} \, dB_t - \frac{\kappa}{f_t(1)} \, dt,$$

where B_t is a standard Brownian motion under $\mathbb{P}$.

We set

$$M_t := R(R-1)^{-\kappa/2} (f_t(R) - f_t(1))^{\kappa/2} f_t(R)^{-1} f_t'(R)^{-1-\kappa/2}.$$

and first prove that $M_{\cdot \wedge \tau_1}$ is a local martingale under $\mathbb{P}$, (which is positive with unit mean by definition), with respect to the filtration $(\mathcal{F}_t)_{t \geq 0}$ generated by $(\xi_t)_{t \geq 0}$.

This follows from Itô's formula. Indeed, for $A, B, C \in \mathbb{R}$, we have that for $t \leq \tau_1$:

$$d(f_t(R))^A$$
$$= A f_t(R)^{A-1} \, df_t(R) + \frac{A(A-1)}{2} d[f_t(R)]$$
$$= \left(\frac{A(A-1)\kappa}{2} - 2A \right) f_t(R)^{A-2} \, dt + \frac{\kappa A f_t(R)^A}{f_t(1) f_t(R)} \, dt - \sqrt{\kappa} A f_t(R)^{A-1} \, dB_t;$$
$$d(f_t'(R))^B = B f_t'(R)^{B-1} \, df_t'(R) = \frac{2B f_t'(R)^B}{f_t(R)^2} \, dt; \quad \text{and}$$

$$d(f_t(R) - f_t(1))^C = C(f_t(R) - f_t(1))^{C-1}\left(-\frac{2}{f_t(R)} + \frac{2}{f_t(1)}\right)dt$$

$$= \frac{2C(f_t(R) - f_t(1))^C}{f_t(1)f_t(R)}\,dt.$$

Thus if $M_t := (f_t(R)^A f_t'(R)^B (f_t(R) - f_t(1))^C$, for $t \leq \tau_1$ we have:

$$dM_t = (f_t(R) - f_t(1))^C d(f_t(R)^A f_t'(R)^B)$$
$$+ f_t(R)^A f_t'(R)^B (f_t(R) - f_t(1))^C \frac{2C}{f_t(1)f_t(R)}\,dt$$
$$= f_t(R)^A (f_t(R) - f_t(1))^C d(f_t'(R)^B)$$
$$+ f_t'(R)^B (f_t(R) - f_t(1))^C d(f_t(R)^A)$$
$$+ f_t(R)^A f_t'(R)^B (f_t(R) - f_t(1))^C \frac{2C}{f_t(1)f_t(R)}\,dt$$
$$= f_t(R)^{A-2} f_t'(R)^B (f_t(R) - f_t(1))^C \left(-2A + 2B + \frac{A(A-1)\kappa}{2}\right)dt$$
$$+ \frac{f_t(R)^A f_t'(R)^B (f_t(R) - f_t(1))^C}{f_t(1)f_t(R)}(\kappa A + 2C)\,dt - \frac{\sqrt{\kappa}AM_t}{f_t(R)}\,dB_t.$$

Hence $M_{\cdot \wedge \tau_1}$ is a local martingale whenever $\kappa A + 2C = 0$ and $-2A + 2B + A(A-1)\kappa/2 = 0$ which is indeed the case when $A = -1$, $B = -1 - \kappa/2$ and $C = \kappa/2$.

In this case, we have

$$M_{t \wedge \tau_1} = 1 + \sqrt{\kappa}\int_0^{t \wedge \tau_1} \frac{M_s}{f_s(R)}\,dB_s = \exp\left(Z_{t \wedge \tau_1} - \frac{1}{2}[Z]_{t \wedge \tau_1}\right)$$

with

$$Z_{t \wedge \tau_1} = \sqrt{\kappa}\int_0^{t \wedge \tau_1} f_s(R)^{-1}\,dB_s,$$

where because $\frac{d}{dt}(f_t(R) - f_t(1)) = \frac{2}{f_t(1)} - \frac{2}{f_t(R)} > 0$, we have $f_t(R) > (R-1)$ for $t < \tau_1$. This means that the quadratic variation is deterministically bounded above on any compact interval of time, so we can apply Novikov's condition to see that $M_{t \wedge \tau_1}$ is in fact a true martingale.

Furthermore, Girsanov's theorem tells us that if we define

$$\left.\frac{d\tilde{\mathbb{P}}}{d\mathbb{P}}\right|_{\mathcal{F}_{\tau_1}} := M_{\tau_1},$$

then the process $\tilde{B}_t = B_t - \int_0^t (\sqrt{\kappa}/f_s(R))\,ds$ will be a Brownian motion under $\tilde{\mathbb{P}}$. Rewriting the expression for $d\xi_t$ in terms of $\tilde{B}$ we get

$$d\xi_t = \sqrt{\kappa}\,d\tilde{B}_t - \frac{\kappa}{f_t(1)}\,dt + \frac{\kappa}{f_t(R)}\,dt$$

B.1 Definitions

for $t \leq \tau_1$, where $\tilde{B}$ is a standard Brownian motion under $\tilde{\mathbb{P}}$. Hence, under $\tilde{\mathbb{P}}$, the law of $\xi_{\cdot \wedge \tau_1}$ is that of the driving function of $SLE_\kappa(\kappa, -\kappa)$ with force points at $(1, R)$. Equivalently, the Radon–Nikodym derivative between the laws of $(\tilde{f}_t)_{t \leq \tilde{\tau}_1}$ and $(f_t)_{t \leq \tau_1}$ (that is, reverse $SLE_\kappa(\kappa, -\kappa)$ and $SLE_\kappa(\kappa)$) up to time τ_1 is equal to M_{τ_1}.

To prove that the total variation distance between the laws of $(\tilde{f}_t)_{t \leq \tilde{\tau}_1}$ and $(f_t)_{t \leq \tau_1}$ converges to 0 as $R \to \infty$, we first claim it suffices to prove that

$$\mathbb{E}(|M_{T \wedge \tau_1} - 1|^2) = \mathbb{E}([M]_{T \wedge \tau_1}) \to 0 \tag{B.3}$$

as $R \to \infty$ for any fixed T. Indeed, suppose (B.3) holds and let $\varepsilon > 0$ be given. We first choose T large enough that $\mathbb{P}(\tau_1 > T) < \varepsilon$. For this T, (B.3) implies that the total variation distance between $(\tilde{f}_t)_{t \leq \tilde{\tau}_1 \wedge T}$ and $(f_t)_{t \leq \tau_1 \wedge T}$ converges to 0 as $R \to \infty$. Therefore, for all R large enough, we can couple the two processes so that they agree up to time $T \wedge \tau_1$ with probability greater than $1 - \varepsilon$. But this implies that for all R large enough, they can be coupled so that they agree up to time τ_1 with probability $1 - 3\varepsilon$. Since $\varepsilon > 0$ was arbitrary, this proves the desired convergence in total variation distance.

To see (B.3), recall that $f_t(R) > R - 1$ for $t \leq \tau_1$. In addition, since $\frac{d}{dt} f'_t(R) = \frac{2 f'_t(R)}{f_t(R)^2} > 0$ we have $f'_t(R) > 1$ for all $t \geq 0$. Thus

$$\frac{M_t}{f_t(R)} \leq R(R-1)^{-\kappa/2} f_t(R)^{\kappa/2 - 2} < \frac{R}{(R-1)^2} < \frac{2}{R}$$

for all $t \leq \tau_1$ and hence, since $[M]_{T \wedge \tau_1} = \kappa \int_0^{T \wedge \tau_1} (M_t / f_t(R))^2 \, dt$,

$$[M]_{T \wedge \tau_1} \leq \frac{4\kappa (T \wedge \tau_1)}{R^2}$$

for any $T \geq 0$. Taking expectations and using the deterministic upper bound $4\kappa T / R^2$, which converges to 0 as $R \to \infty$, proves (B.3).

(3) For this, we claim that for $w \in \mathbb{H}$ with $\Re(w) > 2$ and $z \in [1, 2]$, the process $\Re(\tilde{f}_t(w)) - \tilde{f}_t(z)$ is increasing for $t \leq \tilde{\tau}_a$ (which clearly implies the result). To see the claim, observe that by definition of the reverse flow

$$\frac{\partial (\Re(\tilde{f}_t(w)) - \tilde{f}_t(z))}{\partial t} = \frac{2}{\tilde{f}_t(z)} - \Re\left(\frac{2}{\tilde{f}_t(w)}\right) = \frac{2}{\tilde{f}_t(z)} - \frac{2\Re(\tilde{f}_t(w))}{|\tilde{f}_t(w)|^2},$$

which is positive as long as $\Re(\tilde{f}_t(w)) > \tilde{f}_t(z) > 0$. Since this is true at time 0 for w with $\Re(w) > 2$, it is therefore positive for all $t \leq \tilde{\tau}_a$, and the process $\Re(\tilde{f}_t(w)) - \tilde{f}_t(z)$ is increasing for this range of t. □

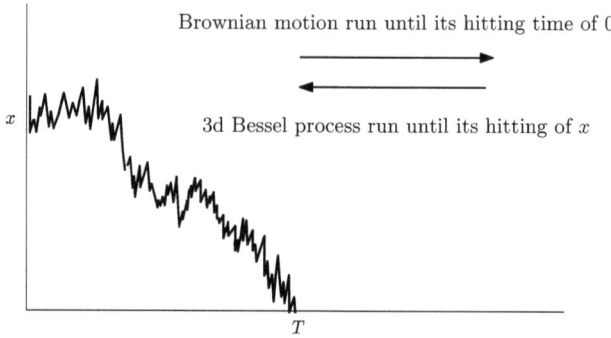

Figure B.3 Illustration of Williams' path decomposition theorem. The classical result says that if X is a Brownian motion started from $x > 0$ and T is its hitting time of zero, then its time-reversal $\hat{X} = (X_{T-t})_{0 \leq t \leq T}$ is distributed as three-dimensional Bessel process, run until its last visit Λ to x.

B.1.1 Symmetries in law for forward/reverse SLE_κ and $\text{SLE}_\kappa(\rho)$

Now, because Brownian motion has time reversal symmetry, the relationship Lemma B.2 between forward and reverse Loewner evolutions has particularly nice consequences for SLE.

More specifically, if $T > 0$ is fixed and $(\xi_t)_{0 \leq t \leq T}$ is $\sqrt{\kappa}$ times a Brownian motion, then $(\hat{\xi}_t)_{0 \leq t \leq T} = (\xi_T - \xi_{T-t})_{0 \leq t \leq T}$ also has the law of $\sqrt{\kappa}$ times a Brownian motion. Consequently:

Lemma B.7 *For any fixed $T > 0$, the curve generated by a reverse SLE_κ run up to time T and the curve generated by a forward SLE_κ run up to time T are equal in law.*

Mind that the *processes* of the previous lemma, defined for all times $t \in [0, T]$, are *not* the same in law. Indeed, we have seen that forward and reverse Loewner evolutions generate hulls via a completely different dynamic. Nonetheless, it is a very useful property that at any fixed time, the laws of the generated hulls are equal.

There are similar consequences for $\text{SLE}_\kappa(\rho)$ processes, but the reversibility properties of solutions to (A.3) are somewhat more complicated. We will explain now what happens in the simplest case of one marked point. Due to Remarks A.6 and B.5, this requires understanding how Bessel processes behave under time reversal.

Remark B.8 (Bessel process properties) Recall that the dimension δ of a Bessel process determines how often it returns to 0: if $\delta \geq 2$, then the Bessel

B.1 Definitions 389

process will almost surely be strictly positive for all positive times; while if $\delta < 2$ then from any starting point it will return to 0 in finite time almost surely.

The following is an extension of a classical result about Brownian motion, due to Williams (see for example Corollary (4.6) in Chapter VII of [RY99] and Figure B.3).

Lemma B.9 (Time reversal of Bessel processes) *Suppose that X is a Bessel process of dimension $\delta \in (0, 2)$ started from $x > 0$, run until its first hitting time T of zero. Then its time reversal $\hat{X} = (X_{T-t}, 0 \le t \le T)$ is a Bessel process of dimension $\hat{\delta} = 4 - \delta \in (2, 4)$, run until its last visit Λ to x.*

The proof of this will boil down to an analogous result for Brownian motion with drift, that we state and prove first.

Lemma B.10 *Let $\mu > 0$. Then the time reversal of a Brownian motion with drift μ, started from 0 and stopped at its last hitting time of $y > 0$, has the law of a Brownian motion with drift $-\mu$, started from y and run up to its last hitting time of 0.*

Proof Let $(X_t)_{t \in \mathbb{R}} = (B_t + \mu t)_{t \in \mathbb{R}}$, where B_t is a standard two-sided Brownian motion with $B_0 = 0$. Then $(\hat{X}_t)_{t \in \mathbb{R}} := (X_{-t})_{t \in \mathbb{R}}$ is equal in law to $(B_t - \mu t)_{t \in \mathbb{R}}$. Define $\tau_0 := \{\inf: s \le 0: X_s = 0\}$ and $\tau_y = \sup\{s \ge 0: X_t = y\}$. Then by the strong Markov property at time τ_0, $(X_{\tau_0+s})_{0 \le s \le \tau_y - \tau_0}$ has the law of a Brownian motion with drift μ, started from 0 and stopped at its last hitting time of y. So, we need to show that the time reversal $(X_{\tau_y-s})_{0 \le s \le \tau_y - \tau_0}$ has the law of a Brownian motion with drift $-\mu$, started from y and run up to its last hitting time of 0.

For this, we use the fact that, by definition of $\hat{X}$,

$$(X_{\tau_y-s})_{0 \le s \le \tau_y - \tau_0} = (\hat{X}_{s-\tau_y})_{0 \le s \le \tau_y - \tau_0} = (\hat{X}_{s+\hat{\tau}_y})_{0 \le s \le \hat{\tau}_0},$$

where $\hat{\tau}_y$ is the first time before 0 that $\hat{X}$ hits y, and $\hat{\tau}_0$ is the last time that $(\hat{X}_{t+\hat{\tau}_y})_{t \ge 0}$ hits 0. Since $\hat{X}$ is equal in law to a two-sided Brownian motion with drift $-\mu$, the law of the process on the right-hand side above is (by the strong Markov property again, but this time for $\hat{X}$) indeed that of a Brownian motion with drift $-\mu$, started from y and run up to its last hitting time of 0. This concludes the proof. □

Proof of Lemma B.9 ([DMS21, Proposition 3.5]) We will make use of the following fact, which is just a rewriting of Lemma 7.20 and Remark 7.21:

- Let $\tau(t) = \inf\{s > 0: [\log(X)]_t > t\}$ and let $Z_t = \log(X_{\tau(t)})$ (recall that

$[M]_t$ denotes the quadratic variation of the continuous semimartingale M). Note that because $\delta \in (0, 2)$, $\tau(t) \uparrow T$ as $t \uparrow \infty$. Then

$$(Z_t)_{t \geq 0} \stackrel{(\text{law})}{=} \left(B_t + \frac{\delta - 2}{2} t \right)_{t \geq 0}, \tag{B.4}$$

where B is a standard Brownian motion with $B_0 = \log x$.

We now want to use this, along with the time reversal symmetry of Brownian motion, to draw a conclusion similar to (B.4) about the time reversal $\hat{X}$ of X, but with the opposite drift (corresponding to a dimension $\hat{\delta} = 4 - \delta$, as claimed). However, there is a slight technical complication that arises, since $\hat{X}_0 = 0$ and so $\log(\hat{X}_0) = -\infty$.

To get around this, we also define for any $\varepsilon < x$, T_ε to be the last time before T that $(X_t)_{t \geq 0}$ hits ε. Then $(Z_t)_{t \in [0, [\log X]_{T_\varepsilon}]}$ is a Brownian motion with drift as in (B.4), started from $\log x$ and stopped at its last hitting time of $\log(\varepsilon)$. This implies (by Lemma B.10) that the time reversal of Z with respect to this time interval is a Brownian motion with drift $-(\delta - 2)/2 = (\hat{\delta} - 2)/2$, started from $\log \varepsilon$ and run up to its last hitting time of $\log x$.

Reversing the argument for (B.4) (i.e., taking the exponential and reparametrising by quadratic variation), this implies that the time reversal of $(X_t)_{t \in [0, T_\varepsilon]}$ is a Bessel process of dimension $(4 - \delta)$, started from ε and run up to its last hitting time of x. Taking a limit as $\varepsilon \to 0$ provides the result. $\square$

As a consequence of this and Remarks A.6 and B.5, we obtain the following:

Corollary B.11 (Symmetries for forward and reverse $SLE_\kappa(\rho)$) *Suppose that $(f_t)_{t \geq 0}$ is the reverse flow for a centred, reverse $SLE_\kappa(\rho)$ process with a single force point at $x > 0$ of weight $\rho < \kappa/2 + 2$. Consider the first time τ that $f_\tau(x) = 0$. Then $H_\tau = f_\tau(\mathbb{H})$ has the same law as $\mathbb{H} \setminus \eta([0, \sigma])$, where η is a forward $SLE_\kappa(\kappa - \rho)$ curve with a force point at 0^+, run until the last time Λ that the centered forward Loewner flow for η sends 0^+ to x.*

Appendix C
Radial Loewner Chains and Radial SLE

C.1 Radial Loewner chains

While chordal Loewner chains describe "locally growing" sets started at one point on the boundary of a domain and targeted at another, *radial* Loewner chains describe growing sets started on the boundary but targeted at a point in the *interior* of the domain. The canonical configuration for chordal Loewner chains is the upper half plane $\mathbb{H}$, with starting point $0 \in \partial \mathbb{H}$ and target point ∞. For radial Loewner chains, things turn out to be nicest if one works in the unit disc $\mathbb{D} \subset \mathbb{C}$ with starting point $1 \in \partial \mathbb{D}$ and target point $0 \in \mathbb{D}$.

Definition C.1 (Radial Loewner chain) Let $(U_t)_{t \geq 0}$ be a continuous process taking values in the unit circle $\partial \mathbb{D}$, with $U_0 = 1$. The *radial Loewner chain driven by* U is the collection of maps $(g_t)_{t \geq 0}$ that solve the *radial Loewner equation*:

$$\frac{\partial g_t}{\partial t}(z) = g_t(z) \frac{U_t + g_t(z)}{U_t - g_t(z)}; \quad g_0(z) = z, \qquad (C.1)$$

for each $z \in \mathbb{D}$ until time $\zeta(z) := \inf_{t>0} g_t(z) = U_t$. If one defines $D_t := \{z \in \mathbb{D} : \zeta(z) > t\}$ for each $t \geq 0$, then g_t is the unique conformal isomorphism

$$g_t : D_t \to \mathbb{D} \text{ with } g'_t(0) = e^t \text{ and } g_t(0) = 0,$$

[SW05]. The hulls generated by U, $K_t := \mathbb{D} \setminus D_t$ for $t \geq 0$, are an increasing family of compact sets in $\mathbb{D}$. With a slight abuse of notation, we will sometimes also refer to $(D_t)_{t \geq 0}$ or $(K_t)_{t \geq 0}$ as the Loewner chain driven by U.

As in the chordal case, continuous non-crossing curves $(\gamma(t))_{t \geq 0}$ in $\overline{\mathbb{D}}$, with $\gamma(0) = 1$, and parametrised so that $-\log \mathrm{CR}(\mathbb{D} \setminus \gamma([0,t]); 0) = t$ for $t \geq 0$, provide examples of radial Loewner chains. That is, when g_t is defined for each t to be the unique conformal isomorphism from $\mathbb{D} \setminus \gamma([0,t])$ fixing 0 and with positive real derivative at 0.

C.2 Radial SLE_κ and $\text{SLE}_\kappa(\rho)$

Definition C.2 (Radial SLE_κ) For $\kappa \geq 0$, radial SLE_κ in $\mathbb{D}$ from 1 to 0 is defined to be the radial Loewner chain driven by

$$(e^{i\sqrt{\kappa}B_t})_{t \geq 0},$$

where B is a standard one-dimensional Brownian motion.

For a general simply connected domain D with a marked boundary point $a \in \partial D$ and interior point $b \in D$, we define the radial SLE_κ in D from a to b to be the random process obtained by taking the image of a radial SLE_κ in $\mathbb{D}$ from 1 to 0 under the unique conformal isomorphism from $\mathbb{D}$ to D sending $1 \mapsto a$ and $0 \mapsto b$.

Radial $\text{SLE}_\kappa(\underline{\rho})$. As with chordal SLE, we can generalise the definition of radial SLE_κ by placing force points on the boundary or in the interior of the domain and keeping track of their evolution under the radial Loewner flow. The definition in the unit disc from 1 to 0 is as follows.

Definition C.3 ($\text{SLE}_\kappa(\rho^1, \ldots, \rho^m)$ in $\mathbb{D}$ from 1 to 0) Let $v^1, \ldots, v^m \in \overline{\mathbb{D}} \setminus \{1\}$ be distinct and $\rho^1, \ldots, \rho^m$ satisfy the condition (A.2). Radial $\text{SLE}_\kappa(\rho^1, \ldots, \rho^m)$ from 1 to 0 in $\mathbb{D}$ with force points at $v^1, \ldots, v^m$ is the radial Loewner chain, whose driving function $(U_t)_{t \geq 0}$ satisfies:

$$U_t = 1 + i\sqrt{\kappa} \int_0^t U_s \, dB_s - \int_0^t \frac{\kappa}{2} U_s \, ds + \sum_i \int_0^t \frac{\rho^i}{2} \hat{\Phi}(V_s^i, U_s) \, ds$$
$$V_t^i = v^i + \int_0^t \Phi(U_s, V_s) \, dt \text{ for } 1 \leq i \leq M.$$
(C.2)

Above we denote $\Phi(u, z) = z\frac{u+z}{u-z}$ and $\hat{\Phi}(u, z) = \frac{\Phi(u,z) + \Phi(1/\overline{u}, z)}{2}$ for $z \in \mathbb{D}$ and $u \in \partial \mathbb{D}$.

When $\underline{\rho}$ satisfies (A.2), the existence and uniqueness of a continuous $(U, V^1, \ldots, V^m)$ satisfying (A.3) up to the first time that $V_t^i = U_t$ for some i with $v^i \in \mathbb{D}$ is proven in [MS16a]. In particular, there is a unique solution for all time when all of the force points v^i are on the boundary $\partial \mathbb{D}$.

As with ordinary radial SLE_κ, we define radial $\text{SLE}_\kappa(\underline{\rho})$ in a domain D from $a \in \partial D$ to $b \in D$, with force points $v^1, \ldots, v^m \in \overline{\mathbb{D}}$, to be the image of $\text{SLE}_\kappa(\underline{\rho})$ in $\mathbb{D}$ from 1 to 0 with force points at $\varphi(v^1), \ldots, \varphi(v^m)$, where φ is the unique conformal isomorphism from D to $\mathbb{D}$ sending a to 1 and b to 0.

Remark C.4 Again we can extend the definition of $\text{SLE}_\kappa(\underline{\rho})$ to include force

C.2 Radial SLE_κ and $\text{SLE}_\kappa(\underline{\rho})$

points located infinitesimally clockwise (respectively anticlockwise) from 1 on $\partial \mathbb{D}$, by taking a limit (in the same way as for chordal $\text{SLE}_\kappa(\underline{\rho})$).

Radial $\text{SLE}_\kappa(\underline{\rho})$ satisfies a very similar collection of properties to chordal $\text{SLE}_\kappa(\underline{\rho})$. Indeed, there is a simple connection between the radial and chordal variants that can be verified using a careful stochastic calculus argument (omitted here).

Lemma C.5 *[SW05] Let D be a simply connected domain, $a, b \in \partial D$ be boundary points, and $c \in D$ be an interior point. Let $\rho^1, \ldots, \rho^m \in \mathbb{R}$ with $\sum_i \rho_i = \kappa - 6$ satisfy (A.2), and let $v^1, \ldots, v^m \in \overline{D}$.*

Suppose that η is a radial $\text{SLE}_\kappa(\underline{\rho})$ from $a \in \partial D$ to $b \in D$ (with force points at $v^1, \ldots, v^m$) stopped at the infimum over t for which c and b are in different connected components of $D \setminus \eta([0, t])$, or an interior force point is swallowed. Let $\tilde{\eta}$ be a chordal $\text{SLE}_\kappa(\underline{\rho})$ from $a \in \partial D$ to $c \in \partial D$ (with force points at $v^1, \ldots, v^m$), stopped at the corresponding time. Then, as curves modulo reparametrisation of time, η and $\tilde{\eta}$ agree in law.

Remark C.6 As already observed, adding a force point of any weight to the target point of a chordal or radial $\text{SLE}_\kappa(\underline{\rho})$ does not effect the law of the curve. So if we start with a given chordal or radial $\text{SLE}_\kappa(\underline{\rho})$, we can add such a force point so that the new weights add up to $\kappa - 6$.

For example, if we want to sample a radial SLE_κ from $a \in \partial D$ to $b \in D$, then we can first run a chordal $\text{SLE}_\kappa(\kappa - 6)$, η_1, in $D =: D_1$ from a to some arbitrary $c_1 \in \partial D$, with the force point at b, up until the first time τ_1 that c_1 and b are separated by η_1. Then, we can run an $\text{SLE}_\kappa(\kappa - 6)$, η_2, in the connected component D_2 of $D \setminus \eta_1([0, \tau_1])$ containing b, from $\eta(\tau_1)$ to some other $c_2 \in \partial D_2$ and with force point at b, and again stop it when η_2 first separates c_2 and b. Iterating this procedure, and reparametrising the concatenated curve so that the conformal radius of b in the to be explored domain is always e^{-t}, we obtain a curve with the law of radial SLE_κ from a to b.

Similar procedures will work to generate radial $\text{SLE}_\kappa(\underline{\rho})$ with non-trivial $\underline{\rho}$. In particular, the following properties hold.

Radial $\text{SLE}_\kappa(\underline{\rho})$: properties.

- Suppose that D is a simply connected domain, $v^1, \ldots, v^m \in \partial D$, and $\underline{\rho}$ satisfies (A.2). Then radial $\text{SLE}_\kappa(\underline{\rho})$ in D from a to b is almost surely generated by a curve γ (that is, there exists a curve $\gamma(t)$ defined for all time such that the connected component of $\mathbb{D} \setminus \gamma([0, t])$ containing 0 is equal to $D_t = \{z \in \mathbb{D} : \tau_z > t\}$ for all t. We will also sometimes refer to the curve γ as "the radial SLE_κ". When there are interior force points, the radial Loewner chain is generated by a continuous curve, until the first time that one of the interior force points is swallowed. Lawler proved in [Law13] that if γ is a radial SLE_κ with target point b, then $\lim_{t\to\infty} \gamma(t) = b$ almost surely. This was extended to the case of $\text{SLE}_\kappa(\underline{\rho})$ with boundary force points and $\underline{\rho}$ satisfying (A.2) in [MS17].

- *Conformal invariance:* follows from the definition of radial $\text{SLE}_\kappa(\underline{\rho})$ in $D \neq \mathbb{D}$ from $a \in \partial D$ to $b \in D$ (see above).
- *Domain Markov property:* suppose that D is a simply connected domain, and γ is a radial $\text{SLE}_\kappa(\underline{\rho})$ from $a \in \partial D$ to $b \in D$, with weights $\underline{\rho} = (\rho^1, \ldots, \rho^m)$ satisfying (A.2) and force points $v^1, \cdots, v^m \in \partial D$. Suppose that T is a bounded stopping time that is measurable with respect to γ. Then, conditionally on $\gamma([0, T])$ and writing D_T for the connected component containing b of $D \setminus \gamma([0, T])$, $\gamma([T, \infty))$ has the law of an $\text{SLE}_\kappa(\underline{\rho})$ from $\gamma(T)$ to b in D_T, with force points at $(V_T^1, \ldots, V_T^m)$.

Target invariance. Finally, we consider the special case when η is a radial $\text{SLE}_\kappa(\kappa - 6)$ from $a \in \partial D$ to $b \in D$ for some domain D, and with force point $c \in \partial D$. Suppose that $\kappa \geq 4$ so that (A.2) holds with $m = 1$ and $\rho^1 = \kappa - 6$. Lemma C.5 then implies that η (which is defined for all time) can be sampled as follows.

- Choose x_1 on ∂D and run a chordal $\text{SLE}_\kappa(\kappa - 6)$, (with force point at c) in D from a to x_1, stopped at the first time that x_1 and b lie in separate connected components of complement of the curve. Reparametrise this curve so that the conformal radius of b in its complement is equal to e^{-t} for all t up to the time that the curve is stopped.
- Repeat the first step with the new domain being the connected component of the complement of the first curve containing b, the new start point being the tip of the curve at the disconnection time, and the new force point being the image of c under the radial Loewner flow generated by the first curve.[1]
- Iterate the above procedure.

In particular, note that the only dependence on b in the above is the choice of exploration domain at "disconnection times". This means that if b' is another point in D, the above procedure (run until b and b' are first separated by the curve) also produces a sample of radial $\text{SLE}_\kappa(\kappa - 6)$ (with the same force point) from $a \in \partial D$ to b' (and stopped when b and b' are first separated).

[1] This is well defined after conformally mapping D to $\mathbb{D}$, a to 1 and b to 0.

C.2 Radial SLE$_\kappa$ and SLE$_\kappa(\rho)$

More precisely, and also applying the Markov property of radial SLE after this separation time, we have the following.

Lemma C.7 (Target invariance of SLE$_\kappa(\kappa-6)$) *Suppose that $\kappa > 4$ and let D be a simply connected domain with $a, c \in \partial D$. For $b_1, b_2 \in D$, one can couple a radial SLE$_\kappa(\kappa-6)$ curve from a to b_1 in D (with force point at c), and from a to b_2 in D (with force point at c) so that they coincide until b_1, b_2 are contained in separate components of the complement of the curve, and afterwards evolve independently.*

In fact, the above lemma means that for given D, a, c, SLE$_\kappa(\kappa - 6)$ can be simultaneously defined towards a countable dense set of target points in D, in such a way that the above description holds for any two given target points. The object created in this manner is referred to as an SLE$_\kappa(\kappa - 6)$ branching tree, or sometimes just a branching SLE$_\kappa$.

Appendix D
Convergence of Random Variables in the Space of Distributions

In this appendix, we prove Lemma 1.34, about the measurability of the convergence event for a sequence of random variables in the space of distributions $\mathcal{D}'_0(D)$, which we restate here for convenience:

Lemma D.1 *Let D be a domain of $\mathbb{R}^d$. Let* Conv *denote the set of sequences in $\mathcal{D}'_0(D)$ which are weak-$*$ convergent. Then* Conv *is a Borel set in $\mathcal{D}'_0(D)^{\mathbb{N}}$ equipped with the product Borel σ-algebra.*

Proof The proof relies on some results in functional analysis, and in particular uses the Schwartz space $\mathcal{S}^*$ of **tempered distributions**, whose definition is as follows. Let $\mathcal{S}$ denote the space of rapidly decaying test functions

$$\mathcal{S} = \{f \in C^{\infty}(\mathbb{R}^d) : \|f\|_j < \infty, \text{ for all } j \geq 1\},$$

where

$$\|f\|_j := \sup_{|\alpha|+|\beta| \leq j} \|x^{\alpha} \partial^{\beta} f\|_{\infty}$$

and we use standard multi index notation for $\alpha = (\alpha_1, \ldots, \alpha_d)$ and $\beta = (\beta_1, \ldots, \beta_d)$. We equip $\mathcal{S}$ with a topology defined by the requirement $f_n \in \mathcal{S}$ converges to $f \in \mathcal{S}$ if and only if $\|f_n - f\|_j \to 0$ for every $j \geq 1$. Thus, the quantities $\|f\|_j$ define seminorms on $\mathcal{S}$ (actually, norms) which together (by definition) generate the topology on $\mathcal{S}$. We define $\mathcal{S}^*$ to be the space of continuous linear functionals on $\mathcal{S}$, equipped once again with the weak-$*$ topology of pointwise convergence: that is, a sequence $x_n^* \in \mathcal{S}^*$ converges to $x^* \in \mathcal{S}^*$ if and only if $x_n^*(x) \to x^*(x)$ for all $x \in \mathcal{S}$.

The advantage of the space $\mathcal{S}$ over $\mathcal{D}_0(D)$ is that it is a (separable) **Fréchet space**, that is, a topological vector space which is locally convex, metrisable and complete. Equivalently, a Fréchet space is one for which there is a countable family of seminorms generating the topology (which is clearly the case for $\mathcal{S}$),

Convergence of Random Variables in the Space of Distributions 397

and which is complete and Hausdorff (which is also straightforward to check in the case of $\mathcal{S}$). The separability of $\mathcal{S}$ is also a standard fact. We will first prove the lemma with $\mathcal{D}'_0(D)$ replaced by $\mathcal{S}^*$, and then explain how to go from $\mathcal{S}^*$ to $\mathcal{D}'_0(D)$. For $\mathcal{S}^*$, the statement boils down to a general fact about Fréchet spaces, which we now introduce.

Fix X a separable Fréchet space and fix a dense countable set $Q = \{x_i\}_{i \geq 1}$ in X. Let X^* be the set of continuous linear functionals on X. Let $\mathrm{Conv}(X^*)$ be the set of converging sequences in X^*: that is,

$$\mathrm{Conv}(X^*) = \{(x_n^*)_{n \geq 1} \in (X^*)^{\mathbb{N}} : x_n^* \to x^* \text{ weak-}^* \text{ for some } x^* \in X^*\}.$$

Let

$$V_j = \{x \in X : \|x\|_k \leq 1/j \text{ for all } 1 \leq k \leq j\},$$

where $(\|\cdot\|_j)_{j \geq 1}$ is a countable family of seminorms generating the topology of X. Then V_j is what is called a countable basis of neighbourhoods of 0: that is, for any neighbourhood V of 0 there exists $j \geq 1$ such that $V \supset V_j$.

Let $K_j = V_j^{\bullet}$ denote the **polar set** of V_j; that is,

$$V_j^{\bullet} = \{x^* \in X^* : |x^*(x)| \leq 1 \text{ for all } x \in V_j\}.$$

Claim

$$\mathrm{Conv}(X^*) = \bigcup_{j \geq 1} \{(x_n^*)_{n \geq 1} \in K_j^{\mathbb{N}}, \text{ such that } x_n^*(x) \text{ converges in } \mathbb{R} \text{ for all } x \in Q\}. \quad (\mathrm{D.1})$$

Proof of (D.1) We have two inclusions to prove. We start by showing that the right-hand side of (D.1) is contained in $\mathrm{Conv}(X^*)$. Let $(x_n^*)_{n \geq 1}$ denote a sequence in X^* and suppose that there is some $j \geq 1$ such that $x_n^* \in K_j$ for all $n \geq 1$, and that $x_n^*(x)$ converges to a limit $\ell(x) \in \mathbb{R}$ for all $x \in Q$. By the Alaoglu theorem, see for example [NB11, Theorem 8.4.1], K_j is compact and furthermore metrisable (as a bounded subset of the dual of a separable space), hence sequentially compact. Let x^* be any weak-* subsequential limit. Then $x^*(x) = \ell(x)$ for all $x \in Q$. Since Q is dense and x^* is continuous, this identifies x^* uniquely. Thus x_n^* converges to x^* in the weak-* sense. Thus $(x_n^*)_{n \geq 1} \in \mathrm{Conv}(X^*)$.

Conversely, suppose $(x_n^*)_{n \geq 1} \in \mathrm{Conv}(X^*)$. Clearly for $x \in Q$, $\langle x_n^*, x \rangle$ converges in $\mathbb{R}$ by definition of the weak-* topology. Therefore it suffices to show that there exists $j \geq 1$ such that $x_n^* \in K_j$ for all $n \geq 1$. We rely on the Banach–Steinhaus theorem in Fréchet spaces [NB11, Theorem 11.9.1], which states that if $(x_n)_{n \geq 1}^*$ is pointwise bounded (that is, if $\sup_{n \geq 1} |x_n^*(x)| < \infty$ for any $x \in X$, which is the case since $x_n^*(x)$ converges in $\mathbb{R}$) then x_n^* is "bounded in the operator norm" (more precisely, equicontinuous): that is, there is some neighbourhood V of 0 such that $x_n^* \in V^\bullet$, the polar set of V. Since $(V_j)_{j \geq 1}$ is basis of neighbourhoods, we can find $j \geq 1$ such that $V_j \subset V$ and thus $V^\bullet \subset K_j$. This concludes the proof of (D.1). □

An immediate consequence of (D.1) is that $\mathrm{Conv}(X^*)$ is a Borel set in $(X^*)^\mathbb{N}$ equipped with the power Borel σ-field. As already mentioned, this applies in particular to the case $X = \mathcal{S}$, $X^* = \mathcal{S}^*$.

To conclude the proof of Lemma 1.34, it remains to reduce the convergence in the sense of distributions to convergence in the Schwartz space $\mathcal{S}^*$ as follows. Let

$$D_k = \{x \in D : \mathrm{dist}(x, \partial D) \geq 1/k\} \cap B(0, k).$$

Fix $(\varphi_k)_{k \geq 1}$ a sequence of test functions with compact support in D such that:

- $\varphi_k \geq 0$,
- $\mathrm{Supp}(\varphi_k) \subset D_{2k+1}$,
- $\varphi_k \equiv 1$ on D_{2k}.

For a distribution $T \in \mathcal{D}'_0(D)$ and $k \geq 1$, let $T\varphi_k$ denote the distribution obtained by setting

$$(T\varphi_k, f) = (T, f\varphi_k).$$

Then note that, given a sequence $(T_n)_{n \geq 1} \in (\mathcal{D}'_0(D))^\mathbb{N}$, we have

$$(T_n)_{n \geq 1} \in \mathrm{Conv} \text{ if and only if } (T_n\varphi_k)_{n \geq 1} \in \mathrm{Conv}, \text{ for all } k \geq 1. \quad (\mathrm{D.2})$$

Indeed, given a test function f with compact support in D, it is always possible to find a $k \geq 1$ such that $\mathrm{Supp}(f) \subset D_{2k}$. On the other hand, for a fixed $k \geq 1$,

$$(T_n\varphi_k)_{n \geq 1} \in \mathrm{Conv} \text{ if and only if } (T_n\varphi_k)_{n \geq 1} \in \mathrm{Conv}(\mathcal{S}^*). \quad (\mathrm{D.3})$$

One implication in (D.3) is trivial: if $(T_n\varphi_k)_{n \geq 1} \in \mathrm{Conv}(\mathcal{S}^*)$ and f is a test function with compact support in D, then clearly $f \in \mathcal{S}$, so $(T_n\varphi_k, f)$ converges as desired. Conversely, if $(T_n\varphi_k)_{n \geq 1} \in \mathrm{Conv}$ and $f \in \mathcal{S}$, then

$$(T_n\varphi_k, f) = (T_n\varphi_k, f\varphi_{k+1}),$$

since $D_{2k+1} \subset D_{2k+3}$. The test function on the right-hand side now has compact support, so the left-hand side converges as desired. Combining together (D.2) and (D.3), we deduce

$$\mathrm{Conv} = \cap_{k \geq 1}\{(T_n)_{n \geq 1} \in \mathcal{D}'_0(\mathbb{D})^\mathbb{N} : (T_n\varphi_k)_{n \geq 1} \in \mathrm{Conv}(\mathcal{S}^*)\}$$

and thus Conv is a Borel set of the product σ-algebra by (D.1), as desired. □

References

[AB22] Yoshihiro Abe and Marek Biskup. Exceptional points of two-dimensional random walks at multiples of the cover time. *Probab. Theory Relat. Fields*, 183(1–2):1–55, 2022.

[ABBS13] Elie Aïdékon, Julien Berestycki, Éric Brunet, and Zhan Shi. Branching Brownian motion seen from its tip. *Probab. Theory Relat. Fields*, 157(1–2):405–451, 2013.

[ABJL23] Élie Aïdékon, Nathanaël Berestycki, Antoine Jego, and Titus Lupu. Multiplicative chaos of the Brownian loop soup. *Proc. Lond. Math. Soc. (3)*, 126(4):1254–1393, 2023.

[ABK13] Louis-Pierre Arguin, Anton Bovier, and Nicola Kistler. The extremal process of branching Brownian motion. *Probab. Theory Relat. Fields*, 157(3–4):535–574, 2013.

[AC15] Omer Angel and Nicolas Curien. Percolation on random maps I: Half-plane models. *Ann. Inst. Henri Poincaré Probab. Stat.*, 51(2):405–431, 2015.

[AF03] Robert A. Adams and John J. F. Fournier. *Sobolev spaces*, Vol. 140 of *Pure and Applied Mathematics (Amsterdam)*. Elsevier/Academic Press, Amsterdam, 2nd ed., 2003.

[AHS17] Juhan Aru, Yichao Huang, and Xin Sun. Two perspectives of the 2D unit area quantum sphere and their equivalence. *Comm. Math. Phys.*, 356(1):261–283, 2017.

[AHS20] Elie Aïdékon, Yueyun Hu, and Zhan Shi. Points of infinite multiplicity of planar Brownian motion: Measures and local times. *Ann. Probab.*, 48(4):1785–1825, 2020.

[AHS24] Morris Ang, Nina Holden, and Xin Sun. Integrability of SLE via conformal welding of quantum surfaces. *Commun. Pure Appl. Math.*, to appear, 2024.

[Ald90] David J. Aldous. The random walk construction of uniform spanning trees and uniform labelled trees. *SIAM J. Discrete Math.*, 3(4):450–465, 1990.

[Ald93] David J. Aldous. The continuum random tree. III. *Ann. Probab.*, 21(1):248–289, 1993.

[Ang03] Omer Angel. Growth and percolation on the uniform infinite planar triangulation. *Geom. Funct. Anal.*, 13(5):935–974, 2003.

[AP22] Juhan Aru and Ellen Powell. A characterisation of the continuum Gaussian free field in arbitrary dimensions. *J. Éc. Polytech. Math.*, 9:1101–1120, 2022.

[ARSar] Morris Ang, Guillaume Remy, and Xin Sun. FZZ formula of boundary Liouville CFT via conformal welding. *J. Eur. Math. Soc.*, to appear, 2023.

[Aru15] Juhan Aru. KPZ relation does not hold for the level lines and SLE_κ flow lines of the Gaussian free field. *Probab. Theory Relat. Fields*, 163(3-4):465–526, 2015.

[Aru20] Juhan Aru. Gaussian multiplicative chaos through the lens of the 2D Gaussian free field. *Markov Process. Relat. Fields*, 26(1):17–56, 2020.

[AS03] Omer Angel and Oded Schramm. Uniform infinite planar triangulations. *Comm. Math. Phys.*, 241(2-3):191–213, 2003.

[Aub98] Thierry Aubin. *Some nonlinear problems in Riemannian geometry*. Springer Monographs in Mathematics. Springer-Verlag, Berlin, 1998.

[AY23] Morris Ang and Pu Yu. Reversibility of whole-plane SLE for $\kappa > 8$. *arXiv preprint arXiv:2309.05176*, 2023.

[Bax00] Rodney J. Baxter. Equivalence of the two results for the free energy of the chiral Potts model. *J. Statist. Phys.*, 98(3-4):513–535, 2000.

[BBG12a] Gaëtan Borot, Jérémie Bouttier, and Emmanuel Guitter. Loop models on random maps via nested loops: The case of domain symmetry breaking and application to the Potts model. *J. Phys. A*, 45(49):494017, 35, 2012.

[BBG12b] Gaëtan Borot, Jérémie Bouttier, and Emmanuel Guitter. A recursive approach to the $O(n)$ model on random maps via nested loops. *J. Phys. A*, 45(4):045002, 38, 2012.

[BBK94] Richard F. Bass, Krzysztof Burdzy, and Davar Khoshnevisan. Intersection local time for points of infinite multiplicity. *Ann. Probab.*, 22(2):566–625, 1994.

[BBM11] Olivier Bernardi and Mireille Bousquet-Mélou. Counting colored planar maps: Algebraicity results. *J. Combin. Theory Ser. B*, 101(5):315–377, 2011.

[BC13] Itai Benjamini and Nicolas Curien. Simple random walk on the uniform infinite planar quadrangulation: Subdiffusivity via pioneer points. *Geom. Funct. Anal.*, 23(2):501–531, 2013.

[Bef08] Vincent Beffara. The dimension of the SLE curves. *Ann. Probab.*, 36(4):1421–1452, 2008.

[Ber96] Jean Bertoin. *Lévy processes*, Vol. 121 of *Cambridge Tracts in Mathematics*. Cambridge University Press, Cambridge, 1996.

[Ber07] Olivier Bernardi. Bijective counting of tree-rooted maps and shuffles of parenthesis systems. *Electron. J. Combin.*, 14(1), 2007.

[Ber08] Olivier Bernardi. A characterization of the Tutte polynomial via combinatorial embeddings. *Ann. Comb.*, 12(2):139–153, 2008.

[Ber15] Nathanaël Berestycki. Diffusion in planar Liouville quantum gravity. *Ann. Inst. Henri Poincaré Probab. Stat.*, 51(3):947–964, 2015.

[Ber17] Nathanaël Berestycki. An elementary approach to Gaussian multiplicative chaos. *Electron. Commun. Probab.*, 22, 2017.

[BGK+24] Guillaume Baverez, Colin Guillarmou, Antti Kupiainen, Rémi Rhodes, and Vincent Vargas. The Virasoro structure and the scattering matrix for Liouville conformal field theory. *Probability and Math. Physics*, to appear, 2024.

[BGRV16] Nathanaël Berestycki, Christophe Garban, Rémi Rhodes, and Vincent Vargas. KPZ formula derived from Liouville heat kernel. *J. Lond. Math. Soc. (2)*, 94(1):186–208, 2016.

[BH19] Stéphane Benoist and Clément Hongler. The scaling limit of critical Ising interfaces is CLE_3. *Ann. Probab.*, 47(4):2049–2086, 2019.

[BHS23] Olivier Bernardi, Nina Holden, and Xin Sun. Percolation on triangulations: A bijective path to Liouville quantum gravity. *Mem. Amer. Math. Soc.*, 289(1440):v+176, 2023.

[BJM14] Jérémie Bettinelli, Emmanuel Jacob, and Grégory Miermont. The scaling limit of uniform random plane maps, *via* the Ambjørn-Budd bijection. *Electron. J. Probab.*, 19, 2014.

[BJRV13] Julien Barral, Xiong Jin, Rémi Rhodes, and Vincent Vargas. Gaussian multiplicative chaos and KPZ duality. *Commun. Math. Phyis.*, 323:451–485, 2013.

[BLR17] Nathanaël Berestycki, Benoît Laslier, and Gourab Ray. Critical exponents on Fortuin-Kasteleyn weighted planar maps. *Comm. Math. Phys.*, 355(2):427–462, 2017.

[BM03] Emmanuel Bacry and Jean-Francois Muzy. Log-infinitely divisible multifractal processes. *Comm. Math. Phys.*, 236(3):449–475, 2003.

[BMR19] Erich Baur, Grégory Miermont, and Gourab Ray. Classification of scaling limits of uniform quadrangulations with a boundary. *Ann. Probab.*, 47(6):3397–3477, 2019.

[BN11] Nathanaël Berestycki and James R. Norris. Lectures on Schramm–Loewner Evolution. Available on the websites of the authors, 2011.

[BPR20] Nathanaël Berestycki, Ellen Powell, and Gourab Ray. A characterisation of the Gaussian free field. *Probab. Theory Relat. Fields*, 176(3–4):1259–1301, 2020.

[BPR21] Nathanaël Berestycki, Ellen Powell, and Gourab Ray. $(1 + \varepsilon)$ moments suffice to characterise the GFF. *Electron. J. Probab.*, 26, 2021.

[BPZ84a] A. A. Belavin, A. M. Polyakov, and A. B. Zamolodchikov. Infinite conformal symmetry in two-dimensional quantum field theory. *Nuclear Phys. B*, 241(2):333–380, 1984.

[BPZ84b] A. A. Belavin, A. M. Polyakov, and A. B. Zamolodchikov. Infinite conformal symmetry of critical fluctuations in two dimensions. *J. Statist. Phys.*, 34(5–6):763–774, 1984.

[Bro89] Andrei Z Broder. Generating random spanning trees. In *FOCS*, Vol. 89, pages 442–447, 1989.

[BSS23] Nathanaël Berestycki, Scott Sheffield, and Xin Sun. Equivalence of Liouville measure and Gaussian free field. *Ann. Inst. Henri Poincaré Probab. Stat.*, 59(2):795–816, 2023.

[BWW18] Nathanaël Berestycki, Christian Webb, and Mo Dick Wong. Random Hermitian matrices and Gaussian multiplicative chaos. *Probab. Theory Relat. Fields*, 172(1–2):103–189, 2018.

[CCM20] Linxiao Chen, Nicolas Curien, and Pascal Maillard. The perimeter cascade in critical Boltzmann quadrangulations decorated by an $O(n)$ loop model. *Ann. Inst. Henri Poincaré D*, 7(4):535–584, 2020.

[CDCH+14] Dmitry Chelkak, Hugo Duminil-Copin, Clément Hongler, Antti Kemppainen, and Stanislav Smirnov. Convergence of Ising interfaces to Schramm's SLE curves. *C. R. Math. Acad. Sci. Paris*, 352(2):157–161, 2014.

[Cer22] Baptiste Cerclé. Liouville conformal field theory on even-dimensional spheres. *J. Math. Phys.*, 63(1):Paper No. 012301, 25, 2022.

[CGRV19] Laurent Chevillard, Christophe Garban, Rémi Rhodes, and Vincent Vargas. On a skewed and multifractal unidimensional random field, as a probabilistic representation of Kolmogorov's views on turbulence. *Ann. Henri Poincaré*, 20(11):3693–3741, 2019.

[Cha84] Isaac Chavel. *Eigenvalues in Riemannian geometry*, Vol. 115 of *Pure and Applied Mathematics*. Academic Press, Inc., Orlando, FL, 1984. Including a chapter by Burton Randol, With an appendix by Jozef Dodziuk.

[Cha06] Pierre Chainais. Multidimensional infinitely divisible cascades. *Eur. Phys. J. B-Condens. Matter Complex Syst.*, 51(2):229–243, 2006.

[Che04] Mu-Fa Chen. *From Markov chains to non-equilibrium particle systems*. World Scientific Publishing Co., Inc., River Edge, NJ, 2nd ed., 2004.

[Che15] Laurent Chevillard. *Une peinture aléatoire de la turbulence des fluides*. PhD thesis, ENS Lyon, (Habilitation à diriger des recherches), 2015.

[Che17] Linxiao Chen. Basic properties of the infinite critical-FK random map. *Ann. Inst. Henri Poincaré D*, 4(3):245–271, 2017.

[CK15] Nicolas Curien and Igor Kortchemski. Percolation on random triangulations and stable looptrees. *Probab. Theory Relat. Fields*, 163(1–2):303–337, 2015.

[CN08] Federico Camia and Charles M. Newman. SLE_6 and CLE_6 from critical percolation. In *Probability, geometry and integrable systems*, Vol. 55 of *Math. Sci. Res. Inst. Publ.*, pages 103–130. Cambridge University Press, Cambridge, 2008.

[Con01] Rama Cont. Empirical properties of asset returns: Stylized facts and statistical issues. *Quant. Finance*, 1(2):223–236, 2001.

[dC16] Manfredo P. do Carmo. *Differential geometry of curves & surfaces*. Dover Publications, Inc., Mineola, NY, 2016. Revised & updated second edition of [MR0394451].

[DDDF20] Jian Ding, Julien Dubédat, Alexander Dunlap, and Hugo Falconet. Tightness of Liouville first-passage percolation for $\gamma \in (0, 2)$. *Publ. Math. Inst. Hautes Études Sci.*, 132:353–403, 2020.

[DG19] Jian Ding and Subhajit Goswami. Upper bounds on Liouville first-passage percolation and Watabiki's prediction. *Comm. Pure Appl. Math.*, 72(11):2331–2384, 2019.

[DG20] Jian Ding and Ewain Gwynne. The fractal dimension of Liouville quantum gravity: Universality, monotonicity, and bounds. *Comm. Math. Phys.*, 374(3):1877–1934, 2020.

[DIM77]	Richard T. Durrett, Donald L. Iglehart, and Douglas R. Miller. Weak convergence to Brownian meander and Brownian excursion. *Ann. Probability*, 5(1):117–129, 1977.
[DKRV16]	François David, Antti Kupiainen, Rémi Rhodes, and Vincent Vargas. Liouville quantum gravity on the Riemann sphere. *Comm. Math. Phys.*, 342(3):869–907, 2016.
[DLG09]	Thomas Duquesne and Jean-François Le Gall. On the re-rooting invariance property of Lévy trees. *Electron. Commun. Probab.*, 14:317–326, 2009.
[DMS21]	Bertrand Duplantier, Jason Miller, and Scott Sheffield. Liouville quantum gravity as a mating of trees. *Astérisque*, 427:viii+257, 2021.
[Dou83]	Adrien Douady. Systèmes dynamiques holomorphes. In *Bourbaki seminar, Vol. 1982/83*, Vol. 105–106 of *Astérisque*, pages 39–63. Soc. Math. France, Paris, 1983.
[DRSV14a]	Bertrand Duplantier, Rémi Rhodes, Scott Sheffield, and Vincent Vargas. Critical Gaussian multiplicative chaos: Convergence of the derivative martingale. *Ann. Probab.*, 42(5):1769–1808, 2014.
[DRSV14b]	Bertrand Duplantier, Rémi Rhodes, Scott Sheffield, and Vincent Vargas. Renormalization of critical Gaussian multiplicative chaos and KPZ relation. *Comm. Math. Phys.*, 330(1):283–330, 2014.
[DS11]	Bertrand Duplantier and Scott Sheffield. Liouville quantum gravity and KPZ. *Invent. Math.*, 185(2):333–393, 2011.
[DSHKS21]	Lorenzo Dello Schiavo, Ronan Herry, Eva Kopfer, and Karl-Theodor Sturm. Conformally invariant random fields, quantum Liouville measures, and random Paneitz operators on Riemannian manifolds of even dimension. *Preprint arXiv:2105.13925*, 2021.
[Dub09a]	Julien Dubédat. Duality of Schramm-Loewner evolutions. *Ann. Sci. Éc. Norm. Supér. (4)*, 42(5):697–724, 2009.
[Dub09b]	Julien Dubédat. SLE and the free field: Partition functions and couplings. *J. Amer. Math. Soc.*, 22(4):995–1054, 2009.
[EK95]	Bertrand Eynard and Charlotte Kristjansen. Exact solution of the $O(n)$ model on a random lattice. *Nuclear Phys. B*, 455(3):577–618, 1995.
[Eva85]	Steven N. Evans. On the Hausdorff dimension of Brownian cone points. *Math. Proc. Cambridge Philos. Soc.*, 98(2):343–353, 1985.
[Eva10]	Lawrence C. Evans. *Partial differential equations*, Vol. 19 of *Graduate Studies in Mathematics*. American Mathematical Society, Providence, RI, 2nd ed., 2010.
[Fal14]	Kenneth Falconer. *Fractal geometry*. John Wiley & Sons, Ltd., Chichester, 3rd ed., 2014. Mathematical foundations and applications.
[FB08]	Yan V. Fyodorov and Jean-Philippe Bouchaud. Freezing and extreme-value statistics in a random energy model with logarithmically correlated potential. *J. Phys. A*, 41(37):372001, 12, 2008.
[FK21]	Johannes Forkel and Jonathan P. Keating. The classical compact groups and Gaussian multiplicative chaos. *Nonlinearity*, 34(9):6050–6119, 2021.
[FQS84]	Daniel Friedan, Zongan Qiu, and Stephen Shenker. Conformal invariance, unitarity, and critical exponents in two dimensions. *Phys. Rev. Lett.*, 52(18):1575–1578, 1984.

[Fri95]	Uriel Frisch. *Turbulence*. Cambridge University Press, Cambridge, 1995. The legacy of A. N. Kolmogorov.
[GH20]	Ewain Gwynne and Tom Hutchcroft. Anomalous diffusion of random walk on random planar maps. *Probab. Theory Relat. Fields*, 178(1–2):567–611, 2020.
[GHM20]	Ewain Gwynne, Nina Holden, and Jason Miller. An almost sure KPZ relation for SLE and Brownian motion. *Ann. Probab.*, 48(2):527–573, 2020.
[GHMS17]	Ewain Gwynne, Nina Holden, Jason Miller, and Xin Sun. Brownian motion correlation in the peanosphere for $\kappa > 8$. *Ann. Inst. Henri Poincaré Probab. Stat.*, 53(4):1866–1889, 2017.
[GHSS18]	Christophe Garban, Nina Holden, Avelio Sepúlveda, and Xin Sun. Negative moments for Gaussian multiplicative chaos on fractal sets. *Electron. Commun. Probab.*, 23, 2018.
[GJSZJ12]	Alice Guionnet, Vaughan F. R. Jones, Dimitri Shlyakhtenko, and Paul Zinn-Justin. Loop models, random matrices and planar algebras. *Comm. Math. Phys.*, 316(1):45–97, 2012.
[GKMW18]	Ewain Gwynne, Adrien Kassel, Jason Miller, and David B. Wilson. Active spanning trees with bending energy on planar maps and SLE-decorated Liouville quantum gravity for $\kappa > 8$. *Comm. Math. Phys.*, 358(3):1065–1115, 2018.
[GKRV21]	Colin Guillarmou, Antti Kupiainen, Rémi Rhodes, and Vincent Vargas. Segal's axioms and bootstrap for Liouville theory. *arXiv preprint arXiv:2112.14859*, 2021.
[GKRV24]	Colin Guillarmou, Antti Kupiainen, Rémi Rhodes, and Vincent Vargas. Conformal bootstrap in Liouville theory. *Acta Mathematica*, 2024.
[GM17]	Ewain Gwynne and Jason Miller. Scaling limit of the uniform infinite half-plane quadrangulation in the Gromov-Hausdorff-Prokhorov-uniform topology. *Electron. J. Probab.*, 22:Paper No. 84, 47, 2017.
[GM21a]	Ewain Gwynne and Jason Miller. Conformal covariance of the Liouville quantum gravity metric for $\gamma \in (0, 2)$. *Ann. Inst. Henri Poincaré Probab. Stat.*, 57(2):1016–1031, 2021.
[GM21b]	Ewain Gwynne and Jason Miller. Convergence of the self-avoiding walk on random quadrangulations to SLE$_{8/3}$ on $\sqrt{8/3}$-Liouville quantum gravity. *Ann. Sci. Éc. Norm. Supér. (4)*, 54(2):305–405, 2021.
[GM21c]	Ewain Gwynne and Jason Miller. Existence and uniqueness of the Liouville quantum gravity metric for $\gamma \in (0, 2)$. *Invent. Math.*, 223(1):213–333, 2021.
[GM21d]	Ewain Gwynne and Jason Miller. Percolation on uniform quadrangulations and SLE$_6$ on $\sqrt{8/3}$-Liouville quantum gravity. *Astérisque*, 429:vii+242, 2021.
[GM21e]	Ewain Gwynne and Jason Miller. Random walk on random planar maps: spectral dimension, resistance and displacement. *Ann. Probab.*, 49(3):1097–1128, 2021.

[GMS19]	Ewain Gwynne, Cheng Mao, and Xin Sun. Scaling limits for the critical Fortuin-Kasteleyn model on a random planar map I: Cone times. *Ann. Inst. Henri Poincaré Probab. Stat.*, 55(1):1–60, 2019.
[GN16]	Evarist Giné and Richard Nickl. *Mathematical foundations of infinite-dimensional statistical models*. Cambridge Series in Statistical and Probabilistic Mathematics, [40]. Cambridge University Press, New York, 2016.
[GP19]	Ewain Gwynne and Joshua Pfeffer. Bounds for distances and geodesic dimension in Liouville first passage percolation. *Electron. Commun. Probab.*, 24:Paper No. 56, 12, 2019.
[GP22]	Ewain Gwynne and Joshua Pfeffer. KPZ formulas for the Liouville quantum gravity metric. *Trans. Amer. Math. Soc.*, 375(12):8297–8324, 2022.
[Gri09]	Alexander Grigor'yan. *Heat kernel and analysis on manifolds*, Vol. 47 of *AMS/IP Studies in Advanced Mathematics*. American Mathematical Society, Providence, RI; International Press, Boston, MA, 2009.
[GRV16]	Christophe Garban, Rémi Rhodes, and Vincent Vargas. Liouville Brownian motion. *Ann. Probab.*, 44(4):3076–3110, 2016.
[GRV19]	Colin Guillarmou, Rémi Rhodes, and Vincent Vargas. Polyakov's formulation of $2d$ bosonic string theory. *Publ. Math. Inst. Hautes Études Sci.*, 130:111–185, 2019.
[GS15]	Ewain Gwynne and Xin Sun. Scaling limits for the critical Fortuin-Kastelyn model on a random planar map III: finite volume case. Preprint arXiv:1510.06346, 2015.
[GS17]	Ewain Gwynne and Xin Sun. Scaling limits for the critical Fortuin-Kasteleyn model on a random planar map II: Local estimates and empty reduced word exponent. *Electron. J. Probab.*, 22:Paper No. 45, 56, 2017.
[HK71]	Raphael Høegh-Krohn. A general class of quantum fields without cut-offs in two space-time dimensions. *Comm. Math. Phys.*, 21:244–255, 1971.
[HMP10]	Xiaoyu Hu, Jason Miller, and Yuval Peres. Thick points of the Gaussian free field. *Ann. Probab.*, 38(2):896–926, 2010.
[HRV18]	Yichao Huang, Rémi Rhodes, and Vincent Vargas. Liouville quantum gravity on the unit disk. *Ann. Inst. Henri Poincaré Probab. Stat.*, 54(3):1694–1730, 2018.
[HS23]	Nina Holden and Xin Sun. Convergence of uniform triangulations under the Cardy embedding. *Acta Math.*, 230(1):93–203, 2023.
[Itō72]	Kiyosi Itō. Poisson point processes attached to Markov processes. In *Proceedings of the Sixth Berkeley Symposium on Mathematical Statistics and Probability (University of California, Berkeley, California Calif., 1970/1971), Vol. III: Probability theory*, pages 225–239. University of California Press, Berkeley, CA, 1972.
[Jan97]	Svante Janson. *Gaussian Hilbert spaces*, Vol. 129 of *Cambridge Tracts in Mathematics*. Cambridge University Press, Cambridge, 1997.
[Jeg20]	Antoine Jego. Planar Brownian motion and Gaussian multiplicative chaos. *Ann. Probab.*, 48(4):1597–1643, 2020.
[Jeg23]	Antoine Jego. Characterisation of planar Brownian multiplicative chaos. *Comm. Math. Phys.*, 399(2):971–1019, 2023.

[Jos02] Jürgen Jost. *Partial differential equations*, Vol. 214 of *Graduate Texts in Mathematics*. Springer-Verlag, New York, 2002. Translated and revised from the 1998 German original by the author.

[JS00] Peter W. Jones and Stanislav K. Smirnov. Removability theorems for Sobolev functions and quasiconformal maps. *Ark. Mat.*, 38(2):263–279, 2000.

[JS17] Janne Junnila and Eero Saksman. Uniqueness of critical Gaussian chaos. *Electron. J. Probab.*, 22:Paper No. 11, 31, 2017.

[JSW19] Janne Junnila, Eero Saksman, and Christian Webb. Decompositions of log-correlated fields with applications. *Ann. Appl. Probab.*, 29(6):3786–3820, 2019.

[Kah85] Jean-Pierre Kahane. Sur le chaos multiplicatif. *Ann. Sci. Math. Québec*, 9(2):105–150, 1985.

[Kah86] Jean-Pierre Kahane. Une inégalité du type de Slepian et Gordon sur les processus gaussiens. *Israel J. Math.*, 55(1):109–110, 1986.

[Kal21] Olav Kallenberg. *Foundations of modern probability*, Vol. 99 of *Probability Theory and Stochastic Modelling*. Springer, Cham, 3rd ed., [2021] ©2021.

[Kem17] Antti Kemppainen. *Schramm-Loewner evolution*, Vol. 24 of *SpringerBriefs in Mathematical Physics*. Springer, Cham, 2017.

[Ken00] Richard Kenyon. The asymptotic determinant of the discrete Laplacian. *Acta Math.*, 185(2):239–286, 2000.

[Ken01] Richard Kenyon. Dominos and the Gaussian free field. *Ann. Probab.*, 29(3):1128–1137, 2001.

[Kiv22] Pax Kivimae. *Random Matrices, Gaussian Multiplicative Chaos, and Complexity*. ProQuest LLC, Ann Arbor, MI, 2022. Thesis (Ph.D.) – Northwestern University.

[KMSW19] Richard Kenyon, Jason Miller, Scott Sheffield, and David B. Wilson. Bipolar orientations on planar maps and SLE_{12}. *Ann. Probab.*, 47(3):1240–1269, 2019.

[KRV19] Antti Kupiainen, Rémi Rhodes, and Vincent Vargas. Local conformal structure of Liouville quantum gravity. *Comm. Math. Phys.*, 371(3):1005–1069, 2019.

[KRV20] Antti Kupiainen, Rémi Rhodes, and Vincent Vargas. Integrability of Liouville theory: Proof of the DOZZ formula. *Ann. of Math. (2)*, 191(1):81–166, 2020.

[KS16] Antti Kemppainen and Stanislav Smirnov. Conformal invariance in random cluster models. II. full scaling limit as a branching SLE. *Preprint arXiv:1609.08527*, 2016.

[Law05] Gregory F. Lawler. *Conformally invariant processes in the plane*, Vol. 114 of *Mathematical Surveys and Monographs*. American Mathematical Society, Providence, RI, 2005.

[Law13] Gregory F. Lawler. Continuity of radial and two-sided radial *SLE* at the terminal point. In *In the tradition of Ahlfors-Bers. VI*, Vol. 590 of *Contemp. Math.*, pages 101–124. Amer. Math. Soc., Providence, RI, 2013.

[LG05] Jean-François Le Gall. Random trees and applications. *Probab. Surv.*, 2:245–311, 2005.

[LG13]	Jean-François Le Gall. Uniqueness and universality of the Brownian map. *Ann. Probab.*, 41(4):2880–2960, 2013.
[LL10]	Gregory F. Lawler and Vlada Limic. *Random walk: a modern introduction*, Vol. 123 of *Cambridge Studies in Advanced Mathematics*. Cambridge University Press, Cambridge, 2010.
[LOS18]	Gaultier Lambert, Dmitry Ostrovsky, and Nick Simm. Subcritical multiplicative chaos for regularized counting statistics from random matrix theory. *Comm. Math. Phys.*, 360(1):1–54, 2018.
[LP16]	Russell Lyons and Yuval Peres. *Probability on trees and networks*, Vol. 42 of *Cambridge Series in Statistical and Probabilistic Mathematics*. Cambridge University Press, New York, 2016.
[LRV17]	Hubert Lacoin, Rémi Rhodes, and Vincent Vargas. Semiclassical limit of Liouville field theory. *J. Funct. Anal.*, 273(3):875–916, 2017.
[LRV22]	Hubert Lacoin, Rémi Rhodes, and Vincent Vargas. The semiclassical limit of Liouville conformal field theory. *Ann. Fac. Sci. Toulouse Math. (6)*, 31(4):1031–1083, 2022.
[LSW01]	Gregory F. Lawler, Oded Schramm, and Wendelin Werner. The dimension of the planar Brownian frontier is 4/3. *Math. Res. Lett.*, 8(4):401–411, 2001.
[LSW03]	Gregory F. Lawler, Oded Schramm, and Wendelin Werner. Conformal restriction: The chordal case. *J. Amer. Math. Soc.*, 16(4):917–955, 2003.
[LSW04]	Gregory F. Lawler, Oded Schramm, and Wendelin Werner. Conformal invariance of planar loop-erased random walks and uniform spanning trees. *Ann. Probab.*, 32(1B):939–995, 2004.
[LSW24]	Yiting Li, Xin Sun, and Samuel S. Watson. Schnyder woods, SLE(16), and Liouville quantum gravity. *Trans. Amer. Math. Soc.*, 2024.
[Mad17]	Thomas Madaule. Convergence in law for the branching random walk seen from its tip. *J. Theor. Probab.*, 30:27–63, 2017.
[MDB00]	Jean-François Muzy, Jean Delour, and Emmanuel Bacry. Modelling fluctuations of financial time series: from cascade process to stochastic volatility model. *Eur. Phys. J. B-Condens. Matter Complex Syst.*, 17(3):537–548, 2000.
[MFC97]	Benoit B Mandelbrot, Adlai J Fisher, and Laurent E Calvet. A multifractal model of asset returns. *Cowles Foundation Discussion Paper, Sauder School of Business Working Paper*, 1164, 1997.
[Mie13]	Grégory Miermont. The Brownian map is the scaling limit of uniform random plane quadrangulations. *Acta Math.*, 210(2):319–401, 2013.
[Mil04]	John Milnor. Pasting together Julia sets: A worked out example of mating. *Experiment. Math.*, 13(1):55–92, 2004.
[Mol96]	George M. Molchan. Scaling exponents and multifractal dimensions for independent random cascades. *Comm. Math. Phys.*, 179(3):681–702, 1996.
[Moo25]	R. L. Moore. Concerning upper semi-continuous collections of continua. *Trans. Amer. Math. Soc.*, 27(4):416–428, 1925.

[MP10]	Peter Mörters and Yuval Peres. *Brownian motion*, Vol. 30 of *Cambridge Series in Statistical and Probabilistic Mathematics*. Cambridge University Press, Cambridge, 2010. With an appendix by Oded Schramm and Wendelin Werner.
[MRV16]	Thomas Madaule, Rémi Rhodes, and Vincent Vargas. Glassy phase and freezing of log-correlated Gaussian potentials. *Ann. Appl. Probab.*, 26(2):643–690, 2016.
[MS16a]	Jason Miller and Scott Sheffield. Imaginary geometry I: Interacting SLEs. *Probab. Theory Relat. Fields*, 164(3–4):553–705, 2016.
[MS16b]	Jason Miller and Scott Sheffield. Imaginary geometry II: Reversibility of $\text{SLE}_\kappa(\rho_1;\rho_2)$ for $\kappa \in (0,4)$. *Ann. Probab.*, 44(3):1647–1722, 2016.
[MS16c]	Jason Miller and Scott Sheffield. Imaginary geometry III: Reversibility of SLE_κ for $\kappa \in (4,8)$. *Ann. of Math. (2)*, 184(2):455–486, 2016.
[MS17]	Jason Miller and Scott Sheffield. Imaginary geometry IV: Interior rays, whole-plane reversibility, and space-filling trees. *Probab. Theory Relat. Fields*, 169(3–4):729–869, 2017.
[MS20]	Jason Miller and Scott Sheffield. Liouville quantum gravity and the Brownian map I: the $\text{QLE}(8/3,0)$ metric. *Invent. Math.*, 219(1):75–152, 2020.
[MS21]	Jason Miller and Scott Sheffield. Liouville quantum gravity and the Brownian map III: The conformal structure is determined. *Probab. Theory Relat. Fields*, 179(3–4):1183–1211, 2021.
[Mul67]	Ronald C. Mullin. On the enumeration of tree-rooted maps. *Canadian J. Math.*, 19:174–183, 1967.
[Mus10]	Giuseppe Mussardo. *Statistical field theory*. Oxford Graduate Texts. Oxford University Press, Oxford, 2010. An introduction to exactly solved models in statistical physics.
[NB11]	Lawrence Narici and Edward Beckenstein. *Topological vector spaces*, Vol. 296 of *Pure and Applied Mathematics (Boca Raton)*. CRC Press, Boca Raton, FL, 2nd ed., 2011.
[NSW20]	Miika Nikula, Eero Saksman, and Christian Webb. Multiplicative chaos and the characteristic polynomial of the CUE: The L^1-phase. *Trans. Amer. Math. Soc.*, 373(6):3905–3965, 2020.
[OS73]	Konrad Osterwalder and Robert Schrader. Axioms for Euclidean Green's functions. *Comm. Math. Phys.*, 31:83–112, 1973.
[OS75]	Konrad Osterwalder and Robert Schrader. Axioms for Euclidean Green's functions. II. *Comm. Math. Phys.*, 42:281–305, 1975. With an appendix by Stephen Summers.
[Pem91]	Robin Pemantle. Choosing a spanning tree for the integer lattice uniformly. *Ann. Probab.*, 19(4):1559–1574, 1991.
[Pit82]	Loren D. Pitt. Positively correlated normal variables are associated. *Ann. Probab.*, 10(2):496–499, 1982.
[Pol70]	A. M. Polyakov. Conformal symmetry of critical fluctuations. *JETP Lett.*, 12:381–383, 1970.
[Pol81]	A. M. Polyakov. Quantum geometry of bosonic strings. *Phys. Lett. B*, 103(3):207–210, 1981.

References

[Pom92] Christian Pommerenke. *Boundary behaviour of conformal maps*, Vol. 299 of *Grundlehren der mathematischen Wissenschaften [Fundamental Principles of Mathematical Sciences]*. Springer-Verlag, Berlin, 1992.

[Pow18] Ellen Powell. Critical Gaussian chaos: Convergence and uniqueness in the derivative normalisation. *Electron. J. Probab.*, 23, 2018.

[Pow21] Ellen Powell. Critical Gaussian multiplicative chaos: A review. *Markov Processes And Relat. Fields*, 27(4):557–606, 2021.

[PW98] James Gary Propp and David Bruce Wilson. How to get a perfectly random sample from a generic Markov chain and generate a random spanning tree of a directed graph. 7th annual ACM-SIAM symposium on discrete algorithms (Atlanta, ga, 1996). *J. Algorithms*, 27(2):170–217, 1998.

[PY82] Jim Pitman and Marc Yor. A decomposition of Bessel bridges. *Z. Wahrsch. Verw. Gebiete*, 59(4):425–457, 1982.

[PY96] Jim Pitman and Marc Yor. Decomposition at the maximum for excursions and bridges of one-dimensional diffusions. In *Itô's stochastic calculus and probability theory*, pages 293–310. Springer, Tokyo, 1996.

[Rem20] Guillaume Remy. The Fyodorov-Bouchaud formula and Liouville conformal field theory. *Duke Math. J.*, 169(1):177–211, 2020.

[RS05] Steffen Rohde and Oded Schramm. Basic properties of SLE. *Ann. of Math. (2)*, 161(2):883–924, 2005.

[RV10a] Rémi Rhodes and Vincent Vargas. Multidimensional multifractal random measures. *Electron. J. Probab.*, 15:no. 9, 241–258, 2010.

[RV10b] Raoul Robert and Vincent Vargas. Gaussian multiplicative chaos revisited. *Ann. Probab.*, 38(2):605–631, 2010.

[RV11] Rémi Rhodes and Vincent Vargas. KPZ formula for log-infinitely divisible multifractal random measures. *ESAIM Probab. Stat.*, 15:358–371, 2011.

[RV14] Rémi Rhodes and Vincent Vargas. Gaussian multiplicative chaos and applications: A review. *Probab. Surv.*, 11:315–392, 2014.

[RV19] Rémi Rhodes and Vincent Vargas. The tail expansion of Gaussian multiplicative chaos and the Liouville reflection coefficient. *Ann. Probab.*, 47(5):3082–3107, 2019.

[RV23] Rémi Rhodes and Vincent Vargas. Lecture notes on Liouville theory and the DOZZ formula. In *Topics in statistical mechanics*, Vol. 59 of *Panor. Synthèses*, pages 185–229. Soc. Math. France, Paris, 2023.

[RY99] Daniel Revuz and Marc Yor. *Continuous martingales and Brownian motion*, Vol. 293 of *Grundlehren der Mathematischen Wissenschaften [Fundamental Principles of Mathematical Sciences]*. Springer-Verlag, Berlin, 3rd ed., 1999.

[Sch00] Oded Schramm. Scaling limits of loop-erased random walks and uniform spanning trees. *Israel J. Math.*, 118:221–288, 2000.

[Sha16] Alexander Shamov. On Gaussian multiplicative chaos. *J. Funct. Anal.*, 270(9):3224–3261, 2016.

[She07] Scott Sheffield. Gaussian free fields for mathematicians. *Probab. Theory Relat. Fields*, 139(3–4):521–541, 2007.

[She09] Scott Sheffield. Exploration trees and conformal loop ensembles. *Duke Math. J.*, 147(1):79–129, 2009.

[She16a] Scott Sheffield. Conformal weldings of random surfaces: SLE and the quantum gravity zipper. *Ann. Probab.*, 44(5):3474–3545, 2016.

[She16b] Scott Sheffield. Quantum gravity and inventory accumulation. *Ann. Probab.*, 44(6):3804–3848, 2016.

[Smi01] Stanislav K. Smirnov. Critical percolation in the plane: conformal invariance, Cardy's formula, scaling limits. *C. R. Acad. Sci. Paris Sér. I Math.*, 333(3):239–244, 2001.

[SS13] Oded Schramm and Scott Sheffield. A contour line of the continuum Gaussian free field. *Probab. Theory Relat. Fields*, 157(1–2):47–80, 2013.

[Ste05] Kenneth Stephenson. *Introduction to circle packing*. Cambridge University Press, Cambridge, 2005. The theory of discrete analytic functions.

[SW05] Oded Schramm and David B. Wilson. SLE coordinate changes. *New York J. Math.*, 11:659–669, 2005.

[SW16] Scott Sheffield and Menglu Wang. Field-measure correspondence in Liouville quantum gravity almost surely commutes with all conformal maps simultaneously. *Preprint arXiv:1605.06171*, 2016.

[SW20] Eero Saksman and Christian Webb. The Riemann zeta function and Gaussian multiplicative chaos: Statistics on the critical line. *Ann. Probab.*, 48(6):2680–2754, 2020.

[VW24] Fredrik Viklund and Yilin Wang. The Loewner-Kufarev energy and foliations by Weil-Petersson quasicircles. *Proc. Lond. Math. Soc. (3)*, 128(2):Paper No. e12582, 62, 2024.

[Wat93] Yoshiyuki Watabiki. Analytic study of fractal structure of quantized surface in two-dimensional quantum gravity. *Progress of Theoretical Physics Supplement*, 114:1–17, 1993.

[Web15] Christian Webb. The characteristic polynomial of a random unitary matrix and Gaussian multiplicative chaos – the L^2-phase. *Electron. J. Probab.*, 20, 2015.

[Wil74] David Williams. Path decomposition and continuity of local time for one-dimensional diffusions. I. *Proc. London Math. Soc. (3)*, 28:738–768, 1974.

[Wil96] David Bruce Wilson. Generating random spanning trees more quickly than the cover time. In *Proceedings of the Twenty-eighth Annual ACM Symposium on the Theory of Computing (Philadelphia, PA, 1996)*, pages 296–303. ACM, New York, 1996.

[Won20] Mo Dick Wong. Universal tail profile of Gaussian multiplicative chaos. *Probab. Theory Relat. Fields*, 177(3–4):711–746, 2020.

[WP21] Wendelin Werner and Ellen Powell. *Lecture notes on the Gaussian free field*, Vol. 28 of *Cours Spécialisés [Specialized Courses]*. Société Mathématique de France, Paris, 2021.

[Zei15] Ofer Zeitouni. Gaussian fields (notes for lectures). Available on the website of the author, 2015.

[Zha08a] Dapeng Zhan. Duality of chordal SLE. *Invent. Math.*, 174(2):309–353, 2008.

[Zha08b] Dapeng Zhan. Reversibility of chordal SLE. *Ann. Probab.*, 36(4):1472–1494, 2008.

Notation and Symbols

Brownian motion

τ_D; hitting time of ∂D, 20

$p_t^D(x, y)$; transition probability for speed two Brownian motion killed when leaving D, 20

$p_t^{\Sigma, g}(\cdot, \cdot)$; heat kernel on a Riemannian manifold (Σ, g), 166

$p_t(x, y)$; transition probability for speed two Brownian motion on $\mathbb{R}^d$, 19

Function spaces

$\bar{\mathcal{D}}'_0(\mathbb{C})$; distributions modulo constants on $\mathbb{C}$, that is, the dual space of $\hat{\mathcal{D}}_0(\mathbb{C})$, 219

$\mathcal{D}_0(D)$; compactly supported smooth functions in D, or test functions, 34

$\hat{\mathcal{D}}_0(\mathbb{C})$; smooth functions with compact support and zero average on $\mathbb{C}$, 219

$\bar{\mathcal{D}}(D)$; smooth functions in D with finite Dirichlet energy, considered modulo constants, 204

$\bar{\mathcal{D}}'_0(D)$; distributions modulo constants on D, 204

$\bar{\mathcal{H}}_{\text{circ}}$; closure of smooth functions on the infinite strip (resp. cylinder) S (resp. C) which have mean zero on vertical segments, 237

$\bar{\mathcal{H}}_{\text{rad}}$; closure of smooth functions modulo constants on the infinite strip (resp. cylinder) S (resp. C) which are on vertical segments, 237

$\bar{H}^1(D)$; Hilbert space closure of $\bar{\mathcal{D}}(D)$ with respect to $(\cdot, \cdot)_\nabla$, 204

Conv; the set of sequences in $\mathcal{D}'_0(D)$ which are weak–$*$ convergent, 35

Harm(U); harmonic functions in U, 47

$\mathcal{D}'_0(D)$; distributions on D, that is, dual space of $\mathcal{D}_0(D)$, 35

$\mathfrak{M}$; completion of $\mathcal{D}(D)$ with respect to the inner product induced by a covariance kernel K, 78

$\mathfrak{M}_0$; signed measures of the form $\rho = \rho^+ - \rho^-$ with $\rho_\pm \in \mathfrak{M}_0^+$, 30

$\mathfrak{M}_0^+$; non-negative measures ρ supported in D with finite integral tested against G_0^D, 30

$\mathfrak{M}_N(\mathbb{D})$; difference of two elements in $\mathfrak{M}_N^+(\mathbb{D})$, 215

$\mathfrak{M}_N(D)$; pushforward of a signed measure in $\mathfrak{M}_N(\mathbb{D})$ under a conformal isomorphism from $\mathbb{D}$ to D, 215

$\mathfrak{M}_N^+(\mathbb{D})$; non-negative Radon measures on $\bar{\mathbb{D}}$ whose restriction to $\mathbb{D}$ is an element of $\mathfrak{M}_0^\mathbb{D}$ and such that integral of m against the Poisson kernel on $\partial \mathbb{D}$ is an element $H^{-1/2}(\partial \mathbb{D})$, 215

$\overline{\text{Harm}}(D)$; harmonic functions on D with finite Dirichlet energy, viewed modulo constants, 205

$\tilde{\mathcal{D}}_0(D)$; test functions $f \in \mathcal{D}_0(D)$ with total integral zero, 204

$G_N^D(x, y)$; choice of Neumann Green function on D, 213

$H^s(\Sigma, g)$; Sobolev space of index s on (Σ, g), 165

$H^{-1}(\hat{\mathbb{C}})$; distributions of the form $\{\varphi + c\,;\, \varphi \in H^{-1}(\hat{\mathbb{C}}, \hat{g}_0), c \in \mathbb{R}\}$, 174

$H^{-1}(\mathbb{S})$; distributions of the form $\{\varphi + c\,;\, \varphi \in H^{-1}(\mathbb{S}, g_0), c \in \mathbb{R}\}$, 174

$H_{\text{loc}}^{-1}(D)$; distributions in element of $H_0^{-1}(U)$ for any $U \Subset D$, 46

$H_0^1(D)$; Sobolev space, completion of $\mathcal{D}_0(D)$ with respect to the Dirichlet inner product, 37

$H_0^s(D)$; Sobolev space of index s in D, 38

$L^2(D)$; square integrable functions in D, 37

Gaussian multiplicative chaos

$\mathcal{M}$; general Gaussian multiplicative chaos measure, 76

$\mathcal{M}_\varepsilon$; approximation of $\mathcal{M}$ at spatial scale ε, 80

$\mathcal{M}_h$; Gaussian multiplicative chaos associated with a field h, 81

$\mathcal{M}_h^\gamma$; Gaussian multiplicative chaos associated with a field h and parameter γ, 81

$\mathcal{V}$; boundary Gaussian multiplicative chaos on the boundary for a field on a domain, 227

$\mathcal{V}_\varepsilon$; approximation of $\mathcal{V}$ at spatial scale ε, 227

$\mathcal{V}_h$; Gaussian multiplicative chaos on the boundary associated with a field h on a domain, 228

$\mathcal{V}_h^\gamma$; Gaussian multiplicative chaos on the boundary associated with a field h on a domain and parameter γ, 228

$\mathfrak{d}$; dimension of the reference measure, 80

σ; reference measure, 80

$\theta_\varepsilon(\cdot)$; mollifier at scale ε, 80

$\xi(\cdot)$; multifractal spectrum function of Gaussian multiplicative chaos, 102

$\mathcal{M}_{h;g}(A)$; Gaussian multiplicative chaos of a field h on a Riemannian manifold (Σ, g), 173

Geometry

$\Delta^{\Sigma, g}$; Laplace operator on Riemannian manifold (Σ, g), 164

$\hat{g}_0$; spherical metric on $\hat{\mathbb{C}}$, 162

$\hat{\mathbb{C}}$; extended complex plane $\mathbb{C} \cup \{\infty\}$, 162

$\mathbb{S}$; unit two-sphere, 162

g_0; spherical metric on $\mathbb{S}$, 162

$R(x; D)$; conformal radius of x in D, 28

R_g; scalar curvature associated to g, 162

v_g; volume form associated with a metric g, 162

Green functions

Γ_N; bilinear form, covariance of the Neumann GFF, 217

Γ_0; bilinear form, doubly integrating against G_0^D, 30

G^c; Green function for whole plane GFF with zero average on the unit circle, 189

$G^{\Sigma, g}(\cdot, \cdot)$; Green function with zero average on (Σ, g), 167

$G_0^D(\cdot, \cdot)$; for Laplacian with zero boundary conditions in D, 21

Inner products

$(\cdot, \cdot)_\nabla$; Dirichlet energy, 37

$(\cdot, \cdot)_g$; L^2 inner product on Riemannian manifold (Σ, g), 165

$(\cdot, \cdot)_s$; H_0^s inner product, 38

Liouville CFT

Δ_α; conformal weights, 159, 185

$\langle \cdot \rangle_{\hat{g}}$; expectation with respect to the Polyakov measure, 175

$\langle V \rangle_{\hat{g}}$; correlation function, 180

$\hat{\mathbf{P}}_{\hat{g}}$; Polyakov measure, 175

$S(\varphi)$; Polyakov action, 162

$V_{\alpha_1, \ldots, \alpha_k}(\mathbf{z})$; vertex operator, 180

Miscellaneous

Υ; special Upsilon function, 200

f^*; the conjugate function $z \mapsto f(\bar{z})$, 219

d_H; Hausdorff dimension, 51

d_M; Minkowski dimension, 120

$\Gamma(s, \mu)$; the Gamma function with parameters s and μ, 185

$\mathcal{T}_\alpha$; α-thick points of the Gaussian free field, 51

d_{TV}; total variation distance between two measures, 220

ν_δ^{BES}; Itô excursion measure for the δ-dimensional Bessel process, 252

$\rho_{z, \varepsilon}$; uniform distribution on the circle of radius ε around z, 49

$A \Subset B$; closure of A is a subset of B, 210

$m_* g$; pushforward of a metric g by a map m, 168

$T_* \mu$; pushforward of a measure μ by a map T, 215

Parameters

κ'; dual parameter value of $\kappa \in (0, 4]$, $\kappa' = 16/\kappa$, 134

γ; Coupling constant (GMC), 61, 76

κ; SLE parameter, 336

q; FK model parameter, 134

Q; parameter in change of coordinates formula for LQG, 72

Planar maps
  $\bar{m}$; refinement map of a map m, 128
  $\mathcal{M}_n$; maps with n edges and one distinguished root edge, 128
  $e^{\dagger}$; dual edge of an edge e, 127
  $m^{\dagger}$; dual map of a map m, 127

Subject Index

Action, 172
  Polyakov action, 174
Background charge, 70
Bessel
  Excursion measure, 278, 280
  Process, 275, 389
  Time reversal, 389
Boltzmann–Gibbs distribution, 214
Boundary Liouville measure. *See* Liouville measure
BPZ equations, 217
Branching $SLE_{\kappa'}(\kappa' - 6)$, 348, 354

Canonical description. *See* Random surface
Cardy embedding, 140
Central charge, 172, 190
Change of coordinates formula, 73
Circle average, 44, 59, 247
Circle packing, 139, 214
Conformal bootstrap, 168, 218
Conformal covariance of Liouville measure, 70
Conformal embedding, 214
Conformal invariance
  of Dirichlet GFF, 43
  of Green function, 14
conformal loop ensemble, CLE, 140
Conformal radius, 19, 58, 59
Conformal weights, 170, 200
Conformal welding, 301, 326
Continuous Random Tree (CRT), 137, 146, 160, 364
Correlation functions, 169, 194
Coupling
  SLE and GFF, 300
Coupling constant ($= \gamma$), 57, 75

Critical exponents, 129
  FK percolation, 166
  Loop-Erased Random Walk, 150
Definite. *See* Non-negative definite
Dirichlet energy, 7
Dirichlet inner product, 29
Discrete excursion, 145
Domain, 9
  Regular, 12, 21
DOZZ formula, 168
Duality
  Bessel processes, 280
  Quantum wedges, 280

Euler's formula, 144

FK model, 137, 141, 156, 160, 210, 214
FKG inequality, 120
Freezing, 113

Gauss–Bonnet theorem, 174
Gauss–Green formula, 28, 42
Gaussian multiplicative chaos, 64, 69, 96, 185
  critical, 69, 113
  supercritical, 113
GFF
  on a compact manifold, 175
  Dirichlet boundary conditions, 24
  Dirichlet–Neumann, 244
  Free boundary conditions, 223
  Neumann, random series, 222
  Neumann, stochastic process, 234
  on the sphere, 175, 187
  Whole plane, 239
  Zero average, 175
Gibbs measure, 172, 209
Girsanov lemma, 64
  for GMC, 94

Subject Index

Green function
  on a compact surface, 178
  Dirichlet boundary conditions, 11
  Discrete, 1
  Neumann, 231, 232
  on the sphere, 182
Hadamard's formula, 306
Hamburgers/cheeseburgers. *See* Sheffield's bijection
Hamiltonian, 172

Imaginary geometry, 350, 352, 357
Insertion, 194, 195
Integration by parts
  Gaussian, 97, 217
Isothermal ball, 127

Kahane's convexity inequality, 96
Karhuhen–Loeve expansion, 132
KPZ relation, 124, 151
KPZ theorem, 122

Liouville Brownian motion, 176
Liouville equation, 211
Liouville field, 212
  Unit volume Liouville sphere, 214
Liouville measure
  Boundary, 248
  Boundary (quantum wedge), 269
  Bulk, 58
  Multifractal spectrum, 105
  Scaling relation, 105
Local time, 278
Loewner chain, 373
Loop-Erased Random Walk, 150

Markov property, 9, 40
  Neumann GFF, 228
  Neumann GFF: boundary, 245
Markov property (of Liouville quantum gravity), 303
Mating of trees, 156, 160
Mating of trees (discrete), 146
Minkowski dimension, 125

Neumann problem, 230
Non-atomicity of Liouville measure, 130
Non-negative definite, 2, 22, 24

Orthogonal decomposition of $H_0^1(D)$, 42

Peanosphere convergence, 159
Peeling, 303
Peyrière measure, 64
Pioneer points, 166
Planar maps, 353

Decorated, 134
Definition, 134
Dual map, 134
Fortuin–Kasteleyn model, 136
Loops, 134
Percolation, 138
Refinement edges, 134
Uniform case, 138
Polyakov action, 172, 174
Polyakov measure, 175

Quantum cones, 73, 272, 298
Quantum discs, 280
Quantum field theory, 169
Quantum length (of SLE), 309
Quantum spheres, 285
Quantum surface, 257
Quantum wedges, 73, 259, 275

Radial decomposition, 56, 261
Random surface, 73
  Canonical description, 258
  Convergence, 258
  Unit circle embedding, 265
  Weights, 289
  Zooming in, 258
Reverse Loewner flow, 304, 382
Riemann uniformisation, 57, 139
Rooted measure, 64, 92, 94

Scalar curvature, 174
scaling exponent, 123
  Euclidean, 123
  Minkowski, 125
  quantum, 123
  quantum Minkowski, 125
Schwartz space, 396
SDE, 275
Seiberg bounds, 168, 195
Sheffield's bijection, 143
SLE
  Chordal, 375
  Chordal with force points, 376
  Radial, 392
  Radial with force points, 392
  Reverse, 384
  Target invariance, 394
Sobolev space, 29, 30
Space-filling SLE
  Definition, $\kappa' \geq 8$, 343
  Definition, $\kappa' \in (4, 8)$, 354
  Markov property, 345
  Reversibility, 347

Stress energy tensor, 217
Subdiffusivity, 167

Tempered distributions, 396
Thick points, 46, 64, 67
Total variation distance, 240, 246
Tutte bijection, 135

Uniform infinite half plane triangulation, 303

Vertex operator, 194

Ward identities, 217
Watabiki formula, 124
Weyl anomaly Formula, 190
Weyl law, 34, 177
Wick's rule, 169
Williams' path decomposition theorem, 389
Wilson's algorithm, 150

Zipper
    Capacity, 316
    Quantum, 317

For EU product safety concerns, contact us at Calle de José Abascal, 56–1°, 28003 Madrid, Spain or eugpsr@cambridge.org.

www.ingramcontent.com/pod-product-compliance
Ingram Content Group UK Ltd.
Pitfield, Milton Keynes, MK11 3LW, UK
UKHW022318120326
468863UK00033B/828